Preventing Weaponization of CNS-acting Chemicals
A Holistic Arms Control Analysis

Preventing Weaponization of CNS-acting Chemicals

A Holistic Arms Control Analysis

By

Michael Crowley
University of Bradford, UK
Email: M.J.A.Crowley1@bradford.ac.uk

and

Malcolm Dando
University of Bradford, UK
Email: M.R.Dando@bradford.ac.uk

ROYAL SOCIETY OF CHEMISTRY

Hardback ISBN: 978-1-83767-134-2
EPUB ISBN: 978-1-83767-816-7
PDF ISBN: 978-1-83767-817-4

A catalogue record for this book is available from the British Library

The Royal Society of Chemistry is a charity, registered in England and Wales, Number 207890, and a company incorporated in England by Royal Charter (Registered No. RC000524), registered office: Burlington House, Piccadilly, London W1J 0BA, UK, Telephone: +44 (0)20 7437 8656.

For further information see our website at www.rsc.org

For general enquiries, please contact books@rsc.org

For EU product safety enquiries, please email books@rsc.org or contact Royal Society of Chemistry Worldwide (Germany) GmbH, Römischer Hof, Unter den Linden 10, 10117 Berlin.

Printed in the United Kingdom by CPI Group (UK) Ltd, Croydon, CR0 4YY, UK

Foreword

Neuroscience is entering a critical phase. The field no longer merely studies the brain; it increasingly offers mechanisms for manipulating core Central Nervous System (CNS) functions. Fuelled by sustained investment and deepening convergence with adjacent fields, neuroscience has shifted from observation to intervention and, in the process, is reshaping how we understand the human brain—not as a black box, but as a complex, modifiable system. The implications are profound: from neurodegenerative disease therapies to enhanced cognitive tools and brain-computer interfaces. Yet, the ability to modulate consciousness, induce compliance, distort perception, or temporarily incapacitate through biochemical or neurological means is increasingly viewed through a second lens: one concerned with its potential weaponisation.

It is within this context that Michael Crowley and Malcolm Dando present their urgent and authoritative volume. *Preventing Weaponization of CNS-acting Chemicals* offers both an empirically grounded survey and a normative intervention. It shows that as we move toward more precise and powerful tools for intervening in brain and CNS functions, the possibility that such tools could be harnessed for coercion, control or weaponization is no longer speculative. It is real—and pressing. More importantly, it outlines how existing governance frameworks—particularly the Biological and Toxin Weapons Convention (BWC) and the Chemical Weapons Convention (CWC)—might be adapted, supplemented or reinterpreted to more adequately respond to these risks.

This book is firmly rooted in precedent. The authors trace how earlier waves of scientific development—especially in neuropharmacology during the Cold War— piqued the interest of military and security actors and led to covert and classified state efforts to develop incapacitating chemical agents. These programmes rarely resulted in field-deployable weapons, but their legacy remains

Preventing Weaponization of CNS-acting Chemicals: A Holistic Arms Control Analysis
By Michael Crowley and Malcolm Dando

Published by the Royal Society of Chemistry, www.rsc.org

salient, particularly as 21st Century neuroscience achieves levels of precision, targeting, and integration that their predecessors could not.

The Russian use of fentanyl derivatives during the 2002 Moscow theatre siege brought these concerns from the margins to the centre of arms control discourse. Crowley and Dando show that while the BWC and CWC nominally cover such agents, interpretive ambiguities, implementation gaps and political divergences have allowed the space for potential misuse to persist. One of the enduring contributions of the book is its sustained focus on that governance space—the 'in-between' areas of chemical and biological weapons control where CNS-acting agents fall. The book's framing of a 'biochemical threat spectrum' challenges us to move beyond categorical thinking. Rather than forcing emerging agents into pre-defined silos of 'chemical' or 'biological', Crowley and Dando suggest it is more productive to think in terms of continua: of agents, mechanisms, and purposes. This conceptual shift is subtle but essential. It invites arms control communities to consider threats based on functional characteristics and application contexts, not just on origins or production methods.

The volume's introduction of a Holistic Arms Control (HAC) approach responds to this complexity. It is not a wholesale replacement of existing regimes, but an analytical and policy framework aimed at building out from them. Drawing from earlier models like 'webs of prevention,' HAC integrates multiple legal, ethical and institutional layers. These include international human rights law, international humanitarian law, the UN drug conventions, and the codes and practices of the scientific and medical professions. Taken together, these overlapping instruments offer a more resilient architecture—one capable of addressing both the long-standing and emerging risks associated with CNS-acting agents.

What distinguishes HAC is its practicality. Crowley and Dando map it onto the structure of the book itself, moving from technical substance to legal instruments to implementation challenges and actor roles. This method enables not just diagnosis but prescription: actionable recommendations for review processes under the BWC and CWC; technical advisory group activities within the OPCW; normative guidance for brain science initiatives like the U.S. BRAIN project or China's Brain Science and Brain-Inspired Intelligence programme; and educational tools for engaging researchers in ethical reflection, responsible science and anticipatory governance. These are not abstract calls. They are rooted in the authors' long experience in arms control diplomacy and civil society engagement.

Indeed, it is worth noting the significant roles that both authors have played over the years. Malcolm Dando's work on dual-use education, verification mechanisms and the sociology of scientific responsibility has left an enduring mark on BWC processes. Michael Crowley's persistent efforts to address riot control agents and the misuse of 'less lethal' weapons have informed numerous international human rights fora. Together, they have consistently argued for a broader conception of arms control—one that recognises the agency of scientists, civil society, and professional bodies, not just states.

Their argument here is unambiguous: existing treaty regimes, while vital, are structurally limited in their ability to anticipate and respond to the weaponisation of neuroscience. The BWC lacks verification provisions and systematic science and technology review. The CWC, though more technically elaborate, is politically constrained and has shown limited willingness to tackle controversial issues such as law enforcement applications of CNS-acting chemicals. The 2021 CSP decision on aerosolised use of these agents was a step forward, but as the authors note, its interpretation and implementation remain unsettled.

What this volume ultimately offers is not certainty, but strategic clarity. It identifies what is known, what is possible, and where the legal and normative gaps lie. It invites the arms control community—not just states parties, but also scientific institutions, international organisations, and advocacy groups—to engage proactively rather than reactively. It acknowledges that science does not stand still, and that regulation must evolve with it.

At a time when the boundaries between science, medicine, security and control are increasingly fluid, *Preventing Weaponization of CNS-acting Chemicals* provides a well-evidenced, sober and much-needed guide. It deserves careful reading, not just within specialist circles but across the broader domains of international security, human rights and scientific governance. In doing so, it reminds us that the protection of the human mind must become a priority of arms control in the 21st century—not just because of what neuroscience can do, but because of what it could undo.

Filippa Lentzos

Preface

In late February 2025, as we were in the final stages of completing this book, life scientists were gathering in Asilomar, California, for a second meeting[1] 50 years after the famous Asilomar seminar which set much of the framework of thinking about biosafety in biological research. The first Asilomar meeting has been subsequently criticised as consciously avoiding the issue of biosecurity, namely how to prevent the intentional misuse of biotechnology to cause harm.[2] Detailed historical research about why this happened is frankly alarming in pointing to the impact of the then-current view that there should be no interference with scientists' ability to freely investigate whatever they thought useful.[3] In contrast the programme for the 2025 meeting[4] had an explicit focus on the role of the scientific community in preventing biological warfare and noted that:

> "Times have changed. The revealing of a Soviet bioweapons program, bioterror attacks by sub-state actors, and the ongoing proliferation of increasingly powerful biotechnology tools have renewed and are amplifying concerns that biotechnology will be used to cause harm. Nations – including China, USA, and Russia – have recently lobbied public accusations of adversarial bioweapons programs.... Continuing ambiguity regarding the origins of the SARS-CoV-2 pandemic has eroded trust at multiple levels. Use of chemical weapons risks weakening moral norms preventing use of biological weapons..."

Times have certainly changed. However, the central 'dual-use' dilemma remains: how can the chemical and life science community and national and international governance structures ensure that benignly-intended scientific advances are not misused by others for creating biological and chemical weapons or for other hostile purposes. These are complex, entangled issues which we seek to explore in detail in our book, but there are two more general points that we need to make here. First, it is obvious that we are in the very

---

Preventing Weaponization of CNS-acting Chemicals: A Holistic Arms Control Analysis
By Michael Crowley and Malcolm Dando

Published by the Royal Society of Chemistry, www.rsc.org

early stages of the current revolution in our understanding of the physiology of living organisms and therefore we can expect decades of further major advances that will be beneficial to human society, but which also may be subject to hostile misuse. Secondly, while some scientists have taken leading parts in these discussions, it is clear that most practicing chemical, life and associated scientists have little awareness or interest in the problem of 'dual use', and they are therefore ill-prepared to assist in preventing such malign applications. Therefore, scientists across the range of life and associated sciences and technologies will need to obtain the necessary capabilities to engage fully in the societal debate about what they should, or should not, do, in full knowledge of the consequences of their actions. This will require a radical revision of, and redoubled energies in, awareness-raising, education and training.

This book is specifically concerned with the potential hostile misuse of benignly-intended neuroscience and neurotechnology, with an emphasis on neuropharmacology and neurotoxicology. We think this focus is needed because most debate about 'dual use' application of the life and associated science and technologies has concentrated on preventing the weaponization of pathogenic microorganisms. While such work has been and continues to be crucial,[5] we show in detail that there is also a long and continuing history of the hostile misuse of modern advances in neuroscience, including military and security development and use of novel weapons utilising pharmaceutical chemicals, toxins and bioregulators to attack the human central nervous system or other core physiological systems. We therefore seek in this publication to use a Holistic Arms Control system of analysis to set out the whole problem: highlighting relevant developments in neuroscience and related chemical and life sciences and associated technologies notably artificial intelligence and machine learning; investigate present and potential future hostile applications; and explore how such misuse might be most comprehensively and effectively addressed through existing regulatory regimes and law, and by a range of different stakeholders, notably including the chemical and life science communities. We hope that this analysis will make a contribution both to the understanding of the very dangerous problems that will arise as we gather more and more detailed information about how our behaviour is underpinned by the operations of the central nervous system, and to the development of effective educational material for future life and associated scientists.

We would like to thank the Leverhulme Trust (under EM-2018-005/10) and the Marmot Trust for partial support of this study and our colleagues at Bradford University and London Metropolitan University for their support of our endeavours over the last 2 years. Finally, we would like to sincerely thank Dr Christopher Timperley and Dr Ralf Trapp for kindly reviewing earlier versions of this book. Of course, any remaining errors are our responsibility alone.

Michael Crowley and Malcolm Dando

References

1. The Spirit of Asilomar and the Future of Biotechnology, February 23–26.
2. M. Cobb, Money and Murder: The dark side of the Asilomar meeting on recombinant DNA, *Nature*, 2025, **638**, 603–608.
3. M. Cobb, The Curious Incident of Crick in the Night-Time and Other Asilomar Enigmas, *J. Hist. Biol.*, 2025, **58**, 49–66.
4. The Spirit of Asilomar and the Future of Biolotechnology. https://www.spiritofasilomar.org/home. (Accessed 1 March, 2025).
5. See, for example, K. P. Adamala, *et al.*, Confronting the risk of mirror life, *Science*, 2024, **386**(6728), 1351–1353.

Acronyms

ABC	Applied Biology Committee
ABC-M7A2/A3	Medetomidine-based hand grenades
AB2K	Afterburner 2000 system
AEs	Autoimmune encephalopathies
ABEO	Advisory Board on Education and Outreach
AFOSR	Air Force Office of Scientific Research
AFU	Armed Forces of Ukraine
AJOB	American Journal of Bioethics
AHI	Anomalous health incidents
AI	Artificial intelligence
ANAOs	Associate National Adhering Organisations
AlphFold	Modelling software
Anti-LGI1	Anti-leucine-rich-glioma-inactivated 1 receptor encephalitis
Anti-NMDA	Anti-*N*-methyl-D-aspartate receptor encephalitis
ARDEC	Armament Research Development Engineering Center
ARPA	Advanced Research Projects Agency
ATT	Arms Trade Treaty
BBB	Blood brain barrier
BCR	B cell receptor
BIOS	Biology-oriented synthesis
BOLD	Blood-oxygen level dependent
BRAIN	Brain Research Through Advancing Innovative Neurotechnologies
BTWC	Biological and Toxin Weapons Convention
BW	Biological warfare/biological weapon
BZ	3-Quinuclidinyl benzilate
C2c2	Class-2 candiate-2 genome editing technology

Preventing Weaponization of CNS-acting Chemicals: A Holistic Arms Control Analysis
By Michael Crowley and Malcolm Dando

Published by the Royal Society of Chemistry, www.rsc.org

CAT	Convention Against Torture
CBDOO-108	Chemical immobilizing agents for nonlethal applications project
CBM	Confidence-building measure
CBP	China brain project
CBW	Chemical and biological weapons
CCCW	Convention on Certain Conventional Weapons
CCKA/B	Cholecystokinin receptors A/B
CCK	Cholecystokinin
CDAB	Chemical Defence Advisory Board
CDC	Center for Disease Control
CDEE	Chemical Defence Experimental Establishment
CESCR	Committee on Economic, Social and Cultural Rights
CFPS	Cell-free protein synthesis
CHEMRAWN	Chemical Research Applied to World Needs
CIA	Central Intelligence Agency
CITNDPS	Convention Against Illicit Traffic in Narcotic Drugs and Psychotropic Substances
CLM	Chemical learning model
CNS	Central nervous system
COHSE	Committee on the Safety of Human Experimentation
CoE	Council of Europe
COVID-19	Coronavirus disease caused by SARS-CoV-2 virus
Cpf1	Genome editing technology
CPS	Convention on Psychotropic Substances
CPT	Committee for the Prevention of Torture and Inhuman or Degrading Treatment or Punishment
CR	Dibenzoxazepine
CRF	Corticotropin-releasing factor
CRISPR/Cas	Genome editing technology
CRPD	UN Convention on the Rights of People with Disabilities
CS	*o*-Chlorobenzylidene malononitrile
CSI	Combined Systems Incorporated
CS	Conditioned stimulus
CSP	Conference of States Parties
CS–US	Conditioned stimulus–unconditioned stimulus
CWA	Chemical warfare agent
CW	Chemical warfare/chemical weapon
CWC	Chemical Weapons Convention
CWPF	Chemical Weapons Production Facility
DARPA	Defense Advanced Research Projects Agency
DCDC	Development, Concepts and Doctrine Centre
DDBJ	DNA Data Bank of Japan
DG	Director-General
DL	Deep learning
DMN	Default mode network

DMSO	Dimethyl sulfoxide
DMT	*N*,*N*-dimethyltryptamine
DNA	Deoxyribonucleic acid
DoD	Department of Defense
DOI	2,5-Dimethoxy-4-iodoamphetamine
DPIC	Death Penalty Information Center
DPRK	Democratic People's Republic of Korea (North Korea)
DR	Dorsal Raphe
DRDE	Defence Research and Development Establishment
DRDO	Defence Research and Development Organisation
DRPC	Defence Research Policy Committee
DREADDs	Designer receptors exclusively activated by designer drugs
DSTL	Defence Science and Technology Laboratory
DTC	Deseret Test Center
EA 3834A	Glycollate chemical agent
EAE	Experimental autoimmune encephalomyelitis
E&O	Education and Outreach
EC	Executive Council
ECBC	Edgewood Chemical Biological Center
ECHR	European Convention on Human Rights
ECtHR	European Court of Human Rights
ECOSOC	Economic and Social Council
ED	Effective dose
EDIS	Electronic Declaration Information System
EEG	Electroencephalogram
EMBL	European Molecular Biology Laboratory
ESMFold	Evolutionary Scale Modelling Software
EU	European Union
FBI	Federal Bureau of Investigation
FET	Future and Emerging Technologies
fMRI	Functional magnetic resonance imaging
GABA	Gamma amino butyric acid
GenBank	Gene Bank sequence database
GGE	Group of Governmental Experts
GPC	General purpose criterion
GPCR	G protein coupled receptors
GP-Write	Genome Project-Write Initiative
HAC	Holistic Arms Control
HBP	Human Brain Project
HIV	Human Immunodeficiency Virus
HLA	Human leukocyte antigen
HPCSA	Health Professions Council of South Africa
HRW	Human Rights Watch
HSE	Herpes simplex virus encephalitis
HTS	High throughput screening

5-HT	5-Hydroxytryptamine
IAEA	International Atomic Energy Agency
IAU	Investigation of alleged use
IBSEN	International Biological Security Education Network
ICA	Incapacitating chemical agent
ICCA	International Council of Chemical Associations
ICESCR	International Covenant on Economic, Social and Cultural Rights
ICCPR	International Covenant on Civil and Political Rights
ICRC	International Committee of the Red Cross
ICTA	International Chemical Trade Association
IED	Improvised explosive device
IgLON5	Immunoglobulin-like Cell Adhesion Molecule 5
IHL	International Humanitarian Law
IHRL	International Human Rights Law
IHU	Imam Hossein University
INCB	International Narcotics Control Board
INLDT	Institute for Non-Lethal Defense Technologies
INSEN	International Nuclear Security Education Network
INS	International Neuroethics Society
IRCT	International Rehabilitation Council for Torture Victims
IRGC	Iranian Republican Guard Corp
ISF	Israeli Security Forces
ISO	International Standards Organisation
ISU	Implementation Support Unit
IUPAC	International Union of Pure and Applied Chemistry
JAG	Judge Advocate General
JNLWD	Joint Non-Lethal Weapons Directorate
JNLWP	Joint Non-Lethal Weapons Program
KGB	Committee for State Security (Russia)
LAE	Lysergic acid ethylamide
LC	Locus coeruleus
LD	Lethal dose
LSD	Lysergic acid diethylamide
MAARS	Modular Advanced Armed Robotic System
MBP	Myelin basic protein
MDMA	3,4-Methylenedioxymethamphetamine ('Ecstasy')
MG	Myasthenia gravis
MHC	Major histocompatibility complex
MI6	Military Intelligence Section 6 (UK Secret Intelligence Service)
miRNA	MicroRNA
MK2	Riot control munition
MKUltra	Secret CIA 'Mind Control' Research Programme
ML	Machine learning
MOOTW	Military Operations Other Than War

mPFC	Medial prefrontal cortex
MPIPG	*N*-methyl-4-piperidyl-isopropyl-phenol glycollate (T3436)
MRI	Magnetic resonance imaging
MS	Multiple sclerosis
MSP	Meeting of States Parties
MUT	Malak Ashtar University
MX	Meeting of Experts
NAOs	National Adhering Organisations
NAS	National Academy of Sciences
NATO	North Atlantic Treaty Organisation
NGO	Non Governmental Organisation
NHP	Non-human primate
NIH	National Institutes of Health
NIJ	National Institute of Justice
NIS	National Intelligence Service (South African)
NKE	Non-kinetic energy
NL	Non-lethal
NLW	Non-lethal weapon
NMDAr	*N*-Methyl-D-aspartate (NMDA) receptor
NORINCO	China North Industries Corporation
NPDs	Neurodegenerative and psychiatric diseases
NREM	Non-rapid eye movement
OC	Oleoresin capsicum
OHCHR	Office of the High Commissioner for Human Rights
OPCW	Organisation for the Prohibition of Chemical Weapons
OR	Opiate receptors
ORF8	Open Reading Frame 8 protein
PA	Panic attack
PAMP	Pathogen-associated molecular pattern
PAVA	Pelargonic acid vanillylamide
PBA	Pharmaceutical based agent
PCC	Posterior cingulate cortex
PCP	Phencyclidine
PD	Panic disorder
PD	Parkinson's disease
PD	Porton Down
PET	Positron emission tomography
PG	Staphylococcal enterotoxin B agent
PLA	(Chinese) People's Liberation Army
PLAN	Principles of proportionality, legality, accountability, and necessity
PPR	Pattern recognition receptor
PRC	Peoples Republic of China
pSAgs	Primary self-antigens
PSIM	Political, security, intelligence and military
PSU	Pennsylvania State University

QNA	QinetiQ North America Inc.
RCA	Riot control agent
RDEC	Research Development and Engineering Center
REM	Rapid eye movement
RENAMO	Mozambican National Resistance
RF	Relevant facilities
RNA	Ribonucleic acid
RoseTTAFold	Modelling software
RSFC	Resting-state functional connectivity
RSTA	Reconnaissance, Surveillance and Target Acquisition
SAB	Scientific Advisory Board
SARS-CoV-2	Severe Acute Respiratory Syndrome Coronavirus 2
SADF	South African Defence Force
S&T	Science and technology
SCN	Suprachiasmatic nucleus
SCND	Single Convention on Narcotic Drugs
SEB	Staphylococcal enterotoxin B
SIRUS	Superfluous injury or unnecessary suffering
SIS	Secret Intelligence Service
SLD	Sublaterodorsal tegmental nucleus
SMG	Shahid Meisami Group
SMILES	Simplified Molecular Input Line Entry Systems
SWAT	Special Weapons And Tactics (Law Enforcement units)
TASS	Russian news agency
T2614	*N*,*N*-diethyl tryptamine
T2619	*N*,*N*-di-isopropyl tryptamine
T2627	*N*-benzyl tryptamine
T2673	*N*-ethyl tryptamine
T2785	Tremoran
TCR	T cell receptor
TFTT	Torture-Free Trade Treaty
TL2636, TL2696, TL2833 and TL3046	Oripavine derivatives
TRC	Truth and Reconciliation Commission
TWG	Temporary Working Group
UAE	United Arab Emirates
UAV	Unmanned aerial vehicle
UGV	Unmanned ground vehicle
UK	United Kingdom
UN	United Nations
UNBP	UN Basic Principles on the Use of Force and Firearms
UNCoC	UN Code of Conduct for Law Enforcement Officials
UNGA	United Nations General Assembly
UNODA	United Nations Office of Disarmament Affairs
UNSC	United Nations Security Council

US	United States
USSR	Union of Soviet Socialist Republics
VERIFIN	Institute in the Department of Chemistry University of Helsinki
vHTS	Virtual high throughput system
VMAT2	Vesicular monoamine transporter-2
VMM	Ventromedial medulla
VX	Nerve agent
WA	Wassenaar Arrangement
WMA	World Medical Association
WHO	World Health Organisation
WIV	Wuhan Institute of Virology
WMD	Weapon of Mass Destruction
WRF	Weapons Review Forum

Contents

Section I: Introduction

Chapter 1 Neuroscience and the Future of the Chemical and Biological Weapons Disarmament and Arms Control Regimes 3

1.1 Advancing and Converging Chemical and Life Sciences 3
1.2 Overview of Advances in Neuroscience 5
1.2.1 Areas of Neuroscience of Potential Dual-use Concern 7
1.2.2 Implications for Monitoring Dual Use Research and Preventing the Weaponization of CNS-acting Chemical Agents 11
1.3 Current Limited Regulation of CNS-acting Agents Under CBW Arms Control and Disarmament Treaties 12
1.3.1 The 1925 Geneva Protocol 13
1.3.2 The Biological and Toxin Weapons Convention 13
1.3.3 The Chemical Weapons Convention 14
1.4 Holistic Arms Control 17
1.5 The Structure of the Book 19
References 22

Preventing Weaponization of CNS-acting Chemicals: A Holistic Arms Control Analysis
By Michael Crowley and Malcolm Dando

Published by the Royal Society of Chemistry, www.rsc.org

Section II: Science and Technology

Chapter 2 Malign Application of Neuroscience and Related Fields 29

2.1 Introduction 29
2.2 Illustrative Examples of Past and Contemporary Malign Application of Neuroscience and Related Fields for Development of CNS-acting Weapons 30
2.2.1 UK Pharmaceutical CNS-acting Chemical Agent Weapons Research 30
2.2.2 South Africa Pharmaceutical CNS-acting Chemical Agent Weapons Development 35
2.2.3 US CNS-acting Chemical Agent Weapons Development 38
2.2.4 Soviet Union/Russia CNS-acting Agent Weapon Development and Use 44
2.2.5 Chinese Pharmaceutic CNS-acting Chemical Agent Weapons Development and Use 47
2.3 Illustrative Examples of Past and Contemporary Dual Use Neuroscience and Related Research of Potential Concern 48
2.3.1 Czech Republic 48
2.3.2 India 49
2.3.3 Iran 50
2.4 Application of CNS-acting Chemicals and Neuroscience for Torture and Ill-treatment 51
2.4.1 Development of Psychopharmacology for Benign Purposes 51
2.4.2 Malign Application of Psychopharmacology 53
2.4.3 Malign Application of CNS-acting Chemicals for Interrogation 60
2.4.4 Contemporary Malign Application of CNS-acting Chemicals for Interrogation 64
2.5 Conclusions 66
References 67

Chapter 3 21st Century Advances in Understanding of Basic Survival Circuits of the Brain 87

3.1 Introduction 87
3.2 The Importance of Neuropeptides 91
3.3 What We Knew About Neuropeptides in 2000 93

3.4 Overview of Neuronal Circuit Research in 2023 95
3.5 Overview of Neuropeptide Research in 2025 100
3.6 Understanding Simpler Circuits 103
3.6.1 Innate Behaviours 103
3.6.2 The Sleep/Wake Cycle 105
3.6.3 Fear Circuits 109
3.7 Simple Circuits and Higher Levels of CNS Functions and Behaviours 113
3.8 Conclusion 115
References 115

Chapter 4 Disrupting Circuits of the Higher Central Nervous System 121

4.1 Introduction 121
4.2 Psychedelic Disruption: Old and New 122
4.3 The Immune System and Disruption of the CNS 128
4.3.1 COVID and Disruption of the Nervous System 128
4.3.2 Basic Immunology 130
4.3.3 Some History of Weapons Research During the Cold War 133
4.3.4 Knowledge of Autoimmune Disease Mechanisms Today 135
4.4 Immunopsychiatry Today 139
4.4.1 Autoimmune Encephalitis 140
4.4.2 Autoimmune Aspects of Neurodegenerative Diseases 144
4.5 The Wider Context of Scientific Advances in Neuroscience 147
4.6 Conclusions 151
References 151

Chapter 5 Advances in Other Science and Technology Relevant to Hostile Misuse of CNS-acting Chemicals 158

5.1 Introduction 158
5.2 Developments in CNS-acting Chemical Agent Discovery and Synthesis 159
5.2.1 Screening for Novel Chemical Agents 159
5.2.2 From Discovery to Rationale Engineering 165
5.2.3 Systems Biology, AI and Rational Drug Design 166
5.2.4 Identifying Retro-synthesis Pathways 167

5.2.5 Biotechnology Techniques Employed for Development and Synthesis of Novel Agents 168
5.2.6 Chemical Synthesis and Production of Biological Agents 173
5.3 Science and Technology Developments in CNS-agent Delivery and Uptake 174
5.3.1 Development of CNS-acting Agent Weapons Employing Limited Area Delivery Mechanisms 174
5.3.2 Developments in CNS-agent Uptake and Targeting the Brain and CNS 178
5.3.3 Dual Use Technologies that Could be Employed for Aerosolised Delivery of CNS-acting Agents 181
5.3.4 Commercial Dual Use Wide Area Delivery Aerosol Devices that Could be Employed for Delivery of CNS-acting Agents 186
5.3.5 Wide Area Aerosolised CNS-agent Weapons in Enclosed Spaces 188
5.3.6 Wide Area RCA Dissemination Devices 190
5.4 Conclusions 199
References 200

Section III: Holistic Arms Control Analysis

Chapter 6 The Biological and Toxin Weapons Convention and Preventing the Weaponization of CNS-acting Chemicals 225

6.1 Introduction 225
6.2 General Obligations Under the BTWC 226
6.3 CNS-acting Chemicals and the Scope of Coverage 227
6.4 Scope of Activities Regulated 232
6.5 Measures to Facilitate Implementation, Verification and Compliance 234
6.5.1 Confidence Building Measures 235
6.5.2 Review of Science and Technology (S&T) 236
6.5.3 Implementation Support Unit 240
6.5.4 Non-compliance Measures 240
6.5.5 Education and Awareness Raising 242
6.6 Conclusions 244
References 245

Chapter 7 The Chemical Weapons Convention and Preventing the Weaponization of CNS-acting Chemicals 253

7.1 Introduction 253
7.2 General Obligations Under the Chemical Weapons Convention 254
7.3 Scope of Coverage 254
7.4 The Definition of Chemical Weapons 255
7.5 Definitional Lacunae and Their Potential Consequences 256
7.6 Regulating CNS-acting Chemical Research, Development and Use 257
7.6.1 Scenario 1: State Research and Development of CNS-acting Weapons Intended for Use in Armed Conflict 257
7.6.2 Scenario 2: State Research and Development of CNS-acting Weapons Intended for Use for Law Enforcement Purposes 258
7.6.3 Scenario 3: State Research and Development of CNS-acting Chemical Agent Weapons for Protective Purposes 260
7.6.4 Scenario 4: Research and Development of CNS-acting Chemicals and Delivery Mechanisms Purportedly for Medical or Other Non-prohibited Purposes, but Which Can Be Employed as Weapons 261
7.7 OPCW 'Culture' and Working Practices 262
7.8 Measures to Facilitate Implementation, Verification and Compliance 262
7.9 Assessing and Responding to Science and Technology Developments 264
7.9.1 The SAB and the Third Review Conference 266
7.9.2 SAB and the Fourth Review Conference 267
7.9.3 SAB and the Fifth Review Conference 268
7.10 Allegations Made by CWC States of Development, Transfer and Use of CNS-acting Weapons Undertaken by Another CWC State Party 270
7.10.1 Allegations by the US 270
7.10.2 Russian Allegations 274
7.11 Conclusions 277
References 280

Chapter 8 International Human Rights Law Constraining the Weaponization of CNS-acting Chemical Agents **291**

8.1 Introduction 291
8.2 Restrictions on the Use of Force 292
8.3 Protection of the Right to Life 294
8.3.1 Application to CNS Acting Chemical Agent Weapons 295
8.4 The Rights to Freedom of Association and Assembly 302
8.5 The Rights to Liberty and Security 303
8.6 The Prohibition Against Torture and Other Ill-treatment 304
8.6.1 Non-consensual Application of CNS-acting Chemical Agents to Detainees in Prison/Law Enforcement Context 305
8.6.2 Torture in Non-custodial Contexts 311
8.7 Health Considerations and the Right to Health 312
8.7.1 US: Application of CNS-acting Chemicals for Law Enforcement Chemical Restraint 314
8.8 Considerations Regarding Means of Delivery and Dispersal of CNS-acting Chemical Agents 315
8.9 Obligations Regarding Design, Review, Training and Monitoring Use of 'Less Lethal' Weapons 317
8.10 Trade Agreements Incorporating IHRL and IHL Criteria 320
8.10.1 Conventional Arms and Dual Use Goods 320
8.10.2 Anti-torture Trade Instruments 324
8.11 Conclusions 327
References 328

Chapter 9 International Humanitarian Law Applicable to CNS-acting Chemical Agent Weapons **350**

9.1 Introduction: Nature of Scope of IHL 350
9.2 Over-arching IHL Obligations Constraining Development and Use of Weapons 352
9.2.1 Discrimination and Distinction 352
9.2.2 Superfluous Injury or Unnecessary Suffering 354
9.2.3 Identification and Protection of Persons Considered *Hors de Combat* 356
9.2.4 Preventing Malign Application of Medicine 358
9.2.5 Respect of International Humanitarian Law 358

9.2.6 The Martens Clause 359
9.3 Reviewing Legality of 'New' Weapons 360
9.3.1 Scope of New Weapons to be Reviewed 360
9.3.2 Criteria for Review 362
9.3.3 Restrictions on the Use of New Weapons 364
9.3.4 Mechanism of Review 365
9.3.5 Article 36 Implementation 367
9.4 Implementation of IHL: Investigation and Enforcement Measures 368
9.5 Conclusions 368
References 369

Chapter 10 Application of the United Nations Drug Control Conventions to CNS-acting Chemical Agent Weapons 381

10.1 Introduction 381
10.2 Single Convention on Narcotic Drugs 382
10.3 The UN Convention on Psychotropic Substances 386
10.4 Civil Society and State Discourse Concerning Potential Relevancy of SCND and CPS to Regulation of CNS-acting Chemical Agent Weapons 388
10.5 Conclusion 392
References 393

Chapter 11 The Scientific and Medical Communities' Role in Preventing Weaponization of CNS-acting Chemical Agents and Broader Misuse of Neuroscience 401

11.1 Introduction 401
11.2 Regulating 'Dual-use' Research 402
11.2.1 'Dual-use' Neuroscience 406
11.3 Oaths, Codes and Pledges for the Chemical and Life Science Communities 412
11.3.1 Ethical Codes for the Life Sciences 413
11.3.2 Ethical Codes for Chemists 416
11.3.3 Constraints on Effectiveness of Codes 419
11.4 Education and Outreach for Chemical and Life Science Communities 420
11.4.1 Education and Outreach to Chemists 420
11.4.2 Education and Outreach to Life Science Communities 424
11.5 Whistle Blowing 426

11.6 Medical Ethics and Ethical Practice 429
11.6.1 Regulating Scientific Research on Human Subjects 429
11.6.2 Upholding IHRL and IHL 431
11.6.3 Prohibiting Medical Involvement in Chemical and Biological Weapons Development 432
11.6.4 Developing Medical Ethical Standards Constraining Involvement in Development and Use of CNS-acting Chemical Agent Weapons 435
11.6.5 Developing and Applying Medical Ethical Standards to Regulation of the Life Sciences 437
11.7 Conclusion 439
References 439

Section IV: Conclusions and Recommendations

Chapter 12 Conclusions and Recommendations 459

12.1 Introduction 459
12.2 HAC Stage One: Weapons, Technology and Advances in Science 460
12.2.1 Weapons Development and Use in the Past 460
12.2.2 Advances in Neuroscience 461
12.2.3 Advances in Other Relevant Science and Technology 465
12.3 HAC Stage Two: Potentially Applicable International Law, Arms Control, Disarmament and Other Instruments and Attendant Control Regimes 468
12.3.1 The Biological and Toxin Weapons Convention (BTWC) 468
12.3.2 The Chemical Weapons Convention (CWC) 469
12.3.3 International Human Rights Law 472
12.3.4 International Humanitarian Law 472
12.3.5 International Drug Controls 473

12.3.6 Further Considerations Regarding International Instruments and Control Regimes 474
12.3.7 The Potential Role of Civil Society 475
12.4 HAC Stage Three: Analysis and Recommendations 477
12.4.1 Chemical Weapons Convention 478
12.4.2 Biological and Toxin Weapons Convention 481
12.4.3 International Humanitarian Law 483
12.4.4 International Human Rights Law 484
12.4.5 UN Drug Control Conventions 485
12.4.6 Chemical, Life Science and Medical Communities 486
References 490

Subject Index **492**

Section I: Introduction

CHAPTER 1

Neuroscience and the Future of the Chemical and Biological Weapons Disarmament and Arms Control Regimes

1.1 Advancing and Converging Chemical and Life Sciences

Science and technology can advance in different ways: by narrowing the focus and digging ever deeper into a particular topic to elucidate its finer details or by bringing together very diverse knowledge and techniques from quite different fields to develop new concepts and methods. Both approaches are necessary for rapid progress, but the second way, which has become known as 'convergence', is more likely to generate unexpected and potentially revolutionary changes. Such convergence appears to be a critical element in facilitating the rapid and far-reaching advances in the chemical, life and associated sciences that, from the 1990s onwards, have increasingly allowed scientists to understand and manipulate living systems in unprecedented ways. Of potential relevance have been the developments and convergences in biotechnology, medicinal chemistry, pharmacology, toxicology, immunology, molecular biology, systems biology, synthetic biology, and neuroscience as well as their interactions with nanoscience, artificial intelligence research and development, and computer science.

Ralf Trapp, the chemical and biological weapons expert, has highlighted the crucial role of convergence in these life science developments, giving one example of the fruits of such convergence as the growth of "'omics'—studies of all constituents collectively of a set of data/biomolecules such as genes

Preventing Weaponization of CNS-acting Chemicals: A Holistic Arms Control Analysis
By Michael Crowley and Malcolm Dando

Published by the Royal Society of Chemistry, www.rsc.org

(genomics), lipids (lipidomics), RNA (transcriptomics), proteins (proteomics), metabolites (metabolomics), the brain's structural and functional connections (connectomics) the cellular systems of an organism (cytomics) and more".[1] He argues that such advances have now made it possible to "edit genetic instructions, providing the tools for correcting genetic defects or directing organisms to produce molecules that are alien to them.... *At the same time, advances in bioinformatics, stimulation and modelling help scientists understand how biological systems work and how their processes can be augmented, interrupted or otherwise interfered with*". (Emphasis added).[2] In addition, capabilities for the biological production of chemical agents are also adding to chemists' already formidable capacity for manufacture by more traditional methods. This augmentation of capabilities is having an effect across the whole of the life sciences and certainly in work on chemical agents that act upon the central-nervous system (CNS) and the broader study of neuroscience which will be the subject of this publication.

Although capable of producing immense societal and health benefits, these capabilities can instead be misused for malign purposes. The need to prevent the potentially devastating consequences of the misuse of such Janus-like knowledge and technology without limiting its benefits for mankind – often called the dual-use problem – were highlighted by leading scientists right from the beginning of this life science revolution.

In 2000, Matthew Meselson, the Thomas Dudley Cabot Professor of the Natural Sciences at Harvard University, stated that:

> "During the century ahead, as our ability to modify fundamental life processes continues its rapid advance, we will be able not only to devise additional ways to destroy life but will also become able to manipulate it – including the processes of cognition, development, reproduction, and inheritance. A world in which these capabilities are widely employed for hostile purposes would be a world in which the very nature of conflict had radically changed. Therein could lie unprecedented opportunities for violence, coercion, repression, or subjugation".[3]

Such concerns were echoed at the time and subsequently by many others in the chemical and life science communities, as well as those engaged in chemical and biological weapons disarmament and arms control. Foremost among the latter being Julian Perry Robison who in 2008 warned that:

> "...Advances in the life sciences, coupled with technologies that allow the analysis and construction of complex biologically active molecules, would eventually make it possible to design a chemical that would interfere with *any* life process that could be understood in molecular terms. And thoughts could also turn to the idea of modifying the genes of factory or vector organisms so as to obtain such chemicals as gene products..."

And he continued:

> "...Because of this convergence of chemistry and biology, we may eventually be able to manipulate at will the processes of development, inheritance, reproduction, locomotion, sensation, cognition and any other process that keeps us working as normal human beings .... *It is the potential for changing human beings and doing so in pursuit of who-knows-what strategy of aggression that makes CBW [chemical and biological weapons] uniquely menacing: weapons, in the extreme, of de-humanization – a risk, therefore, to the whole human species; an existential risk...*"[4] (Emphasis added)

These concerns remain or, if anything, have increased in their pertinence, as our knowledge and capabilities have developed. In 2023, leading life scientists David Gillum, George Poste, Craig Woods, and Rachel Levinson again warned:

> "Life sciences progress is creating both the potential for unprecedented benefits in medicine (and other areas) as well as new risks of harm from accidents, unforeseen consequences, or even deliberate abuse. There has never been a major technological disruption that countries haven't sought to use for industrial or military superiority, or for terrorists and criminals in pursuit of their aims..."

And they pointed to the key issue of precision in our growing understanding of living organisms' physiology:

> "The unifying principle in biotechnology research is understanding the varied genetic regulatory networks and molecular circuitry that encode biological functions in different life forms, from single-cellular micro-organisms to humans. *This knowledge provides the foundation for ever more precise manipulation of biological systems.* [These bring] myriad positive benefits of biotechnology - but the same insights into the organization of the molecular systems creates almost limitless vulnerabilities for their deliberate, targeted disruption".[5] (Emphasis added)

In this publication we focus on one rapidly advancing area of this ongoing revolution, neuroscience, explore the dangers of its malign application and what can be done to mitigate them.

1.2 Overview of Advances in Neuroscience

Neuroscience has been described as:[6]

> "...a large field founded on the premise that all of behavior and all of mental life have their origins in the structure and function of the nervous system..."

And it has been argued further that:

> "...Towards the end of the twentieth century, the study of the brain moved from a peripheral position within both the biological and psychological sciences to become an interdisciplinary field that is now central within each discipline..."

Thus, neuroscience studies the whole of the nervous system from structure to function, development to degeneration, in health and in disease. Whilst it covers the entire central nervous system, one primary area of focus has been on the brain.

There are many different branches of neuroscience. Each focuses on a specific topic, body system, or function including[7] developmental neuroscience describing how the brain forms, grows, and changes; cognitive neuroscience investigating how the brain creates and controls thought, language, problem-solving, and memory; molecular and cellular neuroscience exploring the genes, proteins, and other molecules that guide how neurons function; neurogenetics focusing on inherited changes to neurons, including studies of certain genetic diseases; behavioural neuroscience examining the brain areas and processes underlying how animals and humans act; clinical neuroscience exploring how to treat and prevent neurological disorders and how to rehabilitate patients whose nervous system has been injured; neurophysiology studying the central nervous system itself and how it functions; and sensory neuroscience which examines features of the body's sensory systems and how the central nervous system interprets and processes sensory information.

Relevant key developments in relevant aspects of neuroscience and associated life sciences during the 20th century will be explored in later chapters of this book. For now, it is important to note that despite the discovery of a large number of CNS-acting chemical agents during the 1950s–1980s and their successful employment to ameliorate or treat neurological and psychiatric illnesses, the underlying mechanisms of their action were largely unknown. As was noted in the 1978 second edition of *Principles of Psychopharmacology*:[8]

> "The complexity of the CNS, coupled with inadequate tools for observation, are formidable barriers in gaining information about molecular interactions between CNS drugs and receptor sites..."

Yet, as a result of the contemporary advances in neuroscience we have moved from our previous situation in the 20th century where we could only gain a mechanistic understanding of the neuronal processes underlying behaviour in simple or very specialized systems to the 21st century where we are increasingly able to understand and subsequently manipulate living systems, including the human CNS, in an ever more targeted, controlled and sophisticated fashion.

There is certainly a widely held view that major advances can now be made in our understanding of the neuronal mechanisms underlying our behaviour, including the roles played by bioregulators in such processes, and that the new technologies will allow yet further advances.[9] The obvious advantages of this work, in helping people with brain dysfunctions and injuries, have led States including China,[10] the countries of the European Union,[11] Japan,[12] and the US[13] to institute large-scale funding initiatives to underpin brain research. This optimism about our increasing ability to investigate the interacting neuronal circuits producing human behaviour is clear from the report *BRAIN 2025: A Scientific Vision* for the US National Institutes of Health,[14] which stated that:

> "In considering these goals and the current state of neuroscience, the working group identified the analysis of circuits of interacting neurons as being particularly rich in opportunity, with potential for revolutionary advances..."

The report went on directly to explain:

> "...Truly understanding a circuit requires identifying and characterizing the component cells, defining their specific connections with one another, observing their dynamic patterns of activity as the circuit functions *in vivo* during behavior, and perturbing these patterns to test their significance. It also requires an understanding of the algorithms that govern information processing within a circuit and between interacting circuits in the brain as a whole...".

There appears to be great optimism that significant advances in neuroscience are achievable in our lifetimes. In 2015, one eminent neuroscientist, Ralph Adolphs, suggested that by 2065 amongst the problems that we would be able to solve were: how do we make decisions; why we sleep and dream; how learning and memory work; and what causes psychiatric and neurological illnesses?[15]

1.2.1 Areas of Neuroscience of Potential Dual-use Concern

Undeniably, the growth in neuroscience, and associated technological advances, will be of significant benefit in our understanding and treatment of currently devastating and widespread psychiatric and neurological illnesses, but given its multi-faceted applicability, such knowledge and capabilities could instead be exploited in the development of CNS-acting chemical agent weapons or for other malign manipulation of the CNS and additional core human physiological systems. In 2000, the distinguished scientist George Poste argued that "as we begin to understand the exquisite molecular mechanisms that regulate this remarkable structure called the human body... the ability to understand those circuits means simultaneously we gain the capacity to scramble them".[16] He specifically warned that with such understanding of brain circuits and the capabilities to affect them, malign actors "can engineer a series, a complete spectrum of activity from transient immobilization (...) to catastrophic effects which can be acute or chronic" and so could in effect construct a "brain bomb".[17]

Responding to such concerns, in this book we will examine the contemporary and potential future risks of State-sanctioned malign application of neuroscience and related chemical and life sciences, concentrating on the development of incapacitating chemical agent (ICA) weapons, predominately those employing CNS-acting chemical agents (CNSAC). For the purposes of our study, an "ICA weapon" will be considered to comprise an ICA and associated means of delivery developed with the purported intended purpose to cause prolonged but non-permanent disability or incapacitation of the target. ICA weapons can potentially be developed to target a wide range of core physiological systems beyond the CNS such as the respiratory, circulatory or immune systems. For clarity where such weapons employ chemicals acting on the CNS – producing non-permanent loss of consciousness, sedation, hallucination, incoherence, paralysis, disorientation, or other such effects – they will be termed "CNS-acting chemical agent (CNSAC) weapons".[18] Many putative CNS-acting agents have low safety margins and inappropriate doses cause serious, sometimes permanent, health effects and may result in death. Chemicals of biological origin – notably toxins and bioregulators – and their synthetic analogues have been amongst the substances explored as potential CNS-acting or other types of ICA weapons.[19]

The threats from contemporary and potential future CNSAC weapons development are given added resonance when we consider the previous attempts by certain States, commencing in the Cold War and continuing thereafter, to discover or synthesise and subsequently weaponise CNS-acting chemical agents, and more broadly to manipulate the human CNS for malign purposes. As will be discussed in Chapter 2, it appears that such research and development activities were stimulated by the then-revolutionary advances in neuropharmacology that were taking place in the 1950s, 1960s and thereafter. At first it seems that the Cold War weapons scientists explored all potential pathways and known CNS-acting chemical agents.

The strenuous and comprehensive nature of these investigations was made clear in the 1997 version of the US *Textbook of Military Medicine: Medical Aspects of Chemical and Biological Warfare*. The chapter on 'Incapacitating Agents' stated bluntly that:[20] "[V]irtually every imaginable chemical technique for producing military incapacitation has been tried at some time". The text continued, specifically in regard to the US programme:

> "...Between 1953 and 1973, at the predecessor laboratories to what is now the U.S. Army Medical Research Institute of Chemical Defense, many of these were discussed and, when deemed feasible, systematically tested. Chemicals whose predominant effects were in the central nervous system were of primary interest and received the most intensive study..."

Table 1.1 provides an indication of the range of substances previously explored as potential CNS-acting chemical agent weapons in Cold War and subsequent State weapons programmes.

As the State weapons programmes continued, however, they increasingly narrowed the range of potential weapons candidates to those CNS-acting chemicals that directly interfered with routine lower order systems in the brain such as awareness, and could preferably be delivered as aerosols. The Scientific Advisory Board of the Organisation for the Prevention of Chemical Weapons (OPCW) summarised the outcome of all such investigations, as far as was known from open sources, in its 2012 report to the Third CWC Review

Table 1.1 Summary of selected substances previously explored as potential CNS-acting chemical agent weapons.

Psychedelics	Psychedelics are a group of hallucinogens that have a similar indolamine structure to serotonin (5-hydroxytryptamine, 5-HT) and include LSD (lysergic acid diethylamide) and psilocybin.[21] They exert their effects on the brain through action on 5-HT_{2A} receptors. These effects are dependent on the person, context and dose, but can produce profound alterations in waking consciousness that include:[22] "visual hallucinations, audio-visual synesthesia, and...derealization and depersonalization phenomena". Such substances have a long history of use in human societies, but modern work started with the chance discovery of the effects of LSD by the chemist Albert Hofmann in 1943. Following the discovery of serotonin in the brain of mammals a decade later, this quickly led to considerable civil research based on the assumption that LSD in some way interfered with the normal function of serotonin in the brain. There was military interest in the possible use of LSD as a CNS-acting weapon. The UK, for example, carried out laboratory tests on people briefly in the early 1950s and then more systematically alongside field trials in the early to mid-1960s. However, in 1969, UK officials: "judged LSD unsatisfactory as an operational agent because of the unpredictability of its effects and because it was expensive to produce".[23]
Cannabinoids	Delta-9 tetrahydrocannabinol (THC), the active ingredient of marijuana, was investigated as a potential CNS-acting chemical agent weapon in the 1960s, but it was not found to be sufficiently potent to be of interest.[24] However, since that time our understanding of the brain system affected by THC, the function of that system, and much more powerful agents with similar effects to THC has radically changed. In 1988 it was possible to identify cannabinoid receptors in the brain that were affected by THC and other cannabinoids. Then, in the early 1990s, it was discovered that there were substances in the brain that had a similar activity to THC. These endocannabinoids include anandamide (arachidonoyl ethanolamide),[25] and a variety of chemicals that affect these receptors were developed to further investigate this novel system.

Table 1.1 (*Continued*)

Opioids	Morphine is the prototypical opioid analgesic used in the treatment of moderate to severe pain, however its use is associated with respiratory depression, sedation and addiction. The search for novel opioid narcotic agents that do not cause such side effects is still being actively pursued.[26] Fentanyl, for example, is a synthetic opioid many times more powerful in its effects than morphine, and a large number of fentanyl derivatives have been investigated, although the potential for respiratory depression remains problematic. However, despite advances in analogue synthesis,[27] greater understanding of analogue interactions with the μ receptor in humans and consequent effects upon their targets, the operational use of such fentanyl analogues as CNS-acting weapons by Russian Security forces in 2002 resulted in many deaths and reportedly also immediate and long-term deleterious health effects.[28]
Benzodiazepines	GABA (γ-aminobutyric acid) is a major inhibitory neurotransmitter in the central nervous system. Benzodiazepines enhance the effect of GABA on $GABA_A$ receptors and have therefore found use in the treatment of anxiety and induction of anaesthesia, but they also affect respiration and the blood system. Much research has been carried out to, for example, find faster- and shorter-acting agents with more precise effects, and some of this research may have applicability for CNS-acting weapons development.[29] Benzodiazepines, such as diazepam, are used in drug therapeutic mixtures for treating the convulsions in nerve agent poisoning, so they also have a civil and military medical countermeasure use in this instance.[30]
Alpha 2 adrenoceptor agonists	The *locus coeruleus* neurons in the brain produce the neurotransmitter noradrenaline and have widespread ramifications that function to help induce the alert wakeful state. Alpha 2 adrenoceptors provide inhibitory negative feedback to the *locus coeruleus* neurons when they produce noradrenaline.[31] An agent such as dexmedetomidine, therefore, that mimics the effect of the natural transmitter (*i.e.*, an analogue) can also reduce alertness and wakefulness. It can therefore find medical use in anaesthesia and has also been investigated as a potential CNS-acting weapon.[32]
Neuroleptic anaesthetic agents	Neuroleptic anaesthesia, unlike conventional general anaesthesia, produces a state of unawareness in the patient characterised by unconsciousness and analgesia whilst the patient's muscle tone and reflexes remain largely intact.[33] Consequently, researchers have highlighted the potential application of combinations of neuroleptic anaesthetic agents as CNS-acting weapons, particularly given the possibility of developing a mixture of agents that would produce the neuroleptic state without causing undesired side effects.[34]

Table 1.1 (*Continued*)

Bioregulators	Bioregulators are naturally occurring chemicals produced within living organisms that help to ensure the proper functioning of vital physiological systems in those same living organisms. In mammals bioregulators control vital and diverse functions such as temperature, sleep, blood pressure, heart rate, consciousness, mood and the immune response. However, while they occur naturally in the body at low concentrations, they can be extremely toxic at higher concentrations or if the molecular structure is altered.[35] Bioregulators were previously explored as potential CNS-acting chemical agent weapons, notably by the Soviet Union.[36]

Conference as follows: "[T]he types of chemicals and pharmaceuticals, known to have been considered as [CNS-acting chemical agent weapons] incapacitants from open literature sources, were discussed. Most are centrally acting compounds that target specific neuronal pathways in the brain. All of them emerged from drug programmes undertaken from the 1960s to the 1980s, as far as can be judged by the research that has been published".[37]

According to the 2012 *Brainwaves* study by the UK's Royal Society: "Many different forms of incapacitation were investigated during the Cold War, but with increasing emphasis on rapid action and short duration of effects, contemporary interest has tended to focus on sedative-hypnotic agents that reduce alertness and as the dose increases, produce sedation, sleep, anaesthesia and death".[38]

Although we cannot be certain, it appears that this narrowing of focus was not a deliberate scientific ethical choice to avoid attempting to interfere with the high functions of the human CNS, but instead a forced choice caused by the lack of understanding at the time, notably with regard to the circuits and neuronal receptors in the CNS, and their interactions with potential CNS-acting agents. Subsequently, our knowledge of, and ability to manipulate, the CNS has vastly expanded, opening up further areas of neuroscience research – including molecular neuroscience, genetic neuroscience, drug discovery technology, research into bioregulators and drug delivery to the brain. Such developments could facilitate the discovery of novel CNS-acting agents, the modification of existing CNS-acting agents (including those studied and disregarded by previous weaponeers), the design and synthesis of bespoke CNS-acting agents that can affect with greater and greater specificity, targeted functions of the CNS, and the more effective delivery and uptake of these CNS-acting agents to their targets.

1.2.2 Implications for Monitoring Dual Use Research and Preventing the Weaponization of CNS-acting Chemical Agents

Any contemporary and future search for candidate CNS-acting agents with potential weapons utility – and consequently, the range of research and associated activities of potential dual-use concern – is likely to be informed by the current advances in neuroscience, as well as pharmacology, genomics, and related disciplines which have revolutionised our understanding of the body's bioregulatory systems and notably brain neurotransmitter/neuroreceptor systems. Although such research is at a relatively early stage in the understanding of the ways in which biologically-active-peptides and related chemicals are employed in the body's regulatory systems and specifically the brain's information-processing system, the search for novel CNS-acting agent weapons may prove attractive to certain States as new potential opportunities arise.

Consequently, our awareness of the potential areas of dual-use concern should broaden.

In such circumstances, in addition to work on more efficient new methods of extraction or production of known naturally occurring CNS-acting agents (*i.e.*, neurotoxins and bioregulators) and chemical synthesis of existing pharmaceutical CNS-acting agents, other dual-use research could also raise concern. For example, attempts to modify existing neurotoxins and bioregulators; to design and synthesise novel (more effective) analogues of known agents; to study the structure of known neuroreceptor sub-types and their interaction with relevant neurotoxins, bioregulators and their synthetic analogues; to explore and modify the neuroregulatory systems; or to explore novel packaging and transportation of peptides to potential targets across the blood-brain barrier or within the wider CNS, might raise legitimate concerns. Similarly, studies exploring potential CNS-acting agent or surrogate agent aerosolisation, dispersal, and uptake for which there appears to be little justification for medical, veterinary, or other peaceful purposes need to be closely reviewed and evaluated. In addition to monitoring such processes, it is important to look beyond research directly connected to particular potential candidate CNS-acting agents or means of delivery and also explore the mechanisms by which such dual-use research may be transformed from that undertaken to further knowledge, or for the benign development of drugs or other measures to alleviate neurological and psychiatric illness, to be instead employed in the development of weapons.

Given the radically changing scientific and technological landscape and the corresponding increasing potential threats of malign use, we next explore the current chemical and biological weapons arms control and disarmament treaties and briefly assess their currently limited and circumscribed capabilities to meet the contemporary as well as the growing future threats.

1.3 Current Limited Regulation of CNS-acting Agents Under CBW Arms Control and Disarmament Treaties

The norm against the deliberate use of both poison and disease as weapons of warfare can be traced back to antiquity. These ancient interconnected taboos subsequently became embodied, codified, and expanded in the twentieth century and continue to this day through three international instruments: the 1925 Geneva Protocol, the Biological and Toxin Weapons Convention (BTWC) and the Chemical Weapons Convention (CWC). They establish two distinct yet overlapping regimes prohibiting biological weapons and chemical weapons respectively which, in theory at least, both prohibit toxin weapons, and collectively encompass CNS-acting agents within their scope.

1.3.1 The 1925 Geneva Protocol

Adopted in 1925, the Geneva Protocol prohibits the use in war of "asphyxiating, poisonous or other gases, and of all analogous liquids, materials or devices" and this was subsequently extended to "bacteriological methods of warfare".[39] This prohibition has now become established under customary international law (and is consequently applicable to all States even if they are not party to the Protocol) and applies to both international and non-international armed conflict. Although the range of substances covered by the Geneva Protocol has been contested by certain States, for example with regard to riot control agents, the weaponised employment of CNS-acting chemical agents in armed conflict would clearly breach the Protocol. However, its prohibitions relate solely to the use of such weaponised agents in armed conflict – consequently, the Geneva Protocol does not prohibit the development, production, transfer, or stockpiling of such weapons. Furthermore, it would not prohibit the use of such weapons outside armed conflict, for example for law enforcement purposes.

1.3.2 The Biological and Toxin Weapons Convention

Negotiated in 1972, the BTWC explicitly prohibits the development, production, stockpiling, acquisition or retention of biological and toxin weapons and implicitly bans their use. Unfortunately, the Convention does not formally define "toxins" in its text. However, toxins (and we would argue bioregulators when employed as *de facto* toxins) are encompassed under Article I of the Convention, which declares that:

> "Each State Party to this Convention undertakes never in any circumstances to develop, produce, stockpile or otherwise acquire or retain:
>
> 1. Microbial or other biological agents, or toxins whatever their origin or method of production, of types and in quantities that have no justification for prophylactic, protective or other peaceful purposes.
> 2. Weapons, equipment or means of delivery designed to use such agents or toxins for hostile purposes or in armed conflict".[40]

Article I's wide and essentially open-ended scope – which makes no restriction with regard to lethality – has become known as the General Purpose

Criterion (GPC). It has subsequently been regularly confirmed and applied by the BTWC States Parties to the rapidly advancing life sciences and associated technologies through the Additional Understandings adopted as part of the Final Documents of repeated BTWC Review Conferences. Consequently, those CNS-acting toxic chemicals which are of biological origin (*i.e.*, toxins) or are synthetic analogues of such substances – regardless of their lethality – are covered under the scope of the BTWC, with their weaponization being prohibited. Although the development and use of any CNS-acting toxin weapons in armed conflict is clearly prohibited by the Convention, the BTWC States Parties have not collectively clarified whether such comprehensive prohibitions extend to CNS-acting toxin weapons intended for law enforcement purposes.

1.3.3 The Chemical Weapons Convention

Negotiated in 1993, the CWC, under Article I, explicitly prohibits the development, production, stockpiling, acquisition, retention or use of "chemical weapons".[41] These are consequently defined under Article II:

> "1. 'Chemical Weapons' means [*inter alia*]: a. Toxic chemicals and their precursors, except where intended for purposes not prohibited under the Convention, as long as the types and quantities are consistent with such purposes".[42]

And

> "2. 'Toxic Chemical' means: Any chemical which through its chemical action on life processes can cause death, temporary incapacitation or permanent harm to humans or animals. This includes all such chemicals, regardless of their origin or of their method of production, and regardless of whether they are produced in facilities, in munitions or elsewhere".[43]

It can be seen that through its interacting definitions of "toxic chemicals" and "chemical weapons", the CWC – as with the BTWC – has been designed to incorporate a General Purpose Criterion (GPC). In the case of the CWC, this GPC consequently ensures the Convention a comprehensive and essentially 'open-ended' scope of toxic chemicals encompassed whether they be naturally occurring (*i.e.*, toxins and bioregulators) or synthesised by humans; it covers all such existing agents as well as those that will be discovered or synthesised in the future. Once again, as with the BTWC, the Convention makes no explicit reference to lethality as a determinant in coverage, so the CWC would encompass all CNS-acting chemical agents, including those explored or developed as purportedly "less lethal" weapons.

Although the Convention completely prohibits "chemical weapons", it does not ban the development and employment of certain types of "toxic chemicals" – notably including riot control agents – for "purposes not prohibited" which specifically include "law enforcement including domestic riot control".[44] However, the use of "toxic chemicals" for such "purposes not prohibited" would only be permitted "as long as the types and quantities are consistent with such purposes".[45] At present there are divergent views amongst CWC States Parties on whether the CWC permits development and use of at least certain CNS-acting weapons for law enforcement purposes, and if so what types of weapons and for what law enforcement activities. This issue remains highly contested despite a 2021 Decision by the Conference of States Parties (CSP) clarifying that aerosolised use of CNS-acting chemicals was inconsistent with law enforcement as a 'purpose not prohibited' under the CWC.[46]

The comprehensive regulatory coverage that should be provided collectively by the BTWC and the CWC has become more and more important as the ongoing revolution in the chemical and life sciences has proceeded and the boundaries between chemistry and biology have become increasingly blurred. Such developments have broadened the range of potential agents available to today's weaponeers and have also meant that the clear distinctions previously drawn between chemical and biological weapons have become less useful.

Consequently, rather than thinking of chemical and biological weapons threats as separate, the current authors have been among those arguing that it is more appropriate to conceptualise such agents as lying along a continuous biochemical threat spectrum running from the traditional chemical weapons agents (including nerve, blood, blister and choking agents) on one extreme, through certain pharmaceutical chemicals, toxins, and bioregulators which could all be used as CNS-acting chemical agent (CNSAC) weapons (or indeed other "incapacitating chemical agent weapons") and on to biological weapons agents (including traditional and genetically modified weapons agents).[47] Figure 1.1 shows this biochemical threat spectrum and also indicates how the BTWC and CWC should cover (and where appropriate prohibit) the weaponization of all such substances.

From the preceeding discussion and review of Figure 1.1, it would appear that a range of substances – notably certain CNS-acting chemical agents (and other types of incapacitating chemical agents) – are covered under the scope of the BTWC and the CWC, and presumably thus be afforded an increased level of scrutiny and regulation. However, because of long-standing textual ambiguities in both Conventions and divergent interpretations by States Parties, exacerbated by inconsistencies and failures in implementation by certain States, the apparent overlapping protection of the BTWC and the CWC with regard to these mid-spectrum agents in reality masks a dangerous regulatory gap. Perry Robinson long ago recognized and warned of the consequences of this regulatory lacuna:

> "Such overlap ought to mean, one might think, that the weapons are well controlled, being subject not just to one but to two international disarmament treaties. In the real world, however, that is not the way

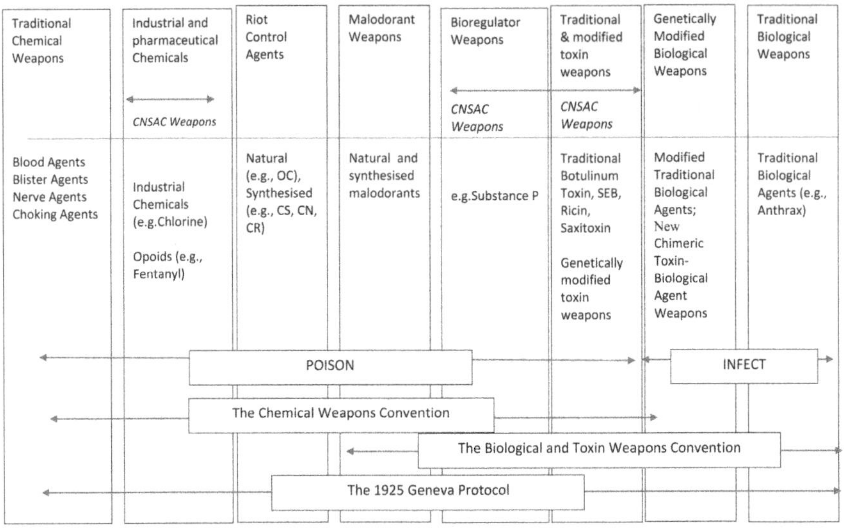

Figure 1.1 The CBW Threat Spectrum.

> it is. That overlap seems simply to have given people involved in implementing one of the two treaties opportunity to relinquish responsibility for anything also covered by the other treaty. The area of overlap thus risks becoming a gulf into which things disappear. It looks like this has been happening to toxins".[48]

Further limitations in the effectiveness of both Conventions arise from the fact that they were negotiated in the Cold War and primarily designed to address then-immediate threats of State-level development and large-scale use on the battlefield of chemical, biological and toxin weapons, and to facilitate the destruction of all extant weapons production capacity and stockpiles.

Consequently, there are serious questions about whether these two Conventions and the associated control regimes are sufficient, in themselves, to prevent and respond adequately to the diverse potential malign applications of today's rapidly advancing and converging chemical and life sciences, the capabilities of which could be employed by a widening array of State and non-State actors. This is particularly relevant when weaponised agents, including CNS-acting chemical agents, are intended to be used by military, security or police forces, beyond the traditional battlefield; notably in counter-insurgency, peacekeeping or law enforcement operations; and where they could be used to facilitate or conduct human rights violations ranging from torture or ill-treatment of detainees, large scale repression of protesters, or even subjugation of entire populations. Alternatively, such agents may instead be developed or acquired and subsequently misused by non-State actors including private military or security companies, armed opposition groups, terrorists or criminal organisations.

Given such limitations in the existing CBW arms control and disarmament architecture, we employ a Holistic Arms Control methodology to examine additional and/or alternative instruments and other measures that could be brought to bear to help to prevent and address the development and use of these CNS-acting chemical agent weapons, and the broader malign application of the chemical and life sciences.

1.4 Holistic Arms Control

For many years the governmental and non-governmental arms-control communities have sought to construct strategies to combat the development, proliferation and subsequent use of chemical and biological weapons by State and non-State actors. Recognising that reliance upon a single disarmament or arms-control agreement alone would not guarantee success, scholars have explored a number of concepts seeking to broaden the range of possible regulatory mechanisms. Utilising and building upon such work, particularly the concepts of "*preventative arms control*"[49] and "*webs of prevention (or protection)*",[50] a "*holistic arms control*" (HAC) framework for regulation has been developed,[51] and in this book it is adapted and applied to frame an exploration of effective strategies to address the threats arising from the misapplication of the rapid advances in neuroscience, and more broadly, the chemical and life sciences, to help to prevent the development of the next generation of CNS-acting chemical agent weapons.

Whilst the proposed HAC analytical framework concentrates upon existing traditional arms-control and disarmament measures (in this case primarily the Biological and Toxin Weapons Convention and the Chemical Weapons Convention), it attempts to identify and explore application of further mechanisms for regulation (notably to include international human rights law, international humanitarian law, and medical and scientific ethical frameworks), and also the nature of the actors involved in such potential regulatory measures (such as the chemical, life science and medical communities). In doing so, HAC attempts to make connections between these different bodies of law and other measures, examining how they interact (or could interact); where synergies exist so that one body of law can interact with another, actors can exchange knowledge, regulatory outcomes, mechanisms and resources, *etc.*

Consequently, HAC can be thought of as a framework intended to facilitate the development of a comprehensive, layered, flexible and strategic approach to link traditional arms control and disarmament with other relevant bodies of law to exploit regulatory and operational synergies between them. In this sense, HAC:

- Is tailored for and unique to the specific type of weapon or technology under consideration rather than for a broad grouping of weapons;
- Is not necessarily limited to a single existing arms control or disarmament treaty, but actively explores and seeks to incorporate States'

existing responsibilities under the full range of relevant international laws and applicable agreements;
- Potentially covers all stages of a weapon's existence, *i.e.*, research and development conducted as part of a weapons programme, mass production, stockpiling, deployment, use, transfer and destruction;
- Seeks to identify the types of permitted and prohibited weapons, acknowledges where existing ambiguity lies and highlights potential mechanisms for resolving such ambiguities;
- Seeks to identify existing constraints upon the permitted use of weapons, *i.e.*, legitimate targets (*e.g.* whether this would include armed combatants, terrorist organisations, criminals, civilians), legitimate types of operations (*e.g.* whether this would include law enforcement, military operations other than war, armed conflict) and how such operations should be conducted (rules of engagement); acknowledges where existing ambiguity lies and highlights potential mechanisms for resolving such ambiguities;
- Seeks to identify measures to facilitate effective national implementation as well as reporting, transparency, verification and enforcement mechanisms;
- Seeks to identify measures for regular "feedback", both "official" (*e.g.*, annual State reporting) and "unofficial" (*e.g.*, civil society briefings to States Parties or regime secretariats, media reports), and a comprehensive review of the relevant control regime(s);
- Explores mechanisms to facilitate regime adaptation to ensure continued effectiveness and relevancy in response to developments including in:
    - the nature of the use/misuse of weapons in practice;
    - science and technology (and can be employed to facilitate regulation of weapons and technologies that have not been invented yet);
    - international law, particularly that limiting types and use of weapons;
- Recognising that States are the prime actors in existing regulatory regimes, allows for and encourages participation by the full range of relevant stakeholders.

HAC consists of the following three stage process:

Stage one: constitutes an examination of the nature of the weapons and technology to be controlled and explores the current and potential future scenarios of application, together with the attendant national and human security concerns of inappropriate use. During this stage the potential relevance of advances in science and technology is assessed.

Stage two: constitutes an analysis of a range of potentially applicable international law; arms control, disarmament and others instruments and attendant control regimes; highlighting strengths, weaknesses and ambiguities in these mechanisms. The potential roles of relevant non-State actors are also explored.

Stage three: following an analysis of information derived from stages one and two, a strategy is developed to strengthen existing mechanisms and/or introduce new mechanisms to facilitate effective regulation or prohibition of the weapon or weapons-related technology of concern.

The manner in which this three-step process is undertaken will depend on, and be constrained by, the experience, expertise, time and resources that a researcher or research body can call upon, as well as the nature of the weapon or weapons-related technology being investigated. It will also necessitate a balance between:

- breadth, *i.e.* the range of potentially applicable instruments, control regimes, international law and other potential regulatory actors examined; and
- depth, *i.e.* the level to which each relevant mechanism is examined.

Due to such considerations Stage Two of our HAC analysis of the potential regulatory mechanisms and measures that could be employed to constrain or prevent weaponisation of CNS-acting chemicals has not been exhaustive, instead we have endeavoured to examine the most relevant and promising regulatory mechanisms and measures currently available. Similarly for Stage Two we have concentrated upon the core prohibitions, constraints and requirements established in the relevant international law and instruments, and have not sought to examine in detail how such high-level obligations are translated and implemented by individual States through national legislation and associated measures.

1.5 The Structure of the Book

The conception, development and organisation of this book have been informed by the HAC analytical framework. Following this introduction (*Section I*), the book consists of eleven chapters divided into four sections.

Section II provides an overview of the relevant advances in neuroscience and the chemical and life sciences more broadly that could be misused for the development of neurological weapons. Chapter Two examines relevant chemical and life science developments from 1945 to the turn of the century and their subsequent misuse in military, security and police spheres. It traces relevant chemical and life science advances from the discovery of a range of CNS-acting chemicals that helped people with mental health problems during the late 1940s through to the completion of the Human Genome Project at the turn of the century and the vision of this genomic knowledge being used to radically advance our understanding of the functions of the CNS. It then surveys previous and contemporary misuse of such knowledge derived in this period, including through case studies of chemical weapons research and development projects targeted at the CNS, and broader core human physiological systems undertaken by UK, South African,

US, Soviet/Russian, and Chinese military, security or police entities sometimes with the involvement of commercial companies. The chapter also explores the nature and scope of previous and contemporary misuse of these scientific developments notably by State officials including those in intelligence and law enforcement bodies to conduct torture and other ill-treatment such as forced interrogation and chemical restraint.

Chapter Three examines the 21st century advances in our understanding of the circuits in the CNS, charting the way in which advances in applications of genomics have been combined with rapid changes in neurotechnology, for example in optogenetics, to open up new ways of exploring the neuronal circuits that underly our behaviour. This chapter covers research in both simple circuits in model animals and complex circuits in human and non-human primates.

Chapter Four then examines the 21st century advances in our understanding of the disruption of circuits in the CNS. It first investigates the implications of the renewed interest in the medical community in using psychedelic drugs to assist people with mental illnesses. It then turns to the implications of the recent discovery of the impact of errors in the immune system on the functions of the CNS. Mental illnesses have complex causes, but one of the barriers previously problematic in elucidating these was the dogma that the CNS was practically isolated from the immune system of the body. However, it is now clear that this dogma was incorrect: the CNS is not isolated from the immune system of the rest of the body, but has its own immune defence system. Also some complex diseases of the CNS such as Parkinson's disease and narcolepsy are caused by specific destruction of key neuronal elements in important underlying circuits probably by autoimmune processes. Moreover, there is clear evidence that dysfunctions of the immune system may underly some psychiatric problems. As in the previous chapter, whilst the focus will be on contemporary CNS disease research of benign intent, the chapter also assesses relevant research trajectories of potential malign application.

Finally, Chapter Five reviews previous and contemporary advances in relevant science and technology, including that relating to artificial intelligence, that could facilitate the development of CNSAC weapons, notably through improved agent discovery, production methods, delivery and targeting of potential CNS-acting chemicals to specific systems in the body. It describes previous and contemporary delivery mechanisms employed in CNSAC weapons and then explores the contemporary development of wide area and remote-control mechanisms promoted for the delivery of riot control agents over large areas or extended distances, that could be adapted for the delivery of alternative payloads potentially including CNS acting chemicals.

Section III of the book provides an analysis of the two key arms control and disarmament treaties relevant to preventing the misuse of the chemical and life sciences, namely the BTWC and the CWC. Chapter Six assesses the current state and future development of the BTWC in the 21st century, providing a summary of the treaty and its application to chemicals of biological origin (*i.e.*, toxins and bioregulators) and their synthetic analogues that affect the CNS. It follows the attempts to strengthen key BTWC provisions in relation to advances in science and technology that are particularly relevant to attacks on the CNS. It explores how this issue was dealt with in the 9th BTWC Review Conference in 2022 and the following implementation process in 2023 and 2024, and assesses the likely developments through to the 10th BTWC Review Conference in 2027.

Chapter Seven assesses the current state and future development of the CWC in the 21st century, providing a summary of the treaty and its application to CNS-acting chemicals. Following the use of fentanyl derivatives by Russian security forces to break the 2002 Moscow Theatre siege this chapter analyses the subsequent State debates on how to deal with CNS-acting chemical agents through to the 2021 CSP Decision and the consequences of that Decision in the context of the general breakdown of consensus decision-making within the OPCW. It explores how this issue was dealt with at the 5th CWC Review Conference in 2023 given the requirement in the 2021 Decision for input from the SAB on this issue at the Review Conference. Particular attention will also be paid to the results of the Temporary Working Group on Biotoxins and its recommendations as to further work required on toxins and bioregulators. The chapter will additionally review likely developments through to the 6th CWC Review Conference in 2028.

Section III also explores the potential application, beyond the BTWC and CWC, of other relevant international law, standards and mechanisms that could facilitate regulation of neuroscience, and the chemical and life sciences more broadly, to prevent their malign application. Chapter Eight surveys potentially applicable international human rights law (IHRL) concentrating upon that relating to the rights to life; to liberty and security; to freedom from torture and other cruel, inhuman or degrading treatment or punishment; to engage in peaceful protest; and to health; together with attendant obligations on the restraint of force. Case studies explore the application of IHRL to the Russian CNS-acting chemical agent weapon used in the Moscow theatre siege; CNS-chemical facilitated torture and ill-treatment; Indian CNS-chemical "narcoanalysis" and other State use of coercive CNS-chemical interrogation; and the misuse of "chemical restraint" in prisons and other places of detention.

Chapter Nine similarly examines obligations derived from international humanitarian law (IHL) – notably the Geneva Conventions and Additional Protocols – to prevent the development and use of CNS-acting chemical weapons in armed conflict. Chapter Ten explores the potential application of the Single Convention on Narcotic Drugs and the UN Convention on Psychotropic Substances to the regulation of CNS-acting chemicals and the prevention of their weaponised use. Whilst this book concentrates on State-centric mechanisms, Chapter Eleven examines the potential roles that can be played to prevent the development and use of CNS-acting chemical weapons by the chemical, life science and medical communities, and informed civil society more broadly. This includes regulation of dual use research; development and implementation of ethical codes by the chemical and life science communities; education and outreach; whistle blowing; and application of medical ethics and ethical practice.

Section IV of the book comprises the concluding Chapter Twelve. This seeks to employ the HAC framework as a means of analysing the contributions of the proceeding chapters and synthesising them into a proposed integrated approach which would prioritise the most appropriate and effective mechanisms to address potential threats arising from the rapid advances in neuroscience and the chemical and life sciences more broadly so as to prevent the development of CNS-acting chemical agent weapons and the malign manipulation of the human CNS. Whilst prioritising the CWC (and OPCW) as the most appropriate and currently most receptive forum for addressing CNSAC and ICA weapons, a range of alternative and/or synergistic measures employing the BTWC, IHRL, IHL and the UN Drug Conventions are also proposed, and increasing engagement of the scientific and medical communities highlighted.

References

1. R. Trapp, Convergence of Chemistry and Biology, and Nanotechnology, in *Preventing Chemical Weapons; Arms Control and Disarmament as the Sciences Converge*, ed. M. Crowley, M. Dando and L. Shang, Royal Society of Chemistry, London, 2018, ch. 7, p. 195.
2. R. Trapp, Convergence of Chemistry and Biology, and Nanotechnology, in *Preventing Chemical Weapons; Arms Control and Disarmament as the Sciences Converge*, ed. M. Crowley, M. Dando and L. Shang, Royal Society of Chemistry, London, 2018, ch. 7, p. 195.
3. M. Meselson, Averting the hostile exploitation of biotechnology, *CBW Conv. Bull.*, 2000, **48**, 16–19.
4. J. Perry Robinson, Bringing the CBW conventions closer together, *CBW Conv. Bull.*, 2008, **80**, 1–4.
5. D. Gillum, G. Poste, C. Woods and R. Levinson, Biotech promises miracles, But the risks call for more oversight, Bulletin of Atomic Scientists, 2023. https://thebulletin.org/2023/08/biotech-promises-miracles-but-the-risks-call-for-more-oversight/.
6. L. R. Squire *et al.*, *Fundamental Neuroscience*, Academic Press, Elsevier, Amsterdam, Preface, 4th edn, 2013, p. xix.
7. US, National Institute of Health, What are some different areas of neuroscience? https://www.nichd.nih.gov/health/topics/neuro/conditioninfo/areas#:~:text=Developmental%20neuroscience%20describes%20how%20the,that%20guide%20how%20neurons%20function.
8. F. J. Petrasek, Chemistry and structure – activity relationships of psychotropic drugs part I general topics, in *Principles of Psychopharmacology*, ed. W. G. Clarke and J. del Giudice, Academic Press, New York, 1978, ch. 7, pp. 134–140.

9. R. Adolphs, The unsolved problems of neuroscience, *Trends Cognit. Sci.*, 2015, **19**(4), 173–175.
10. M. Dando, *Neuroscience and the Problem of Dual Use: Neuroethics in the New Brain Projects*, Springer Nature, Switzerland, 2020, pp. 149–172.
11. P. Frackowiak and H. Markham, The future of human cerebral cartography: a novel approach, *Philos. Trans. R. Soc., B*, 20152015, **370**, 171.
12. H. Okano, A. Miyawaki and K. Kasai, Brain/MINDS: brain mapping project in Japan, *Philos. Trans. R. Soc., B*, 20152015, **370**, 310.
13. L. A. Jorgenson and W. T. Newsome, *et al.*, The BRAIN Initiative: developing technology to catalyse neuroscience discovery, *Philos. Trans. R. Soc., B*, 20152015, **370**, 164.
14. Brain Research through Advancing Innovative Neurotechnologies (BRAIN) Working Group, *Report to the Advisory Committee to the Director, NIH: Brain 2025: A Scientific Vision*, National Institutes of Health, Washington, D. C., 2014, p. 5.
15. R. Adolphs, The unsolved problems of neuroscience, *Trends Cognit. Sci.*, 2015, **19**(4), 175.
16. G. Poste, *Advances in biotechnology: promise and peril*, Second National Symposium on Medical and Public Health Response to Bioterrorism, Center for Biosecurity, Washington, DC, 28–29 November, 2000. www.upmc-biosecurity.org/pages/events/2nd_symposia/transcripts/*trans*_post.html (no longer active).
17. G. Poste, *Advances in biotechnology: promise and peril*, Second National Symposium on Medical and Public Health Response to Bioterrorism, Center for Biosecurity, Washington, DC, 28–29 November, 2000. www.upmc-biosecurity.org/pages/events/2nd_symposia/transcripts/*trans*_post.html (no longer active).
18. Certain chemical agents like organophosphorus and carbamate nerve agents that act principally on the CNS, are not included, for ease of distinction, within the definition of CNSAC weapons because they are not intended to reversibly incapacitate their targets.
19. (a) For further discussion see: M. R. Dando, *The Chemical and Biological Nonproliferation Regime after the Covid-19 Pandemic: Dealing with the Scientific Revolution in the Life Sciences*, Palgrave Macmillan, Springer Nature, Cham, Switzerland, 2023, pp. 15–27; (b) M. Crowley, *Chemical Control: Regulation of Incapacitating Chemical Agent Weapons, Riot Control Agents and their Means of Delivery*, Palgrave, Basingstoke, 2016.
20. J. S. Ketchum and F. R. Sidell, Incapacitating agents, in *Textbook of Military Medicine: Medical Aspects of Chemical and Biological Warfare*, ed. F. R. Sidell, E. T. Takafuji and D. R. Franz, Office of the Surgeon General, Department of the Army, United States, Washington D.C., 1997, vol. 287–306, ch. 11, pp. 291–294.
21. D. E. Nichols, *Pharmacol. Rev.*, 2016, **68**, 264.
22. D. E. Nichols, *Pharmacol. Rev.*, 2016, **68**, 270.

23. A. Maclean, *Historical Survey of the Porton Down Service Volunteer Programme 1939–1989*, Part IV, Human studies with incapacitating agents, Ministry of Defence, London, 2006, p. 119.
24. J. S. Ketchum, *Chemical Warfare Secrets Almost Forgotten*, Private Publication, Santa Rosa, California, 2006, pp. 35–42.
25. J. S. Meyer and L. F. Quenzer, *Psychopharmacology: Drugs, the Brain, and Behavior*, Sinauer Associates, Sunderland, Mass, 2005, pp. 331–332.
26. Royal Society, *Neuroscience, conflict and security*, Royal Society, London, 2012, p. 47.
27. L. Dosen-Micovic, Molecular modelling of fentanyl analogs, *J. Serb. Chem. Soc.*, 2004, **69**, 843–854.
28. (a) European Court of Human Rights (ECtHR) (2011), Finogenov and others v. Russia, application numbers 18299/03 and 27311/03, Judgement, 20th December 2011; (b) M. Wheelis Human impact of incapacitating chemical agents in: International Committee of the Red Cross, Expert Meeting: Incapacitating chemical agents, implications for international law, Montreux, Switzerland, 24th–26th March 2010; (c) D. Levin and V. Selivanov Medical and Biological Issues of NLW Development and Application, Proceedings of the 5th European Symposium on Non-Lethal Weapons, 11th–13th May 2009, Ettlingen, Germany, European Working Group on Non-Lethal Weapons. V23, p. 7. (d) 4 Years Later, Moscow Hostages Suffering, CBS Evening News, 21st October 2006; (e) J. R. Riches, *et al.*, Analysis of clothing and urine from Moscow theatre siege casualties reveals carfentanil and remifentanil use, *J. Anal. Toxicol.*, 2012, **36**(9), 647–656.
29. Royal Society, *Neuroscience, conflict and security*, Royal Society, London, 2012, pp. 47–48.
30. See, for example: C. M. Timperley and M. Bird, *et al.*, 1,1′-(Propane-1,3-diyl)bis(4-tert-butylpyridinium) di(methanesulfonate) protects guinea pigs from soman poisoning when used as part of a combined therapy, *Med. Chem. Commun.*, 2012, **3**, 352–356.
31. Royal Society, *Neuroscience, conflict and security*, Royal Society, London, 2012, p. 48.
32. (a) J. Lakoski, W. Murray and J. Kenny, *The advantages and limitations of calmatives for use as a non-lethal technique*, College of Medicine Applied Research Laboratory, Pennsylvania State University, 3rd October 2000, p. 37; (b) *Recommended Operating Procedures for Analysis in the Verification of Chemical Disarmament (VERIFIN Blue Book)*, ed. P. Vanninen, Ministry for Foreign Affairs of Finland/University of Helsinki, Chemicals, 2023, ch. 3, pp. 73–74.
33. Royal Society, *Neuroscience, conflict and security*, Royal Society, London, 2012, pp. 48–49.
34. See for example: J. Lakoski, W. Murray and J. Kenny, *The advantages and limitations of calmatives for use as a non-lethal technique*, College of Medicine Applied Research Laboratory, Pennsylvania State University, 3rd October 2000, p. 37.

35. M. Crowley and M. Dando, *Toxin and Bioregulator Weapons*, Palgrave MacMillan, 2022.
36. M. Crowley and M. Dando, *Toxin and Bioregulator Weapons*, Palgrave MacMillan, 2022, pp. 110–113.
37. OPCW, *Report of the Scientific Advisory Board on Developments in Science and Technology for the Third Special Session of the Conference of the States Parties to Review the Operation of the Chemical Weapons Convention*, RC-3/DG-1, OPCW, The Hague, 29 October 2012, paragraph 12.
38. Royal Society, *Brain Waves Module 3: Neuroscience, conflict and security*, The Royal Society, London, February, Section 5.2.1 Neuropharmacological Agents as Incapacitating Chemical Agents, 2012, p. 46.
39. *Protocol for the Prohibition and Use in War of Asphyxiating, Poisonous or Other Gases, and of Bacteriological Methods of Warfare* (Geneva Protocol), 17 June 1925.
40. ICRC, Treaties, *States Parties and Commentaries: Convention on the Prohibition of the Development, Production and Stockpiling of Bacteriological (Biological) and Toxin Weapons and on Their Destruction (BTWC)*, 1972. Available at https://ihl-databases.icrc.org/applic/ihl/ihl.nsf/INTRO/450?OpenDocument, Article 1.
41. Organisation for the Prohibition of Chemical Weapons (OPCW), Convention on the Prohibition of the Development, Production, Stockpiling and Use of Chemical Weapons and on their Destruction (Chemical Weapons Convention [CWC]), 1993. Available at http://www.opcw.org, Article I.
42. Organisation for the Prohibition of Chemical Weapons (OPCW), Convention on the Prohibition of the Development, Production, Stockpiling and Use of Chemical Weapons and on their Destruction (Chemical Weapons Convention [CWC]), 1993. Available at http://www.opcw.org, Article II.1.
43. Organisation for the Prohibition of Chemical Weapons (OPCW), Convention on the Prohibition of the Development, Production, Stockpiling and Use of Chemical Weapons and on their Destruction (Chemical Weapons Convention [CWC]), 1993. Available at http://www.opcw.org, Article II.2.
44. Organisation for the Prohibition of Chemical Weapons (OPCW), Convention on the Prohibition of the Development, Production, Stockpiling and Use of Chemical Weapons and on their Destruction (Chemical Weapons Convention [CWC]), 1993. Available at http://www.opcw.org, Article II.9.a-d.
45. Organisation for the Prohibition of Chemical Weapons (OPCW), Convention on the Prohibition of the Development, Production, Stockpiling and Use of Chemical Weapons and on their Destruction (Chemical Weapons Convention [CWC]), 1993. Available at http://www.opcw.org, Article II.1.a.
46. Conference of States Parties, *Decision: Understanding regarding the aerosolised use of central nervous system-acting-chemicals for law enforcement purposes*, C-26/DEC.10. OPCW, The Hague, 1 December 2021.

47. (a) For further discussion see: G. Pearson, *Relevant Scientific and Technological Developments for the First CWC Review Conference: The BTWC Review Conference Experience*, CWC Review Conference Paper No. 1, University of Bradford, August, 2002; (b) P. Aas, The threat of mid-spectrum chemical warfare agents, *Prehospital and Disaster Med.*, 2003, **18**(4), 306–312; (c) M. Crowley and M. R. Dando, *Toxin and Bioregulator Weapons*, Palgrave MacMillan, 2022, pp. 197–234; (d) *Recommended Operating Procedures for Analysis in the Verification of Chemical Disarmament (VERIFIN Blue Book)*, ed. P. Vanninen, Ministry for Foreign Affairs of Finland/University of Helsinki, Chemicals, 2023, ch. 3.
48. J. Perry Robinson, Bringing the CBW conventions closer together, *CBW Conv. Bull.*, 2008, **80**, 2.
49. See for example: J. Altmann, Preventive arms control: concept and design, in Altmann, *Journal Military nanotechnology: potential applications and preventive arms control*, Routledge, London, 2006.
50. See for example: *A Web of Prevention: Biological Weapons, Life Sciences, and the Governance of Research*, ed. C. McLeish and B. Rappert, Earthscan, London, 2007.
51. (a) For development and previous application see: M. Crowley, *Chemical Control: Regulation of Incapacitating Chemical Agent Weapons, Riot Control Agents and their Means of Delivery*, Palgrave, Basingstoke, 2016; (b) *Preventing Chemical Weapons*, ed. M. Crowley, M. Dando and L. Shang, 2018; (c) M. Crowley and M. R. Dando, *Toxin and Bioregulator Weapons*, Palgrave MacMillan, 2022.

Section II: Science and Technology

CHAPTER 2

Malign Application of Neuroscience and Related Fields

2.1 Introduction

This chapter will examine the concerted attempts by a number of States to develop CNS-acting chemical agent weapons and more broadly explore the potential malign manipulations of the CNS. Although the majority of such State CNS weapons development programmes have long ago been shut down, certain States continue to undertake troubling research and development activities in this area today. The past and present record of State interest in CNS-acting chemical agent weapons and broader weaponization of neuroscience should be a concern for the future as our understanding of the CNS continues to expand rapidly. This chapter sets out to demonstrate this growing danger by following the broad understanding of dual use suggested in the EU Human Brain Project's[1] *Opinion on 'Responsible Dual Use'*. Therefore, we agree that dual use concerns should encompass potential misuse in *political, security, intelligence* as well as *military* spheres. Thus this chapter explores the misuse of relevant chemical and life scientific advances from the discovery of a range of CNS-acting chemicals that helped people with mental health problems during the late 1940s through to the completion of the Human Genome Project at the turn of the 21st century as this new genomic knowledge began to be used to radically advance our understanding of the functions of the CNS.[2] To do this, the chapter surveys the previous and contemporary misuse of knowledge derived in this period, including five case studies of Chinese, Soviet/Russian, Apartheid South African, UK, and US CNS-acting chemical agent weapons research and development projects targeted at the CNS (and other core human physiological

Preventing Weaponization of CNS-acting Chemicals: A Holistic Arms Control Analysis
By Michael Crowley and Malcolm Dando

Published by the Royal Society of Chemistry, www.rsc.org

systems). Further examples of dual use research of concern and potential attempts at activities related to weaponization in the Czech Republic, Iran and India are also highlighted. The chapter then explores the nature and scope of previous and contemporary misuse of these scientific developments and CNS-acting chemicals notably by State institutions and agencies including security and law enforcement bodies to conduct torture and other ill-treatment including forced interrogation and chemical restraint.

2.2 Illustrative Examples of Past and Contemporary Malign Application of Neuroscience and Related Fields for Development of CNS-acting Weapons

From the late 1940s onwards military, security or police entities and related State policy-making bodies of certain countries explored the potential utility of CNS-acting weapons. States that reportedly conducted research and attempted development of CNS-acting weapons or acquired such weapons at some stage prior to the signing of the CWC in 1993 included: Albania, China, Iraq, Israel, (Apartheid) South Africa, the Soviet Union, the UK, the US and Yugoslavia.[3] There are indications from open source information that research related to CNS-acting chemical agent weapons development continued after the coming into force of the CWC and may still be taking place in certain States. Details of the historical CNS-acting chemical agent weapons research and development programmes that have been made public are partial and of varying reliability. And information concerning contemporary activities is scarce. However, despite these limitations the following case studies give some indication of the range of CNS-acting chemical agents that have been explored to date. These examples are discussed in relation to the evidence available for different countries rather than under specific scientific or technological topics.

2.2.1 UK Pharmaceutical CNS-acting Chemical Agent Weapons Research

According to official national records, the UK undertook research into and attempted development of pharmaceutical CNS-acting agent weapons during the Cold War period. This programme of activities appears to have been initiated in the mid-1950s and to have been halted in the early 1970s.[4] It was centred at the Chemical Defence Experimental Establishment (CDEE) at Porton Down, Wiltshire. Whilst there appears to have been some research related to the use of CNS-acting chemical agents for intelligence purposes, the over-riding focus was on the development of CNS-acting weapons for military purposes. In its incapacitating agent programme, the CDEE appeared to distinguish between weaponised chemical agents that had incapacitating physical effects and those that had incapacitating psychological effects, though it appears that certain potential CNS-acting agents fell within both designations.

In light of the 21st century's ongoing revolutionary advances in the chemical, life and associated sciences, explored in this publication, it is interesting to note UK government experts' recognition of then-contemporary technological limitations and how these could be overcome. The UK's approach was set out in a discussion paper for the 15th Tripartite Conference in 1960: "ideally, the best possible method for preparing a new agent with a given action would be to design a molecule which would have this specific type of action". However, the paper noted that "knowledge of structure–activity relationships is not sufficiently exact for this to be possible". Thus, it argued, another approach would be a wide-ranging literature survey, which might "perhaps almost by chance" throw up promising leads.[5] Three lines of investigation were subsequently employed: systematic review of the scientific literature; liaison with industrial laboratories and universities; and research stimulated at the CDEE itself.[6] In comparison to the massive scale of current toxic chemical screening programmes for the pharmaceutical industry, the UK screening capacity was relatively modest. For example, it was reported that from August 1961 to December 1963, 240 new compounds had been received for testing as potential incapacitating agents; whilst the CDEE annual report covering 1 July 1963–30 June 1964 stated that 150 compounds had been screened during this period with 10–15% coming from industry or universities.[7] As it was deemed impracticable for all these substances to be directly tested on humans, in 1960 the CDEE established a range of physical and behaviour tests for animal work to screen new substances which would then potentially be the subject of human trials.

2.2.1.1 Psychological Incapacitating Agents[8]

The CDEE research concentrated upon three classes of chemicals as potential psychological incapacitating agents: lysergic acid derivatives, glycollates and tryptamines.

Two derivatives of lysergic acid were investigated as potential CNS-acting weapons in human trials by the CDEE: lysergic acid diethylamide (LSD) and lysergic acid ethylamide (LAE). Only a very short series of human tests was conducted with LAE, which was thought to be about one tenth the potency of LSD. In 1954 human tests – sponsored by the UK's Secret Intelligence Service (SIS) [which is otherwise known as MI6] – were undertaken to establish the possible utility of LSD as a 'truth drug'; these tests showed that LSD had no such utility and so these trials ceased. In 1956 it was recognised that LSD might have an important role against "individuals, key groups or perhaps in controlling hostile crowds".[9] The Defence Research Policy Committee (DRPC) agreed in 1956 that a small programme of work on such drugs should proceed at CDEE and this was later expanded. The programme of human tests with LSD started in the autumn of 1961 and ended in January 1965. This included the application of LSD by oral, intravenous injection and subsequently inhalation in laboratory conditions. This was followed by three field tests – Moneybags, Recount and Small Change – undertaken to establish how LSD could affect troops under simulated operational conditions. The first field test, Moneybags, was held in late November/early December 1964, with soldiers from 41 Commando Royal Marines undertaking an exercise reproducing an internal security operation similar to those the men had experienced in "[B]orneo, Cyprus and elsewhere". The results showed that 70 minutes after receiving an oral dose of 200 μg of LSD, the soldiers became progressively disorganised, ill-disciplined and incapable of taking orders. Soldiers walked aimlessly in the open and would have been exposed to enemy fire. Communications between the soldiers were chaotic: "[B]y the time the recipient had been 'woken up' enough to realise that someone was talking to him, the speaker had forgotten what he was supposed to be saying". Two soldiers fed their sandwiches to imaginary birds, one soldier was found climbing a tree and another soldier spent 25 minutes trying to cut down a tree with a spade. It was concluded that the dose of LSD could render military operations impossible, and that the unit would have been "annihilated by modest enemy action" within an hour. Furthermore, the experienced infantry officer observing the soldiers conducting the same exercise the following day without LSD found their performance to be "well-nigh perfect", indicating that impairment was transitory.

However, the results from the subsequent field tests – Recount and Small Change – were not as positive, with unaffected men being able to exert a considerable stabilising influence on the unit. It was subsequently concluded that LSD "is of doubtful chemical warfare value". LSD investigations stopped by December 1968, apart from a little work with animals on antidotes which continued until September 1969. Members of the Chemical Defence Advisory Board (CDAB), at its meeting in January 1969, judged LSD unsatisfactory as an operational agent because of the unpredictability of its effects and because it was expensive to produce. Furthermore, a UK expert, Dr F. W. Beswick, explained that LSD was not a practical agent because "there were problems of dissemination, the 100% effective doses by inhalation was relatively high, and the material was expensive".[10]

The effects of glycollates on human behaviour include complete dissociation from surroundings, unintelligible muttering, extreme restlessness and exploratory activity, with food, drink and sleep requirements being ignored by the subjects. The onset of symptoms was usually marked by euphoria, drowsiness, a dry mouth, slurred speech, expanding pupils and uncontrolled muscle twitching. Two glycollates were investigated by CDEE in human studies: the first was BZ, a long-lasting glycollate which had been developed and standardised as a weapon by the US. The CDEE began human tests with BZ in August 1962 and these ran until the summer of 1967.

In November 1967 CDEE examined "*N*-methyl-4-piperidyl-isopropyl-phenyl glycollate" (MPIPG) – which was given the name T3436. It was similar in action to BZ but its effects were of shorter duration. Human tests with T3436 began in January 1968, initially with subjects being given T3436 doses at

0.5 μg kg^{-1}, with doses gradually increasing in subsequent trials until February 1971 when clear incapacitation was observed at doses of 9–11 μg kg^{-1}. Just as with LSD, consideration was given to conducting tests with a "realistic" dissemination method – aerosol – prompting a series of inhalation tests. T3436 by inhalation was reputedly more effective than T3436 given orally. Human tests inhaling T3436 ran from June 1971 to July 1972. All men receiving a dose of 6 μg kg^{-1} showed changes in behaviour: 5 reported visual or auditory hallucinations. All the men inhaling a dose of 8 μg kg^{-1} were affected, with two possibly being incapacitated. However, despite these results, human testing of glycollates appears to have been halted by 1973.

Tryptamines are derivatives of indole. They were known about in 1956; academic references note that one of the tryptamines (*N*,*N*-dimethyl tryptamine) caused hallucinations in humans at doses of 1 mg kg^{-1} body weight when injected intra-muscularly. Four simple tryptamines were examined in human tests by the CDEE: T2614 (*N*,*N*-diethyl tryptamine), T2619 (*N*,*N*-diisopropyl tryptamine), T2627 (*N*-benzyl tryptamine) and T2673 (*N*-ethyl tryptamine). However, the effects produced by all four were only marginally incapacitating and the majority of short duration (less than 3 hours), and consequently rejected as potential CNS-acting weapons.

2.2.1.2 Physical Incapacitating Agents[11]

The CDEE research concentrated upon three classes of chemicals as potential physical incapacitating agents: tremoram; pyrexal and oripavine derivatives.

Interest in tremoram as a possible incapacitating agent was provoked by an external paper published in 1956. Intra-venous injections of tremoram produced muscle tremors, muscular weakness and general incapacitation in a number of species. Work was conducted at the CDEE in 1962 with animals having tremoram (referred to as T2785) administered by parenteral injection. The lowest effective dose was around 30 μg kg^{-1}. Marked tremor was usually present but at lower doses tremoram almost always induced profuse salivation and diarrhoea. From this work the signs of tremoram poisoning in man were expected to be salivation, diarrhoea and possibly Parkinson-like tremor. Human tests with tremoram were conducted at the CDEE, but the details are obscure.

Pyrexal is a lipopolysaccharide prepared from *Salmonella abortus equi.* It induces the signs and symptoms usually associated with fever: headache, shivering, sweating, and aching of limbs and the back. Reports of its being used in the UK, either as therapy or in scientific investigations, appeared from 1956 onwards. These reports suggested that an intra-venous dose of less than 1 μg induced feverish symptoms in man. Some reports indicated pyrexal was effective when inhaled. The CDEE conducted human tests with commercial pyrexal samples in 1959 and 1960 through intravenous injection and nasal sprays. CDEE scientists determined in 1961 that pyrexal acted as a potential harassing agent (*i.e.*, a possible riot control agent [RCA]) rather than as an incapacitating agent, with the main effect, which was transient, being a headache which, while severe, was not enough to prevent tasks from being completed. Pyrexal was quite inactive when applied to the mucosa of the nose and upper respiratory tract. Consequently, no further work with it was conducted by the CDEE.

In February 1961, the CDEE was informed by a commercial pharmaceutical company of its development of a new type of powerful synthetic analgesic derived from oripavine, such derivatives having similar effects to morphine but being 1000 to 10 000 times more powerful. The company had patented five derivatives and gave CDEE samples of each. In low doses, the oripavine derivatives produce euphoria, nausea and vomiting and lethargy. Following CDEE animal testing of these derivatives it was observed that: death occurred from respiratory depression and that such respiratory effects could be reversed by a known antidote called nalorphine; the duration of effects induced varied between animal species (short-lived in small animals but lasting as long as 18 hours in larger animals); and that ratio between the lethal dose and the effective dose was found to be 4 in rhesus monkeys, 20 in cats, 130–200 in rats, rabbits and mice and 1000 in pigs.

CDEE undertook human tests with four of the oripavine derivatives, identified as TL2636, TL2696, TL2833 and TL3046, with the majority of work being conducted with TL2636. The main series of human tests with TL2636 was conducted in 1961 and 1962. These included laboratory tests with TL2636 given orally, by intra-venous injections and also *via* inhalation through cigarettes injected with TL2636. The conclusions drawn from these human tests with TL2636 were that it incapacitates in low doses, with the estimated effective oral dose being 4 $\mu g\,kg^{-1}$; it appeared to be eight times more effective when inhaled than when ingested; and that no deleterious mental effects were observed. However, it was found to have a very low safety margin (*i.e.*, the ratio of lethal to incapacitating dose) of between 5–10. This, in combination with limitations in its supply, suggested TL2636 was "of doubtful value as an incapacitating agent". However, despite these conclusions human tests in the laboratory resumed in 1967 and 1968. Also, in 1968, TL2636 was field tested. It was given to a contingent of those soldiers participating in Small Change – the last in a series of LSD field exercises, as an "active placebo". It was assessed to have reduced the military effectiveness of soldiers then conducting simulated anti-terrorist sweeps by half. However, the majority of soldiers (with the exception of one who was hospitalised) were able to resist the effects and complete their duties. Consequently, it was concluded that a dose of TL2636 sufficiently large to guarantee certain incapacitation in the field would carry an "unacceptably high risk of fatal outcome".

The UK ended its active research and development of CNS-acting chemical agent weapons for military purposes decades prior to the coming into force of the CWC. In a paper for the Offensive Evaluation Committee, a UK expert, Fisher, stated: "[O]n general grounds I think it unlikely that... a pure incapacitator agent will emerge. Any chemical agent, a small dose of which is capable of profound disturbance of bodily or mental function, is certain to be able to cause death in large dose... and no attack with a chemical warfare agent is likely to be designed with the primary objective of avoiding overhitting".[12]

2.2.2 South Africa Pharmaceutical CNS-acting Chemical Agent Weapons Development

In 1981 the South African Defence Force (SADF) established a clandestine chemical and biological weapons research and development programme code-named 'Project Coast'. In 1982 Delta G Scientific and Roodeplaat Research Laboratories were established as the front company facilities of Project Coast for chemical and biological research, respectively.[13] Its objectives included research and development of defensive materials and capability for the SADF and also biological and chemical weapons research and development. This included research and development of riot control agents, notably CR, and incapacitating chemical agent weapons. Although significant details and documents concerning Project Coast were made public, particularly during South Africa's Truth and Reconciliation Commission (TRC) hearings in 1998 and also during the trial of Dr Wouter Basson (the former project officer) in 1999 and 2000, there is still continuing uncertainty regarding certain Project Coast activities, their purposes and results, due to its clandestine nature.

In 1983 Project Coast explored the potential utilisation of a range of illegal recreational drugs as potential incapacitating chemical agent weapons including LSD, cannabis and a hypnotic sedative called methaqualone (mandrax), in part utilising stocks confiscated by the South African police. Between 1983–1989 Delta G Scientific undertook research firstly into the potential employment of a powdered cannabis formulation for use in grenades and when that proved unsuccessful, development and synthesis of cannabinoid analogues as potential incapacitating weapons; however, by the closure of Project Coast no cannabinoid formulation had been produced for weaponization.[14]

Delta G Scientific manufactured 1000 kilograms of methaqualone together with various analogues. Basson stated that he and a Special Forces operator devised the formulas for the weaponization according to which the methaqualone was compacted into tablet form before being aerosolised. The tablets varied in size and the intention was to weaponize a maximum of 500 to 2000 projectiles, including hand grenades, 81-millimetre mortar bombs and 155-millimetre projectiles. Basson stated that the payload of a hand grenade was 350 grams and an 81-millimetre mortar approximately 500 grams.[15] According to a document provided to the TRC by the former Surgeon General Knobel, methaqualone was placed in mortars to be tested on animals and humans. The results were, however, "disappointing" and Knobel stated that production halted in 1988. Thereafter "a far more active analogue was developed through further research" which was envisaged to overcome these "disadvantages".[16] However Gould and Folb note that Knobel's account is not verifiable from any other source and it is not known whether methaqualone was successfully weaponised.[17] It seems doubtful that this drug is potent enough to have had the desired effect.

Between February 1992 and January 1993 production of MDMA (Ecstasy) took place at Delta G Scientific under the code-name Baxil. Whilst the SADF and the Surgeon-General have consistently maintained that the MDMA produced at Delta G was intended for use as a crowd control agent, the organic chemist responsible for its manufacture, doubted its efficacy for crowd control and was concerned that it was to be used for purposes other than that stated. 912 kilograms of 90-96% pure MDMA in crystalline form had been produced when Project Coast closed.[18] The TRC found that "there was no scientific basis for thinking that it would be an appropriate, safe or sensible form of crowd control".[19] No documents indicate that research was done on its efficacy as a crowd control agent despite claims by Basson's defence team that Swartklip Products did tests on its weaponization.[20]

Delta G Scientific also explored the development of BZ weapons. According to Basson, Project Coast funds were used to purchase a "large amount" (5000 kilograms) of BZ in a joint operation with Gen. Peter Regli, head of Swiss intelligence. In addition, 1000 kilograms of a locally produced BZ variant were manufactured and intended for weaponization in 1993. Basson testified that BZ was used to fill hand grenades, 81-millimetre mortars and 155-millimetre projectiles, though in the case of the latter, the process went only as far as filling the canisters. He said that pyrotechnical testing and weaponization of the BZ took place in what was known as the Pilot Plant at Speskop (the headquarters of the South African Special Forces). In addition, Basson also stated that cocaine was combined with BZ and used to fill certain weapon prototypes. There does not appear to have been any largescale production of BZ weapons. There were no reports of the use of BZ internally in South Africa.[21] There was one case of a possible employment of a BZ weapon against Mozambique armed forces (see below), although this was never confirmed.

Basson was subsequently charged in 2007 by the Health Professions Council of South Africa with six counts of professional misconduct as a result of his activities at Project Coast from 1981 to 1993. Although two counts were dropped, one of the remaining counts against him was of providing unidentified "disorientation substances for over-the-border kidnapping ('grab') exercises, where the substances were used to tranquilise the person to be kidnapped".[22] Dr Basson did not dispute the basis of the charges, indicating that "disorientation substances", presumably CNS-acting chemical agents, were developed and subsequently used by the South African security forces as described. Basson instead presented arguments as to why his conduct was not "unprofessional". On 18 December 2013, the HPCSA's Professional Conduct Committee concluded that the "conduct under discussion was in breach of established ethical rules" and that "the breaches of medical ethics amount to unprofessional conduct".[23]

South Africa signed the Chemical Weapons Convention in January 1993. Project Coast activities were subsequently terminated, and all remaining stockpiles of BZ, MDMA, methaqualone and cocaine were subsequently reported as being destroyed during 1993-4, although certain questions of material accounting remain.[24]

2.2.2.1 Possible Use of CNS-acting Weapons

On 27 January 1992, Mozambique formally alleged that chemical weapons were used in an attack on its forces during South African operations against the Mozambican National Resistance (RENAMO) and requested that a UN team of experts be sent to Mozambique to investigate the incident.[25] According to the testimonies of Mozambican Government soldiers present, the alleged chemical weapon attack took place on 16 January 1992, against Mozambican Government forces operating close to the South African border in the Ngungue region. After examining a deserted RENAMO base, the Mozambican Government force came under small arms fire. An explosion occurred overhead which produced dark smoke that dissipated. The Mozambican Government forces withdrew from the area, but after 15 minutes there were problems keeping the soldiers moving and control was lost. Soldiers were disorientated and confused and their vision was affected. Between 18 and 27 January 1992, 28 soldiers were admitted to Maputo hospital, four were reported dead, two were wounded during *"uncontrollable shooting"* and 38 were reported missing.[26] Three separate investigations were undertaken by teams of South African,[27] Swedish[28] and UK[29] experts – the latter two at the request of the Mozambican Government. The three investigations came to three differing conclusions. According to a subsequent report by Mozambique, an (unnamed) physician working for the International Committee of the Red Cross who examined some of the hospitalised soldiers "concluded that the victims were exposed to a non-conventional weapon and of toxic nature".[30]

The UN Secretary General appointed a team of experts from Sweden, Switzerland and the UK to conduct an investigation in Mozambique.[31] The investigation took place from 23 to 27 March 1992. The subsequent report, which was submitted to the UN Secretary-General on 1st April 1992,[32] was somewhat contradictory in nature. On the one hand the report stated that: "[T]he signs and symptoms of the casualties are consistent with the use of an atropine-like chemical warfare agent". However, it also noted that these symptoms "are also consistent with severe heat stress. The use of an atropine-like agent does not remove the possibility of heat stress; rather, it increases an individual's susceptibility to it".[33]

The report also stated that: "[I]t is improbable that a chemical attack would be planned using a single round [as reported by the victims interviewed]. However, it cannot be excluded that the limited quantity of agent that could be delivered would have had an effect which could have been exacerbated by local climatic conditions and limited water supply".[34] The report concluded that: "[I]n the current absence of analytical data, we cannot conclude that a chemical warfare agent was used in the attack".[35]

Gould and Folb have stated that the UN report presented "two interpretations of the symptoms, reflecting the opinions of Heiner Staub [from Switzerland] on the one hand and J P Thompson [from the UK] on the other. Staub believed that the symptoms experienced by the troops were the result of dehydration and resultant heat stress. Thompson contended that the symptoms were consistent with exposure to a centrally acting atropine-like agent".[36]

Further information of relevance to the Ngungue incident was revealed through the publication of a previously classified document prepared by General Steyn of the SADF, based on investigations carried out by the South African National Intelligence Service (NIS) into the activities of certain SADF members.[37] As part of its wide-ranging determinations, the Steyn Report claimed that: "[A]llegedly the chemical attack on Frelimo soldiers in Mozambique was a practical training session. A small unmanned reconnaissance bomber was located shortly before the attack on Komatipoort. The toxic substance used in the attack was manufactured and stored by Protechnik".[38] The Steyn Report noted the allegation was "Confirmed and the individuals involved are known".[39]

Gould has concluded: "[A]ll those consulted have... agreed that something strange happened [in Mozambique] and it raises the spectre that the SADF may, as claimed by the Steyn report, have seen an opportunity to test BZ on human subjects. Given the amount of time which has passed since the incident it is unlikely ever to be resolved satisfactorily".[40]

2.2.3 US CNS-acting Chemical Agent Weapons Development

2.2.3.1 Research and Development of Pharmaceutical CNS-acting Chemical Agent Weapons

The United States of America (US) has a long history of research and attempted development of weapons employing incapacitating chemical agents (ICAs), notably CNS-acting chemical agents, dating back to the 1950s.[41] According to the 1997 US Army textbook, *Medical Aspects of Chemical and Biological Warfare*:

> "Virtually every imaginable chemical technique for producing military incapacitation has been tried at some time. Between 1953 and 1973, at the predecessor laboratories to what is now the U.S. Army Medical Research Institute of Chemical Defense, many of these were discussed and, when deemed feasible, systematically tested. Chemicals whose predominant effects were in the central nervous system were of primary interest and received the most intensive study..."[42]

The authors went on to suggest that almost all such agents could be put into one of four classes: stimulants, depressants, psychedelics and deliriants. Stimulants include, for example, amphetamines and cocaine, depressants include barbiturates, and psychedelics include LSD. Delirium, "an incapacitating syndrome, involving confusion, hallucinosis, disorganized speech and behavior," can be produced by a wide variety of drugs. But as the text pointed out: "[F]rom this large number of possibilities, chemical compounds in a single subgroup – the 'anticholinergics' – are regarded as most likely to be used as military incapacitating agents".[43]

The US studies that were carried out on such deliriant agents have been described in some detail,[44] and one, BZ (3-quinuclidinyl benzilate),[45] was weaponized by the US. BZ interferes with the operation of the acetylcholine neurotransmitter in the brain by binding to muscarinic receptors for the transmitter.[46] However, as there are many types of such receptors, the effects of BZ are complex and the behaviour produced was unpredictable.[47] Approximately 60 000 kilograms (130 000 pounds) of BZ were manufactured and the agent was weaponised in two munitions – the 175-lb M44 generator cluster and the 750-lb M43 cluster bomb[48] – that entered the US arsenal in 1964.[49] Although the US military proposed initiating use of BZ along with CS in the Vietnam war,[50] BZ was never utilised.

From 1955 to 1975, the US military also conducted research into a wide range of other potential CNS-acting chemical agents, many of which were tested on US military volunteers.[51] In the late 1970s the US army conducted some advanced development including work on a pilot plant for production of the glycollate EA 3834A and a filling facility for an XM96 66mm CNS-acting agent rocket warhead, although these munitions do not appear to have entered the US arsenal.[52] From the late 1970s onwards there is no evidence of further advanced large-scale CNS-acting agent weapons development. BZ was removed from the US arsenal; the stockpile of BZ munitions and agent was eventually destroyed between 1988 and 1990.[53]

It does appear that there was research into CNS-acting agents and suitable dispersal mechanisms continued for military and law enforcement purposes, even after signing and coming into force of the Chemical Weapons Convention. In the law enforcement sphere, from 1992 until at least 1997, the National Institute of Justice (NIJ) funded Lawrence Livermore National Laboratory to undertake CNS-acting agent weapon research including fentanyl-family based delivery targeting individuals.[54]

In December 1999 the US Army's Edgewood Chemical Biological Center (ECBC) solicited research proposals for a three phase project – CBD00-108 Chemical Immobilizing Agents for Non-lethal Applications – the objective of which was to: "demonstrate the feasibility of a safe, reliable chemical immobilizing agent(s) for non-lethal (NL) applications in appropriate military missions and law enforcement situations".[55] Under Phase 1 of this programme researchers would "conduct an analysis of promising new chemical immobilizing agents or combinations of agents" including "recent breakthroughs in the pharmacological classes such as anaesthetics/analgesics, tranquilizers, hypnotics and neuromuscular blockers".[56] Researchers would "conduct a toxicological test program" with the "most promising" new immobilizing agents to fill data gaps and consequently "establish the mode of immobilization, the effective dose(age) for immobilization, onset time and duration of effects, and safety ratio in the most appropriate animal species".[57]

Under Phase 2, input would be gathered from potential military and law enforcement users on the desired performance/operational characteristics; and the implications of the Chemical Weapons Convention for proposed scenarios of use would be determined. Following the selection of the "optimum scenario(s) of use" a series of "non-human primate and clinical tests" would be conducted "to establish safety and performance characteristics".[58] Subsequently an "appropriate delivery technique" for example, "an aerosol generator for dissemination for the inhalation route of entry, or a dart for injection in the intra-muscular route of entry" would be designed and demonstrated. Under Phase 3, dual-use applications of the technology were to be analysed.[59] The potential military uses cited included: "meeting US and NATO objectives in peacekeeping missions; crowd control; embassy protection; rescue missions; and counter-terrorism".[60] The potential law enforcement uses highlighted included: "hostage and barricade situations; crowd control; close proximity encounters... to halt fleeing felons; and prison riots".[61] In June 2000, ECBC awarded a contract for Phase 1 of this project to OptiMetrics, Inc. The company completed Phase 1 by November 2002.[62] It is not known whether Phase 2 or 3 were ever undertaken and if so, when or by whom.

In 2000, the Applied Research Laboratory and the College of Medicine at Pennyslvania State University published the results of its literature survey and analysis of bio-medical research into a range of pharmaceutical agents including sedative-hypnotic agents, anaesthetic agents, skeletal muscle relaxants, opioid analgesics, anxiolytics, antipsychotics, antidepressants and selected drugs of abuse which attempted to identify "calmatives with potential use as non-lethal techniques".[63] The report identified ten classes of pharmaceutical agents and 32 representative agents or agent combinations has having a "high potential for further consideration as a non-lethal technique"[64] (as summarised in the Table 2.1).

In 2001 the NIJ funded a three-phase project on 'non-lethal' weapons at the Institute for Non-Lethal Defense Technologies (INLDT) at Pennsylvania State University (PSU). Phase two of the project was to "conduct an investigation of controlled exposure to calmative-based oleoresin capsicum".[65] Although publicly available information regarding this project is scarce, it apparently involved the combination of a CNS-acting chemical agent with the chemical irritant oleoresin capsicum in order to produce more profound effects.[66] In 2007 the NIJ awarded a grant to Penn State University to "explore the potential of operationalizing calmatives and to examine possible pharmaceuticals, technologies and legal issues".[67]

Table 2.1 Indicative drug classes and pharmaceutical chemical agents considered to have potential utility as CNS-acting agents. Data from ref. 64.

Drug class	Selected compounds	Site of action
Benzodiazepines	Diazepam Midazolam Etizolam Flumazenil (antagonist)	GABA receptors
Alpha$_2$ adrenergic receptor agonists	Clonidine, Dexmedetomidine Fluparoxan (antagonist)	Alpha$_2$-adrenergic receptors
Dopamine D3 receptor agonists	Pramipexole Cl-1007 PD 128907	D3 receptors
Selective serotonin reuptake inhibitors	Fluoxetine Sertraline Paroxetine WO-09500194	5-HT transporter
Serotonin 5-HT$_{1A}$ receptor agonists	Buspirone Lesopitron Alnespirone MCK-242 Oleamide WAY-100, 635	5-HT$_{1A}$ receptor
Opioid receptors and Mu agonists	Morphine Carfentanil Naloxone (antagonist)	Mu opioid reception
Neurolept anaesthetics	Propofol Droperidol and fentanyl combination, Phencyclidines	GABA receptors DA, NE and GABA receptors Opioid receptors
Corticotrophin-releasing factor receptor antagonists	CP 154,526 (antagonist) NBI 27914 (antagonist) CRF-BP (binding protein)	CRF receptor
Cholecystokinin B receptor antagonists	CCK-4 Cl-988 (antagonist) Cl-1015 (antagonist)	CCKB receptor

Despite such US research activities, there is no evidence of any completed development or production of CNS-acting chemical agent weapons for either military or law enforcement purposes.

2.2.3.2 *Research of Toxin and Bioregulator Incapacitating Chemical Agents*

Whilst US incapacitating chemical agent (ICA) weapon research and development activities concentrated upon certain classes of pharmaceutical chemicals that affected the CNS; the US also explored the potential utility of employing toxins and bioregulators as ICAs, some of which primarily affected core physiological systems beyond the CNS. During the 1960s, staphylococcal enterotoxin B (SEB), which incapacitates through severe nausea, diarrhoea and intestinal cramp, was extensively investigated as an ICA weapon in the US CBW programme.[68] SEB has an inhalation ED_{50} (Effective Dose for 50% of the population affected) in humans estimated to be at least three orders of magnitude smaller than the corresponding LD_{50} (Lethal Dose for 50% of the population) of nerve agents such as sarin or VX.[69] The 1997 United States *Textbook of Military Medicine* noted, for example, that during the 1960s, Staphylococcal enterotoxin B (SEB):[70]

> "...was especially attractive as a biological agent because much lower quantities were needed to produce the desired effect than were required with synthetic chemicals. The dose that is incapacitating for 50% of the human population exposed... was found to be 0.0004 $\mu g\,kg^{-1}$, and the dose that is lethal for 50% of the human population exposed...was estimated to be approximately 0.02 $\mu g\,kg^{-1}$, both by the inhalation route."

In 1968, during DTC Test 68-50, dry-agent spray tanks filled with SEB (referred to by the US as Agent PG) mounted on US *Phantom* strike aircraft allowed the agent to be released over caged monkeys and other animals at sea off Eniwetok Atoll in the Marshall Islands. The test data reportedly indicated a 30% casualty rate over an area of 2400 km^2.[71] In addition to such large-scale testing, US documents indicate that by the beginning of 1970, the US had the production capacity to make 600 pounds of SEB per month[72]—although there are no confirmed publicly available records of the total agent quantities produced. As far as is known, SEB was never employed as a weapon, and all US research, development, and production of such offensive toxin weapons ended with the moratorium announced by President Nixon on 14 February 1970, and SEB stocks for offensive weapons were subsequently destroyed.[73]

The US ICA/CNS-acting chemical agent weapons programme continued during the 1970s and 1980s. Whilst there is evidence that this programme explored a wide range of pharmaceutical chemicals, it is unknown whether research extended to toxins and bioregulators during this period. In the 1990s, the US conducted dual-use research, including on bioregulators and bioregulatory systems, applicable to the study and potential development of incapacitating chemical agent weapons.[74] Certain activities were carried out for purportedly defensive purposes, under the auspices of the Chemical and Biological Defense Command (CBDCOM) and its largest component, the Edgewood Research Development and Engineering Center (RDEC). An analysis by Dando of RDEC research paper abstracts published and reported at its annual conferences indicated interest in the μ-opioid receptors, analgesic opioids and their synthetic analogues, notably fentanyl;[75] and a second line of research into the locus coeruleus (LC) system of the mammalian brain that is required for alert activity, associated alpha2 adrenergic receptors, sedative adrenergic agonists notably noradrenaline, and their synthetic analogues notably medetomidine.[76] Further related research, including into the LC and its neurons, was undertaken by the US Air Force Office of Scientific Research (AFOSR) notably under its Neuroscience Programme.[77]

These themes and lines of research were again taken up in the highly influential joint report on "calmatives" by researchers at the Applied Research Laboratory and the College of Medicine at Pennsylvania State University published in October 2000. Although concentrating on pharmaceutical chemicals, the study also identified two bioregulatory peptides and associated bioregulatory systems of interest.[78] The first bioregulator discussed was corticotropin-releasing factor (CRF), a peptide that triggers the release of stress-related hormones. Concentrated in the hypothalamus of the brain, CRF binds to two classes of receptors that are closely linked to feelings of anxiety and stress. In principle, receptor antagonist peptides or synthetic analogues that block CRF receptors in the brain could induce a "calm behavioral state".[79] The second bioregulatory peptide examined was cholecystokinin (CCK) which is present in the body in several biologically active forms. There are two classes of CCK receptors: CCK-A receptors concentrated in the gastrointestinal tract and CCK-B receptors in the central nervous system. Synthetic CCK analogues that stimulate CCK-B receptors trigger panic attacks, disruption of memory, and increased sensitivity to pain; synthetic CCK antagonists that block these receptors induce a tranquil state and could therefore serve as potential calmative agents.[80]

2.2.3.3 US Rejection of CNS-acting Weapons Development

Despite its long-standing interest in research related to CNS-acting chemical agent weapons development, in 2013 senior US representatives explicitly stated that no attempts to develop weapons employing such chemical agents were currently taking place. At the Third CWC Review Conference, the US Acting Under Secretary for Arms Control and International Security, Ms Rose Gottemoeller, declared that: "the development, production, acquisition, stockpiling, or use of incapacitating chemical agents—or any other toxic chemicals—in types and quantities inconsistent with purposes not prohibited by the Chemical Weapons Convention, is clearly prohibited by Article I of the Convention".[81] Ms Gottemoeller highlighted concerns that "illicit programmes could possibly be concealed under the guise of a legitimate treaty purpose, such as law enforcement" and further warned that States Parties "must all be vigilant to ensure that incapacitating chemical agents and other technologies do not jeopardise the twin goals of the Convention—the destruction of all chemical weapons and the prevention of the re-emergence of chemical weapons".[82] Furthermore, in a statement to the OPCW Executive Council meeting held in May 2013, the US Ambassador, Dr Robert Mikulak confirmed "very clearly and directly... that the United States is not developing, producing, stockpiling, or using incapacitating chemical agents".[83] This position has subsequently been repeatedly reiterated by US officials. To date the scope of this suspension with regard to the range of toxic chemicals has yet to be clarified; consequently it is unclear whether it explicitly covers bioregulator and toxin research in these areas.

2.2.4 Soviet Union/Russia CNS-acting Agent Weapon Development and Use

2.2.4.1 Pharmaceutical CNS-acting Agent Research and Weaponization

There are indications that, prior to the coming into force of the CWC, the USSR and subsequently the Russian Federation conducted research into and development of pharmaceutical CNS-acting chemical agent weapons. Recent archival research has uncovered documents indicating that in the late 1940s and 1950s, the Soviet authorities in Ukraine used CNS-acting chemical agent weapons to combat an insurgency waged by Ukrainian guerillas. These included Neptune-47, a sleep-inducing paralytic chemical added to food that allowed Soviet agents to capture leaders and section heads of the underground; Neptune 7/93 which was used against bunkers occupied by Ukrainian insurgents; and Typhoon, a fast-acting soporific, used in the 1950s in order to overcome resistance from insurgents' bunkers.[84]

In a presentation to the International Committee of the Red Cross (ICRC) 2012 expert meeting, Perry Robinson reported that "[I]n the Soviet Union... the Central Committee and the Council of Ministers adopted a resolution in May 1971 on the building of production capacity for 'non-lethal' chemicals resulting from the Foliant programme".[85]

According to a 2012 paper by Riches, Read, Black, Cooper and Timperley (scientists working at the UK's Defence Science and Technology Laboratory), "[R]ussian military research on fentanyls occurred before 1999".[86] In support of their assertion, the authors cited the following extract from a publication by General Antonov, a former director of the Military Chemical Institute in Shikhany:

> "...the action of analgesics is a knock-out blow—personnel subject to an attack of forces only a few minutes after the beginning of a chemical attack will lose their capacity to stand, not to mention move about. In severe cases people will enter an 'unconscious state' and 'carfentanil is one of the most active substances of the entire group of the studied derivatives of fentanyl. It manifests its activity for different pathways of entry into the organism, including inhalation of vapours or aerosol...'"[87]

Following the collapse of the Soviet Union, the Russian Federation clearly maintained stockpiles of at least certain CNS-acting weapons and the capability to use them. In October 2002, Russian security forces used a CNS-acting chemical weapon against armed Chechen separatists holding 900 hostages in a Moscow theatre. Although the bulk of the hostages were freed, more than 120 died from the effects of the chemical agents, and an undetermined number of additional hostages have suffered long-term damage or died prematurely in the years following the siege.[88]

Treatment of the hostages who had been poisoned was delayed and compromised by the refusal of the Russian authorities to state publicly what type of CNS-acting chemical had been used in the theatre for 4 days after the siege had ended.[89] On 30 October 2002 the Russian Health Minister, Yuri Shevchenko, identified the incapacitating agent as “a mixture of derivative substances of the fast action opiate Fentanyl”.[90] Mr Shevchenko further stated that: “[I] officially declare: chemical substances which might have fallen under the jurisdiction of the international convention on banning chemical weapons were not used during the special operation”.[91] However, the Minister refused to be more precise about the chemicals used even on 11 December 2002 when faced with a parliamentary question. He said it was a “State secret”.[92] Although Russia has to date refused to fully identify the CNS-acting chemicals employed, analysis by the UK’s Defence Science and Technology Laboratory (UK DSTL) indicated that the weapon comprised two analgesic-anaesthetics, carfentanil and remifentanil.[93] Russian scientists have subsequently undertaken research potentially related to CNS-acting weapons.[94] This has included computer modelling of the employment of an aerosolised chemical “calmative” against groups of individuals in enclosed spaces.[95]

Furthermore, there are indications that Russian scientists – including those associated with or then working at Russian defence establishments – investigated the properties of a range of potential CNS-acting chemical agents. The 2012 paper by Riches, Read, Black, Cooper and Timperley, from the UK DSTL asserted that: “[S]cientific papers published by Russian military officers indicate an interest in fentanyls extending back 12 years: opioid receptor studies,[96,97] fentanyl analysis[98] and synthesis of fentanyl precursors”.[99,100] The current authors have identified additional papers published from 2006–2012 by Russian life scientists indicating continued research relating to opiate receptors (ORs) and their interaction with OR ligands.[101–105]

To date, Russia has provided no public information as to whether stockpiles of weaponised pharmaceutical CNS-acting chemical agents are currently held by Russian authorities, nor whether research into the development and employment of such weapons is taking place, and if so, for what purposes. In its *2021 CWC Compliance Report*, the US raised its concerns that “[R]ussia has a PBA [pharmaceutical-based agent] program intended for purposes inconsistent with the CWC”. It highlighted Russia’s previous acknowledgement that its “special operations forces employed ‘derivatives of fentanyl’ to resolve the October 2002 Dubrovka theater hostage crisis”, expressing its concerns that “Russia is pursuing these types of agents for offensive purposes, which would be a violation of Article I [of the CWC]”.[106]

2.2.4.2 *Bioregulator and Toxin CNS-acting Weapons and Other ICA Weapons*

According to Alibek, Soviet scientists under Biopreparat's Project Bonfire undertook research which was "intended to develop a new kind of toxin weapon"[107] based on bioregulatory peptides that "could damage the nervous system, alter moods, trigger psychological changes, and even kill".[108] From 1978, Project "Bonfire" scientists attempted to incorporate short DNA sequences that coded for the production of bioactive peptides into the genomes of bacteria and viruses. Their aim was to create genetically engineered microbes that would manufacture toxic substances and cause serious illness and death.[109]

One area of investigation under Project Bonfire and Project Factor concerned the application of then-advanced genetic engineering techniques to the identification, investigation, and subsequent manipulation of the gene encoding myelin basic protein (MBP). According to Popov and Alibek, molecular biologists at the Soviet Academy of Sciences examined the properties of the naturally occurring MBP which maintains the correct structure of the body's myelin and is thought to be important in the process or myelination (coating the nerve fibres of the CNS with myelin sheaths, and insulating and protecting them). Although essential to the normal functioning of the CNS, MBP when present in excessive amounts triggers an autoimmune reaction that damages the myelin sheaths and impedes the transmission of nerve impulses. This research was subsequently taken up by scientists at the State Research Center for Applied Microbiology in Obolensk, who inserted the gene for MBP into the bacterium *Yersinia pseudotuberculosis*. When guinea pigs were subsequently infected with the engineered bacteria, the microbes multiplied in the host releasing MBP and triggering an autoimmune reaction that caused death by paralysis in about 2 weeks.[110]

According to Alibek, these results were considered a breakthrough: "[A] new class of weapons had been found. For the first time, we would be capable of producing weapons based on chemical substances produced naturally by the human body. They could damage the nervous system, alter moods, trigger psychological changes, and even kill".[111]

In other bioregulator research, conducted under Project Factor, Soviet scientists at the State Research Center for Virology and Biotechnology (Vector) near Novosibirsk engineered viruses by inserting genes coding for neuroactive peptides that affected brain function. Some of these experiments involved opiate-like neurotransmitters called enkephalins and endorphins, which in high concentrations can induce neurological and psychiatric disorders.[112] Popov, who worked at Vector from 1976 to 1986 and then at the Obolensk centre until 1992, stated that Vector had worked on endorphins, enkephalins, and other neuromodulating peptides potentially capable of altering human cognition and emotions. According to Popov, "[I]t has been discovered that personalities could be adjusted with these agents. For example, you could make people more aggressive. Or you could create feelings of insomnia, where people wanted to sleep, but would never feel tired".[113]

The Third Directorate of the Ministry of Health, under Project Flute, was also active in this area, with a network of biomedical research facilities that developed toxins, bioregulators, and other substances for use against individual human targets.[114] According to Alibek, this network included the Moscow Institute of Applied Molecular Biology (which later became the Russian Scientific Center of Molecular Diagnostics and Treatment) which studied "various biological substances in order to find those that could kill or cause irreversible mental damage" and the Moscow Institute of Immunology which studied "regulatory peptides with toxic properties capable of triggering both reversible and irreversible changes in the neural and immune systems".[115]

According to Alibek, "[T]he mood-altering possibilities of regulatory peptides were of particular interest to the KGB", possibly for use in coercion or interrogation of prisoners. In addition, the KGB was also interested in their application for covert assassination operations as such bioregulator weapons could not be traced by pathologists and victims would appear to have died of natural causes. "[W]hat intelligence service would not be interested in a product capable of killing without a trace?"[116] However, whilst whistle blower testimony has described Soviet bioregulator research and attempted development activities, there is no publicly available evidence that bioregulator weapons were produced or employed for interrogation, covert assassination operations or in armed conflict.

The important point here, as was stressed by Petro and his colleagues in 2003,[117] is that weaponeers were increasingly not restricted to modifying serendipitously-discovered potential agents but were becoming capable of designing agents to attack specific targets within living organisms.

2.2.5 Chinese Pharmaceutic CNS-acting Chemical Agent Weapons Development and Use

From at least the mid-1990s, Chinese State-owned companies developed and promoted a weapon incorporating an unidentified CNS-acting chemical for use in law enforcement operations against individuals. In 1995, marketing materials distributed internationally by China North Industries Corporation (NORINCO) promoted the "[B]BQ-901 anaesthetic gun system". This weapon discharges a projectile, with an effective range of 40 metres, which on impact injects a liquid incapacitating chemical agent into the target.[118] According to NORINCO "[T]he Model BBQ-901 Anaesthetic system is a fine unlethal [sic] special weapon system for SWAT units and other special usage.... It can be used for reconnaissance and capture of criminals in a concealed place. It is also used as a riot control weapon to subdue the ruffians and maintain public order".[119] The essentially similar if not identical, "[B]BQ-901 narcosis gun" was promoted by a second Chinese company, State 9616 Plant, at the

Asia Pacific China Police Expo held in Beijing in 2004 and again in 2006.[120] Although no further marketing materials from NORINCO or State 9616 Plant have come to light, photographs of this weapon being displayed in May 2011 by the Chinese Peoples' Liberation Army garrison in Hong Kong were subsequently published in 2012.[121]

In June 2013 Radio Free Asia broadcast a report alleging that Chinese police had employed "anaesthesia guns" to suppress ethnic Hui villagers resisting a large-scale enforced house demolition operation in Xiangdong village, Hualong County, Qinghai Province. According to a reported eyewitness "[T]he special police beat people and used anaesthesia guns. The villagers who were shot were unable to stand up. After fainting, they were taken away and detained". According to this report, a spokesperson from the Public Security Brigade of the Hualong County Public Security Bureau denied that anaesthetic guns had been used in this operation.[122]

China has provided no public information regarding its stockpiles, if any, of CNS-acting weapons nor the specific purposes of their intended employment. To date, China has made no statement clarifying whether any Chinese entity has conducted or is conducting research activities related to the development of CNS-acting weapons targeting individuals or groups. In its *2021 CWC Compliance Report*, the US reported "[S]cientists at a Chinese military institute have expressed interest in military applications of PBAs [pharmaceutical based agents] and are engaged in research involving the synthesis, characterization, and testing of PBAs with potential dual-use applications".[123]

2.3 Illustrative Examples of Past and Contemporary Dual Use Neuroscience and Related Research of Potential Concern

2.3.1 Czech Republic

From 2005–2007, Czech scientists published papers describing their investigations over several years relating to a range of pharmaceutical chemicals including various opioids, ketamine, medetomidine and midazolam, specifically highlighting their potential utility as, in their words, "pharmacological non-lethal weapons".[124] In a paper presented at the 3rd Ettlingen European Symposium on Non-Lethal Weapons in 2005 Fusek, Hess and Schreiberova reported administering rhesus monkeys with various "pharmacological cocktails" in order to determine which combinations and doses resulted in "fully reversible immobilization".[125] The paper also described how "[F]ully reversible analgesic sedation was... tested in man", utilising the triple combination of dexmedetomidine, midazolam and fentanyl given to patients undergoing surgery, and a second combination of dexmedetomidine, midazolam and ketamine which was tested on ten nurses.[126] The researchers also investigated a number of alternative means of agent delivery, including *trans*-buccal and sub-lingual administration in cats and rhesus monkeys; and conjunctival, nasal, sub-lingual and *trans*-dermal administration in rabbits.[127] They also explored inhalation administration, which was initially tested on rats with subsequent inhalation experiments utilising human "volunteers", which included children.[128]

In 2007, in a paper authored by Hess, Schreiberová, Fusek, Málek and Votava, which was presented at the 4th Ettlingen European Symposium on Non-Lethal Weapons, the researchers described how they "decided to test new combinations [of drugs] for suppression or complete abolition of aggressive behaviour" in macaque monkeys.[129] The researchers stated that: "[A]ll tested combinations resulted in macaques in reduction or complete loss of aggressiveness. Optimal combinations were naphtylmedetomidine + dextrorotatory isomer of ketamine + hyaluronidase. The onset of effect was rapid and we achieved complete manipulability of the animal with low motoric sedation".[130] Furthermore, the researchers argued that: '*the results can be used to pacify aggressive people during* medical treatment (mental disease), *terrorist attacks and during [sic] production of new pharmacological nonlethal weapons.*'[131] (Emphasis added).

Research by Czech scientists into such chemicals continued after 2007, but additional papers contained no explicit reference to their potential application as so-called "pharmacological non-lethal weapons". The Czech Republic CWC National Authority subsequently investigated Czech research activities and in 2014 correspondence to the current authors stated that "[T]here was no connection [between] the research [and] creation of any sort of weapons or devices which could be used for military or police purposes".[132] With regard to the Czech "pharmacological non-lethal weapons" papers, the National Authority stated "research programmes had justifiable medical goals, but their reporting in public media exceeded actual results of the research thus creating a false impression of possible development of some sort of chemical weapons".[133]

2.3.2 India

A review of publicly available scientific papers from 2005 till 2013 indicated that researchers at the Indian Defence Research & Development Establishment (DRDE) of the Defence Research & Development Organisation (DRDO) undertook wide-ranging research into the synthesis, aerosolization and bioefficacy of fentanyl and/or its analogues over this time period.[134] The Indian CWC National Authority subsequently investigated these research activities and in 2014 correspondence to the current authors stated that: "[I]ndia does not hold any stockpiles of weapons involving ICAs... does not conduct research in order to develop weapons ICAs... [nor] ... conduct research related to weaponisation of incapacitating agents" for defensive purposes such as the development of counter-measures.[135] In its response, the Indian CWC National Authority did confirm that "DRDE/DRDO undertakes research into fentanyl and its analogues" but that this was "only for purposes of characterization including its detection and protection aspects".[136]

A review of publicly available scientific papers from 2016 till 2019 indicated that Indian military and military-related and funded institutions and scientists searched for, identified, and characterised toxins and other substances from a broad range of indigenous Indian stinging and other poisonous plants.[137] Contemporary scientific papers published by these researchers described their work as being associated with the potential "formulation of novel future bio-threat agents for self-defence"[138] or alternatively of potentially being "exploited for the development of novel multi-system targeted agents for self-defence".[139] Some of the toxins investigated as potential agents are highly toxic and could be lethal; others though not as toxic could still potentially be employed as "less lethal" toxin weapons attacking the CNS and/or other core physiological systems. In scientific publications describing their research, authors from Indian military institutions claimed that such activities were permitted under the BTWC,[140] although this interpretation appears to be a misreading of the Convention. The Indian authorities did not respond to the current authors' previous requests for clarification as to the purposes of such research.

2.3.3 Iran

Iranian research scientists including those at Imam Hossein University (IHU) (an academic institution run along military lines and controlled by the Iranian Revolutionary Guard) and Malek Ashtar University of Technology (affiliated with the Iranian Ministry of Defence) have undertaken dual use research potentially related to CNS-acting agent weapons, published in scientific papers from 2005 until 2023.[141] This research explored the synthesis of fentanyl precursors[142] and an analogue;[143] structural activity relationships of fentanyl and its analogues;[144] the synthesis of medetomidine,[145] attempted generation of stable long lasting aerosols of this agent;[146] and investigated the phase behaviour of this and other potential CNS-acting agents.[147]

Further information indicating related activities of concern has also been made public. In its *CWC Compliance Reports*, from 2019 until 2023, the US has repeatedly alleged that in 2014, the IHU chemistry department sought "kilogram quantities of medetomidine—*a sedative it has researched as an incapacitant*—from Chinese exporters". (Emphasis added).[148] According to the US, the IHU chemistry department has "little history of veterinary or even medical research", and the quantities sought (over 10 000 effective doses) were "inconsistent with the reported end use of research".[149]

In its *2024 CWC Compliance Report,* the US described how on 23 September 2023 the X (formerly Twitter) account of Lab Dookhtegan (a hacker group opposed to current Iranian regime) posted confidential documents detailing an Iranian military university's development of grenades meant to disseminate medetomidine which included information on the production and testing of prototype weapons.[150]

When questioned in 2014 by the current authors about relevant IHU dual use research then in the public domain, the Iranian CWC National Authority stated that the lead researcher was "interested in advance [of] academic and scientific chemical issues that [are] not prohibited by the Chemical Weapons Convention" and that the "academic research is financed by [the] ministry of science and technology and is 'solely [for] scientific purposes'".[151] The Iranian CWC National Authority also stated that:

> "The Islamic Republic of Iran strongly believes that deploying and using any kind of chemical substances, RCA or ICA, as [a] matter of warfare is against the letter and spirit of the CWC. We also support any effort in the OPCW to clarify the issue and to list them in the toxic chemical schedules."[152]

There are indications that Iran's policy and practice regarding development and use of ICA/CNS-acting weapons is currently different from that as stated in 2014. In addition to the continuing dual use activities of concern, Iran voted against the OPCW CSP-26 Decision claryifing that aerosolised use of CNS-acting chemical agents would not be consistent with law enforcement purposes as set out in the CWC.[153] Iran (together with China, Russia and Syria) also declared the decision was an "ultra vires act which goes beyond the powers and functions of the Policy-Making Organs of the OPCW, so could not have any legal effect(s) on the States Parties' rights and obligations under the Convention".[154]

2.4 Application of CNS-acting Chemicals and Neuroscience for Torture and Ill-treatment

2.4.1 Development of Psychopharmacology for Benign Purposes

Psychopharmacology can be described as the study of drugs that affect mood, behaviour and cognition. It aims to develop an understanding of the actions and effects of such drugs, which may include treatments for psychiatric disorders as well as substances that can be abused.[155] The key stages in its development have been examined by Mitchell and Hadzi-Pavlovic and are summarised[156] in Box 2.1.

Prior to its revolutionary advances following World War Two, the main method of treating many people with severe mental illnesses was to confine them to asylums where they could be looked after as well as possible, and wider society could also be protected from them. The often-serendipitous discovery of key agents such as chlorpromazine, imipramine and lithium helped to render patients more able to cope in the world and, in many western countries, to a large reduction of the number of people confined to

Box 2.1 Eras of psychopharmacology.[156]

The Premodern Era

Before the revolution in psychopharmacology

The Era of Discovery, Expectation and Promise

From the late 1940s to the mid-1960s when "most of the seminal psychotropic drug discoveries were made and the basic pharmacological actions of some of these agents elucidated." The era was "characterised by a seismic shift in paradigms of mental illness".

The Era of Expansion and Consolidation

From the mid-1960s to the early years of the 21st century when "there was a dramatic increase in the number of new psychopharmacological agents and incremental growth in understanding of the underlying pharmacology".

The Era of Disillusionment and Retreat

During the following years of this century during which there was "a dawning recognition of the profound limitations in the efficacy and effectiveness of these agents" and "the growing scientific impasse of our failure to identify the underlying pathophysiology of the various psychiatric disorders" which precluded "the possibility of aetiologically targeted and tailored therapies".

asylums. The term 'neuroleptic' was introduced, for example, to describe the effect of chlorpromazine in "reducing but not paralysing" nervous activity. Mitchell and Hadzi-Pavlovic noted that:[157]

> "Chlorpromazine appeared to be most effective in schizophrenia, particularly in the acute phase.... For the first time, admission rates to US psychiatric hospitals declined..."

Chlorpromazine's introduction in the early 1950s was quickly followed by the discovery of the anti-depressant effects of imipramine and iproniazid later in the 1950s. There also rapidly evolved what appeared to be a scientific explanation of the reasons for the success at least of some of these new agents. Following the discovery during the 1920s and 1930s, that whilst transmission of information within neurons was electrical for the most part transmission of information between neurons was chemical, major investigations led to the discovery of amine neurotransmitters and important aspects of their functions. For example:[158]

> "... Brodie discovered that reserpine depleted brain levels of serotonin and noradrenaline. As reserpine was known to cause suicidal depression, these findings implicated noradrenaline and serotonin in psychiatric disease for the first time, leading to the amine hypothesis for the affective disorders."

So, there were certainly successful advances in treating patients and during the second era from the mid-1960s considerable expansion of the use of such chemical agents in medical practice. Of course, as with all drugs, care was needed to keep to the minimal effective dose and to monitor for the inevitable side effects of the drugs used and if necessary to change the drug or halt drug use. However, despite the success of these novel drugs, there were few new types of drugs developed, and for the most part advances were made by modifications of known agents or by finding new uses for known agents.

After the turn of the century, in the third era of psychopharmacology, concerns grew over the failure of drug companies to report negative findings, and at the over-prescribing of chemical agents for less severe illnesses when a non-drug approach, or a combined drug and non-drug approach, worked well.[159] Consequently "the compounding evidence of the limited efficacy and effectiveness of psychotropics led to a crisis of confidence in these medicines". Fundamentally, however, the problem that became clear was that the efforts to define the various mental illnesses through identification of set of symptoms that could be recognised by different practitioners did not actually relate directly to distinct illnesses with different causes.[160] Therefore, it was not possible to know how the available medications worked to effect change. Not surprisingly, the new State-level brain research projects that began to be initiated in 2014 focused on a different approach by attempting to discover the basic neuronal circuits that underlie our behaviour with the hope of discovering how these became modified when illnesses arose.[161]

2.4.2 Malign Application of Psychopharmacology

Evidence of the detention of large numbers of political prisoners and other sane detainees in mental institutions at some stage during a period from the 1960s until the 1990s, has been documented in China and the Soviet Union. In certain institutions, forcible administration of psychotropic substances for non-therapeutic reasons was documented.[162] Although apparently at lower levels, such abusive practices have continued. In the last decade, UN, regional and non-governmental human rights monitors have reported the enforced application of psycho-active chemicals, to subdue, intimidate or punish political prisoners or detainees in for example China, Egypt,[163] Iran, Russia[164] and Uzbekistan.[165]

2.4.2.1 Malign Application of Psychopharmacology in Soviet/Russian Psychiatric Institutions

It is clear from the documentary evidence obtained that psychiatry was systematically misused in the Soviet Union to falsely diagnose, incarcerate and abusively "treat" sane individuals and that such abusive treatment utilised the knowledge gained in the post-Second World War psychopharmacology revolution. Human rights organisations, medical associations and academics have documented how, from the 1950s to the 1990s, large numbers of political prisoners and other sane individuals were forcibly detained in psychiatric institutions in the Soviet Union.[166] According to Amnesty International these individuals were detained "as a direct result of their political or religious beliefs and with no medical justification".[167] One leading scholar, Reddaway, emphasised that despite the large knowledge gaps, rough speculative estimates can be made for the "number of dissidents held in the dozen or so high-security psychiatric prisons, yielding a figure of between 10 000 and 15 000 in all between the late 1950s and the late 1980s. But the number held in the hundreds of ordinary mental facilities cannot as yet even be estimated".[168]

According to Amnesty International there were nine major maximum security Special Psychiatric Hospitals in the Soviet Union but considerable documentation was only available on six of them situated in Kazan, Sychyovka, Leningrad, Chemaykhovsk, Dnepropetrovsk, and Oryol. These hospitals were under the direct control of the Ministry of Internal Affairs rather than the Ministry of Health and many of these hospitals had formerly operated as prisons. These institutions had formidable external walls and barbed wire and were guarded by armed military personnel. The dissidents were kept in cells, subject to strict discipline, and the orderlies who directly dealt with them were convicted criminals recruited from corrective institutions. These orderlies were allowed to physically attack the inmates and the whole context was one of intimidation.

This then was the situation within which drugs were administered to the dissenters. The Amnesty International report mentions the use of one drug – sulphazin – which had been given up in other countries because of the lack of any beneficial results for patients, being administered in these hospitals despite it causing intense pain. Other forms of directly harmful physical treatments were also employed including inappropriate use of the novel drugs available at that time. The report notes that:[169]

> "A recognized form of treating schizophrenia and certain other psychiatric disorders involves the administration of depressant drugs, including haloperidol, aminazin and triftazin. Even administered in carefully-regulated doses and with close consideration of the patient's condition, such drugs have extremely unpleasant side-effects: 'toxic inflammation of the liver, elevated intraocular pressure, fluctuations of arterial pressure, increased tension and cramping of the muscles, headaches, malaise, depressive moods, dryness of the mouth'."

The report comments that such drugs were used indiscriminately and with little regard for inmates' needs.

The effects of such treatment on the victims were devastating. Vladimir Gershuni, an ordinary Moscow worker, is quoted in the Amnesty Report writing from the Oryol Institute in 1970:[170] "[T]his medicine makes me feel more awful than anything I have experienced before; you no sooner lie down than you want to get up, you no sooner take a step than you're longing to sit down, and if you sit down, you want to walk again.... Everyone here has their life made miserable by triftazin".

The wife of Leonid Plyushch managed to visit him in the Dnepropetrovsk Special Psychiatric Hospital in October 1973. He had until August of that year continued his mathematical work, but then he was treated with haloperidol. His wife found after 2 months of this treatment that:[171]

> "...it was impossible to recognise him. His eyes were full of pain and misery, he spoke with difficulty and brokenly.... His efforts at self-control was evident as from time to time he closed his eyes, trying to carry on a conversation and to answer questions. But his inner strength was exhausted, finished..."

When she was able to visit again in November 1973 her husband was in an even worse condition and asked her not to send him any more scientific books as he was not capable of reading them. When the use of haloperidol was stopped his condition improved, but later he was treated with large doses of triftazin and became incapable even of replying to the letters from his wife.

Clearly the psychiatrists who controlled this system were involved with the security authorities and knew exactly what they were doing and why. Professor A. Snezhnevsky, Director of the Institute of Psychiatry of the Soviet Academy of Medical Sciences, developed a novel set of classifications of mental illnesses and associated diagnostic criteria that were used to incarcerate and 'treat' the dissidents:[172]

> "...patients with sluggish schizophrenia could present as ostensibly sane yet show minimal but clinically relevant personality changes that could pass unnoticed to the untrained eye.... Symptoms referred to as part of the 'negative axis' included conflict with authorities, poor social adaptation and pessimism, and were themselves sufficient for a formal diagnosis of 'sluggish schizophrenia with scanty symptoms'..."

With this kind of classification and criteria anyone could fall victim to the system if that was thought necessary by the security authorities.

A closer look at the drugs involved shows precisely why they were so devastating when used in this manner. While the initial discoveries were made in the West, psychiatrists were quickly interested in the new drugs and by 1966 the Soviet government had listed ten such drugs that were to be given priority by the pharmaceutical industry.[173] Aminazin, triftazin and haloperidol are all now classified as first-generation typical anti-psychotics.[174] While in proper medical practice they do affect critical dopamine receptors involved in schizophrenia, they also affect other receptors in different neuronal circuits and this is the reason for the side effects of the drugs. Aminazine (chlorpromazine) is rated as having low potency while haloperidol and triftazine (trifluoperazine) are rated as having high potency. Common side effects for chlorpromazine include[175] "dizziness, drowsiness, dry mouth... hypotension... sedation, photosensitivity to sunlight". Common side effects for haloperidol include[176] "[B]lurred vision, drowsiness, dry mouth, dystonia, hallucinations, impotence, insomnia... visual disturbances," and for trifluoperazine common side effects are[177] "[B]lurred vision, dizziness, drowsiness, dry mouth, dystonia... insomnia, motor restlessness". In general, this class of drugs carries the warning that[178] they "[C]an impair mental/physical abilities and lower seizure threshold".

How then did medical doctors end up treating dissidents who were not mad, with these novel drugs? Robert van Voren has noted that "most experts agree that the core group of psychiatrists who developed this concept did so on the orders of the Party and the Soviet secret service KGB". Additionally, van Voren takes a more nuanced view of those Soviet psychiatrists who were not in this core group:[179] "for many Soviet psychiatrists this seemed a very logical explanation because they could not explain to themselves otherwise why somebody would be willing to give up his career, family, and happiness for an idea or conviction that was so different from what most people believed or forced themselves to believe". So, they avoided difficult questions and potential dangers for themselves and their families by not standing out against what was happening to the dissidents.

2.4.2.2 Malign Application of Psychopharmacology in Chinese Psychiatric Institutions

Evidence of the detention of large numbers of political prisoners and other sane detainees in mental institutions at some stage during a period from the 1960s until the present day has been recorded in China. The complicated history of such misuse of psychiatry in China has been documented by a range of scholars and human rights organisations notably Human Rights Watch (HRW) and the Geneva Initiative on Psychiatry (GIP) in their 2002 report, *Dangerous Minds: Political Psychiatry in China Today and its Origins in the Mao Era*.[180] Based on a large number of documents compiled from China's official literature on psychiatry and the law, the HRW/GIP report estimated that "somewhere in excess of 3000 'political cases' (broadly defined) have been dealt with by Chinese forensic psychiatric examiners countrywide over the past two decades [from 1980–2000], and moreover that the great majority of these were subjected, as a result, to some form and duration of forced psychiatric custody and treatment".[181]

The lead author of this report Robin Munro, in a substantial 2000 paper on "Judicial Psychiatry in China and its Political Abuses" stated that:[182]

"...Chinese authorities have ... a longstanding record of the misuse of psychiatry for politically repressive purposes, one that resembles in all key respects that of the former Soviet Union, and one, moreover, that may well have exceeded in scope and intensity the by now thoroughly documented abuses that occurred in the latter country prior to 1990..."

And he suggested that:

"...It seems likely that these abuses are confined mainly to those working within the sub- specialist domain of forensic psychiatry, a small and still secretive field of which most regular Chinese psychiatrists may have little direct knowledge or experience..."

And, as in the Soviet Union, the extent of the abuse reflected the prevailing political situation and the level of authoritarianism at the time (see Box 2.2).

Box 2.2 Stages in the evolution of abuse in China.[182]

1. From 1949 through the 1950s

   "During which Soviet influences predominated;"

2. The Cultural Revolution decade (1966–1976)

   "When political psychiatry reached its absurd apogee;"

3. The 1980s

   "When the reform era of Deng Xiaoping seems to have meant, for forensic psychiatry, a partial return to the orthodoxies of the pre-Cultural Revolution period;"

4. The 1990s

   "Which appeared to see a significant decrease in politically-directed psychiatry in China;"

5. The end of the 1990s

   "A substantial resurgence of abusive practices, notably in the case of Falun Gong detainees."

The similarities to Soviet practice are clear, for example, in the entanglement of criminal and medical classifications.

As Munro points out:[183]

> "...it should be noted that the persons in question were in most cases arrested on criminal charges — but for activities not held to be crimes under international legal standards — prior to being committed for forensic psychiatric evaluation. If truly mentally disturbed, they should not have fallen within the scope of the psychiatric-criminal justice system, but should rather have been given appropriate treatment by the regular mental healthcare system."

And in the arbitrary definitions used for the confinement of dissidents:[184]

> "...China's expansive definition of the key legal determinant of involuntary psychiatric committal, namely 'social dangerousness.' Whereas under international standards, the applicable scope of the 'dangerousness' criterion is mainly restricted to situations where mentally ill people pose a direct physical danger either to themselves or to others, in China it is applied also to those, such as certain types of dissidents, whom the government regards as posing a political threat to 'social order.'"

Notable also is the use made of special high security facilities:[185]

> "...in 1987 the Chinese government for the first time decided to establish a nationwide system of high-security facilities for 'dangerously mentally ill offenders.' These, the equivalent of the USSR's Special Psychiatric Hospitals run by the Interior Ministry, were to be uniformly designated as 'Ankang' (Peace and Health) institutions, and were to be directly administered and run by the Ministry of Public Security..."

And there was the use of the then-newly available drugs. For example, one person described his treatment in the 1960s as follows:[186]

> "...I was subjected to numerous bouts of drug interrogation...and was *fed large quantities of chlorpromazine*. They treated me like an experimental object and it was all a disguised form of physical torture. It was extremely painful, and by the end, I was left trembling and sweating all over and my memory had started to go." (Emphasis added).

Human rights organisations and investigative journalists have detailed how the misuse of such CNS-acting drugs continued decades afterwards, notably in the Ankang institutions. In 2005, Human Rights Watch[187] and the German newspaper,[188] *Die Zeit*, worked together to highlight testimonies of former Ankang prisoners. Wang Wanxing was forcibly detained for 13 years in Beijing Public Security Bureau's Ankang Hospital for the Custody and Treatment of Mentally Ill Offenders following his brief, one-man pro-democracy demonstration in Tiananmen Square on the eve of the third anniversary of the 4 June 1989 crackdown. Throughout his detention at the Beijing Ankang, Wang was forced to take chlorpromazine three times a day. Despite close monitoring by nursing staff, Wang, unable to cope with the severe mental and physical side effects, developed a method of concealing the pills in his mouth until after the inspection and then secretly disposing of them.[189]

Qiu Jinyou, a rural rights activist, was detained at the Hangzhou Ankang for 208 days and subjected to frequent electric shock treatment by officials, in an effort to make him reveal the names of fellow petitioners against corruption in his village. As a result of the frequent psychiatric medication he was given, Qiu said that his hair began falling out and he suffered from spasms and tremors, nervous despair, memory loss, forgetfulness and other symptoms.[190]

Ms. Meng Xiaoxia was forcibly held in Xi'an Ankang for 10 years in total, without any medical or psychiatric examination, for protesting against her employer's failure to discipline a fellow worker who had beaten her unconscious. She was given electric shock treatment, with electrodes placed on her forehead, on three occasions, and she was also subjected to insulin coma therapy – a treatment that is elsewhere almost unknown today due to its dangerous side effects.[191]

Despite subsequent purported reforms to its repressive Ankang system of enforced detention of political activists in psychiatric institutions, recent evidence from human rights organisations indicates that this system remains in place. In its 2022 report, Safeguard Defenders claim to have uncovered "Some 109 hospitals, from 21 provinces, municipalities or regions across China [involved in the Ankang system] . . . indicating that the political abuse of psychiatry is geographically widespread and routinely practiced in China".[192]

The report highlights the cases of 99 individuals who were subjected to forced psychiatric hospitalization for political reasons between 2015 and 2021.[193] In a large majority of the cases studied, individuals were forced to take medication either orally or by injection against their will. For example, political activist, Zhang Jilin, was arrested in January 2019 after giving a speech in a public square in Chongqing, a municipality in Southwest China, calling for the removal of Xi Jinping, the General Secretary of the Chinese Communist Party. He was subsequently held for 28 days in a psychiatric institution in early 2019, where he was force-fed drugs three times a day. If he tried to resist, medical staff resorted to injecting him. As he stated:

> "...They didn't tell me what kind of medication I was taking, but I suspect it was two different kinds, one to make me feel euphoric and

another to ... sedate me... Each time I took the medication, I felt dizzy and nauseous. A couple of times in the beginning I vomited, and there was blood in my vomit as well."[194]

Whilst medical staff normally refused to divulge what medication was being administered, a few 'patients' who continued to be prescribed drugs after release were able to identify their medication. The majority were anti-psychotics such as Thioridazine, Perphenazine, Risperidone, Aripiprazole and Clozapine. The side effects of such drugs, as we have noted, include drowsiness, anxiety, memory loss, difficulty in thinking, tremors and insomnia.[195]

2.4.3 Malign Application of CNS-acting Chemicals for Interrogation

2.4.3.1 Cold War Research into CNS-acting Chemicals

During the Cold War, military, intelligence and law enforcement bodies in certain States including the UK, US and Soviet Union sponsored dedicated research into CNS-acting chemicals (such as LSD, sodium thiopental, sodium amytal or scopolamine) with the intention of developing so-called "truth drugs" or "truth serums" to be employed against detainees without their consent, for the purposes of interrogation.[196] While the majority of such activities remain classified an intimation of the chemical agents explored and research undertaken can be gleaned from the limited documents that have come to light into the US programme.

2.4.3.1.1 Cold War Case Study: US Intelligence Sponsored Research into Malign Application of CNS-acting Chemicals. At the height of the Cold War, from the early 1950s till the mid-1970s, the US Central Intelligence Agency (CIA) oversaw a huge network of clandestine research on human subjects under code names including Project MKUltra, Project BLUEBIRD and Project ARTICHOKE. Although much of the documentation was either destroyed or remained classified for decades, an increasing amount is now being uncovered which shows this research included the use of CNS-acting chemical agents, hypnosis, isolation, and/or sensory deprivation. This research took place in a variety of medical settings including prison hospitals, addiction centres and juvenile detention facilities, often without the informed consent of those tested, and with the involvement of leading figures in the behavioural sciences.[197]

Perhaps the most notorious research programme, for which we have information to date, is Project MKUltra. Initiated in 1953 and terminated in 1973, it incorporated human experimentation with the objective of developing procedures that could be used during interrogations to weaken people and force confessions. MKUltra tested and utilised numerous methods to manipulate its subjects' cognisance including the covert administration of high doses of CNS-acting chemicals without the subjects' informed full consent.[198] In his opening remarks to a Congressional Hearing into Project MKUltra, Senator Kennedy, Chair of the Subcommittee on Health and Scientific Research of the Committee on Human Resources, stated that: "The Central Intelligence Agency drugged American citizens without their knowledge or consent. It used university facilities and personnel without their knowledge. It funded leading researchers, often without their knowledge".[199]

We lack a full understanding of Project MKUltra because many of the relevant documents were deliberately destroyed before the first Congressional investigations of the project in the 1970s. However, by chance some of the financial records were not erased and this allowed Admiral Stansfield Turner, then Director of the CIA, to provide a more detailed insight into the nature of the sub-projects within the overall project in 1977.[200] In his Statement to Congress the Admiral noted that 149 different sub-projects had been identified within the newly discovered material. Some of these sub-projects that he identified are set out in the categories he used in Box 2.3.

Box 2.3 MKULTRA sub-projects.[200]

1. ***Research into the effects of behavioural drugs and/or alcohol:***
   - 17 subprojects probably not involving human testing;
   - 14 subprojects definitely involving tests on human volunteers;
   - 19 subprojects probably including tests on human volunteers. While not known, some of these subprojects may have included tests on unwitting subjects as well;
   - 6 subprojects involving tests on unwitting subjects.
2. ***Research on hypnosis:***
   - 8 subprojects, including 2 involving hypnosis and drugs in combination.
3. ***Acquisition of chemicals or drugs:***
   - 7 subprojects.
4. ***Aspects of magicians' art useful in covert operations: e.g., surreptitious delivery of drug-related materials:***
   - 4 subprojects.
5. ***Studies of human behaviour, sleep research, and behavioural changes during psychotherapy:***
   - 9 subprojects.
6. ***Library searches and attendance at seminars and international conferences on behavioural modification***
   - 6 subprojects.
7. ***Motivational studies, studies of defectors, assessment, and training techniques***
   - 23 subprojects.

8. ***Funding mechanisms for MKULTRA external research activities:***
   - 3 subprojects.
9. ***Research on drugs, toxins, and biologicals in human tissue; provision of exotic pathogens and the capability to incorporate them in effective delivery systems:***
   - 6 subprojects.

In summary he suggested that for the most part, this new data was just additional data to what was known to Congress already from previous investigations. However, he also added that:

> "The most significant new data discovered are, first, the names of researchers and institutions who participated in the MKULTRA project.... We are now in possession of the names of 185 non-government researchers and assistants who are identified in the recovered material dealing with the149 subprojects. The names of 80 institutions where work was done of with which people were affiliated are also mentioned..."

And he continued: "The institutions include 44 colleges or universities, 15 research foundations or chemical or pharmaceutical companies and the like, 12 hospitals or clinics (in addition to those associated with universities) and 3 penal institutions".

That listing of institutions connected with MKULTRA gives a good idea of the huge scale of the project. To explore the manner in which psychiatry and psychopharmacology were misused, it is useful to examine one of the CIA-funded subprojects.

2.4.3.1.2 The Montreal Experiments. The Montreal Experiments were carried out in the 1950s and 1960s by the then eminent psychiatrist Dr Donald Cameron (a previous President of the Psychiatric Association from 1961–1966) with financing from the CIA's MKULTRA project.[201] The CIA funding was provided through a front company called the "Society for the Investigation of Human Ecology" and it is not clear whether Cameron knew of the real source, but it is quite easy to identify the link from the application made for funding found in the Project MKULTRA financial files for Subproject 68 where a proposal for the use of "Psychic Driving", Cameron's method of trying to cure Schizophrenia, is set out and publications on this topic are mentioned.[202] The Canadian Encyclopedia notes that the Canadian Government also funded this work, the purpose of which it describes as follows:[203]

> "...The experiment's purpose was originally and officially listed as an effort to find a cure for schizophrenia. However, in reality, the project conducted illegal human experimentation to determine whether drugs

and psychological techniques could be used for the purposes of mind control..."

Several hundred patients are reported to have been subject to such treatment in these experiments.[204] Some of the victims and their families have, over the years, obtained some compensation, but non-disclosure and confidentiality agreements have restricted the release of more information.

Cameron described what he was doing in a 1962 co-authored paper, "The Depatterning Treatment of Schizophrenia", published in the *Journal Comprehensive Psychiatry*, the official journal of the American Psychopathological Association.[205] The paper explained that over the previous 30 years various treatments for schizophrenia had been developed, but that the extensive experience of the authors led them to seek a more powerful method of treatment. This essentially consisted of keeping the victims in drug-induced sleep and the repeated use of electroshocks and various drugs. The net result was to reduce the person through a series of stages that could be gauged by reference to the loss of their memories. As the authors stated:[206]

> "In the first stage of disturbance of the space-time, there are marked memory deficits, but it is possible for the individual to maintain a space-time image. In other words, he knows where he is, how long he has been there and how he got there. In the second stage, the patient has lost his space-time image, but clearly feels that there should be one.... In the third stage, there is not only a loss of the space-time image but loss of all feeling that should be present..."

The text goes on to state that:

> "...During this stage the patient may show a variety of other phenomena, such as loss of a second language or all knowledge of his marital status. In more advanced forms, he may be unable to walk without support, to feed himself, and he may show double incontinence..."

The authors state that "[At] this stage all schizophrenic symptomatology is absent". They called this process "depatterning", which was a development of the earlier idea of Psychic Driving in which drugged sleeping patients were subject to continuous repetitive auditory verbal input. If the process was of any help to the patients is surely open to question, but in the middle of the Cold War with concerns about how brainwashing might be used, the interest of the CIA in funding such work should not be unexpected.

In the grant application of 1957 found in the financial records the nature of the intended illegal research is quite clear:

> "...After considerable experimentation, we have developed a procedure which in the most successful case has produced behavioral changes lasting up to two months.

The procedure requires:[207]

1. The breaking down of ongoing patterns of the patient's behaviour by means of particularly intensive electroshock (depatterning),
2. The intensive repetition (16 hours a day for 6 or 7 days) of the prearranged verbal signal.
3. During this period of intensive repetition the patient is kept in partial sensory isolation.
4. Repression of the driving period is carried out by putting the patient, after conclusion of the period, in continuous sleep for 7 to 10 days".

Furthermore, amongst the specific objectives stated in the application was the question:[208]

> "Can we develop better methods of inactivating the patient during the period of driving (exposure to repetition) and at the same time maintain him at a high level of activity, by physiological and chemical agents...Amongst the chemical agents we propose to explore ...are the following (used either singly or in combination): Artane, Anectine, Bulbocapnine, Cuare...We propose to use LSD 25 and other similar agents as a means of breaking down the ongoing patterns of behavior."

Klein, in assessing the Montreal experiments, argues that they provided the basis for the development and use of techniques of torture by the United States, stating that:[209] "This, it turns out, was the true purpose of MKULTRA: not to research brainwashing (that was a mere side project), but to design a scientifically based system for extracting information from 'resistant sources.' In other words, torture".

So, during the Cold War, in both Western democratic countries (notably the US) and in the Communist States (notably China and the Soviet Union), the last century saw attempts to use the increasing understanding of the brain for malign purposes and this warns us of the need to be aware of such possibilities in the future.

2.4.4 Contemporary Malign Application of CNS-acting Chemicals for Interrogation

Human rights organisations have reported contemporary cases of the forced application by certain State intelligence operatives, law enforcement officials or (*para*) medical attendants of usually unidentified drugs, with apparent psychoactive properties, to prisoners and other detainees in order to facilitate interrogation. These practices have been reported infrequently; normally they appear to be conducted in a limited and *ad hoc* manner for example in Iran,[210] Turkmenistan,[211] and in the US detention centre at

Guantanamo Bay.[212] However in India, chemically facilitated interrogation has become institutionalised within the law enforcement system.

2.4.4.1 Iran

Human rights organisations have documented the forced administration of unidentified chemical substances having "mind-and mood-altering" effects, apparently to facilitate interrogation by Iranian law enforcement officials. Prisoners describe "los[ing] control over themselves, laughing or crying uncontrollably or on occasions feeling that they had unlimited strength" or of such chemicals exacerbating "feelings of fear and anxiety".[213] Such practices were reported against those detained during and following the nationwide protests of 2009,[214] 2017[215] and 2019.[216] Amnesty International has documented the case of Ali Abdoli, a civil society activist arrested on 18 November 2019 after participating peacefully in protests in Shiraz, Fars Province, who was held in a secret detention facility known as "Pelak-e 100" (Number 100) run by the Ministry of Intelligence. Here he was subjected to a range of torture techniques including repeated forced administration of unidentified pills that had a sedative and intoxicating effect, making Mr Abdoli sleepy and confused, and affecting his ability to think clearly. It was in this state of disorientation that his interrogators tried to force him to sign self-incriminating 'confessions'. When he was subsequently transferred to Adelabad prison in Shiraz, he suffered convulsions and experienced other chemical withdrawal symptoms.[217]

2.4.4.2 India

Indian law enforcement agencies currently employ certain CNS-acting chemicals to augment interrogation.[218] In this technique, known in India as narcoanalysis, sodium pentothal has been administered intravenously to prisoners who then enter various stages of anaesthesia.[219] According to proponents of narcoanalysis, whilst in the hypnotic stage of anaesthesia, the subject becomes less inhibited and is more likely to divulge information, which they would normally try to suppress in the conscious state. However, the potential utility and acceptability of narcoanalysis have been and continue to be contested internationally and within the Indian academic legal community and medical profession. Despite being known as "truth drugs" or "truth serums" the principle utilitarian concern of administering such CNS-acting chemicals is that there is no guarantee that the person will divulge the truth; some subjects may retain an ability to deceive even in the hypnotic state, whilst others may become extremely suggestible to questioning. This is especially worrying, since investigators may frame questions in a manner prompting misleading or incriminatory responses. Clearly subjects questioned whilst in a hypnotic state are not in a clear state of mind and the statements made are not truly voluntary.[220]

In May 2010 the Indian Supreme Court ruled that the administration of CNS-acting chemicals to detainees during interrogation without their consent was illegal as it constituted cruel, inhuman or degrading treatment.[221] The Court however left open the possibility of admitting into evidence information obtained under narcoanalysis when the subject consented to such practices, potentially undermining the prohibition. Indeed, despite this ruling, narcoanalysis has become institutionalised as part of the Indian law enforcement and judicial system. In a 2012 interview with UK newspaper, *The Guardian*, Mr V. H. Patel, then Deputy Director of the Directorate of Forensic Sciences (DFS) in Gujarat, Western India, stated that he had personally conducted narcoanalysis in nearly 100 cases. He estimated that his laboratory received requests for narcoanalysis three to four times a month.[222] In 2017, the *Times of India* reported that the DFS in Gandhinagar had performed 42 narco-analysis tests in the first 6 months of 2017 against the annual average of 60 tests.[223] In 2018, narcoanalysis capacity within India further increased with the establishment of a Forensic Sciences Laboratory facility in Dehli.[224] Although there are no publicly available official statistics comprehensively detailing the current frequency of narcoanalysis use by Indian law enforcement agencies, the employment of this technique continues to be reported in the Indian media.[225]

2.5 Conclusions

In its 2008 study of *Emerging Cognitive Neuroscience and Related Technologies* the US National Research Council[226] argued that the search for means of enhancing cognitive performance could also lead to what it called a "degradation market" in which means of *reducing* cognitive functions could also develop on the back of the benignly-intended research. As they noted:

> "The development of a more humane way of fighting a war by using cognitive weapons could dramatically change the degradation market.... The fear that this approach to fighting war might be developed will be the justification for developing countermeasures to possible cognitive weapons. This escalation might lead to innovations that would cause this market area to expand rapidly..."

And, specifically, it noted that:

> "...*If a particularly effective degradation product is developed that has few side effects, escalation of this market will be self-fulfilling*". (Emphasis added)

Moreover, the authors continued immediately to add that "The concept of torture could also be altered by products in this market". This perspective reflects the point that the members of the EU Human Brain Project made in regard to dual use: that this concept must not just be considered in relation to military misuse, but to several other areas of potential misuse as well.

As we have shown in this chapter, such diverse misuse has occurred in the recent past in relation to advances in our understanding of the brain, and there have clearly also been interactions between these diverse forms of misuse. For example, the search for new means of law enforcement through the use of chemical agents has overlapped with the search for militarily effective novel 'non-lethal' or 'less lethal' weapons employing similar agents. So, as we now move to investigate what has happened to neuroscience since the turn of the century and what kinds of novel insights and applications are becoming possible, we have to keep this broad understanding of the dual or rather multiple uses of such knowledge and technology, and the interactions of the differing forms of potential misuse in mind.

References

1. Human Brain Project, *Opinion on 'Responsible Dual Use'*, Danish Board of Technology Foundation, Copenhagen, 2018.
2. N. C. Andreasen, *Brave New Brain: Conquering Mental Illness in the Era of the Genome*, Oxford University Press, Oxford, 2001.
3. (a) In addition to country case studies detailed in this Chapter, see also J. Perry Robinson, Incapacitating chemical agents in context: an historical overview of States' policy, in ICRC, "Incapacitating chemical agents" Law enforcement, human rights law and policy perspectives, Montreux, Switzerland, 2013, pp. 89–96; (b) Royal Society, *Brain Waves Module 3: Neuroscience, conflict and security*, 2012, pp. 10–13; (c) M. Dando and M. Furmanski, Midspectrum incapacitant programs, in *Deadly Cultures: Biological Weapons Since 1945*, ed. M. Wheelis, L. Rózsa and M. Dando, Harvard University Press, Hartvard, 2006.
4. (a) A. Maclean, Historical survey of the Porton Down volunteer programme, Ministry of Defence, Part 4: Human studies with incapacitating agents, June 2006, pp. 109–142; (b) M. Dando and M. Furmanski, Midspectrum incapacitant programs, in *Deadly Cultures: Biological Weapons Since 1945*, ed. M. Wheelis, L. Rózsa, and M. Dando, Harvard University Press, Cambridge, 2006; (c) R. Evans, *Gassed: British Chemical Warfare Experiments on Humans at Porton Down*, House of Stratus, London, 2000; (d) J. Perry Robinson, Disabling Chemical Weapons: Some Technical and Historical Aspects (paper presented to the Pugwash Study Group on Implementation of the CBW Conventions), Den Haag/Noordwijk, 27th–29th May', 1994.
5. UK, Biology Committee, *The UK Approach to New Agents*, PRO, WO 195/15031, 1960.
6. UK, Offensive Evaluation Committee, *Minutes of the 9th Meeting*, 3 January, 1962. PRO, WO 195/15329, 1962.
7. M. Dando and M. Furmanski, Midspectrum incapacitant programs, in *Deadly Cultures: Biological Weapons Since 1945*, ed. M. Wheelis, L. Rózsa and M. Dando, Harvard University Press, Hartvard, 2006, pp. 239–240.

8. Unless otherwise stated the information summarised in this section derives from, A. Maclean, *Historical survey of the Porton Down volunteer programme*, Ministry of Defence, June, Part 4: Human studies with incapacitating agents, 2006, pp. 109–142, And this report published by the UK MoD contains full details of the primary sources.
9. Psychochemical agents OAD/TA2800/257/58 14 April 1958. [U], as cited in A. Maclean, *Historical survey of the Porton Down volunteer programme*, Ministry of Defence, June, Part 4: Human studies with incapacitating agents, 2006, ch. 13, pp. 109–142, footnote 27.
10. UK, Applied Biology Committee, *Minutes of the 7th Meeting*, PRO, WO 195/16855, 4 December 1968.
11. Unless otherwise stated the information summarised in this section derives from A. Maclean, *Historical survey of the Porton Down volunteer programme*, Ministry of Defence, June, Part 4: Human studies with incapacitating agents, 2006, pp. 109–142, And this report published by the UK MoD contains full details of the primary sources.
12. R. Fisher, *An Outline of CW and Its Problems*, PRO, WO 195/16641, 1968.
13. C. Gould and P. Folb, The South African Chemical and Biological Warfare Program: An Overview, *Non-proliferation Rev.*, 2000, 10–23, Fall-Winter.
14. C. Gould and P. Folb, *Project Coast: Apartheid's Chemical and Biological Warfare Programme*, United Nations Institute for Disarmament Research (UNIDIR), 2002, pp. 1–294.
15. C. Gould and P. Folb, Project Coast: Apartheid's Chemical and Biological Warfare Programme, United Nations Institute for Disarmament Research (UNIDIR), 2002, p. 127.
16. Correspondence from Gen. D. P. Knobel to J. Swanepoel, 11 January 1992. Document made available to the TRC and public during the TRC hearing into chemical and biological warfare, Cape Town, June,1998, as cited in C. Gould and P. Folb, Project Coast: Apartheid's Chemical and Biological Warfare Programme, United Nations Institute for Disarmament Research (UNIDIR), 2002, p. 125.
17. C. Gould and P. Folb, Project Coast: Apartheid's Chemical and Biological Warfare Programme, United Nations Institute for Disarmament Research (UNIDIR), 2002, p. 125.
18. C. Gould and P. Folb, Project Coast: Apartheid's Chemical and Biological Warfare Programme, United Nations Institute for Disarmament Research (UNIDIR), 2002, p. 128.
19. Final report of the Truth and Reconciliation Commission, *Special Investigation into Project Coast: South Africa's Chemical and Biological Warfare Programme*, 1999, vol. 2, ch. 6, as cited by C. Gould and P. Folb, Project Coast: Apartheid's Chemical and Biological Warfare Programme, United Nations Institute for Disarmament Research (UNIDIR), 2002, p. 128.
20. C. Gould and P. Folb, Project Coast: Apartheid's Chemical and Biological Warfare Programme, United Nations Institute for Disarmament Research (UNIDIR), 2002, p. 128.

21. C. Gould and P. Folb, Project Coast: Apartheid's Chemical and Biological Warfare Programme, United Nations Institute for Disarmament Research (UNIDIR), 2002, pp. 128–129.
22. HPCSA, Before the Professional Conduct Committee of the Medical and Dental Professions Board, in the matter of Dr W. J. Basson, accused, Charge sheet and annexures, charge 5, 2007.
23. HPCSA, Professional Conduct Committee: Concerning Dr W.Basson, Reasons of the Committee, Issued by the HPCSA, 18 December 2013, 2013.
24. C. Gould and P. Folb, Project Coast: Apartheid's Chemical and Biological Warfare Programme, United Nations Institute for Disarmament Research (UNIDIR), 2002, pp. 223-30.
25. Letter dated 27th January 1992 from the Permanent Representative of Mozambique to the United Nations Addressed to the Secretary General, General Assembly, Security Council document A/47/87, S/23490, 29th January 1992.
26. For further descriptions of the Ngogunue incident see reports by the South African, Swedish, UK and UN investigation missions cited below.
27. B. Davy, *Chemical Incident Verification Mission, Mozambique 22nd-24th January 1992*, 1992. 29th January. The South African team, comprising nine experts and four support staff, was joined by a Mozambican five-member delegation during the mission.
28. A. Andersson and S. Persson, *The final report given by the experts appointed by ASDI [Swedish International Development Cooperation Agency] to assist the government of Mozambique in order to investigate the alleged use of chemical warfare agent(s) in the Ngungue Incident*, 3rd March. The Swedish two-person team conducted its mission from 5th to 11th February 1992, 1992.
29. J. Thompson, *Report of an investigation into the alleged use of chemical weapons in Mozambique, January 1992*. Chemical and Biological Defence Establishment, 17th February. The UK two-person team conducted its mission from 27th to 31st January 1992, 1992.
30. Government of Mozambique, *Report on South African Verification Mission* (from Mozambican perspective), 30th January 1992, Maputo.
31. The experts were Dr Sven-Ake Persson, Professor of Toxicology and Research Director of the Swedish National Defence Research Establishment, Mr Heiner Staub, Chemical Engineer, from the Swiss NC-Laboratory Defence Technology and Procurement Agency, and Dr J. P. Thompson, Head of Human Studies at the UK Chemical and Biological Defence Establishment. Dr Persson and Dr Thompson had previously conducted investigations into this incident on behalf of Sweden and the UK respectively. See: United Nations Security Council, Note by the Secretary General, Report of the Mission dispatched by the Secretary-General to investigate an alleged use of chemical weapons in Mozambique, S/24065, 12th June 1992. Paragraphs 3 and 4, pp. 1–2.
32. United Nations Security Council, Note by the Secretary General, Report of the Mission dispatched by the Secretary-General to investigate an alleged use of chemical weapons in Mozambique, S/24065, 12th June 1992.

33. United Nations Security Council, Note by the Secretary General, Report of the Mission dispatched by the Secretary-General to investigate an alleged use of chemical weapons in Mozambique, S/24065, 12th June 1992, paragraph 32, p. 5.
34. United Nations Security Council, Note by the Secretary General, Report of the Mission dispatched by the Secretary-General to investigate an alleged use of chemical weapons in Mozambique, S/24065, 12th June 1992, paragraph 33, p. 5.
35. United Nations Security Council, Note by the Secretary General, Report of the Mission dispatched by the Secretary-General to investigate an alleged use of chemical weapons in Mozambique, S/24065, 12th June 1992, paragraph 36, p. 5.
36. C. Gould and P. Folb, *Project Coast: Apartheid's Chemical and Biological Warfare Programme*, United Nations Institute for Disarmament Research (UNIDIR), 2002, p. 136.
37. (a) Republic of South Africa, National Intelligence Service, Steyn Report: Staff paper prepared for the Steyn Commission on alleged dangerous activities of SADF components, December 1992, p. 14, as cited in C. Gould, South Africa's chemical and biological warfare programme 1981–1995, PhD thesis, Rhodes University, August 2005, p. 230; (b) For further discussion of the Steyn Report allegations see: S. Burgess and H. Purkitt, *The Rollback of South Africa's Biological Warfare Programme*, USAF Institute for National Security Studies, USAF Academy, Colorado, 2001; (c) C. Gould, South Africa's chemical and biological warfare programme 1981 – 1995, PhD thesis, Rhodes University, August 2005, pp. 53–54. (d) See also: J. Bale, South Africa's Project Coast: death squads, covert State-sponsored poisonings, and the dangers of CBW proliferation, *Democracy Secur.*, 2006, **2**, 27–59. (e) Nuclear Threat Initiative, South Africa Profile, Chemical Chronology, http://www.nti.org/e_research/profiles/SAfrica/Chemical/index.html (accessed 11th January 2012, now unavailable).
38. Republic of South Africa, National Intelligence Service, *Steyn Report*, December 1992, p. 14, as cited in C. Gould, South Africa's chemical and biological warfare programme 1981 – 1995, PhD thesis, Rhodes University, August 2005, p. 230.
39. Republic of South Africa, National Intelligence Service, *Steyn Report*, December 1992, p. 14, as cited in C. Gould, South Africa's chemical and biological warfare programme 1981 – 1995, PhD thesis, Rhodes University, August 2005, p. 230.
40. C. Gould, *South Africa's chemical and biological warfare programme 1981 – 1995*, PhD thesis, Rhodes University, August 2005, p. 237.
41. (a) For further discussion of historical U.S. CNS-acting chemical research, see for example: M. Dando and M. Furmanski, Midspectrum incapacitant programs, in *Deadly Cultures: Biological Weapons Since 1945*, ed. M. Wheelis, L. Rózsa, and M. Dando, Harvard University Press, Cambridge, Mass., 2006; (b) N. Davison, 'Off the Rocker' and 'On

the Floor': The Continued Development of Biochemical Incapacitating Weapons, Bradford Science and Technology Report No. 8, August 2007; (c) M. Furmanski, Historical military interest in low-lethality biochemical agents, in *Incapacitating Biochemical Weapons*, ed. A. Pearson, M. Chevrier and M. Wheelis, 2007; (d) A. Pearson, Late and post-cold war research and development of incapacitating biochemical weapons, in *Incapacitating Biochemical Weapons*, ed., A. Pearson, M. Chevrier and M. Wheelis, Lexington Books, 2007, p. 74; (e) J. Perry Robinson, Disabling Chemical Weapons, A Documented Chronology of Events, 1945–2011, (copy provided by author), 20 November 2012.

42. J. S. Ketchum and F. R. Sidell, Incapacitating agents, in *Military Aspects of Chemical and Biological Warfare*, ed. F. R. Sidell, E. T. Takafuji and D. R. Franz, Office of the Surgeon General, U.S. Army, Washingon D.C., 1997, pp. 287–305.
43. J. S. Ketchum and F. R. Sidell, Incapacitating agents, in *Military Aspects of Chemical and Biological Warfare*, ed. F. R. Sidell, E. T. Takafuji and D. R. Franz, Office of the Surgeon General, U.S. Army, Washingon D.C., 1997, p. 294.
44. J. S. Ketchum, *Chemical Warfare: Secrets Almost Forgotten*, Private publication, United States, 2006, ISBN: 1-4243-0080-0
45. Given the attempts to weaponise this chemical agent, BZ is listed under Schedule 2.a. of the Chemical Weapons Convention. Two of its immediate precursors, 3-Quinuclidinol and Benzilic Acid are also listed (both under Schedule 2.b.). See: OPCW, Chemical Weapons Convention, Annex on Chemicals, B. Schedules of Chemicals, Schedule 2, 1993.
46. M. S. Gitler, R. De la Cruz, B. R. Zeeberg and R. C. Reba, [3H]QNB displays in vivo selectivity for the m^2 subtype, *Life Sci.*, 1994, **55**, 1493–1508.
47. M. R. Dando, *A New Form of Warfare: The Rise of Non-Lethal Weapons*, Brassey's, London, 1996, pp. 90–94.
48. According to a US Army munitions manual, the M43 cluster munition was designed for: "aerial delivery of 57 M138 10-pound BZ incapacitating agent bombs on selected targets to temporarily incapacitate enemy personnel. Inhaling BZ causes temporary slowing of mental and physical activity, disorientation, and hallucinations among exposed personnel". [*Technical manual, US Army, equipment data sheets, chemical weapons and munitions*, TM 43-0001-26-2, Department of the Army, Washington, DC, 29th April 1982.]
49. M. Furmanski, Historical military interest in low-lethality biochemical agents, in *Incapacitating Biochemical Weapons*, ed. A. Pearson, M. Chevrier and M. Wheelis, Lexington Books, Lanham, MD, 2007, p. 54.
50. HQ U.S.M.A.V., Command History 1964, volume 35, p. 133, as cited in: M. Furmanski, Historical military interest in low-lethality biochemical agents, in *Incapacitating Biochemical Weapons*, ed. A. Pearson, M. Chevrier and M. Wheelis, Lexington Books, Lanham, MD, 2007, p. 54.

51. (a) US Army Department List of Agents Used on Human Volunteers, 12th August 1975, tabled in evidence at the joint hearing of Senate Judiciary Subcommittee on Administrative Practice and Procedure and Senate Labor and Public Welfare Subcommittee on Health, 10 and 12 September and 7 November 1975; (b) Biomedical and Behavioral Research, 1975: Human-Use Experimentation Programs of the Department of Defense and Central Intelligence Agency, p. 1142, as cited in: J. Perry Robinson, Disabling Chemical Weapons, A Documented Chronology of Events, 1945–2011, (copy provided by author), 20 November 2012. Annex I contains an excerpted table of the wide-range of chemical and biological agents tested over this period. In addition, Annex 4 of Perry Robinson's Chronology also contains a list of ICAs held in a CIA store. This was taken from the inventory reproduced in: US Senate, Select Committee to Study Governmental Operations with respect to Intelligence Activities, hearings, 16-18 September 1975, Unauthorized Storage of Toxic Agents, pp. 12 & 192–197.
52. J. Perry Robinson, *Disabling Chemical Weapons, A Documented Chronology of Events, 1945–2011*, (copy provided by author), 20th November 2012, chronology reference: 761000, p. 85.
53. M. Dando and M. Furmanski, Midspectrum incapacitant programs, in *Deadly Cultures: Biological Weapons Since 1945*, ed. M. Wheelis, L. Rózsa and M. Dando, Harvard University Press, Harvard, 2006
54. (a) A. Pearson, Late and post-cold war research and development of incapacitating biochemical weapons, in *Incapacitating Biochemical Weapons*, ed., A. Pearson, M. Chevrier and M. Wheelis, Lexington Books, 2007, p. 74; (b) N. Davidson, *Non-lethal weapons*, Palgrave, 2009, pp. 116–118.
55. United States Army, Topic CBD00-108, Chemical Immobilizing Agents for Non-lethal Applications, Small Business Innovation Research Solicitation, CBD 00.1, December 1999 http://web.archive.org/web/20070213055848/http://www.acq.osd.mil/osbp/sbir/solicitations/sbir001/cbd001.htm (accessed 12th June 2014).
56. United States Army, Topic CBD00-108, Chemical Immobilizing Agents for Non-lethal Applications, Small Business Innovation Research Solicitation, CBD 00.1, December 1999 http://web.archive.org/web/20070213055848/http://www.acq.osd.mil/osbp/sbir/solicitations/sbir001/cbd001.htm (accessed 12th June 2014).
57. United States Army, Topic CBD00-108, Chemical Immobilizing Agents for Non-lethal Applications, Small Business Innovation Research Solicitation, CBD 00.1, December 1999 http://web.archive.org/web/20070213055848/http://www.acq.osd.mil/osbp/sbir/solicitations/sbir001/cbd001.htm (accessed 12th June 2014).
58. United States Army, Topic CBD00-108, Chemical Immobilizing Agents for Non-lethal Applications, Small Business Innovation Research Solicitation, CBD 00.1, December 1999 http://web.archive.org/web/20070213055848/http://www.acq.osd.mil/osbp/sbir/solicitations/sbir001/cbd001.htm (accessed 12th June 2014).

59. United States Army, Topic CBD00-108, Chemical Immobilizing Agents for Non-lethal Applications, Small Business Innovation Research Solicitation, CBD 00.1, December 1999 http://web.archive.org/web/20070213055848/http://www.acq.osd.mil/osbp/sbir/solicitations/sbir001/cbd001.htm (accessed 12th June 2014).
60. United States Army, Topic CBD00-108, Chemical Immobilizing Agents for Non-lethal Applications, Small Business Innovation Research Solicitation, CBD 00.1, December 1999 http://web.archive.org/web/20070213055848/http://www.acq.osd.mil/osbp/sbir/solicitations/sbir001/cbd001.htm (accessed 12th June 2014).
61. United States Army, Topic CBD00-108, Chemical Immobilizing Agents for Non-lethal Applications, Small Business Innovation Research Solicitation, CBD 00.1, December 1999 http://web.archive.org/web/20070213055848/http://www.acq.osd.mil/osbp/sbir/solicitations/sbir001/cbd001.htm (accessed 12th June 2014).
62. United States Army, CBD,26 Phase I Selections from the 00.1 Solicitation http://web.archive.org/web/20101123094323/http://dodsbir.net/selections/abs001CBD.htm (accessed 12th June 2014).
63. J. Lakoski *et al.*, The advantages and limitations of calmatives for use as a non-lethal technique, *College of Medicine Applied Research Laboratory*, Pennsylvania State University, 2000.
64. J. Lakoski *et al.*, The advantages and limitations of calmatives for use as a non-lethal technique, *College of Medicine Applied Research Laboratory*, Pennsylvania State University, 2000.
65. National Institute of Justice, Grant No. 2001-RD-CX-K002. Details from NIJ Research Portfolio available December 2006 at: http://nij.ncjrs.org/portfolio/ as cited in N. Davison, 'Off the Rocker' and 'On the Floor': The Continued Development of Biochemical Incapacitating Weapons, Bradford Science and Technology Report No. 8, August 2007 p. 24.
66. (a) In February 2003, a presentation by the Senior Program Manager for the NIJ Less-Than-Lethal Technology Program, indicated that the project had been reviewed by a liability panel and that work was progressing at Pennsylvania State University; (b) J. Cecconi, Less-Than-Lethal Program. Presentation to the 2003 National Institute of Justice Annual Technology Conference, 2003; (c) as cited inN. Davison, 'Off the Rocker' and 'On the Floor': The Continued Development of Biochemical Incapacitating Weapons, Bradford Science and Technology Report No. 8, August 2007, 2007, p. 24.
67. (a) For further discussion of these and other CNS-acting chemical research projects together with original sources, see: M. Crowley, *Chemical Control*, Palgrave Macmillan, 2016, pp. 31–35; (b) N. Davison, *Non-lethal' Weapons*, Palgrave Macmillan, 2009, pp. 111–125.
68. US Army Test and Evaluation Command, *Joint CB Technical Data source Book, Volume VI, Toxin Agents, Part Two: Agent PG*. DTC 73 - 33, US Army Deseret Test Center, Fort Douglas, Utah, June 1982.

69. J. Perry Robinson, Bringing the CBW Conventions Closer Together, *CBW Conv. Bull.*, 2008, **80**(1–4), 3.
70. R. G. Ulrich *et al.*, Staphylococcal enterotoxin B and related pyrogenic toxins, in *Medical Aspects of Chemical and Biological Warfare*, ed. F. R. Sidell *et al.*, Borden Institute, Washington, D. C., 1997, ch. 31, pp. 621–630.
71. J. Perry Robinson, Bringing the CBW Conventions Closer Together, *CBW Conv. Bull.*, 2008, **80**(1–4), 3.
72. Interdepartmental Political-Military Group in response to NSSM 85, *Report to the National Security Council: US Policy on Toxins*, January 1970. Declassified by the US Department of State, August 2007 in Document 177 in Department of State, Office of the Historian, *Foreign Relations of the United States: Nixon-Ford Administrations*, volume E-2: Documents on Arms Control and Nonproliferation, 1969–1972, p. 5.
73. (a) Office of the White House Press Secretary, Key Biscayne, FL, Statement on Toxins, February 14 in, 1970 Document 189. Available at www.state.gov/r/pa/ho/frus/nixon/e2/83628.htm. (b) J. Tucker and E. Mahan, President Nixon's Decision to Renounce the U.S. Offensive Biological Weapons Program. Case Study 1, Center for the Study of Weapons of Mass Destruction, National Defense University, October,. pp. 13–17; (c) M. Dando and M. Furmanski, Midspectrum incapacitant programs, in *Deadly Cultures: Biological Weapons Since 1945*, ed. M. Wheelis, L. Rózsa, and M. Dando, Harvard University Press, Cambridge, 2006 pp. 36–39.
74. M. R. Dando, *A New Form of Warfare: The Rise of Non-Lethal Weapons*, Brassey's, London, See Chapter 8: An Assault on the Brain, 1996, pp. 157–168.
75. M. R. Dando, *A New Form of Warfare: The Rise of Non-Lethal Weapons*, Brassey's, London, See Chapter 8: An Assault on the Brain, 1996, pp. 158–160.
76. M. R. Dando, *A New Form of Warfare: The Rise of Non-Lethal Weapons*, Brassey's, London, See Chapter 8: An Assault on the Brain, 1996, pp. 161–168.
77. M. R. Dando *A New Form of Warfare: The Rise of Non-Lethal Weapons*, Brassey's, London, See Chapter 8: An Assault on the Brain, 1996, pp. 166–167.
78. (a) J. Lakoski *et al.*, The advantages and limitations of calmatives for use as a non-lethal technique, *College of Medicine Applied Research Laboratory*, Pennsylvania State University, 2000; (b) for further discussion see J. Tucker, The body's own bioweapons, *Bull. At. Sci.*, 2008, **64**(1), 16–22. (c) M. Crowley, *Chemical Control*, Palgrave Macmillan, 2016.
79. J. Lakoski *et al.*, The advantages and limitations of calmatives for use as a non-lethal technique, *College of Medicine Applied Research Laboratory*, Pennsylvania State University, 2000, pp. 39–42.
80. J. Lakoski *et al.*, The advantages and limitations of calmatives for use as a non-lethal technique, *College of Medicine Applied Research Laboratory*, Pennsylvania State University, 2000, pp. 42–45

81. United States of America, *Statement by Rose E. Gottemoeller*, Acting Under Secretary for Arms Control and International Security, at the Third Review Conference, Third Special Session, RC-3/NAT.45, 8th–19th April 2013. OPCW, The Hague, 9th April 2013.
82. United States of America, *Statement by Rose E. Gottemoeller*, Acting Under Secretary for Arms Control and International Security, at the Third Review Conference, Third Special Session, RC-3/NAT.45, 8th–19th April 2013. OPCW, The Hague, 9th April 2013.
83. United States of America, *Statement by Ambassador Robert P. Mikulak*, United States Delegation to the OPCW, at the Seventy-Second Session of the Executive Council, EC-72/NAT.8, 6th May 2013.
84. V. Viatrovych *et al.*, *Enemy Archives*, McGill-Queen's University Press, 2023, p. 48, as cited in R. Petersen, *A sense of place*, CBRNe World, 2024, pp. 40–41.
85. J. Perry Robinson Incapacitating chemical agents' in context: an historical overview of states policy, in *Incapacitating chemical agents" Law enforcement, human rights law and policy perspectives*, Montreux, Switzerland, 24th–26th April 2012, p. 92.
86. J. R. Riches, *et al.*, Analysis of clothing and urine from Moscow theatre siege casualties reveals carfentanil and remifentanil use, *J. Anal. Toxicol.*, 2012, **36**(9), 647–656.
87. General Antonov publication: English Translation LN72-96, Progress Publisher, Moscow, Russia, 1994. Chemical weapons at the turn of the century; as cited in J. R. Riches, *et al.*, Analysis of clothing and urine from Moscow theatre siege casualties reveals carfentanil and remifentanil use, *J. Anal. Toxicol.*, 2012, **36**(9), 647–656.
88. (a) European Court of Human Rights (ECtHR), Finogenov and others v. Russia, application numbers 18299/03 and 27311/03, Judgement, 20th December 2011; (b) M. Wheelis, Human impact of incapacitating chemical agents in: International Committee of the Red Cross, Expert Meeting: Incapacitating chemical agents, implications for international law, Montreux, Switzerland, 24th–26th March 2010; (c) D. Levin and V. Selivanov, Medical and Biological Issues of NLW Development and Application, Proceedings of the 5th European Symposium on Non-Lethal Weapons, 11th–13th May 2009, Ettlingen, Germany, European Working Group on Non-Lethal Weapons. V23, p. 7. (d) 4 Years Later, Moscow Hostages Suffering, CBS Evening News, 21st October 2006.
89. See for example: Human Rights Watch, *Independent Commission of Inquiry Must Investigate Raid on Moscow Theater: Inadequate Protection for Consequences of Gas Violates Obligation to Protect Life*, Human Rights Watch, 30 October 2002.
90. ITAR-TASS, from Moscow in English, 2112 hrs GMT 30th October 2002, as in FBIS-SOV-2002-1030, Russian experts discuss use of Fentanyl in hostage crisis; as cited by J. Perry Robinson, Categories of Challenge now facing the Chemical Weapons Convention, 52nd Pugwash CBW

Workshop, 10 Years of the OPCW: Taking Stock and Looking Forward, Noordwijk, The Netherlands, 17th–18th March 2007.

91. S. Alison [from Moscow for Reuters], 1257 hrs ET 30th October 2002, 'Russian confirms siege gas based on opiate fentanyl', as cited in J. Perry Robinson, Categories of Challenge now facing the Chemical Weapons Convention, 52nd Pugwash CBW Workshop, 10 Years of the OPCW: Taking Stock and Looking Forward, Noordwijk, The Netherlands, 17th–18th March 2007.
92. Amnesty International, Rough Justice: *The law and human rights in the Russian Federation*, AI Index EUR 46/054/2003, October 2003, p. 53.
93. J. R. Riches, *et al.*, Analysis of clothing and urine from Moscow theatre siege casualties reveals carfentanil and remifentanil use, *J. Anal. Toxicol.*, 2012, **36**(9), 647–656.
94. For further discussion, see: M. Crowley and M. R. Dando, *Down the Slippery Slope? A Study of Contemporary Dual-Use Chemical and Life Science Research Potentiallty Applicable to Incapacitating Chemical Agent Weapons*, University of Bradford and University of Bath, 2014, pp. 42–50.
95. V. Klochikhin *et al.* M. Principles of modelling of the scenario of calmative application in a building with deterred hostages. *Proceedings of the 3rd European Symposium on Non-Lethal Weapons*, Ettlingen, Germany, 10–12th May 2005, V17, Pfinztal: Fraunhofer ICT.
96. N. E. Kuzmina and V. S. Kuzmin, Development of concepts on the interaction of drugs with opioid receptors, *Russ. Chem. Rev.*, 2011, **80**, 145–169.
97. F. S. Dukhovich, *et al.*, The agonist paradox: Agonists and antagonists of acetylcholine receptors and opioid receptors, *Chem. Biodiversity*, 2005, **2**, 354–366.
98. V. A. Zlobin, *et al.*, Analysis of opiates by HPLC with indirect spectrophotometric detection, *Pharm. Chem. J.*, 2000, **34**, 279–280.
99. A. V. Panfilov, *et al.*, Sodium borohydride in reductive amination reactions, *Pharm. Chem. J.*, 2000, **34**, 76–78.
100. A. V. Panfilov, *et al.*, Reactions of sodium borohydride in acetic acid: Reductive amination of carbonyl compounds, *Pharm. Chem. J.*, 2000, **34**, 371–373.
101. N. E. Kuz'mina, *et al.*, Geometric parameters as a criterion for assessment of the bioactive conformations of opiate receptor ligands, *Russ. Chem. Bull.*, 2006, **55**(9), 1516–1522.
102. N. E. Kuz'mina, *et al.*, General model of the opiate pharmacophore, *Russ. Chem. Bull.*, 2008, **57**(6), 1277–1284.
103. N. E. Kuz'mina, *et al.*, General model of the opiate pharmacophore, *Russ. Chem. Bull.*, 2008, **57**(6), 1285–1298.
104. N. E. Kuzmina and I. N. Stankov, Role of saturability of noncovalent interactions in the analysis of the three-dimensional structure—opioid activity relationship, *Russ. Chem. Bull.*, 2012, **61**(6), 1207–1214.

105. N. E. Kuzmina, Method for estimating a compound's opiate activity based on a versatile three-dimensional model of nonselective opiate pharmacophore, *Russ. J. Bioorg. Chem.*, 2012, **38**(5), 507.
106. United States, Compliance with the Convention on the Prohibition of the Development, Production, Stockpiling, and Use of Chemical Weapons and on Their Destruction [2021 CWC Compliance Report]. April 2 2022, p. 17. For similar analysis see also: US, *2018 CWC Compliance Report*, 15 April 2019, p. 10; US *2019 CWC Compliance Report*, June 2020, p. 13; US *2021 CWC Compliance Report*.
107. K. Alibek and S. Handelman, *Biohazard: The Chilling True Story of the Largest Covert Biological Weapons Program in the World—Told from the Inside by the Man Who Ran It*, Random House, New York, 1999, p. 154.
108. K. Alibek and S. Handelman, *Biohazard: The Chilling True Story of the Largest Covert Biological Weapons Program in the World—Told from the Inside by the Man Who Ran It*, Random House, New York, 1999, p. 164.
109. K. Alibek and S. Handelman, *Biohazard: The Chilling True Story of the Largest Covert Biological Weapons Program in the World—Told from the Inside by the Man Who Ran It*, Random House, New York, 1999, pp. 154–155.
110. (a) Interview: S. Popov, *J. Homeland Secur.*, 2000 (updated November 19, 2002). Available from https://web.archive.org/web/20110719234905/http://www.homelandsecurity.org/journal/Interviews/PopovInterview_001107.html; (b) K. Alibek and S. Handelman, *Biohazard: The Chilling True Story of the Largest Covert Biological Weapons Program in the World—Told from the Inside by the Man Who Ran It*, Random House, New York, 1999, pp. 163–167.
111. K. Alibek and S. Handelman, *Biohazard: The Chilling True Story of the Largest Covert Biological Weapons Program in the World—Told from the Inside by the Man Who Ran It*, Random House, New York, 1999, p. 164.
112. J. Gilsdorf and R. Zilinskas, New considerations in infectious disease outbreaks: the threat of genetically modified microbes, *Clinical Infectious Diseases*, 2005, **40**(8), 1 and 1160–1165.
113. Interview: S. Popov, *J. Homeland Secur.*, 2000, 11(updated November 19, 2002).
114. (a) K. Alibek and S. Handelman, *Biohazard: The Chilling True Story of the Largest Covert Biological Weapons Program in the World—Told from the Inside by the Man Who Ran It*, Random House, New York, 1999, pp. 171–172 and p. 302; (b) See also V. J. Birstein, *The Perversion of Knowledge: The True Story of Soviet Science*, Westview Press, Boulder, Colorado, 2001, p. 107, According to Alibek, this network included the Moscow Institute of Applied Molecular Biology (which later became the Russian Scientific Center of Molecular Diagnostics and Treatment).
115. K. Alibek and S. Handelman, *Biohazard: The Chilling True Story of the Largest Covert Biological Weapons Program in the World—Told from the Inside by the Man Who Ran It*, Random House, New York, 1999, p. 302.

116. K. Alibek and S. Handelman, *Biohazard: The Chilling True Story of the Largest Covert Biological Weapons Program in the World—Told from the Inside by the Man Who Ran It*, Random House, New York, 1999, p. 164.
117. J. B. Petro, *et al.*, Biotechnology: Impact on biowarfare and biodefense, *Biosecur. Bioterrorism*, 2003, **1**(3), 161–168.
118. China North Industries Corporation (NORINCO), undated *Security, Anti-Riot Weapons and Ammunition*. Brochure distributed at MILIPOL security exhibition, Paris, France, 1995, p. 11.
119. China North Industries Corporation (NORINCO), undated *Security, Anti-Riot Weapons and Ammunition*. Brochure distributed at MILIPOL security exhibition, Paris, France, 1995, p. 11.
120. State 9619 Plant (undated) Brochures distributed during Asia Pacific China Police Expo 2004 (23rd–26th June 2004), and Asia Pacific China Police Expo 2006 (24th–27th May 2006) held at Beijing Exhibition Centre, Beijing, China.
121. G. Arthur, *New Equipment in Hong Kong, Defence Review Asia*, 19th March, 2012, p. 35, http://www.defencereviewasia.com/articles/153/NEW-EQUIPMENT-IN-HONG-KONG (accessed 8th January 2015).
122. Radio Free Asia, *Thousands of people demolished the Hui village, police with anesthesia guns against the Hui*, 7 June 2013, Available at https://www.rfa.org/cantonese/news/demolition-06072013094102.html (accessed 17 September 2020) [original in Cantonese, translation by google translate].
123. United States, Compliance with the Convention on the Prohibition of the Development, Production, Stockpiling, and Use of Chemical Weapons and on Their Destruction [2021 CWC *Compliance Report*], April 2022, p. 11.
124. For more extensive discussion including details of original Czech research papers and correspondence with Czech authorities see, M. Crowley and M. R. Dando, *Down the Slippery Slope? A Study of Contemporary Dual-Use Chemical and Life Science Research Potentiallty Applicable to Incapacitating Chemical Agent Weapons*, University of Bradford and University of Bath, 2014, pp. 20–29.
125. L. Hess, J. Schreiberova and J. Fusek, Pharmacological Non-Lethal Weapons, Proceedings of the 3rd European Symposium on Non-Lethal Weapons, 10th–12th May 2005, Ettlingen, Germany, European Working Group on Non-Lethal Weapons, Pfinztal: Fraunhofer ICT, V23.
126. L. Hess, J. Schreiberova and J. Fusek, Pharmacological Non-Lethal Weapons, Proceedings of the 3rd European Symposium on Non-Lethal Weapons, 10th–12th May 2005, Ettlingen, Germany, European Working Group on Non-Lethal Weapons, Pfinztal: Fraunhofer ICT, V23, pp. 8–9.
127. L. Hess, J. Schreiberova and J. Fusek, Pharmacological Non-Lethal Weapons, Proceedings of the 3rd European Symposium on Non-Lethal Weapons, 10th–12th May 2005, Ettlingen, Germany, European Working

Group on Non-Lethal Weapons, Pfinztal: Fraunhofer ICT, V23, pp. 10–14.
128. L. Hess, J. Schreiberova and J. Fusek, Pharmacological Non-Lethal Weapons, Proceedings of the 3rd European Symposium on Non-Lethal Weapons, 10th–12th May 2005, Ettlingen, Germany, European Working Group on Non-Lethal Weapons, Pfinztal: Fraunhofer ICT, V23., pp. 11–12.
129. L. Hess, J. Schreiberová, J. Málek, M. Votava and J. Fusek, Drug-Induced Loss of Aggressiveness in the Macaque Rhesus, *Proceedings of 4th European Symposium on Non-Lethal Weapons*, 21st–23rd May 2007, Ettlingen, Germany, European Working Group on Non-Lethal Weapons, Pfinztal: Fraunhofer ICT, V15, p. 6.
130. L. Hess, J. Schreiberová, J. Málek, M. Votava and J. Fusek, Drug-Induced Loss of Aggressiveness in the Macaque Rhesus, *Proceedings of 4th European Symposium on Non-Lethal Weapons*, 21st–23rd May 2007, Ettlingen, Germany, European Working Group on Non-Lethal Weapons, Pfinztal: Fraunhofer ICT, V15, p. 7.
131. L. Hess, J. Schreiberová, J. Málek, M. Votava and J. Fusek, Drug-Induced Loss of Aggressiveness in the Macaque Rhesus, *Proceedings of 4th European Symposium on Non-Lethal Weapons*, 21st–23rd May 2007, Ettlingen, Germany, European Working Group on Non-Lethal Weapons, Pfinztal: Fraunhofer ICT, V15, p. 7
132. Czech Republic, National Authority of the CWC, *Reply to the University of Bradford, Re: Request for information concerning research potentially related to incapacitating chemical agents*, 14th July 2014, p. 4.
133. Czech Republic, National Authority of the CWC, *Reply to the University of Bradford, Re: Request for information concerning research potentially related to incapacitating chemical agents*, 14th July 2014, p. 5.
134. (a) See for example: P. K. Gupta, *et al.*, A convenient one-pot synthesis of fentanyl, *J. Chem. Res.*, 2005, 452–453. (b) L. Manral and P. K. Gupta, *et al.*, Thermal behaviour of fentanyl and its analogues during flash pyrolysis, *J. Therm. Anal. Calorim.*, 2009, **96**(2), 531–534. (c) L. Manral, *et al.*, Effect of exposure to fentanyl aerosol in mice on breathing pattern and respiratory variables, *Drug Chem. Toxicol.*, 2009, **32**(2), 108–113. (d) For more extensive discussion including details of original Indian research papers and correspondence with Indian authorities see, M. Crowley and M. R. Dando, Down the Slippery Slope? A Study of Contemporary Dual-Use Chemical and Life Science Research Potentiallty Applicable to Incapacitating Chemical Agent Weapons, University of Bradford and University of Bath, 2014, pp. 29–34.
135. Correspondence to Dr M.Crowley, BNLWRP, from R. K. Singh, Deputy Chief of Mission, Embassy of India, The Hague, forwarding the response of the Indian CWC National Authority, 22nd July 2014.
136. Correspondence to Dr M.Crowley, BNLWRP, from R. K. Singh, Deputy Chief of Mission, Embassy of India, The Hague, forwarding the response of the Indian CWC National Authority, 22nd July 2014.

137. For more extensive discussion including details of original Indian research papers see, M. Crowley and M. Dando, *Toxin and Bioregulator Weapons*, Palgrave MacMillan, 2022, pp. 79–85.
138. S. M. Gupta, *et al.*, Phytochemical analysis of indian stinging plants: an initiative towards development of future novel biothreat agents for self-defence, *Proc. Natl. Acad. Sci., India, Sect. B*, 2018, **88**(2), 819–825.
139. S. M. Gupta, *et al.*, Himalayan toxic plants of defense importance, *ACTA Sci. Med. Sci.*, 2018, **2**(3), 44–48.
140. S. M. Gupta, *et al.*, Himalayan toxic plants of defense importance, *ACTA Sci. Med. Sci.*, 2018, **2**(3), 820.
141. (a) For more extensive discussion including details of original Iranian research papers and correspondence with Iranian authorities see, M. Crowley and M. R. Dando, Down the Slippery Slope? A Study of Contemporary Dual-Use Chemical and Life Science Research Potentiallty Applicable to Incapacitating Chemical Agent Weapons, University of Bradford and University of Bath, 2014, pp. 29–34; (b) and also M. Giveh and the Good ISIS Team, The Islamic Republic's work on pharmaceutical based agents, Institute for Science and International Security, 25 November 2024.
142. (a) H Fakhraian, *et al.*, Two-step protic solvent-catalyzed reaction of phenylethylamine with methyl acrylate, *Org. Prep. Proced. Int.*, 2005, **37**(6), 579–584. (b) H. Fakhraian, Improved procedure for the preparation of 1-(2-phenethyl)-4-piperidone, *Org. Prep. Proced. Int.*, 2008, **40**(3), 307–310.
143. M. J. Maghizadeh, *et al.*, Improved method for the total synthesis of thiofentanyl, *J. Heterocycl. Chem.*, 2023, **60**(6), 950–956.
144. T. Nezamoleslam, *et al.*, Structure-activity relationship, atomic electron density and conformational investigation of fentanyl analogues, *J. Passive Def. Sci. Technol.*, 2010, **1**, 23–32.
145. (a) H. Fakhraian, *et al.*, Synthesis and enantiomeric resolution of medetomidine, *Chirality*, 2015, **27**(2), 141–148. (b) H. Toulabi and H. Fakhraian, An easy method and two steps for synthesis of the medetomidine with high yield, *Iran. J. Org. Chem.*, 2014, **6**(3), 1487–1491. (c) The synthesis of medetomidine and its naphtalenic derivative, *Org. Chem. J.*, 2015 cited by M. Giveh, and the Good ISIS Team (2024) but without author attribution.
146. (a) H. Kamranpey, Aerosolisation of medetomidine hydrochloride as an incapacitating agent, *J. Pass. Def. Sci. Technol.*, 2011, **1**, 51–56. (b) M. S. Abazari, Investigating the phase behavior of medetomidine hydrochloride, ketamine hydrochloride and sevoflurane in the presence of ethanol and propellant, *Pass. Def. Sci. Technol.*, 2013, **3**, 65–70.
147. M. S. Abazari, Investigating the phase behavior of medetomidine hydrochloride, ketamine hydrochloride and sevoflurane in the presence of ethanol and propellant, *Pass. Def. Sci. Technol.*, 2013, **3**, 65–70.
148. See for example United States, *Compliance with the Convention on the Prohibition of the Development, Production, Stockpiling, and Use of*

Chemical Weapons and on Their Destruction, April, Condition (10)(C) Report, 2022, p. 15.
149. See for example United States, *Compliance with the Convention on the Prohibition of the Development, Production, Stockpiling, and Use of Chemical Weapons and on Their Destruction*, April, Condition (10)(C) Report, 2022, p. 15.
150. United States, *Compliance with the Convention on the Prohibition of the Development, Production, Stockpiling, and Use of Chemical Weapons and on Their Destruction*, Condition (10)(C) Report, April 2024. See Chapter 5 of this book for further discussion.
151. Correspondence to Dr M. Crowley, BNLWRP, from Dr H. Farajvand, Secretary of the National Authority for the CWC, Ministry of Foreign Affairs of the Islamic Republic of Iran, 15th July 2014.
152. Correspondence to Dr M. Crowley, BNLWRP, from Dr H. Farajvand, Secretary of the National Authority for the CWC, Ministry of Foreign Affairs of the Islamic Republic of Iran, 15th July 2014.
153. OPCW, *Decision: Understanding regarding the aerosolised use of central nervous system-acting-chemicals for law enforcement purposes.* CWC Conference of States Parties, 26th Session. C-26/DEC.10, 2021.
154. OPCW, Joint Statement on behalf of 4 Delegations delivered by the Delegation of the Islamic Republic of Iran at the 26th Session of the Conference of States Parties of the OPCW under the Subitem 26.1 "Any Other Business" on the Draft Decision entitled "Understanding Regarding the Aerosolized Use of Central Nervous System-Acting Chemicals for Law Enforcement Purposes", 29 November 2021.
155. See British Association for Psychopharmacology, https://www.bap.org.uk/#:~:text=Psychopharmacology%20is%20the%20study%20of,substances%20that%20can%20be%20abused. (accessed 27 January 2024).
156. Adapted from P. B. Mitchell and D. Hadzi-Pavlovic, Psychopharmacology, in *Psychiatry: Past, Present and Prospect*, ed. S. Bloch *et al.*, Oxford University Press, Oxford, 2014, ch. 18, pp. 335–354, p. 335.
157. P. B. Mitchell and D. Hadzi-Pavlovic, Psychopharmacology, in *Psychiatry: Past, Present and Prospect*, ed. S. Bloch *et al.*, Oxford University Press, Oxford, 2014, ch. 18, p. 340.
158. P. B. Mitchell and D. Hadzi-Pavlovic, Psychopharmacology, in *Psychiatry: Past, Present and Prospect*, ed. S. Bloch *et al.*, Oxford University Press, Oxford, 2014, ch. 18, p. 345.
159. P. B. Mitchell and D. Hadzi-Pavlovic, Psychopharmacology, in *Psychiatry: Past, Present and Prospect*, ed. S. Bloch *et al.*, Oxford University Press, Oxford, 2014, ch. 18, p. 350.
160. G. E. Berrios, Defining and classifying mental illness, in *Psychiatry: Past, Present and Prospect*, ed. S. Bloch *et al.*, Oxford University Press, Oxford, 2014, ch. 10, pp. 180–195.

161. M. R. Dando, *Neuroscience and the Problem of Dual Use: Neuroethics in the New Brain Projects*, Springer Nature, Switzerland, Chapter 1 Modern Neuroscience, 2020, pp. 3–16.
162. See for example: Amnesty International, *Prisoners of Conscience in the USSR: Their Treatment and Conditions*, AI Publications, London, 1975.
163. Human Rights Watch, "*We Are in Tombs*", Abuses in Egypt's Scorpion Prison, 28 September 2016.
164. (a) Amnesty International, Activist forcibly detained in psychiatric hospital, EUR 46/3996/2016 Russian Federation, 9 May 2016; (b) Amnesty International, Activist released from psychiatric hospital: Dmitry Vorobyovsky, EUR 46/4041/2016, 13 May 2016.
165. (a) Human Rights Watch, Uzbekistan: Brutal Police Attack On Activist, 4 June 2015; (b) Human Rights Watch, Uzbekistan: Happy Birthday, Elena Urlaeva!, 9 February 2017.
166. (a) See for example:British Medical Association, *Medicine Betrayed: The Participation of Doctors in Human Rights Abuses*, Zed Books, London, 1992, pp. 64–72; (b) Amnesty International, *Prisoners of Conscience in the USSR: Their Treatment and Conditions*, AI Publications, London, 1975; (c) S. Block and P. Reddaway, *Soviet Psychiatric Abuse: The Shadow Over World Psychiatry*, Gollancz, London, 1984.
167. Amnesty International, *Prisoners of Conscience in the USSR: Their Treatment and Conditions*, AI Publications, Londonm, 1975, p. 104.
168. (a) P. Reddaway, *The Dissidents: A Memoir of Working with the Resistance in Russia, 1969–1990*, The Brookings Institute, Washington, D.C., 2020, pp. 148–149. (b) See also S. Bloch and P. Reddaway, *Russia's Political Hospitals*, Futura Publications, London, 1978.
169. Amnesty International, *Prisoners of Conscience in the USSR: Their Treatment and Conditions*, AI Publications, Londonm, 1975, p. 130.
170. Amnesty International, *Prisoners of Conscience in the USSR: Their Treatment and Conditions*, AI Publications, Londonm, 1975, p. 131.
171. Amnesty International, *Prisoners of Conscience in the USSR: Their Treatment and Conditions*, AI Publications, Londonm, 1975, pp. 131–132.
172. D. Ougrin, *et al.*, Psychiatry in post-communist Ukraine; dismantling the past, paving the way for the future, *Psychiatr. Bull.*, 2006, **30**, 456–459.
173. B. Zajicek, The Psychopharmacological Revelution in the USSR: Schizophrenia Treatment and the Thaw in Soviet Psychiatry, 1954–64, *Med. Hist*, 2019, **63**(3), 249–269.
174. T. Padder, *Practical Guide to Psychiatric Medicines*, MTP Publishing, Maryland, 2nd edn, 2022, pp. 122–133.
175. T. Padder, *Practical Guide to Psychiatric Medicines*, MTP Publishing, Maryland, 2nd edn, 2022, p. 123.
176. T. Padder, *Practical Guide to Psychiatric Medicines*, MTP Publishing, Maryland, 2nd edn, 2022, p. 125.
177. T. Padder, *Practical Guide to Psychiatric Medicines*, MTP Publishing, Maryland, 2nd edn, 2022, p. 130.

178. T. Padder, *Practical Guide to Psychiatric Medicines*, MTP Publishing, Maryland. 2nd edn, 2022, p. 122.
179. R. van Voren, Political abuse of psychiatry – an historical perspective, *Schizophr. Bull.*, 2010, **36**(1), 33–35.
180. Human Rights Watch and Geneva Initiative on Psychiatry, *Dangerous Minds: Political Psychiatry in China Today and its Origins in the Mao Era*, 2002.
181. Human Rights Watch and Geneva Initiative on Psychiatry, *Dangerous Minds: Political Psychiatry in China Today and its Origins in the Mao Era*, 2002, p. 144.
182. R. Munro, Judicial psychiatry in china and its political abuses, *C. J. Asian Law*, 2000, **14**(1), 1–101.
183. R. Munro, Judicial psychiatry in china and its political abuses, *C. J. Asian Law*, 2000, **14**(1), 5.
184. R. Munro, Judicial psychiatry in china and its political abuses, *C. J. Asian Law*, 2000, **14**(1), 4.
185. R. Munro, Judicial psychiatry in china and its political abuses, *C. J. Asian Law*, 2000, **14**(1), 5.
186. R. Munro, Judicial psychiatry in china and its political abuses, *C. J. Asian Law*, 2000, **14**(1), 23–24.
187. Human Rights Watch, *Political Prisoner Exposes Brutality in Police-Run Mental Hospital*, Human Rights Watch, China, 2 November 2005.
188. G. Blume, Electric shocks against the freedom virus, *Die Zeit*, 2 November 2005, vol. 3. https://www.zeit.de/2005/45/Gulag_HL.
189. Human Rights Watch, *Political Prisoner Exposes Brutality in Police-Run Mental Hospital*, Human Rights Watch, China, 2 November 2005.
190. Human Rights Watch, *Political Prisoner Exposes Brutality in Police-Run Mental Hospital*, Human Rights Watch, China, 2 November 2005.
191. Human Rights Watch, *Political Prisoner Exposes Brutality in Police-Run Mental Hospital*, Human Rights Watch, China, 2 November 2005.
192. Y. Mou and D. Gardner, Drugged and detained: China's psychiatric prisons, *Safeguard Defenders*, August 2022, p. 12.
193. Y. Mou and D. Gardner, Drugged and detained: China's psychiatric prisons, *Safeguard Defenders*, August 2022, pp. 12–16.
194. Y. Mou and D. Gardner, Drugged and detained: China's psychiatric prisons, *Safeguard Defenders*, August 2022, p. 37.
195. Y. Mou and D. Gardner, Drugged and detained: China's psychiatric prisons, *Safeguard Defenders*, August 2022, p. 37.
196. J. Perry Robinson, *Disabling Chemical Weapons: A Documented Chronology of Events*, 1945–2003, Unpublished monograph, 1 November 2003.
197. (a) For an overview see: J. Marks, *The Search for the 'Manchurian Candidate': The CIA and Mind Control: The Secret History of the Behavioral Sciences*, W. W. Norton & Company, New York, US, 1979; (b) and CIA, Behavior Control Experiments Focus of New Scholarly Collection, Briefing Book, Digital National Security Archive, https://nsarchive.gwu.edu/briefing-book/dnsa-intelligence/2024-12-23/cia-behavior-control-

experiments-focus-new-scholarly (accessed 26 January 2025); (c) For access to original unclassified documents see: Digital National Security Archive (DNSA): CIA and the Behavioral Sciences: Mind Control, Drug Experiments and MKULTRA, https://proquest.libguides.com/dnsa/64.

198. (a) For an overview see: J. Perry Robinson, Disabling Chemical Weapons: A Documented Chronology of Events, 1945–2003, Unpublished monograph, 1 November 2003; (b) US Senate, Committee on Intelligence and Human Resources Subcommittee on Health and Scientific Research, Joint Hearing, Project MKULTRA, the CIA's Program of Research in Behavioral Modification, 3 August 1977; (c) United States Senate, Nineth-Fifth Congress, First Session, 3 August 1977; (d) L. J. Meier, Mind Control: Past and Future, Issue 2025-01, Carr Center for Human Rights Policy, Harvard Kennedy School, Harvard University, 13 January 2025.
199. US Senate, Committee on Intelligence and Human Resources Subcommittee on Health and Scientific Research, Joint Hearing, Project MKULTRA, the CIA's Program of Research in Behavioral Modification, 3 August 1977, p. 4.
200. Admiral Stansfield Turner, Director of Central Intelligence, Prepared Statement on Project MKULTRA, The CIA's Program of Research in Behavioral Modification. Senate Hearing on MKULTRA. Joint Hearing of the Select Committee on Intelligence and the Subcommittee on Health and Scientific Research of the Committee on Human Resources, United States Senate. Nineth-Fifth Congress, First Session, August 3 1977.
201. N. Klein, *The Shock Doctrine*, Penguin Books, London, 2007, pp. 26–27.
202. Anonymous, Application for grant to study the effects upon human behavior of the repetition of verbal signals, section 68–37, *MKULTRA Subproject 68. The Black Vault*, National Security Archive, George Washington University, 1957, pp. 1–7. Accessed 21 March, 2023.
203. T. C. Noakes, Montreal MKULTRA Experiments, *The Canadian Encyclopedia*, 2021, https://www.thecanadianencyclopedia.ca/en/article/mkultra. Accessed 21 March 2023.
204. A. Kassam, The toxic legacy of Canada's CIA brainwashing experiments: 'They strip you of your soul'. *The Guardian*, London, 3 May 2018. Accessed 1 February, 2023.
205. D. E. Cameron and J. G. Lohrenz, The depatterning treatment of schizophrenia, *Compr. Psychiatry*, 1962, **3**(2), 65–76.
206. D. E. Cameron and J. G. Lohrenz, The depatterning treatment of schizophrenia, *Compr. Psychiatry*, 1962, **3**(2), 67.
207. Anonymous, Application for grant to study the effects upon human behavior of the repetition of verbal signals, section 68-37, *MKULTRA Subproject 68. The Black Vault*, National Security Archive, George Washington University, 1957, pp. 3–4.
208. Anonymous, Application for grant to study the effects upon human behavior of the repetition of verbal signals, section 68 37, *MKULTRA*

Subproject 68. The Black Vault, National Security Archive, George Washington University, 1957, p. 5.
209. N. Klein, *The Shock Doctrine*, Penguin Books, London, 2007, pp. 38–46, Chapter 1 subsection titled "The Science of Fear".
210. (a) Amnesty International, Iran: From protest to prison: Iran one year after the election, MDE 13/062/2010, June 2010, pp. 38–39; (b) Freedom From Torture, We will make you regret everything: Torture in Iran since the 2009 elections. Medical Foundation for the Care of Victims of Torture, March 2013. www.bit.ly/3dGpr3w.
211. (a) Amnesty International, Turkmenistan: Individuals continue to be at risk of violations in Turkmenistan, AI Index: EUR 61/001/2009, 12 February 2009; (b) Human RightsWatch, Turkmenistan: Open Letter from a Coalition of Human Rights Organizations, 17 July 2006.
212. (a) J. Warrick, Detainees Allege Being Drugged, Questioned. U.S. Denies Using Injections for Coercion', Washington Post, 22 April 2008; (b) UN Commission on Human Rights, Situation of Detainees at Guantánamo Bay, 27 February 2007, UN doc. E/CN.4/2006/120. The report was prepared jointly by: the Chairperson-Rapporteur of the Working Group on Arbitrary Detention, the Special Rapporteur on the independence of judges and lawyers, the Special Rapporteur on torture and other cruel, inhuman or degrading treatment or punishment, the Special Rapporteur on freedom of religion or belief, and the Special Rapporteur on the right of everyone to the enjoyment of the highest attainable standard of physical and mental health. For details of the US Government investigation into this issue; (c) see: US Department of Defense, Deputy Inspector General for Intelligence, Investigation of Allegations of the Use of Mind-Altering Drugs to Facilitate Interrogations of Detainees, 23 September 2009. Obtained under the US Freedom of Information Act by the NGO Truthout, https://www.documentcloud.org/documents/395950-pentagon-inspector-generals-report-investigation.html (accessed on 1 September 2012); (d) J. Kaye and J. Leopold, DoD Report Reveals Some Detainees Interrogated While Drugged, Others "Chemically Restrained." Truthout, 11 July 2012.
213. Freedom From Torture, *We will make you regret everything: Torture in Iran since the 2009 elections*. Medical Foundation for the Care of Victims of Torture, March 2013. www.bit.ly/3dGpr3w.
214. (a) Amnesty International, Iran: From protest to prison: Iran one year after the election, MDE 13/062/2010, 9 June 2010, 2010; (b) Freedom From Torture, We will make you regret everything: Torture in Iran since the 2009 elections. Medical Foundation for the Care of Victims of Torture, March 2013. www.bit.ly/3dGpr3w.
215. Centre for Human Rights in Iran, *Silencing the Streets, Deaths in Prison: The December 2017 Crackdown in Iran*, 19 February 2018.
216. Amnesty International, *Iran: Trampling humanity – mass arrests, disappearances and torture since Iran's 2019 November protests*, MDE 13/2891/2020, 2 September 2020.

217. Amnesty International, *Iran: Trampling humanity – mass arrests, disappearances and torture since Iran's 2019 November protests*, MDE 13/2891/2020, 2 September 2020, p. 44.
218. Subhojyoti Acharya, 'Is Narco Analysis a Reliable Science? – Present Legal Scenario In India', *Legal Service India*, 19 February 2008. Available at http://www.legalserviceindia.com/article/l176-Narco-Analysis.html; Mumbai attacks: Militant kept in underwear to prevent suicide, *Daily Telegraph*, 8 December 2008.
219. H. Pidd, Indian police still using truth serum, The Guardian, 5 July 2012.
220. S. Bada Math, Supreme Court judgment on polygraph, narco-analysis & brain-mapping: A boon or a bane, *Indian J. Med. Res.*, 2011, **134**(1), 4–7.
221. Supreme Court of India, Criminal Appellate Jurisdiction, *Selvi v. State of Karnataka & Anr.*, Criminal Appeal Number 1267 of 2004, 5 May 2010.
222. H. Pidd, Indian police still using truth serum, The Guardian, 5 July 2012.
223. *The Times of India/TNN*, Directorate of Forensic Sciences does record number of narco analyses in two months, 12 July 2017.
224. NDTV/Press Trust of India, Delhi's Forensic Lab To Have Psychological Autopsy Facility: Officials Forensic Science Laboratory in Delhi's Rohini will be the second lab in the country to provide facility for narco-analysis, according to a senior official, 2 September 2018.
225. (a) See for example: CBI puts Kotkhai suspects under lie detection, narco tests, The Indian Express, 3 October 2017; (b) NDTV, Journalist Gauri Lankesh's Murder: Accused To Undergo Narco Test, 13 March 2018; (c) Financial Express, Unnao case: CBI court allows narco test, brain mapping of truck driver, cleaner, 9 August 2019; (d) NDTV, Aaftab Poonawala's Post-Narco Session Completed: Report, 2 December 2022; (e) Times of India, Shastri, P. Custodial assault case: 5 Odisha cops brought to DFS for narcoanalysis, psychoanalysis, 4 October 2024.
226. Committee on Military and Intelligence Methodology for Emerging Neurophysiological and Cognitive/Neural Science Research in the Next Two Decades, *Emerging Cognitive Neuroscience and Related Technologies*, National Research Council, Washington D.C., 2008, p. 133.

CHAPTER 3

21st Century Advances in Understanding of Basic Survival Circuits of the Brain

3.1 Introduction

The fourth edition of *Principles of Neural Science* opens by stating[1] that "[T]he goal of neural science is to understand the mind – how we perceive, move, think and remember". It then emphasises that behaviour is examined at the level of the individual cells – the neurons – of the nervous system in order to find answers to five basic questions:

> "...How does the brain develop? How do nerve cells in the brain communicate with one another? How do different patterns of interconnections give rise to different perceptions and motor acts? How is communication between neurons modified by experience? How is communication altered by diseases?"

Our concerns here are mainly with the neuronal circuits that directly underly our behaviours and what goes wrong with them in diseases or potentially as the result of malign manipulation, but we will also have to touch upon some of the wider and longer-term interests of neuroscientists. In this chapter we focus on our rapidly developing knowledge of the basic regulatory and instinctive circuits of the lower reaches of the brain before turning, in the next chapter, to how very recent advances are also opening up routes to understand the more complex behaviours that are underpinned by neuronal circuits within the higher reaches of the brain.

Preventing Weaponization of CNS-acting Chemicals: A Holistic Arms Control Analysis
By Michael Crowley and Malcolm Dando

Published by the Royal Society of Chemistry, www.rsc.org

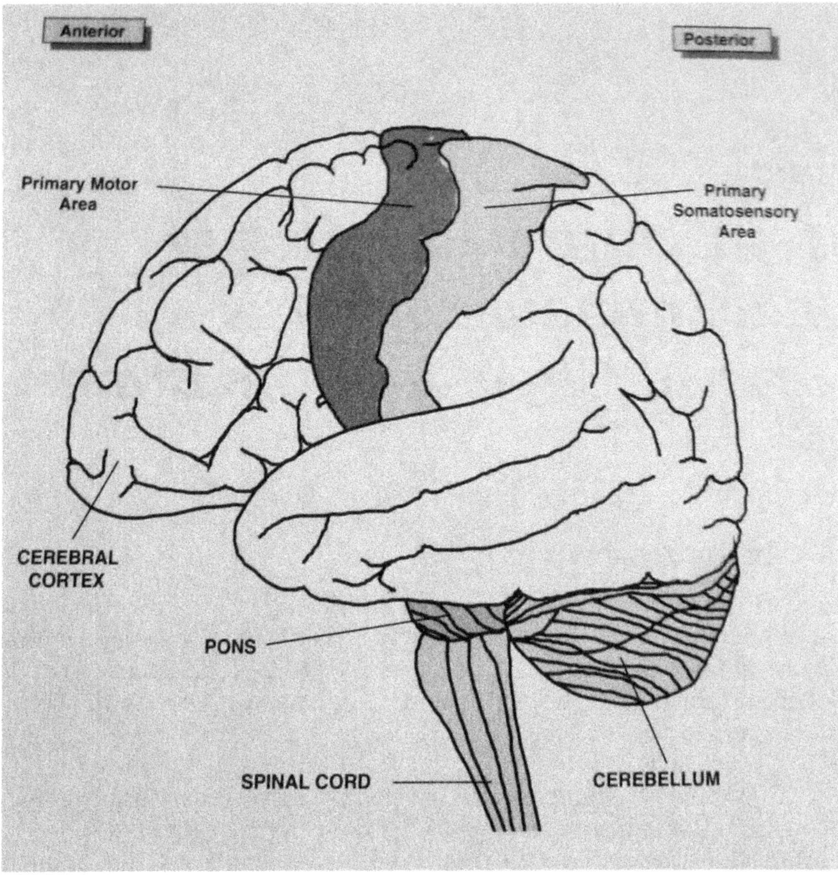

Figure 3.1 A lateral view of the brain. Adapted from ref. 2 with permission from Brassey's (UK) Ltd, Copyright 1996.

When we look at standard pictures of lateral views of the brain[2] what we see are mainly the massive convoluted cerebral hemispheres as shown in Figure 3.1.

It is in these structures where most of our executive decision making is thought to take place. However, these expanded hemispheres of the brain in human beings obscure the important structures of the rest of the forebrain, the midbrain, and the hindbrain as shown in Figure 3.2.

It is important to recognise that these underlying structures and their component groups of cells also have significant functions in the production of our behaviours, in moving information from the sense organs to the higher centres and passing information from the higher centres back down, for example, through the spinal cord to the muscular systems of the body. They also have influential inputs to the operations of the higher centres when basic regulatory functions, that are located in the lower reaches of the brain, require such modifications. It is useful to think of the brain in

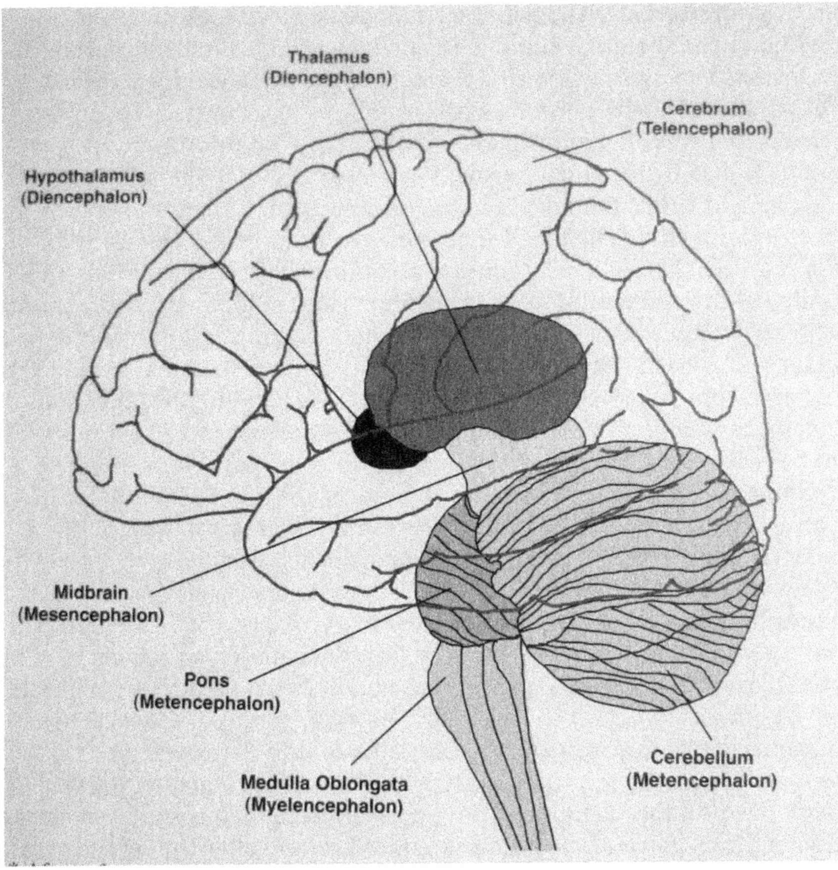

Figure 3.2 Brain structures. The labelling of the major structures of the CNS is based on the primary and secondary vesicles of the developing brain. Forebrain (Telencephalon and Diencephalon), Midbrain (Mesencephalon) and Hindbrain (Metencephalon and Myelencephalon). Some of the brain regions in these major structures are also illustrated. Adapted from ref. 2 with permission from Brassey's (UK) Ltd, Copyright 1996.

simplified terms as a complex information transmission system that has developed over evolutionary time to enable the organism to function in an integrated manner. So, there are afferent inputs of information to the central nervous system from sensory organs in the muscles and viscera as well as from the special sense organs focused on exterior events. This afferent information is subject to processing in the central nervous system and when necessary, results in efferent motor output to the muscles of the body or interior organs such as the heart.[3]

What is important for us here is the focus on the individual neurons, that we have known are the key cells involved in information transmissions in the brain for a little over a century. These specialised cells receive information through dendrites (input projections) and send information out through their axon projections. The communications between such neurons, as we have known for about a century are mainly by chemical means, despite the information transmissions within the neurons being by electrical means. So, within the individual neurons, information is transmitted by pulses of electrical changes. Sometimes the transmission of information between neurons is also by electrical means, but overwhelmingly the pulses of electrical changes cause the release of chemical transmitters at the end of the axons, and it is these chemicals that affect receptors on the dendrites of the following cells. The last century saw a continuing increase in our understanding of these chemical neurotransmitters and of the neuronal receptors that react by excitation or inhibition to these chemicals.[4] Particularly important also is to understand that during this century there has been a very rapid development of new technical means of investigating the operations of the nervous system in humans, and in a wide variety of model species[5] that can be used to understand the causes of our behaviour because of evolutionary continuities between different organisms. Furthermore, these advances have been accelerated by the initiation of a number of specific State-level brain research programmes over the last decade.[6]

We will not be able to survey all the advances that could be of dual-use concern here, but, as has been made clear in our first two chapters, previous advances in our understanding of these neurotransmitter operations have allowed malign misuse to take place – and not on a small scale (see Tables 1.1 and 2.1 for details of such misuse). The first of these chemical neurotransmitters to be discovered in the early 20th century was acetylcholine. This also functions as the transmitter at the neuromuscular junction for our voluntary muscles, and disruption of its normal operations is the basis of the action of the chemical nerve agents discovered in and after the Second World War. Other relatively simple chemicals such as glutamic acid (the major transmitter that causes excitation of affected cells within the brain) and gamma-aminobutyric acid (the major neurotransmitter that causes inhibition of affected cells within the brain) were also discovered in the last century as were the catecholamines, dopamine and noradrenaline, and serotonin.[7] These last four neurotransmitters were of particular interest as they were found to be produced by neurons that were located in small groups in the lower reaches of the brain but which had widespread inputs to the higher parts of the brain – in short suggesting that they have a *neuromodulatory* function at many different synapses rather than a neurotransmitter function at single synapses.

While the investigations at the end of the last century established that such neurotransmitters can affect different types of receptors and therefore produce different responses in neurons affected, the number of this type of neurotransmitter is not large. As noted by Solomon Snyder:[8]

> "...Neurotransmitters occur in chemical classes: the amines, amino acids, peptides and gases. As best we can guess, numbers of amines, amino acids, and gases are notably constrained. By contrast, there could exist many more peptides than are presently known..."

But he added:

> "...Identifying most of these should not be an insuperable challenge. With presently available technology, one should be able to define the universe of biologically active peptides in all tissues and, especially, in the brain. I suspect the number will be finite, perhaps no more than two hundred..."

As we certainly do not yet know anything like 200 neuropeptides in detail, this is clearly also an area where considerable new information may well become available in the coming decade or two. It is therefore an additional issue that is worth considering in relation to future dual-use possibilities.

3.2 The Importance of Neuropeptides

Advances in science and technology relevant to the potential development and misuse of biochemical agents that attack the central nervous system (CNS) have continued apace in this century. Such advances could be focused at a variety of different levels *i.e.*, on a particular behaviour, the underlying neuronal circuits, the neurons involved in a key circuit, the receptors on such neurons, the agent that could be used to attack and thus affect the behaviour and the production, dissemination and delivery of the agent.[9] While new agents could be novel synthetic chemicals produced *de novo,* there is also growing concern about how our understanding of the natural peptides[†] that influence the operations of the CNS might be misused in the development of biological and chemical weapons that target the human brain and broader CNS. Such concerns have been highlighted by several States in previous meetings of the Biological & Toxin Weapons Convention (BTWC) and the Chemical Weapons Convention (CWC). For example, in its contribution to the background information document on new scientific developments relevant to the Convention, prepared for the 9th BTWC Review Conference in 2022, the UK noted:[10]

> "A scientific advancement that has seen much progress since the eighth Review Conference in 2016 is peptide synthesis technology. This technology is rapidly advancing, and the emerging fields of peptide-based drugs and biomaterials are increasing accessibility of custom peptide synthesis and driving down its cost. Peptide synthesis equipment is available at all scales, although currently from a limited number of manufacturers globally. However, companies providing custom peptide synthesis services are numerous and globally widespread..."

[†] Defined by the *Compact Oxford English Dictionary* (Soanes, C. and Hawker, S. Third Edition, 2008, Oxford University Press, page 753) as "A chemical compound consisting of two or more linked amino acids."

And it went on to point out that:

"...peptide synthesis can be applied to many research applications such as, research into peptide vaccines, peptide drugs, peptide-based biomaterials, target validation, epitope mapping and structure and activity studies. Some recent developments of note include:

- Increased purity of commercially available amino acid building blocks increases yield and purity of peptides, which in turn allows for synthesis of longer peptides.
- The use of microwave or infrared heating and flow-based synthesis equipment has sped up the process to a matter of minutes, rather than hours, for each amino acid addition in a peptide chain. This can also increase the purity of the product".

And it concluded that such advances converging with other developments could be of concern for States Parties to the Convention:

"Advances in a number of technology areas, including synthetic biology, bioautomation, AI, machine learning, cloud-based laboratories, materials science, quantum and energetics are rapidly converging, which may consequently increase potential risks of misuse or misapplication of bioscience research. These convergences may also give rise to questions regarding the distinction between weapons of mass destruction and conventional weapons..."

Similarly, the report of the OPCW Scientific Advisory Board (SAB) for the 2023 5th CWC Review Conference stated that:[11]

"...The technical barriers for the artificial design and chemical synthesis of dangerous bacteria or a viral genome, *toxin, or bioregulator chemical*, are falling quickly due to advances in synthetic biology technologies that have reduced the cost and increased accessibility. Special attention should be paid to the commercialisation of benchtop DNA synthesisers that could be misused to covertly print dangerous sequences that code for the bioproduction of toxic chemicals. In addition, international biological academic competitions such as the International Genetically Engineered Machine and the rapid expansion of amateur biological groups have made it easy to find and utilise relevant professional knowledge and skills." (Emphasis added).

Indeed, the SAB was of the opinion that these issues were so serious that a special Temporary Working Group (TWG) should be established to study current developments concerning CNS-acting chemicals stating that:

"The SAB recommends that the Director-General consider establishing a TWG to study current developments concerning

CNS-acting chemicals, and it is further recommended that this TWG include the participation and expertise of other relevant international institutions (*e.g.*, the United Nations Commission on Narcotic Drugs) . . ."

As we will discuss in later chapters, the establishment of such a TWG would seem to us to be an essential step in view of the rate of scientific and technological change now and in the foreseeable future. In this chapter we concentrate on developments in our understanding of neuronal circuits underlying basic and regulatory behaviours that might be of interest to CNS-acting chemical weapons developers and in particular also what we know of the involvement of key neuropeptides in the operation of such circuits.

3.3 What We Knew About Neuropeptides in 2000

At the turn of the century, the journal *Peptides* published a review of all the studies on peptides that it had published over the last 20 years since its founding.[12] The review focused on the observable effects of the external administration of the peptides on seven categories of behaviour: activity levels, addiction, ingestion, learning and memory, nociception (pain), social and sexual behaviours, and simple stereotype behaviours (*e.g.*, catalepsy – sudden weakness in the muscles). We can therefore obtain a reasonable idea of the level of knowledge of the circuits involved in producing these behaviours and the investigative technologies available to the researchers from this review.

In regard to activity levels, for example, they measured motor activity by the distance moved in an enclosed space by an animal and concluded that:

"Despite the wealth of data, there is no unifying mechanism yet known to underlie activity level. Some peptides appear to work though common neural substrates such as the ventral striatum[‡] and globus pallidus[§] to produce increased activity, whereas others may be independent of such pathways. Still others may regulate activity without directly acting on the brain at all, but rather through peripheral control of metabolism."

‡ "The primary limbic portion of the striatum that includes the nucleus accumbens and related parts of the caudate nucleus and putamen." Loring, D. W (Ed.) *INS Dictionary of Neuropsychological and Clinical Neurosciences.* Oxford University Press, Oxford. Second Edition, 2015. Pages 379 – 380.

§ "A wedge-shaped nucleus medial to the putamen from which most of the efferents from the basal ganglia arise." Loring, D. W (Ed.) *INS Dictionary of Neuropsychological and Clinical Neurosciences.* Oxford University Press, Oxford. Second Edition, 2015. Page 172.

In regard to addiction, the reviewers concluded that it was clear that peptides were important and that drugs of addiction acted at various brain areas and receptors:

> "...Common mechanisms of addiction appear to come into play with multiple drugs, however. The mesolimbic dopamine system[¶] is the best known, but other, peptide-mediated systems also are critical to addiction. Opioid peptide systems within and independent of the mesolimbic system play a role in drug-use reinforcement. CRF [Corticotropin-Releasing Factor, a hormone having a role in the body's stress response] and other stress-peptides are critical in withdrawal behaviors, and other peptide neurotransmitters may modulate transmission within the dopamine system. In short, the system is hooked on peptides."

The system was also obviously very complex. The same could be said for social and sexual behaviours as they concluded, in part, that:

> "...Emotional responsiveness provoked in the amygdala[||] is known to be critical to social interactions, and thus the actions of peptides may influence social behaviors in this area. Memory and pain also tie in strongly with social behaviors..."

What stands out in this review is the lack of reference to studies of individual receptors in identified neurons within the key circuits.

A similar global review of research papers 5 years later in the same journal again conveys the idea of developing general concepts, but also the complexity and the difficulties of unravelling the nature of the circuits involved in producing the behaviours:[13]

> "Over the years, studies by many researchers as well as our own work ... have illustrated some novel concepts. These include the idea that hypothalamic peptides can act 'up' as well as 'down' (*i.e.*, that neurons in the lower parts of the brain can produce peptides that act on neurones in the higher reaches of the brain and *vice versa*), that peptides in the periphery of the body can act on the brain, that one peptide can exert more than one action, that peptide actions can persist longer than their blood half-lives (*i.e.*, that while the peptide can be rapidly destroyed by enzymes in the blood it actions on neurons affected can persist for much longer), that increased peptide doses can result in decreased effects, that peptides can cross the blood–brain barrier (BBB) [that forms a barrier to

---

¶ A major group of dopamine producing neurons that originate in the ventral tegmental area and ascend to innervate various parts of the limbic system such as the amygdala and hippocampus.

|| "Almond-shaped collection of nuclei in the anterior medial temporal lobe, just beneath the uncus, that play a role in memory and emotional control." Loring, D. W (Ed.) *INS Dictionary of Neuropsychological and Clinical Neurosciences.* Oxford University Press, Oxford. Second Edition, 2015. Page 22.

many substances access to the brain], and that a peptide's name does not restrict its actions..."

The different situation in 2022 can be seen in an Editorial in the same journal on "Peptides as mediators of stress responses" which notes that:[14]

> "In this special issue we focused on 'diversity' and 'harmony/- disruption' of stress responses involving *various peptides*, covering *model organisms* [fish, mouse, rat, genetically modified animals, and human] ...*modern techniques* [...genetically manipulated animals expressing GPCR-based sensors...], and stress responses and resilience to stress..." (Emphases added)

This looks like a very different field of research from that in 2000. Before investigating specific examples of modern understanding of significant neuronal circuits and the role of specific peptides in the modulation of such circuits, it is therefore necessary first to describe the general state of research in this area two decades after the turn of the century.

3.4 Overview of Neuronal Circuit Research in 2023

The editors of a special issue of the journal *Pharmacology, Biochemistry and Behavior* summarised the current situation in neuroscience succinctly as follows:[15]

> "*The last two decades of neuroscience research have witnessed an explosion of technological advances, arming scientists with myriad new tools to probe and manipulate brain function.* These technical advances, which span from fluorescence microscopy to opto- and chemo-genetics, fiber photometry, calcium imaging, and tissue clearing and expansion, have rapidly sculpted the neuroscience research landscape and fuelled new discoveries of the neural circuitries underlying brain function and behavior..." (Emphasis added)

The aim of the special issue was also clearly set out:

> "...The marriage of modern circuit neuroscience techniques with established pharmacological tools is therefore more vital than ever in the discovery of the next generation of treatments for human brain disorders. *This special issue will highlight recent research that has successfully married neural circuitry manipulations with pharmacological and biochemical techniques to elucidate the complexities of neural function underlying maladaptive behaviors and neuropsychiatric disease...*" (Emphasis added)

The special issue looked to the future in order to inspire new investigations into the translation of circuit neuroscience and pharmacological manipulations to the next generation of treatments for human brain disease.

New technologies developed in this century have therefore radically changed capabilities to access, assess and affect the CNS.

The potential scale of the likely change in capabilities was well illustrated by a paper by Ralph Adolphs, a prominent neuroscientist, in 2015.[16] He divided the unsolved problems of neuroscience into four categories: problems that are solved, or soon will be; problems that we should be able to solve in the next 50 years; problems that we should be able to solve, but who knows when; and problems we may never be able to solve. The interesting set of problems now, nearly 10 years after Adolphs published his estimate, are those he thought should be solved within 50 years. These are shown in Box 3.1.

What is of particular relevance is the question of what causes psychiatric and neurological illness, as knowledge of such causes might be put to malign use. The same might well be said of knowledge of how learning and memory work, why we sleep, and how we make decisions. There is little doubt about the possibilities of such successful research objectives being achieved, as major States and the European Union (EU) have initiated large scale brain research projects with these objectives in mind.[17] While most have not given much public attention to the possibilities of dual use, the EU Human Brain Project has explicitly investigated the dangers of dual-use and argued that steps have to be taken to reduce such dangers.[18] Indeed, while it may be inconvenient, such possibilities raise the question posed by David Relman in 2014 of whether there might be some experiments we should not do.[19]

What we are dealing with, therefore, is clearly an intended revolution in technologies for investigating the functions of the brain. This point was made obvious in the full name of the United States BRAIN Initiative – Brain Research Through Advancing Innovative Neurotechnologies. The initiative began in 2014 and had expended about 2.4 billion US dollars by early 2022. It was expected to have expended a total of over 5 billion dollars by 2026. The initiative has been guided by a major report in 2014, and a follow up report in 2019[20] that had updated the plans for the initiative. An update on the state of the BRAIN initiative and its current plans was given by the Director in a

Box 3.1 Problems solvable in 50 years.[16]

How do circuits of neurons compute?
What is the complete connectome of the mouse brain?
How can we image a mouse brain at cellular and millisecond resolution?
What causes psychiatric and neurological illnesses?
How do learning and memory work?
Why do we sleep and dream?
How do we make decisions?
How does the brain represent abstract ideas?

2022 article in the journal *Cell*.[21] He stated that continuing support would be provided for investigating neural circuits in non-human and human models in seven priority areas of funding (see Box 3.2).

Additionally, there were to be three large transformative projects that were the focus of his article. These transformative projects would:

> "...*(1) build a comprehensive atlas of cell types in the human brain, (2) develop and scale up the tools necessary to complete a microconnectivity map of an entire mammalian brain, and (3) provide tools for precision access to the identified cell types to allow interrogation and modulation of neural circuits...*" (Emphasis added)

Box 3.2 Objectives of the US BRAIN project.[21]

Priority Area 1. Discovering Diversity: Identify and provide experimental access to the different brain cell types to determine their roles in health and disease.

Priority Area 2. Maps at Multiple Scales: Generate circuit diagrams that vary in resolution from synapses to the whole brain.

Priority Area 3. The Brain in Action: Produce a dynamic picture of the functioning brain by developing and applying improved methods for large-scale monitoring of neural activity.

Priority Area 4. Demonstrating Causality: Link brain activity to behaviour by developing and applying precise interventional tools that change neural circuit dynamics.

Priority Area 5. Identifying Fundamental Principles: Produce conceptual foundations for understanding the biological basis of mental processes through the development of new theoretical and data-analysis tools.

Priority Area 6. Advance Human Neuroscience: Develop innovative technologies to understand the human brain and treat its disorders; create and support integrated human brain research networks.

Priority Area 7. From the Brain Initiative to the Brain. Integrating new technological and conceptual approaches produced in goals #1–6 to discover how dynamic patterns of neural activity are transformed into cognition, emotion, perception, and action in health and disease.

Transformative projects

...Recognizing progress and promise, BRAIN 2.0 is poised to take on a set of new grand challenges. Such "transformative projects," some examples of which we have provided in this report, involve complex and multiscale lines of inquiry and will require new technological, scientific, and organizational inventions...

These projects he suggested would "enable the eventual development of novel interventions – including molecular and gene-editing therapies – for human circuit disorders". Looking in more detail at these projects can give us an overview of the scale of transformation of capabilities that brain researchers have in mind to achieve over the next few years.

In regard to building a comprehensive human brain cell atlas, the Director argued that to understand the neurobiological basis of behaviour we need an account of the cell types and their properties, and the ways that they communicate with each other to process information, and thus generate specific behaviours. So, he noted that:[22]

> "...Much work is needed to generate a complete catalog or "parts list" to base deeper mechanistic studies on neural circuit function on and to reveal the developmental and evolutionary principles underlying cellular diversification in the brain..."

In his view this would include the genes and proteins expressed by all different cell types, their physiological and morphological characteristics and the connections that they make within and outside the nervous system. Moreover, all of this data would need to be available in a systematic classification system. In regard to developing a microconnectivity map of human brain he argued that:[23]

> "Mapping the connections across an entire mammalian brain at nanometer resolution promises to revolutionize the field of neuroscience by providing a ground-truth "wiring diagram" of this highly complex biological system..."

That knowledge, he noted, would allow us to reveal circuit motifs, to identify key neurons for further studies as well as identify new cell types based on long range connectivity. Furthermore, he made an analogy with the progress of the Human Genome Project, arguing that "we are now at the level of sequencing genes and loci (local circuits), whereas new scalable technologies are needed to sequence and assemble entire genomes (whole brains)". In his view, such progress would open up presently-unimaginable ways to probe neural circuit functions.

Perhaps most importantly, the third transformative project of the BRAIN initiative aims to develop further means of access to identified cell types within the brain. As the Director noted:[24]

> "Together with the revolutions in optogenetics**, chemogenetics††, and gene-editing technologies, rapid progress in generating molecularly

** Optogenetics is a technique used to control the activity of identified neurons by the use of light probes.

†† Chemogenetic approaches such as designer receptors exclusively activated by designer drugs (DREADDs) permit the remote control of neuronal function by systemic drug administration.

grounded brain cell atlases opens up exciting new possibilities for gaining genetic access to neural circuits for the purposes of interrogating and modulating their function…"

He went on to state specifically that "the goal is to invent, refine, and validate new technologies for building an expanded toolkit or 'armamentarium' for brain cell access that can be used to probe neural circuits across multiple species with greater precision…"

In the Director's view with these developments in technology it should:[25]

> "…*be possible, for example, to target a cell type—defined by its molecular, physiological, and connectivity properties—and perturb its activity optogenetically, chemogenetically, or through gene editing to test its role in generating a specific output or behavior…*" (Emphasis added)

Interestingly, from our point of view here, the article ends with a concern about the potential misuse of such capabilities. In the Director's view:

> "…As technologies in this realm advance—likely at a rapid but unpredictable pace—we as a scientific community must take stock not only of the value of these future applications, *but also of the potential for misuse or harm*, even if unintentional…" (Emphasis added)

Finally, he also noted that there are other efforts going on around the world that have similar objectives.

So, when considering the future developments of neuroscience capabilities, we have to keep in mind worldwide neuroscience. In that regard it is important to understand that the novel technologies are widely available and detailed explanations of how to use them are in the open literature, for example for optogenetics and chemogenetics,[26] fiber photometry,[27] and quantification of animal behaviour,[28] for application in an ever-widening range of animal models including non-human primates.[29] Moreover, the results of these extensive endeavours will not be long in coming. For example, in early 2023 *Nature* published a report of work on a map of the brain of a fly under the title "Gigantic map of fly brain is a first for a complex animal: Fruit fly 'connectome' could lead to deeper understanding of neurological diseases" The reporter noted that:[30]

> "…The map shows all of the 3016 neurons and 548 000 synapses that were tightly packed in a young *Drosophila*'s brain, which is smaller than a poppy seed.
>
> The map is a milestone in understanding how the brain processes the flow of sensory information and translates it into action…"

The results produced some very interesting findings such as that 93% of the neurons were paired with partners on the other side of the brain, but most of the unpaired neurons were a specific kind of neuron known to be important in learning and memory. The synapses between the neurons could be classified into just four different types, and the wiring diagram showed that the brain was a nested structure with multiple layers interconnected by pathways of varying lengths and that 41% of the neurons formed "recurrent loops" providing feedback to upstream partners. According to the report, previously, the only brain connectomes available were from very simple organisms, but in contrast the young fly can exhibit complex behaviours including "navigating landscapes, processing smells and weighing the risks and benefits of an action". Consequently, having this map of its brain opens up possibilities of investigating how connectivity deficits affect its behaviour, and potentially using it as a model of human degenerative diseases like Alzheimer's and Parkinson's.

Technological advances have also had a major impact on our understanding of the role of neuropeptides in the modulation of neuronal circuits, so we need also to briefly review these advances before turning to specific examples of research that might be misused in the future.

3.5 Overview of Neuropeptide Research in 2025

As Theodosius Dobzhansky famously pointed out in the 1960s and the 1970s,[31] "Nothing in Biology Makes Sense Except in the Light of Evolution". This viewpoint seems particularly appropriate in regard to neuropeptides as they clearly arose along with nervous systems amongst the cnidarians (sea anemones, jellyfish and the like) early on in the evolution of animal species and have important roles in both the invertebrates and the vertebrates – including mammals.[32] Disentangling these roles in various species is difficult, however, as the neuropeptides and their receptors only evolved slowly and often display deeply conserved features, whereas closely related species can demonstrate very different behaviours. The answer to this apparent paradox lies in the way in which neurons can shift their "expression of GPCRs [G protein coupled receptors for the neuropeptides] and the content of monoamines and neuropeptides," and thus reconcile the deep homology of neuromodulation with the variations of behaviour.[33] So the same neuropeptide can produce different behaviours in a variety of species because the receptors it affects can change over time and there can be different other neurotransmitters at the same time. Nevertheless, the deep homology of the neuropeptides means that considerable advantages can be gained in our understanding of the roles of these modulators through the study of animal models.

An example of the level of understanding that is becoming available concerns the fly *Drosophila* that was discussed previously in relation to recent work on the microanatomy of the nervous system. Current understanding of the role of neuropeptides in controlling the behaviour of this insect was described in a long 2022 paper[34] titled "Endocrine cybernetics: neuropeptides as molecular switches in behavioural decisions". The authors argue that the context dependent modulation of the fly's behaviour is "largely based on non-synaptic signalling with neuropeptides" and that there is a hierarchy of neuropeptide neurons with a small number that "arborize [branch] widely in multiple areas of the brain to orchestrate or modulate global activity" through to local peptidergic neurones (neurons that produce neuropeptides but only have branches that stay within a short distance of the cell) that are restricted to modulation of specific circuits. In addition, throughout this paper, the authors identify and provide extensive analysis of the multiple neuropeptides involved in controlling the fly's behaviour. It has also become possible to investigate in detail how such modulation operates at the level of the individual neurones involved. For example, neuroscientists, employing other model organisms, have analysed how the output of simple systems such as the crustacean stomatogastric ganglion is modified by input from identified modulatory neurones.[35]

Recently it was estimated[36] that "over 6000 neuropeptides have been identified belonging to more than 60 neuropeptide families found across nearly 500 different species," and that "over 200 different neuropeptides from 47 different neuropeptide families have been identified in humans" both within the central and peripheral nervous system. Some of these neuropeptides such as oxytocin and vasopressin are ancient and are highly conserved with essential functions for survival "in reproduction, water homoeostasis and feeding". Many other neuropeptides evolved more recently, for example:[37]

> "...Current evidence suggests that opioids[‡‡] are a vertebrate novelty as there is no evidence of opioid signalling system in invertebrates. Similarly, bradykinin is only found in vertebrates and plays a crucial role in inflammation and is also involved in fluid homeostasis, pain and blood pressure..."

The production of neuropeptides by a cell is complex. The usual process of protein production from DNA and RNA gives rise to a large precursor protein which is then split in further processing to produce one or more active neuropeptides. These are then transported to the axonal nerve endings in dense core vesicles for release when prolonged increases in calcium levels caused by electrical activity in the axon lead to fusion of the vesicles to the nerve membrane and thus to release of the active peptide. Moreover, differences in processing can lead to different forms of the neuropeptide and the neuropeptide may bind to different receptor subtypes. Moreover, as neuropeptides can affect different internal signalling cascades within the affected cells following the attachment to their receptors:

> "...neuropeptide signalling systems can be very complex as they involve several GPCR subtypes linked to multiple signalling cascades and are expressed in various tissues with different functions."

[‡‡] Compounds that derive from or mimic those found in opium that may share the same addictive properties or physiological effects.

Box 3.3 Examples of mammalian neuropeptides.[36]

Bradykinin
Cholecystokinin
Endothelin
Ghrelin
Neuropeptide Y
Orexin/hypocretin
Insulin
Somatostatin
Substance P
Vasopressin

Additionally, while some neuropeptides are essential for survival many play a modulatory role and gene knockout experiments will detect little impact,[38] neuropeptides may function at very low concentrations and can often act in combination with other small-molecule neurotransmitters.[39] Therefore, a neuropeptide that is essential for survival will cause death of the animal if the gene producing the peptide is removed in a gene knockout experiment, but often the effect of the neuropeptide will be difficult to demonstrate, as it might act at very low concentrations and maybe in combination with other neurotransmitter chemical types. Some examples of well-known mammalian neuropeptides are listed in Box 3.3.

While the first neuropeptide (Substance P) was discovered in the 1930s and its structure was determined in the 1970s, the complex nature of neuropeptides and their functions has led to the development of more and more novel techniques to determine the structure, localisation and functions of these signalling molecules.[40] One important technique that has been useful is mass spectrometry.[§§] This has enabled the quantitative analysis of neuropeptides even given the difficulties of working with neuropeptides as:[41]

> "...Their exact function is difficult to characterize, however, due to dependence on concentration, post- translational modifications, and the presence of other comodulating neuropeptides. Mass spectrometry enables sensitive, accurate, and global peptidomic analyses [detection and estimation of all the peptides in a sample] that can profile neuropeptide expression changes to understand their roles in many biological problems, such as neurodegenerative disorders and metabolic function."

[§§] The Wikipedia definition of mass spectrometry (MS) is "an analytical technique that is used to measure the mass-to-charge ratio of ions. The results are presented as a mass spectrum, a plot of intensity as a function of the mass-to-charge ratio. Mass spectrometry is used in many different fields and is applied to pure samples as well as complex mixtures."

And again, detailed information on the range of available technologies is widely available.[42] Given such technological developments it is to be expected that considerable advances are being made in our understanding of neuronal circuits and their manipulation. It is to these subjects that we now turn.

3.6 Understanding Simpler Circuits

3.6.1 Innate Behaviours

The question is to what extent has new technological capability enabled us to better understand the neural underpinnings of behaviour. For example, now we cannot just "knock in" or "knock out" genes, but also identify particular neurons genetically and modify their functions by optical or chemical means. Additionally, we can much more closely and automatically observe the behavioural effects of such manipulations. In short have these technological advances opened up radically new means of understanding the operations of nervous systems? In an extensive review of the neural underpinnings of innate behaviour (*i.e.*, a behaviour that is genetically determined and performed without prior experience) a large group of authors argued that:[43]

> "Behaviors are the most easily visualized readouts of the brain's inner workings. As such, they provide the most direct gateways to probe the brain's function. Spontaneously emerging *innate behaviors are of particular interest as these behaviors require no prior training and are stable among different individuals of the same species.*" (Emphasis added)

The authors suggest that farmers may have unknowingly used such innate behaviour patterns for breeding since ancient times. Certainly, recently scientists interested in the behaviour of farm and other domestic animals have intensively studied such innate behaviours[44] using in particular Panksepp's ideas on the evolution of emotions in vertebrates.[45] We shall return to Panksepp's work later, but the authors of this extensive review based their analyses on the work of the innovative ethologists[¶¶] who were awarded the Nobel Prize in Physiology and Medicine in 1973, stating that:

> "...innate or instinctual behaviors first gained scientific notary in the 1900s from field studies by Dr. Karl von Frisch, Dr. Nikolaas Tinbergen, and Dr. Konrad Lorenz. *They observed many highly complex and species-specific behaviors across various species, including bees, fish, birds, and mammals. The majority of these behaviors relate to the survival of individuals and the propagation of the species...*" (Emphasis added)

[¶¶] Scientists who study the behaviour of animals in their natural environment.

They note that the genetic accessibility and individual stability of such behaviours had provided ideal topics for the investigation of neuronal circuits by traditional methods such as electrical stimulation. However, they now note that such research on the neuronal control of such behaviours:

> "...has gained tremendous momentum in recent years, aided by the development of new tools that permit monitoring and functional manipulation of genetically defined populations of neurons in model organisms..."

And that:

> "...*As a result, recent studies are beginning to unravel the complex neural circuits that underlie various innate behaviors, including feeding, sleep, urination, prey-predation interactions, and conspecific social interactions*..." (Emphasis added)

So, it is to such innate behaviours and their control, that we can look to answer our question about the impact of the newly developed neurotechnologies.

One key question that can begin to be investigated, when there is some detailed knowledge of the circuits underlying behaviours, is how an animal decides which particular behaviour to undertake at any one time. As another recent review pointed out:[46]

> "Nervous systems must not only generate specific adaptive behaviors, such as reproduction, aggression, feeding, and sleep, but also select a single behavior for execution at any given time, depending on both internal states and external environmental conditions. Despite their tremendous biological importance, the neural mechanisms of action selection remain poorly understood..."

They then review work in the last decade on the fruit fly *Drosophila melanogaster* and compare the findings for this invertebrate with similar work on mice with the aim of seeing whether there are any similarities in the control circuits and their modification by neuropeptides. They were able to come to some interesting conclusions, as set out in Box 3.4.

The authors are certain that the available capabilities will continue to deliver similar and more detailed insights into how these neuronal circuits produce innate behaviours. As they note:

> "...*Future studies on how a small number of neurons control sex, fight, feeding, sleep, as well as selection among these innate behaviors, in both sexes, would provide invaluable insights for understanding individual innate behaviors, action selection among them, and their sexual dimorphism*..." (Emphasis added)

Box 3.4 General principles underlying action selection.[46]

"First, action selection among innate behaviors involves distinct selection models.... selection between sex and feeding involves feed-forward regulation, while selection between sex and sleep uses reciprocal inhibition."

"Second, a prominent feature of decision neurons...is their capacity to integrate multiple sensory inputs and social experiences. Furthermore, subsets of these decision neurons may have distinct functions...and interact with each other to orchestrate closely-related behaviors."

"Third, sexually dimorphic neurons may control the sex-specific prioritization of innate behaviors.... a higher activation level is required to induce aggression in female flies, which is consistent with the phenomenon that females are less aggressive than males."

"Fourth, in addition to circuit configurations, *neuromodulators play important roles in action selection.* In particular, neuromodulators may respond to changes in environmental cues or internal states, and switch a behavioural choice rapidly by acting on their receptors to promote one behavior and/or inhibit another." (Emphasis added)

"Finally, independent studies in flies and mice have demonstrated comparable circuit configurations for action selection among innate behaviours, involving pC1 neurons in flies and VMHvl neurons in mice..."

Against that background of highly likely rapid advances in our understanding of neuronal circuits and their modulation,[47] we can investigate what knowledge is being developed in relation to neuronal functions that might be of most interest to future CNS-acting chemical agent weapons developers.[48]

3.6.2 The Sleep/Wake Cycle

An animal has to be awake in order to carry out its essential tasks for survival and reproduction, but sleep is also a persistent characteristic of animals and seems essential for their survival. Therefore, studies of the sleep-wake cycle can be carried out in many model organisms. We will concentrate here on studies of vertebrates as there appears to be considerable continuity in the organisation of the cycle within this group including human beings. Since the 1930s it has been clear that there is a particular region of the brainstem that is essential for wakefulness. If that region is cut off from the higher regions of the brain by experimental lesion or accident then coma results from the lack of input from the lower region. A recent paper reviewed almost a century of investigations of the mechanism involved and stated that:[49]

> "…Lesion studies in experimental animals and in human patients have converged on a region of the upper brainstem tegmentum where damage causes coma. This region is the only location in the nervous system where a single, focal lesion can cause coma. It contains genetically diverse neurons with widely varying patterns of connectivity, but it remains unclear which are necessary for wakefulness."

The authors argue that there are two main possible explanations:

> "…One hypothesis is that many cell groups work in concert to produce wakefulness …. In this view, wakefulness is an emergent property of a synergistic, interconnected network of brainstem neurons. Damage to any part of this system causes at most a small or transient decrease in wakefulness, but damage to the entire system is required to produce coma."

The other possibility:

> "…is that some undefined set of upper-brainstem neurons sustains wakefulness. These neurons may or may not have been already described, but their role in arousal has not yet been tested. Under this hypothesis, when this group of cells is completely disconnected from the forebrain, coma results. Other neurons still may promote arousal in certain circumstances and eliminating them would impair a specific mode of arousability…"

The authors suggest that there are reasons to favour the second hypothesis, but that the jury is still out. However, they also argue that the continued application of modern techniques will lead towards an answer and thus to the better elucidation of the mechanism of the whole sleep-wake cycle.

It has been known since the middle of the last century that the human sleep/wake cycle is complex with sleep not being a single phenomenon[50] but having different characteristics in EEG (electroencephalogram) recordings[‖‖] in repeated stages of Rapid Eye Movement (REM)[***] and Non-Rapid Eye movement (NREM)[†††] sleep during the night (Table 3.1).

[‖‖] An electroencephalogram (EEG) measures electrical activity in the brain.

[***] During REM sleep there are random rapid movement of the eyes, low muscle tone and the probability of dreaming.

[†††] During NREM sleep a person falls asleep and then moves from a light sleep into a deep sleep.

Table 3.1 Characteristics of awake and sleep states.[50]

States	Characteristics EEG	Sensations	Thoughts	Movements
AWAKE	Fast, low voltage	Normal, externally generated	Normal, logically organized	Normal and voluntary
SLEEP				
NREM	Slow, high voltage	Usually absent	If present logical, but repetitive	Infrequent, involuntary
REM Sleep	Fast, low voltage looks like a normal awake EEG	Internally generated vivid, dreams often present	Illogical	Most muscles paralysed, but rapid eye movements occur

Each cycle of REM and NREM takes about 90 minutes and the cycle repeats about five times. Motor activity during REM sleep is suppressed for protective purposes so that dreams, which occur mainly during these periods, are not acted out. In the mammalian brain the suprachiasmatic nucleus (SCN) is a tiny region in the hypothalamus, situated directly above the optic chiasm. It is responsible for controlling circadian rhythms, and the neuronal and hormonal activities it generates appear to regulate many different body functions in a 24-hour cycle. It is the source of the circadian timing of the sleep/wake cycle, having neurons that oscillate naturally without environmental input. In diurnal animals like humans, SCN neurons promote wakefulness, by facilitating and inhibiting, respectively, the activity of the arousal-promoting and sleep-promoting regions. Essentially then in normal operation two sets of neurons in opposition function like a flip/flop switch so that quite rapid transitions are possible between the sleep and awake states. The cycle can also be modulated by sleep-promoting substances like adenosine being released into the CNS during wakefulness.

The neurons of the sleep/wake cycle produce classical neurotransmitters[‡‡‡] such as noradrenaline, dopamine, gamma-aminobutyric acid (GABA) and acetylcholine and various neuropeptides that cause the excitation and inhibition in their interactions. In humans and rodents these neuropeptides can be divided into three categories:[51]

> "...(i) Galanin, melanin-concentrating hormone, and vasoactive intestinal polypeptide are sleep-promoting peptides. It is also noticeable that vasoactive intestinal polypeptide particularly increases REM sleep. (ii) Orexin and neuropeptide S have been shown to induce wakefulness.

‡‡‡ Classical neurotransmitters are simple chemicals such as acetylcholine, amino acids such as GABA and catecholamines such as dopamine. They are different from neuropeptides that consist of strings of amino acids.

(iii) Neuropeptide Y and substance P may have a bidirectional function as they can produce both arousal and sleep-inducing effects..."

Orexin (also known as hypocretin) was only discovered in the late 1990s but has been the subject of considerable research because if the orexin producing neurons are destroyed, usually in humans by an autoimmune interaction,[52] Type I Narcolepsy results and the person suffers from attacks of cataplexy (sudden loss of muscle tone).[53] Clearly, such effects might be of interest to those with malign intent as well as the benignly-intended scientists who would like to find better ways of dealing with this lifelong disease.

Orexin neurons, that for historical reasons are also known as hypocretin secreting (hypocretinergic) neurons, have been subject to considerable study and it is well known that they:[54]

"...discharge during active wakefulness or the transition from sleep to waking in association with changes in muscle tone, and virtually stop firing during sleep.... Furthermore, hypocretinergic neurons are active...during wakefulness with motor activity related to the motivation to explore a new environment, but not during quiet wakefulness or forced locomotion..."

Given that orexin is a neuropeptide that causes excitation in other neurons it is assumed that it primarily acts by activating other excitatory elements to cause arousal and maintain the wake state. However, more detailed recent investigation show that its actions are more complex, the authors of one recent study argued that:[55]

"...we show using *in vivo* optogenetics,[§§§] *in vitro* optogenetic-based circuit mapping, and single-cell transcriptomics that orexin neurons also *contribute to arousal maintenance through indirect inhibition of sleep-promoting neurons* of the ventrolateral preoptic nucleus.[¶¶¶] Activation of this subcortical circuit rapidly drives wakefulness from sleep by differentially modulating the activity of ventrolateral preoptic neurons..." (Emphasis added)

It is important to add that it has become increasingly possible to access, assess and investigate the effects of manipulation of the orexins in freely moving rodents and thus to examine the role of this and other neuropeptides in natural conditions.[56] It has to be understood that given the importance, for example, of sleep disorders[57] and anaesthesia[58] we can expect such studies to continue at pace. Again, it is clear that the details of the

[§§§] Optogenetics is a technique used to control the activity of identified neurons by the use of light probes.

[¶¶¶] The ventrolateral preoptic nucleus is a small cluster of neurons situated in the anterior hypothalamus that is active during sleep.

necessary technologies for such work are widely available.[59] It is therefore necessary to look ahead and ask what is on the horizon that may impact our understanding of the sleep/wake cycle in the near future. One such possibility concerns the growing interest in microRNAs (miRNAs). To produce a protein the DNA of the gene is transcribed into messenger RNA which is then used to produce the protein. MicroRNAs can control the production of the protein by binding to the messenger RNA as explained by one recent article:[60]

> "...short (20 to 25 nucleotides), noncoding regulatory RNA molecules that exert their effects on target genes by complementary binding of the miRNA target sequence...leading to either mRNA degradation or translational arrest..."

Moreover, the authors suggest that therapeutics based on such mRNAs are emerging, and their article demonstrates that:

> "...intracellular miR-137 levels in Hcrt [orexin] neurons regulate Hcrt [orexin] expression with downstream effects on wakefulness. Specifically, lowering of miR-137 levels increased wakefulness in mice. We further show that the miR-137:Hcrt interaction is conserved across mice and humans...and that the MIR137 locus is genetically associated with sleep duration in humans..."

It is likely that a great deal more research work will be done in regard to this aspect of the regulation of the sleep/wake cycle and this will probably have positive real world applications in medicine. However, such research might also be subject to misuse with potential malign application in the development of CNS-acting chemical agent weapons intended to cause unconsciousness or lessened awareness of the targeted individuals or groups.

3.6.3 Fear Circuits

Studies of the innate circuits underlying responses to threatening circumstances, for example Pavlovian conditioning in rodents,[61] has usually been called work on fear. However, Joseph LeDoux has warned that this terminology should be used with caution as while the circuits in the lower parts of the brain certainly initiate the defensive responses in such circumstances (for example, freezing), in humans the feeling of fear and anxiety involves higher reaches of the brain.[62] While animals may or may not harbour feelings as well as defensive responses, the higher centres of human brains are so different that it seems unlikely that such feelings are very similar. So, LeDoux suggests dropping the current use of 'fear' to describe the innate circuits and instead to call them defensive survival circuits. This is reasonable in our view, but given the predominant use of the fear terminology in the literature we will stick with the current terminology while acknowledging the force of LeDoux's argument.

In 2016 Bianca Silva and her colleagues attempted to summarise the results of decades of work on these fear circuits. They suggested that a useful way of viewing these circuits was as three different functional units:[63]

> "...a detection unit, responsible for gathering sensory information signalling the presence of a threat; an integration unit, responsible for incorporating the various sensory information and recruiting downstream effectors; and an output unit, in charge of initiating appropriate bodily and behavioral responses to the threatful stimulus..."

And additionally, they suggest in parallel "the experience of innate fear also instructs a learning process leading to the memorization of the fearful event". Clearly threats are detected by different sensory systems – visual, olfactory, auditory, but more surprising is they conclude that also "the detection, *integration, and output units* processing acute fear responses to different threats tend to be harboured in distinct brain circuits, memory encoding of these threats seems to rely on a shared learning system". (Emphasis added). While the circuits underlying predator fear, and conspecific fear (fear of an animal of the same species) have been less investigated than the standard Pavlovian fear circuit underlying foot shock learning (a system of learning involving shocks to the animals feet), the data indicates that the foot-shock fear circuit's "neural integration is mediated by a dedicated amygdala pathway... that does not seem to overlap with conspecific or social fear processing structures".

Thus, there remains a great deal to be learnt about these ecologically relevant natural circuits. Of particular interest, the author's point out that:

> "...the investigation of the neural correlates of fear in rodents have classically been limited to the sensory-motor processing of defensive responses to a threat and only few studies have tried to uncover the brain circuits generating an 'emotional state of fear'..."

The authors then suggest that going beyond this limitation in the future would be important as "this would pave the way to understand how the circuits identified in the mouse relate to the brain correlates of human emotion of fear". Their overall conclusion is that the earlier studies they reviewed had provided fundamental insights into the function of fear circuits, but the studies had crucial limitations such as their cellular resolution and therefore they looked forward to the new technologies that were becoming increasingly available being used to clarify these details.

That possibility is well illustrated by a different section of their review in which they analysed work relating to defensive actions taken in response to internal sensory signals rather than external signal inputs. For example, it was clear that obstruction of respiration "invoke intense innate responses across animal species" and they argue that this has important medical implications:[64]

> "The neural circuits of fear induced by suffocation signals have attracted particular attention for their *relevance to panic disorder*, where hyper-responsivity of this system has been proposed to represent a major contributor to this pathology…" (Emphasis added)

Panic disorder (PD) has a lifetime prevalence of 3.5% (5% in women and 2% in men)[65] and is[66] an "anxiety disorder characterised by spontaneous and recurrent panic attacks (PAs)". Panic attacks are:

> "…characterized by multiple physical symptoms including a pounding or racing heart, sweating, dyspnea [shortness of breath], weakness or dizziness, feeling hot or cold chills, tingling or numbness in the hands, chest pain, or stomach pain…"

They are clearly a very unpleasant experience.

Jens Rehfeld described his experience of such an attack in 1968 when he was studying the effect of an intestinal peptide (CCK-4) on hormone secretion by intravenous injection of a bolus of this peptide on himself and a colleague. As he puts it:[67]

> "Half a minute after the injection of CCK-4, we experienced the beginning of what was to be a full-blown panic attack. The symptoms were intense anxiety, with a fear of dying, and a strange sense of the world sliding away, accompanied by palpitations, sweating, and faintness. The attack peaked after 5–8 min, and then gradually disappeared during the following 15–20 min. None of us had experienced such an attack before…" (Emphasis added)

At that time this outcome was most unexpected, but the reason for the panic attack has been clarified since then. The peptide was CCK-4, a type of Cholecystokinin, and as Rehfeld has recently noted "[T]oday, nobody questions CCK-4 as a robust panicogenic peptide, that is and has been a reliable tool in the study of panic disorder in man, and anxiety in most mammals…"

Cholecystokinin (CCK) is produced and released by intestinal cells and is involved in modulating intestinal mobility,[68] but this neuropeptide has also been found to be abundantly expressed in neurons in the brain. There are different types of CCK as a 115-amino-acid precursor is split into smaller active fragments. CCK-8 is the most abundant neuropeptide, but other fragments such as CCK-5 are also active. These peptides affect two different types of receptor CCK-A and CCK-B. CCK-B receptors are more abundant in the brain than in the periphery of the body and these receptors have equal affinity for CCK-8 and CCK-4. The CCK neuropeptides are frequently co-located in neurons with various classical small neurotransmitters such as dopamine. The authors of this 2003 research suggested that the "acute anxiogenic effects of CCK appear to be the result of activation of CCK-B receptors in the basolateral amygdala".

Interestingly in studies of mice which have been genetically modified to remove the gene for cholecystokinin[69] it has been found that they are not much affected in terms of food intake and "these results imply that the functions of CCK related to energy homeostasis are likely redundant with other signals such that CCK is not necessary for many of its actions". However, very much in contrast "CCK-deficient mice appeared more anxious and had impaired spatial memory and passive avoidance". So, the authors concluded that the central neurotransmitter effects of CCK related to anxiety and panic were much more important for normal functioning than the peripheral effects. More recent work with modern technology has begun to elucidate the central role of CCK neuropeptides in more detail. Feng and his colleagues studied mice in a standard Pavlovian fear conditioning experiment, but made the task for the mice more difficult by altering the process. As they explain, fear conditioning is standardly:[70]

> "...performed by pairing a neutral stimulus (conditioned stimulus, CS), such as a tone, with a punishing stimulus (unconditioned stimulus, US), such as a shock. The CS–US pair elicits fear behaviors.... Canonical delay fear conditioning is performed by terminating the CS and US at the same time. However, CS and US do not necessarily occur simultaneously in nature, and the brain has evolved mechanisms to associate temporally distinct events..."

There is clearly involvement of CCK in fear-related mental disorders such as panic disorder but they argue that "it remains elusive the involvement of CCK in fear conditioning and the formation of cue-specific fear memory, which is possibly the neural foundation of these disorders". Therefore, 'trace fear conditioning' is used to study the mechanisms involved when the animal is faced with more difficult circumstances. So, the authors' aim was to use loss-of-function and gain-of-function experiments (either removing or adding genes to experimental animals) to investigate the role of CCK producing neurons in the entorhinal cortex‖‖ in trace memory formation. They found that entorhinal cortex to lateral amygdala (LA) projections****

‖‖ Part of the cortex involved in odour processing and memory and the principle input station to the hippocampus. Loring, D. W (Ed.) *INS Dictionary of Neuropsychological and Clinical Neurosciences.* Oxford University Press, Oxford. Second Edition, 2015. Page 138.

**** "Almond-shaped collection of nuclei in the anterior medial temporal lobe, just beneath the uncus, that play a role in memory and emotional control." Loring, D. W (Ed.) *INS Dictionary of Neuropsychological and Clinical Neurosciences.* Oxford University Press, Oxford. Second Edition, 2015. Page 22.

"modulate neuroplasticity [alter the activity] in the LA and contribute to the formation of trace fear memory". And they conclude that:

> "...Our findings add a novel insight into the participation of the neuropeptide CCK in the formation of the trace fear memory. As various mental disorders, including anxiety, depression, and post traumatic stress disorder (PTSD), are highly correlated with hyper-activation and dysfunction of the amygdala and the fear memory circuitry, our finding supports CCK and its receptors as potential new targets for future therapeutic applications for these disorders."

It is surely evident that further work will continue to elucidate these mechanisms involving the ubiquitous CCK neuropeptide[71] and such research is likely to have positive real world medical applications to cure or treat a variety of anxiety related disorders. However, such research might also be subject to malign application in the development of CNS-acting chemical agent weapons intended to cause anxiety, fear and panic in the targeted individuals or groups.

3.7 Simple Circuits and Higher Levels of CNS Functions and Behaviours

A collection of articles on *Neural Circuits of Innate Behaviors* published in 2020[72] gives a good idea of the range of different behaviours and circuits being studied at the present time (Box 3.5).

So, investigations using modern techniques to research fear, aggression, feeding, reward, social interactions, parental care, special navigation as well as the sleep-wake cycle were underway at that time. Moreover, it is easy to find ongoing investigations of these subjects using modern techniques being reported in the scientific literature: for example, on parental care,[73] and aggression.[74] The argument has also certainly been made by Panksepp

Box 3.5 Articles on neural circuits of innate behaviours.[73]

1. Neural circuits underlying innate fear.
2. Neurobiology and neural circuits of aggression
3. Neural regulation of feeding behaviour
4. Neural circuits for reward
5. Neural response and behavioural modulation in social interactions
6. Neural circuit mechanisms that underlie parental care
7. Special navigation
8. Neural circuits for sleep-wake regulation

Box 3.6 Examples of Panksepp's basic circuits.[76]

SEEKING: Circuits that produce euphoric and expectant responses.
FEAR: Circuits that deal with threats and dangers.
RAGE: Circuits that produce furious responses.
LUST: Circuits that generate sexual desire and attachments.
CARE: Circuits for maternal care.
PLAY: Circuits for rough-and-tumble interactions.

and his colleagues that disruption of these basic circuits can lead to serious difficulties in carrying out more complex behaviour.[75] Panksepp argued that there are circuits in the lower parts of the brain that are so essential for the survival of species that they were highly conserved throughout the mammals (Box 3.6), and therefore studies of model animals could be highly informative in relation to human behaviour.

Our question here is slightly different: we want to know what help such investigations into simple behaviours might give to any CNS-acting chemical agent weapons developers who become interested again[76] in assaulting more complex human behaviour that involves operations of the higher reaches of the CNS?

An insight into how this could come about was shown in the Award of the 11th Breakthrough Prize for the Life Sciences in 2023. The award stated that:[77]

> "Another neurodegenerative disease, *narcolepsy*, was little understood before Emmanuel Mignot and Masashi Yanagisawa, running separate labs and pursuing different research programs, converged on a new understanding of its causes. *They showed that central to the disease is the protein orexin (also called hypocretin), which ordinarily regulates wakefulness.* In some animals, such as dogs, narcolepsy is caused by a mutation affecting the neural receptor that orexin binds to; while in humans, the disease is triggered by the immune system attacking the cells that produce orexin (probably "mistaking" it for a viral particle) . . ." (Emphasis added)

The mechanism by which the loss of orexin leads to Type 1 Narcolepsy with the problem of disrupted sleep, and specifically attacks of cataplexy, is becoming increasingly understood and would surely be of potential interest to a future developer of CNS-acting chemical agent weapons.[78] However, most importantly the statement on the Breakthrough Prize continued:[79]

> ". . .Mignot and Yanagisawa's discoveries have led to treatments shown to relieve the symptoms of narcolepsy, as well as enabling the design of sleep-inducing drugs. *They revealed that narcolepsy is a neurodegenerative disease with autoimmune origins, and raise the possibility that other*

neurodegenerative diseases may be caused by selective loss of neurons. And they shed light on a central mechanism of sleep and waking, an area of behavior that still holds many mysteries..." (Emphases added)

Yanagisawa's work includes an analysis of the neuronal circuit that induces a hibernation-like state in rodents[80] – a subject which is known to have previously interested certain States, notably the US, in regard to potential offensive uses.[81] However, it is Mignot's work with many other colleagues that has elucidated the autoimmune mechanism long suspected as the cause of the destruction of the orexin producing cells, and the possibility of this being the cause of other major immune-mediated neurological diseases that is of particular interest here,[82] and which will be the central subject of the next chapter.

3.8 Conclusion

This chapter began with a brief review of the recent development of neuroscience over the last 120 years since the discovery that the key information processing cells – the neurons – were separate,[83] and then that while information transfer within neurons was by electrical means, between neurons information transfer was predominantly by the release and reception of chemical substances – neurotransmitters. As our understanding of such neurotransmitters and their modes of operation has advanced it has led to the beneficial development of means of assisting people with mental illnesses and also, unfortunately, efforts to develop and use weapons that degrade the normal functions of the brain.

The chemical neurotransmitters that were first found were relatively simple chemicals, but it has become increasingly clear that there are also many neuropeptides that have neurotransmitter and modulatory functions (effecting many neurons simultaneously) within the brain. Such neuropeptides have attracted attention recently in regard to advances that may be of concern in regard to their weaponization as CNS-acting chemical agent weapons, and therefore the chapter goes on to describe these concerns and the rapid advances now being made in our understanding of their functions. The chapter then discusses advancing understanding of some of the simpler basic – survival – neuronal circuits that are located in lower regions of the brain. The chapter ends by linking this advancing understanding of such basic circuits with research that is casting more light on the operations of the executive functions of the higher reaches of the brain which is the subject of Chapter 4.

References

1. E. R. Kandel *et al.*, *Principles of Neural Science*, McGraw-Hill, New York, 4th edn, 2000.

2. For an extended simple account of the structure and function of the brain see M. R. Dando, *A New Form of Warfare: The Rise of Non-Lethal Weapons*, Brassey's, London, 1996, ch. 6, The Human Nervous System, pp. 95–115.
3. For an extended description see Chapter 2 The Structure and Function of the Brain in M. R. Dando, *Neuroscience and the Future of Chemical and Biological Weapons*, Palgrave Macmillan, Basingstoke, 2015.
4. See for example M. R. Dando, The new biological weapons: threat, proliferation and control, in *Neuroscience and the Future of Chemical-Biological Weapons*, ed. L. Rienner and M. R. Boulder Dando, Palgrave Macmillan, Basingstoke, 2015.
5. M. Carter and J. Shieh, *Guide to Research Techniques in Neuroscience*, Elsevier, Amsterdam, 2nd edn, 2015.
6. M. R. Dando, *Neuroscience and the Problem of Dual-Use; Neuroethics in the New Brain Research Projects*, Springer/Nature, Cham, Switzerland, 2020.
7. J. S. Meyer and L. F. Quenzer, *Psychopharmacology: Drugs, the Brain and Behavior*, Sinauer, Sunderland, 2nd edn, ch. 5–8, 2013.
8. S. Snyder, The brain harbors many neurotransmitters, in *Think Tank: Forty Neuroscientists Explore the Biological Roots of Human Experience*, ed. D. J. Linden, Yale University Press, New Haven, 2018, pp. 88–96.
9. See Figure 3.1 "The New Threat Spectrum" in M. R. Dando, *The Chemical and Biological Nonproliferation Regime after the Covid-19 Pandemic: Dealing with the Scientific Revolution in the Life Sciences*, Palgrave Macmillan/Springer Nature, Cham, Switzerland, 2023, p. 32.
10. Implementation Support Unit, Background Document: New Scientific and Technological Developments Relevant to the Convention. BWC/CONF.IX/7. United Nations, Geneva, 13 December. pp. 13 and 14.
11. Organisation for the Prohibition of Chemical Weapons, *Report by the Director General: Report of the Scientific Advisory Board on Developments in Science and Technology to the Fifth Special Session of the Conference of the States Parties to Review the Operation of the Chemical Weapons Convention*. RC-5/DG.1. OPCW, The Hague, 2023, pp. 27 and 8.
12. R. N. McLay, *et al.*, Effects of peptides on animal and human behavior: A review of studies published in the first twenty years of the journal Peptides, *Peptides*, 2000, **22**, 2181–2255.
13. Y. Yu, *et al.*, Effects of peptides, with emphasis on feeding, pain, and behavior: A 5-year (1999-2003) review of publications in Peptides, *Peptides*, 2003, **25**, 2257–2289.
14. Y. Ueta and K.-H. Herzig, Peptides as mediators of stress responses, *Peptides*, 2022, **153**, 170795.
15. C. Jordan, *et al.*, Pharmacology in the age of circuit neuroscience: Illuminating the neural mechanisms of reward, drug use and addiction and enlightening the future of translational research, *Pharmacol., Biochem. Behav.*, 2021, **206**, 173187. The quotations are from the website introduction to the set of articles and dated 17 June, 2022.

16. R. Adolphs, The unsolved problems of neuroscience, *Trends Cogn Sci*, 2015, **19**(4), 173–175.
17. M. R. Dando, *Neuroscience and the Problem of Dual Use: Neuroethics in the New Brain Projects*, Springer Nature, Cham, Switzerland, 2020.
18. Human Brain Project, *Opinion on 'Responsible Dual Use': Political, Security, Intelligence and Military Research of Concern in Neuroscience and Neurotechnology*, Danish Board of Technology Foundation, Copenhagen, 2018.
19. D. Relman, Inconvenient Truths" in the Pursuit of Scientific Knowledge and Public Health, *J. Infect. Dis.*, 2014, **209**, 170–172.
20. Advisory Committee to the NIH Director Brain Initiative Working Group 2.0, *The Brain Research through Advancing Innovative Neurotechnologies (BRAIN) Initiative 2.0: From Cells to Circuits*, Towards Cures. National Institutes of Health, Washington, D.C., 2019.
21. J. Ngai, BRAIN 2.0: Transforming neuroscience, *Cell*, 2022, **185**, 4–7.
22. J. Ngai, BRAIN 2.0: Transforming neuroscience, *Cell*, 2022, **185**, 4.
23. J. Ngai, BRAIN 2.0: Transforming neuroscience, *Cell*, 2022, **185**, 5–6.
24. J. Ngai, BRAIN 2.0: Transforming neuroscience, *Cell*, 2022, **185**, 6.
25. J. Ngai, BRAIN 2.0: Transforming neuroscience, *Cell*, 2022, **185**, 7.
26. K. Vlasov, *et al.*, Optogenetics and chemogenetics, *Methods Enzymol.*, 2018, **603**, 181–196.
27. Y. Wang, *et al.*, A selective review of recent advances in the study of neuronal circuits using fibre photometry, *Pharmacol., Biochem., Behav.*, 2021, **201**, 173113.
28. Y. Hu, *et al.*, *LabGym*: Quantification of user-defined animal behaviors using learning-based holistic assessment, *Cell Rep. Methods*, 2023, **3**, 100415.
29. T. Aida and G. Feng, The dawn of non-human primate models for neurodevelopment disorders, *Curr. Opin. Genet. Dev.*, 2020, **65**, 160–168.
30. M. Naddal, Gigantic map of fly brain is a first for a complex animal: Fruit fly 'connectome' could lead to deeper understanding of neurological diseases, *Nature*, 2023, **615**, 571.
31. See the entry in Wikipedia on "Nothing in Biology Makes Sense Except in the Light of Evolution".
32. C. J. P. Grimmelikhuijzen and F. Hauser, Mini-review: The evolution of neuropeptide signalling, *Regul. Pept.*, 2012, **177**, S6–S9.
33. P. S. Katz and J. L. Lillvis, Reconciling the deep homology of neuromodulation with the evolution of behavior, *Curr. Opin. Neurobiol.*, 2014, **29**, 39–47.
34. D. R. Nassel and M. Zandawala, Endocrine cybernetics: neuropeptides as molecular switches in behavioural decisions, *Open Biol.*, 2022, **12**, 220174.
35. D. M. Blitz, Neural circuit regulation by identified modulatory projection neurons, *Front. Neurosci.*, 2023, **17**, 1154769.
36. H. C. Mendel, *et al.*, Neuropeptide signalling systems – an unexplored target for venom drug discovery, *Biochem. Pharmacol.*, 2020, **181**, 114129.

37. H. C. Mendel, *et al.*, Neuropeptide signalling systems – an unexplored target for venom drug discovery, *Biochem. Pharmacol.*, 2020, **181**, 2.
38. H. C. Mendel, *et al.*, Neuropeptide signalling systems – an unexplored target for venom drug discovery, *Biochem. Pharmacol.*, 2020, **181**, 3.
39. L. E. Eiden, *et al.*, Neuropeptides and small-molecule amine transmitters: cooperative signalling in the nervous system, *Cell. Mol. Life Sci.*, 2022, **79**, 492.
40. K. DeLaney, *et al.*, New techniques, applications and perspectives in neuropeptide research, *J. Exp. Biol.*, 2018, **221**(3), jeb151167.
41. C. S. Sauer, *et al.*, Developing mass spectrometry for the quantitative analysis of neuropeptides, *Expert Rev. Proteomics*, 2021, **18**(7), 607–621.
42. K. S. Given, *et al.*, Emerging approaches for decoding neuropeptide transmission, *Trends Neurosci.*, 2022, **45**, 12.
43. W. Xiao, *et al.*, Neural circuits of innate behaviors, *Life Sci.*, 2022, **65**(3), 466–499.
44. See for example, T. Grandin and M. J. Deesing, Behavioral genetics and animal science, in *Genetics and Behavior of Domestic Animals*, ed. T. Grandin, Academic Press, Elsevier, London, 2022, pp 1–48.
45. J. Pankesepp and L. Biven, *The Archaeology of Mind: Neuroevolutionary Origins of Human Emotions*, W. W. Norton, New York, 2012.
46. X. Jiang and Y. Pan, Neural control of action selection among innate behaviors, *Neurosci. Bull.*, 2022, **38**(12), 1541–1558.
47. (a) See for example: L. Luo, Architecture of neuronal circuits, *Science*, 2021, **373**(6559), eabg7285. (b) R. J. V. Roberts, *et al.*, Evolution of central neural circuits: state of the art and perspectives, *Nat. Rev. Neurosci.*, 2022, **23**, 725–743. (c) J. Mei, *et al.*, Informing deep neural networks by multiscale principles of neuromodulatory systems, *Trends Neurosci.*, 2022, **45**(3), 237–250.
48. (a) J. B. Tucker, The body's own bioweapons, *Bull. At. Sci.*, 2008, **64**(1), 16–22 and 56–57. (b) V. M. Borzenkov, *et al.*, Additive synthesis or regulatory peptide in vivo: introduction of the vaccine strain of Francisella tularensis producing Beta-endorphin, *Bull. Exp. Biol. Med.*, 1993, **116**, 942–945.
49. F. S. Grady, *et al.*, A century of searching for the neurons necessary for wakefulness, *Front. Neurosci.*, 2022, **16**, 930514.
50. M. R. Dando, *Neuroscience and the Problem of Dual Use: Neuroethics in the New Brain Projects*, Springer Nature, Cham, Switzerland, 2020, ch. 1, Section 1.3 Sleep Research, pp. 7–13.
51. Y.-C. Shen, *et al.*, Roles of neuropeptides in sleep-wake regulation, *Int. J. Mol. Sci.*, 2022, **23**, 4599.
52. A. Iranzo, Sleep and neurological diseases, *Neuropsychopharmacology*, 2020, **45**, 129–140.
53. M. R. Dando, *The Chemical and Biological Nonproliferation Regime after the Covid-19 Pandemic: Dealing with the Scientific Revolution in the Life Sciences*, Palgrave Macmillan/Springer Nature, Cham, Switzerland, 2023, p. 48.

54. G. Vanini and P. Torterolo, Sleep-wake neurobiology, in *Cannabinoids and Sleep, Advances in Experimental Medicine and Biology*, ed. J. M. Monti *et al.* Springer Nature, Switzerland, 2021, vol. 1297, pp. 65–82.
55. R. De Luca, *et al.*, Orexin neurons inhibit sleep to promote arousal, *Nat. Commun.*, 2022, **13**, 4163.
56. T. Tsunematsu, Elucidation of neural circuits involved in the regulation of sleep/wakefulness using optogenetics, in *Optogenetics, Advances in Experimental Medicine and Biology*, ed. H. Yawo *et al.* Springer Nature, Singapore, 2021, vol. 1293, pp 391–406.
57. R. E. Brown, *et al.*, Translational approaches to influence sleep and arousal, *Brain Res. Bull.*, 2022, **185**, 140–161.
58. Q. Yang, *et al.*, Neural substrates for the regulation of sleep and general anesthesia, *Curr. Neuropharmacol.*, 2022, **20**, 72–84.
59. (a) P. J. Shiromani and J. H. Peever, New neuroscience tools that are identifying the sleep-wake circuit, *Sleep*, 2017, **40**(4), DOI: 10.1093/sleep/zsx032. (b) M. Carter *et al.*, *Guide to Research Techniques in Neuroscience*, 3rd edn, Academic Press, London, 2022.
60. A. Holm, *et al.*, The evolutionarily conserved miRNA-137 targets the neuropeptide hypocretin/orexin and modulates the wake to sleep ratio, *Proc. Natl. Acad. Sci. U. S. A.*, 2022, **119**(17), e2112225119.
61. M. R. Dando, *The Chemical and Biological Nonproliferation Regime after the Covid-19 Pandemic: Dealing with the Scientific Revolution in the Life Sciences*, Palgrave Macmillan/Springer Nature, Cham, Switzerland, 2023, pp. 35–44.
62. J. E. LeDoux, As soon as there was life, there was danger: the deep history of survival behaviours and the shallower history of consciousness, *Phil. Trans. R Soc. B*, 2021, **377**, 20210292.
63. B. A. Silva *et al.*, The neural circuits of innate fear: detection, integration, action, and memorization, *Learning and Memory Open Access*, Cold Spring Harbor Laboratory Press, 2016, pp. 544.
64. B. A. Silva *et al.*, The neural circuits of innate fear: detection, integration, action, and memorization, *Learning and Memory Open Access*, Cold Spring Harbor Laboratory Press, 2016, pp. 547.
65. J. F. Rehfeld, Cholecystokinin and panic disorder: reflections on history and some unsolved questions, *Molecules*, 2021, **26**, 5657.
66. P. Kyriakoulis and M. Kyrios, Biological and cognitive theories explaining panic disorder: A narrative review, *Front. Psych.*, 2023, **14**, 957515.
67. J. F. Rehfeld, Cholecystokinin and panic disorder: reflections on history and some unsolved questions, *Molecules*, 2021, **26**, 3.
68. S. Rotzinger and F. J. Vaccarino, Cholecystokinin receptor subtypes: role in the modulation of anxiety-related and reward-related behaviours in animal models, *J. Psych. Neurosci*, 2003, **28**(3), 171–181.
69. C.-M. Lo, *et al.*, Characterization of mice lacking the gene for cholecystokinin, *Am. J. Physiol. Regul. Integr. Comp. Physiol.*, 2008, **294**, R803–R810.

70. H. Feng, *et al.*, The entorhinal cortex modulates trace fear memory formation and neuroplacticity in the mouse lateral amygdala *via* cholecystokinin, *eLife*, 2021, **10**, e69333.
71. J. F. Rehfeld, Cholecystokinin – from local gut hormone to ubiquitous messenger, *Front. Endocrinol.*, 2017, **8**, 47.
72. H. Wang, Neural circuits of innate behaviors, in *Advances in Experimental Medicine and Biology*, Springer Nature, Singapore, 2020, vol. 1284.
73. Y. Liu, *et al.*, Molecular and cellular mechanisms of the first social relationship: A conserved role of 5-HT from mice to monkeys, upstream of oxytocin, *Neuron*, 2023, **111**, 1468–1485.
74. T. Yang, *et al.*, Hypothalamic neurons that mirror aggression, *Cell*, 2023, **186**, 1195–1211.
75. J. Pankesepp and L. Biven, *The Archaeology of Mind: Neuroevolutionary Origins of Human Emotions*, W. W. Norton, New York, 2012.
76. See, for example, R. Khatchadourian, Operation Delirium, The New Yorker, December 2012, 1 – 37, https://www.newyorker.com/madazine/2012/12/17/operation-delirium.
77. Breakthrough Prize, Winners of the 2023 Breakthrough Prizes in Life Sciences, Mathematics and Fundamental Physics Announced, San Francisco, 2022, vol. 22, p. 3, breakthroughprize.org.
78. M. R. Dando, *The Chemical and Biological Nonproliferation Regime after the Covid-19 Pandemic: Dealing with the Scientific Revolution in the Life Sciences*, Palgrave Macmillan/Springer Nature, Cham, Switzerland, 2023, pp. 44–48.
79. Breakthrough Prize, Winners of the 2023 Breakthrough Prizes in Life Sciences, Mathematics and Fundamental Physics Announced, San Francisco, 2022, vol. 22, p. 3, breakthroughprize.org.
80. T. M. Takahashi, *et al.*, A discrete neuronal circuit induces a hibernation-like state in rodents, *Nature*, 2020, **583**, 109–114.
81. M. Crowley and M. R. Dando, *Toxin and Bioregulator Weapons: Preventing the Misuse of the Chemical and Life Sciences*, Palgrave Macmillan, Springer Nature, Cham, Switzerland, 2022, p. 177.
82. (a) R. S. Liblau, *et al.*, The immunopathogenesis of narcolepsy type 1, *Nat. Rev. Immunol.*, 2023, DOI: 10.1038/s41577-023-00902-9. (b) H. M. Ollila, *et al.*, Narcolepsy risk loci outline role of T cell autoimmunity and infectious triggers in narcolepsy, *Nat. Commun.*, 2023, **14**, 2708.
83. B. Ehrlich, *The Brain in Search of Itself: Santiago Ramon Y Cajal and the Story of the Neuron*, Farrar, Straus and Giroux, New York, 2022.

CHAPTER 4

Disrupting Circuits of the Higher Central Nervous System

4.1 Introduction

We began this book by referring, particularly in Chapter 2, to the ideas that the early Cold War weaponeers had of disrupting the operations of a wide range of central nervous system (CNS) functions through the use of chemical and biological agents as illustrated in Figure 1.1. However, they appear eventually, because of their lack of knowledge of the CNS, to have retreated mainly to developing and using relatively simple means of sedation as shown in Tables 1.1 and 2.1. In the last chapter we examined how benignly-intended advances in neuroscience have increased our knowledge of the basic regulatory and instinctive functions within the CNS with the implication that such mechanisms might become the subject of misuse in the future. As far back as 2008 Julian Perry Robinson raised such concerns[1] about the impact of scientific and technological change on the Chemical Weapons Convention (CWC) and the Biological and Toxin Weapons Convention (BTWC) in the potential perception of utility for new so-called "non-lethal weapons". He returned to this issue in another paper during the same year, noting that:[2]

> "A ...major source of new utility for chemical weapons is the propensity of knowledge newly gained in the life sciences to suggest novel modes of attack that could be the basis for militarily or politically attractive forms of new weapon. *For example, if a new molecule is discovered that can exert novel disabling effects on the human body at low doses, attempts to weaponize it may well ensue...*" (Emphasis added).

Preventing Weaponization of CNS-acting Chemicals: A Holistic Arms Control Analysis
By Michael Crowley and Malcolm Dando

Published by the Royal Society of Chemistry, www.rsc.org

He then discussed the way that the discovery of LSD in the 1940s led quickly to military efforts to weaponize it in the 1950s, and then added:

> "...*Or if a hitherto unknown molecular pathway serving a process of life comes to be identified, chemical agents capable of interfering with that pathway might also become identifiable* and then form the basis for a novel weapon..." (Emphasis added).

As he argued, the science itself is not the challenge to the regime as there are many other considerations that determine the utility of a weapon, but such a discovery "would be a step towards it; and many such discoveries can be envisaged". In making that particular statement he referred back to the influential US National Academies' *Lemon-Relman Report* in 2006 which had argued that the coming threats could extend beyond just the manipulation of pathogens focused on in the 2004 report of the Fink Committee. The purpose of this chapter is to examine the current status of Perry Robinson's concerns, particularly in regard to the discovery of new pathways of networks, neurons, and receptors,[3] that operate in the higher reaches of the CNS in the control of more complex types of behaviour.

While there are many different areas of chemical and life science potentially applicable, we will focus on two fields of modern civil research into the higher CNS where misuse of benignly-intended work seems, in the light of historical chemical weapons development, to be quite possible. These two fields of dual-use research are the renewed interest in medical applications of psychedelic drugs, and important novel medical investigations of neuroimmunology processes as potential causes of psychiatric disorders. Again, we do not argue that the results of such research will necessarily be subject to misuse in the different fields that concerned the members of the EU Human Brain Project, but that this possibility needs to be kept in mind in order to help protect this civil research from developing into a degradation market.

4.2 Psychedelic Disruption: Old and New

We noted in previous chapters that there had been serious efforts during the East-West Cold War of the previous century to develop means of attacking the CNS through the use of psychedelics such as LSD.[†] In his recent book[4] with Brigid Moss, Professor David Nutt noted that in the middle of the last century there was also use of such chemicals for recreational purposes by civil society, and extensive research by the medical profession seeking new means of helping people suffering from mental illnesses. When the political decision was taken in the late 1960s to make use of such drugs illegal it did

[†] There is huge complexity in this area of research. There are an enormous number of natural, and particularly synthetic, chemicals within different classes of chemical that exert some of these kinds of effects on the brain. Moreover, any one chemical can affect different types of receptors or, more likely, different subtypes of the same class of receptor.

not prevent continued use by some members of civil society, but it did make further medical research very difficult. Therefore, the number of published research papers dropped dramatically and medical research and publication on this topic remained limited until very recently.[5]

In the meantime, of course, other forms of neuroscience research did not stop and over the last decades we have learnt much more about the pharmacology of the bioregulators of the CNS and its neuronal receptor systems, but also, through convergence with the application of other technologies,[6] about the large-scale functional neuronal networks within the brain. It has been possible for some time to assess the structure and activity of the brain by a variety of methods. One such method is magnetic resonance imaging (MRI) which measures the consumption of oxygen as an estimate of neuronal activity. As explained in a 2019 review by Uddin *et al.*:[7]

> "Large-scale brain networks have been successfully delineated using an approach referred to as resting-state functional connectivity (RSFC). *This approach examines synchronized patterns of spontaneous oscillations in blood-oxygen level dependent (BOLD) signal measured at rest with MRI . . .*" (Emphasis added)

This data was then examined using various mathematical techniques, as the text continues:

> ". . .Some of the earliest studies using RSFC to delineate macro-scale functional brain networks used ICA [independent component analysis]. ICA is a model-free approach that decomposes neuroimaging datasets into a set of independent one-dimensional time series and associated three-dimensional spatial maps that describe the temporal and spatial characteristics of the underlying signals. . ."

Essentially, such work demonstrates that the activities of different large groups of neurons in different parts of the brain are active at the same time. These networks have been found to function, for example, with regard to vision, attention, and control.[8] As yet, there is not a consensus on the nomenclature, and some room for disagreement on the number of such networks, but it seems likely that eventually just a small number of such large-scale networks will be recognised in the brain.

One very surprising early result from such studies was the discovery of the Default Mode Network (DMN) which operates when we are at rest. Writing in the *Annual Review of Neuroscience* in 2015, Marcus Raichle noted that:[9]

> "In 2001 we used positron emission tomography (PET) measurements of regional blood flow and oxygen consumption to show, by established metabolic criteria for activation, that areas consistently exhibiting activity reductions during task performance were not activated in the

resting state. Our article was titled, "'A Default Mode of Brain Function'..."

A recent systematic review in 2023 by Gattuso *et al.* confirmed this stating that:[10]

Neuroimaging studies have shown that when an individual is at rest, multiple interconnected brain regions are activated above baseline, and, conversely, there is decreased activity in these regions during task-dependent attention.... These regions amalgamate to create 4 functional hubs: the medial prefrontal cortex (mPFC), the posterior cingulate cortex (PCC), precuneus, and the angular gyrus, together referred to as the Default Mode Network (DMN)..." (Emphasis added)

The review goes on to assert[11] that activities in the DMN have been "implicated in a variety of complex cognitive functions such as theory of mind,[‡] self-referencing,[§] memory and rumination" and that "altered DMN function has been implicated in a range of neuropsychiatric and neurodegenerative conditions such as depression, attention deficit hyperactivity disorder, schizophrenia, anxiety" and so on.

From our point of view here, the important point is that research has conclusively shown that psychedelics affect the operation of the DMN:[12]

"...Functional magnetic resonance imaging (fMRI) studies involving psychedelics show that the modulation of the DMN acutely decreases connectivity and blood flow within nodes of this network ... These changes are paralleled by magneto/electroencephalography studies showing neuronal desynchronization of alpha power [¶](itself previously shown to be highly correlated with DMN connectivity) ..."

Not only does the connectivity within the DMN decrease, but connectivity between the DMN and other large-scale networks increases[13] and these modulations are at least correlated with the well-known consequences of the psychedelics on people's thoughts and behaviours. In such circumstances it is not surprising that medical professionals are researching, and achieving some success, in using psychedelics to help people with problems such as depression.

The review by Gattuso *et al.* concluded that:[14]

"This systematic review provides evidence to support the notion that classical psychedelics are capable of modulating the DMN, which is

[‡] The theory of mind is the idea that we understand other individuals by considering that they have mental states in the same way that we do.
[§] Self-referencing is thinking about yourself.
[¶] Detected oscillations in the frequency range of 8–12 Hz.

correlated with ego dissolution, increased brain entropy, and improved mental health and well-being. *Our review of the data shows psychedelics are a valuable tool for investigating this network and considers the potential therapeutic implications of this effect. Psychedelics are showing promise as tools to aid our understanding of the brain and mind in greater detail...*" (Emphasis added)

So, while the Cold War chemical weapons' weaponeers knew that psychedelics disrupted the functions of the higher reaches of the brain, they had little idea of how this happened. Now we know that there is a Default Mode Network and that it is disrupted by psychedelics. Moreover, we have modern means to investigate how such disruptions are caused. What then do we know about how the mechanism by which psychedelics affect the brain?

A 2022 review by Johansen *et al.* of some of the medical research being carried out on classic psychedelics defined them precisely as agonists[‖] of serotonin** receptors in the brain stating that:[15]

"Classic psychedelics are a group of psychoactive drugs that includes lysergic acid diethylamide (LSD), psilocybin, *N*,*N*-dimethyltryptamine (DMT) and mescaline. When ingested or administered, these compounds can elicit altered states of perception, cognition and emotion..."

And the authors continue to explain that:

"...*Despite having distinct characteristics, the classic psychedelic compounds are often grouped together as they all act as agonists of serotonin 5-hydroxytryptamine (5-HT) 2A receptors. It is through this receptor agonism that classic psychedelics primarily exert their psychoactive effects.*" (Emphasis added)

There is overwhelming evidence for this assertion in the relevant research papers.[16] So, in order to understand what is now known of the mechanism of action of these classic psychedelics we need first to look at the serotonin system in the brain.

The link of serotonin to the modulation – fine tuning rather than triggering or inhibiting – of behaviours arose early and has been evolutionarily conserved in the animal kingdom. It has been well described in both vertebrates and invertebrates.[17] Indeed it has been suggested to be "the

‖ An agonist is a drug or other substance that can bind to a receptor inside a cell or on its surface, and consequently causes the same action as the substance that would normally bind to that receptor.

** Serotonin (5-hydroxytryptamine (5-HT)) is a monoamine neurotransmitter carrying messages between nerve cells in the brain and CNS and throughout the body's peripheral nervous system.

prototypical neuromodulator fundamentally involved in the adaptation of all organisms across the animal kingdom".

The serotonin system in mammals was one of a number of such systems with cell bodies in the brain stem and ramifications into the higher parts of the brain described first by Swedish researchers in the 1960s (see Chapter 3). The axons from these serotonin producing neurons were found to ramify extensively through virtually all of the forebrain.[18] There have been found to be at least 15 different subtypes of receptors for serotonin and some of these have been clearly linked to problems that can be dealt with by drugs.[19] For example, the 5-HT_{1D} receptor is found on blood vessels, particularly those leading into the brain. The exact cause of migraine is not known, but is linked to increased blood flow into the brain. Thus, it has been shown that drugs targeted at this receptor are useful in preventing migraine attacks and stopping them once started. The $5HT_{2A}$ receptors, as described above, are of particular interest because of the affinity that they have for psychedelics, and also because of where neurons with these receptors are located, namely in areas of the default mode network (DMN) of the brain.

In 2022 one of the key people involved in the renewal of research on the medical possibilities of using psychedelics, Felix Vollenweider, and his colleague John Smallridge published a review titled *Classic Psychedelic Drugs; Update on Biological Mechanisms*.[20] Their review covered classic psychedelics such as:

> "...the *indoleamines* such as *N,N*-Dimethyl-tryptamine (DMT) contained in plants, psilocybin and its active metabolite psilocin contained in several mushroom species, the *phenylalkylamines* such as mescaline contained in several cacti, and *synthetic 'amphetamines'* such as 2,5-Dimethoxy-4-iodoam-phetamine (DOI) and 2,5-dimethoxy-4-bromoamphetamine, and the *semisynthetic ergolines* such as lysergic acid diethylamide (LSD)..." (Emphases added)

Despite the different structures of these chemicals, they all affect the $5HT_{2A}$ receptor and produce what these authors describe as "profound alterations in perception, cognition, emotion, and self-consciousness" that have been "used by humans for millennia for spiritual and medicinal purposes". This comprehensive review is set out in several sections as presented in Box 4.1.

The section of particular interest here is concerned with the "Functional Network Models of Psychedelic States". The authors note that altered information processing in the feedback loops between the cortex and the thalamus has been proposed as one mechanism underlying the psychedelic experience. The thalamus gates external and internal input to the cortex but is also subject to input from the cortex and this input is in turn subject to modulation by serotonin producing neurons in the lower parts of the brain.

Box 4.1 Sections of the Vollenweider and smallridge review paper.[20]

Introduction
Phenomenology and Predictors of Psychiatric States
Neurobiology of Psychedelics
  Receptor activation and pharmacological effects of psychedelics
  Neuroplastic effects of psychedelics
Functional Network Models of Psychedelic States
  Thalamic gating model
  Neural entropy model
Effects of Psychedelics on Brain Network Integration
Neural Correlates of Altered Self- and Emotion-Processing in Psychedelic States
  Self-processing
  Emotional processing
Conclusion

The model suggests that the psychedelics activate pyramidal neurons of layer V of the cortex directly *via* their $5HT_{2A}$ receptors and that these layer V neurons are also indirectly activated *via* pyramidal neurons in deeper cortical layers (being activated *via* their $5HT_{2A}$ receptors that causes the neurons to produce glutamate at other excitatory receptors on the layer V pyramidal neurons). The net effect of the output from these over excited neurons to the thalamus is to cause network changes in the information processing system.[21]

The important point, in terms of potential dual use concerns, is that such hypotheses are becoming open to practical testing,[22] for example, in a study of what happens to the DMN when the Designer Receptors Exclusively Activated by Designer Drugs (DREADDs) technology[23] was used to inactivate neurons in the amygdala.[24] This technology involves the use of genetic modifications that allow particular neuronal receptors to be activated or deactivated through the use of drugs that specifically act only on these receptors. Thus, the researchers were able to show that the de-activation resulted not only in disrupted communication between the cortex and the amygdala but also in disruption of cortical communication "across multiple functional brain systems". In such circumstances it seems certain that increasingly precise understanding of the mechanisms involved in the production of psychedelic experiences and of the ability to manipulate them for medical purposes will arise.[25] Unfortunately, that may also enable manipulation by those with malign intent, *i.e.*, to construct CNS-acting chemical agent weapons targeted at the DMN. We turn now to another much more recently discovered source of these benign and malign possibilities.

4.3 The Immune System and Disruption of the CNS

In the Spring of 2023, the widely read journal *Scientific American* published a special edition of articles on our current understanding of the brain. This special edition included an article by Sutherland and Shafer reviewing the problems caused by long COVID – the persistent illness that continues for some time after the initial respiratory difficulties caused by the viral infection.[26] As the authors noted, given the huge numbers of people infected around the world by the virus even the lower-range estimate of 1 in 10 people suffering the persistent illness indicated that many millions of people must have been affected. This finding provoked widespread interest in mechanisms of immune disruption of the higher nervous system that already had been advancing rapidly in recent decades and could obviously be of concern in regard to misuse.

4.3.1 COVID and Disruption of the Nervous System

Surprisingly, the authors of the *Scientific American* article went on to point out that although COVID infection is initially a respiratory disease:[27]

> "The most common, persistent and disabling symptoms of long COVID are neurological.
>
> Some are easily recognised as brain – or nerve – related. Many people experience cognitive dysfunction in the form of difficulty with memory, attention, sleep and mood..."

The authors then suggest that "the Sars-CoV-2 virus is new, but post-viral syndromes are not". In fact, as the article details, there has been a great deal of previous work on neurological disorders following viral infections such as HIV, and, of course, the potential use of viral agents that attack the nervous system has long been a concern in relation to biological warfare. What is particularly interesting from our perspective here is that the authors also note that one of the possible routes to the damage to the nervous system resulting from Sars-CoV-2 and other viral infections, including the CNS, is through *autoimmune processes* in which the immune defence system turns to attack the body, and autoimmune diseases are also known to be the probable cause of diseases of the nervous system such as Myasthenia Gravis and Narcolepsy.

Another special issue of *Scientific American* on the *New Science of the Immune System* contained an article on "Autoimmune Disease by Numbers" by Bender *et al.* which suggested that while definitions are still not completely agreed, about 80 diseases are now known to be caused by autoimmunity problems.[28] A large table sets out the numbers of diseases related to 10 different body systems, for example, 12 different diseases are shown to be associated with the endocrine system. Those associated directly with the nervous system and their prevalence are listed in Box 4.2.

However, the increasing concerns about an autoimmune cause for long COVID come against a background of numerous publications before the pandemic arguing that autoimmune processes could be at the root of even a range of psychiatric disorders. As one such 2019 article by Jeppesen and Benros noted:[29]

> "The notion of immunological pathways playing a role in the etiology of a subset of psychiatric disorders has received increasing interest in the last decades. One of the findings that has spiked interest herein, is the apparent link between autoimmune diseases and psychiatric disorders..."

The article continued elaborating on the evidence, stating that:

> "...This is supported by genetic findings associating immune-related genetic markers with schizophrenia and clinical studies finding increased levels of inflammatory markers in patients with psychosis. Several large-scale epidemiological studies have found positive associations between autoimmune disease and psychosis..."

While the 2006 *Lemon-Relman Report* noted that consideration of dual use issues should focus on the host's immune defence system, this point has not

Box 4.2 Autoimmune diseases of the nervous system and their prevalence.[28]

A. *Between about 1 in 1000 and about 1 in 10 000*
Multiple Sclerosis
Narcolepsy
Myasthenia Gravis

B. *Between about 1 in 10 000 and about 1 in 100 000*
Guillain-Barre syndrome
Hashimoto's encephalopathy

C. *Between about 1 in 100 000 and about 1 in a Million*
Acute disseminated encephalomyelitis

D. *Less than about 1 in a Million*
PANDAS
Miller Fisher syndrome
Autoimmune inner ear disease
Vogt-Koyanagi-Harada syndrome
Morvan's syndrome
Stiff-person syndrome
Issacs' syndrome/acquired neuromyotonia
Rasmussen's encephalitis

been given much focus in the discussion of the threat in subsequent decades and certainly not in the context of provoking autoimmune processes that affect higher cognitive functions. However, as life scientists continue to unravel more and more aspects of infections causing autoimmune processes affecting CNS functions it clearly needs careful consideration now. To investigate this further we need to understand the basics of immunological mechanisms.

4.3.2 Basic Immunology

While human and animal physiology has many different specialist fields of research, it is important not to lose sight of the fact that, for both humans and animals, individual bodies are one integrated system. Therefore, while we study the nervous system in neuroscience and the hormonal system in endocrinology, these systems are intimately linked and both affect and are affected by the other through their respective chemical communication systems. In addition, what is now becoming clear is that the immune system is also intimately linked to these systems in a similar way through its chemical communication system. Clearly, it is necessary to think about the integrity of human and animal organisms in a holistic way.

No organism is going to last long if it does not have defence systems against invading organisms and therefore it is no surprise to find, for instance, that more and more evidence is piling up to demonstrate that *innate immune* defences against viruses are pervasive through the animal kingdom.[30] However, the system in the vertebrates is more complex than in the invertebrates and has an additional multicomponent *adaptive immune* system that is not present in this form in invertebrates. So, we need to briefly discuss both the innate and the adaptive systems.

The innate system has physical barriers to invading organisms such as the skin, mucus and saliva, but if an invader enters the body the innate system reacts first. A 2019 review by Pavon *et al.* summarised this as follows:[31]

> "...The principal components of innate immunity are the physical and chemical barriers, as well as phagocytic cells, with constitutive receptors [Pattern Recognition Receptors PPRs] that recognize the PAMPs [pathogen-associated molecular patterns], release soluble molecules, and are in charge of orchestrating inflammatory reactions".

The adaptive immune responses that arose later in evolution are mediated by various lymphocytes – immune cells made in the bone marrow and found in the blood and in lymph tissue.[††] This part of the system can recognise different microbes and molecules with very high specificity and has the

[††] There are three main categories of lymphocytes in the vertebrate immune system: innate lymphoid cells involved in mucosal immunity and homeostasis; T lymphocyte cells involved in cell-mediated and cytotoxic adaptive immunity; and B lymphocyte cells that are involved in antibody-driven adaptive immunity.

Table 4.1 Some characteristics of the innate and adaptive immune systems.[31]

Characteristic	Innate	Adaptive
Specificity	Recognizes general patterns	Specific patterns recognised
Diversity	Limited range recognised	An enormous range recognised
Receptors	PRRs from germline	T and BCRs from somatic recombinations[a]
Memory	No	Yes
Response	Fast but limited	Slower but more potent
Other components	Physical and chemical barriers	Antibodies

[a] TCRs (T cell receptors), BCRs (B cell receptors).

capability to recall such reactions and thus increase its response to repeated invasions. The main components of the adaptive immune system are "cell membrane bound T cell receptors (TCR) and B cell receptors (BCR) and soluble molecules released by them such as cytokines[‡‡] and antibodies". And the main functions of this adaptive system "are to eliminate the intra- and extracellular foreign microorganisms that innate immunity cannot eliminate, and to boost innate mechanisms such as phagocytosis". Some of the main characteristics of the innate and adaptive immune systems are described and compared in Table 4.1.

So, while the detailed mechanisms by which the immune defence system operates are very complex, what we can see from Table 4.1 is that it has these two very different components: the fast but limited innate component and the slower but much more sophisticated adaptive component that can recognise a much greater diversity of threats and remember the characteristics of these to better deal with them in future. That, of course, is the reason that we vaccinate ourselves against major threats like COVID so that we will be better protected by our adaptive immune system if we are infected by the virus.

Early in the last century, experiments demonstrated that foreign tissues and tumours transplanted into the body that did not survive in the periphery did survive in the brain. This led to the view that while the body outside the brain had these immune defences, the brain was "immune privileged" and itself could not elicit immune responses.[32] This viewpoint has been increasingly overturned, particularly over the last few decades[33] and "[a] contemporary view considers the immune system as acting in an integrated way with other physiological systems, especially the nervous system, in the sense of maintaining homeostasis or body balance", and that "cytokines and leukocytes are indeed major effectors of the immune-to-brain communication, and this cross-talk is established by three independent

[‡‡] Substances, such as interferon or interleukin which are secreted by cells of the immune system and have an effect on other cells.

pathways: the humoral, neural and cellular (leukocyte) routes". In short and in contradiction to the earlier concept of "immune privileged", the CNS is intimately connected with the immune system. Moreover, it has also become clear that the brain has its own resident immune cells.[34]

We are all familiar with these complex interactions when we exhibit sickness behaviour following an infection. Such depression of behaviour is thought to help us overcome the infection and has been described as shown by:[35]

> "...physiological (*e.g.*, fever, alterations in blood composition), behavioral (*e.g.*, decreased locomotion, food and water intake), and hormonal (*e.g.*, secretion of hormones from the hypothalamic-pituitary-adrenal [HPA] axis) changes..."

The mechanisms involved in sickness behaviour have been shown to include changes in CNS monoamine levels as "[B]oth experimental studies and clinical trials have demonstrated that the decrease in synaptic availability of serotonin was highly correlated with the development of the cytokine-induced depressive symptoms".

Although it is not necessary to go into great detail here, it is important to give an idea of the complexity of the adaptive immune system and how it can malfunction. The major histocompatibility complex (MHC) locus[§§] in our DNA contains some 200 genes that code for highly varied MHC proteins found on and within cells that help the immune system recognise foreign substances (*i.e.*, antigens). MHC molecules are highly variable and can attach to peptides from diverse pathogens and present them on the cell surface. This allows the pathogens to be recognised by the T cells and destroyed. Specific T cells recognise and destroy particular pathogens, and they can produce copies of themselves to increase the defence against this pathogen and remember the pathogen's characteristics. Clearly, this could go wrong if material that is not from a foreign body is attached to the MHC molecule and marked for destruction. The ability of the immune system to avoid reactions against itself is termed (central and peripheral) *tolerance*:[36]

> "...*Central tolerance* largely occurs in the primary lymphoid organs (thymus and bone marrow) and is a process *whereby lymphocytes that strongly recognise self-antigens presented by MHC molecules are removed before they enter circulation....* If self-reactive T cell do leave the primary lymphoid organs, they can be further controlled *by peripheral tolerance mechanisms through which mature Self-reactive T cell are rendered functionally inactive (anergic)...*" (Emphases added)

If this system works well all is in order, but it can go wrong in a variety of different ways. A simple illustration of such a malfunction was described in a

[§§] Known in humans as the human leukocyte antigen (HLA) system.

series of stages as follows.[37] While a T cell receptor is usually said to be specific for one antigen it may not be quite that precise and thus also able to recognise closely related molecules. So, stage one is that a pathogen infects the body and it has an antigen very similar to a self-antigen found in cells in the body. Stage two is that T cells that connect to both the pathogen's antigen and the self-antigen are activated. Stage three is that the T cells begin to kill both the cells infected by the pathogen and healthy cells. B cells are then activated to produce autoantibodies that connect to healthy cells and mark them for death in stage four and in stage five the activated B cells and T cells turn into memory cells and the autoimmune response becomes a chronic autoimmune disease.

4.3.3 Some History of Weapons Research During the Cold War

Although the understanding of the immune system was limited during the early Cold War, some of the potential chemical weapons agents explored by certain States certainly affected the immune system. For example, in the early 1970s the Stockholm International Peace Research Institute (SIPRI) produced its original series of studies on *The Problem of Chemical and Biological Warfare*. Volume II of the series dealt with *CB Weapons Today* and its contents give a good idea of what was known at that stage after the early decades of the Cold War. The section of the study on incapacitating agents is of particular interest here, as some of the agents researched clearly manipulated the immune system.[38] The objective of the work was described as "to find substances capable of reducing military effectiveness for lengthy periods without endangering life or causing permanent injury, and to do so at dosages comparable with the effective doses of existing CW [chemical weapons] agents". And the SIPRI authors added "[I]n the early days of the US incapacitating-agent programme, there seemed to be many mechanisms of incapacitation which new CW agents might be able to exploit", and the text proceeded to set some of the mechanisms out:

> "...Hypotension was one such mechanism, for at that time the drug firms were starting to have impressive success in finding new therapeutics for hypertension. In healthy people, fainting is an early effect of suddenly lowered blood pressure, particularly if the subject is standing up...."

Emesis was another example:

> "...for a retching and vomiting soldier would not be an effective one. Besides the harassing vomiting agents such as adamsite, a great many emetics are known...."

And:

"The disturbance of body temperature was a third example.... A raised temperature can lead to incapacitating heat stroke and heat exhaustion. Milligram.... Atropine...can produce heat exhaustion by interfering with heat loss. Several bacterial endotoxins are amazingly potent fever-inducers in man..."

Similarly:

"...A lowered body temperature can lead to prolonged sedation or unconsciousness, and a variety of drugs are known...that can induce 'artificial hibernation.'"

The list ends by stating that further examples:

"...included inhibition of labyrinthine reflexes resulting in loss of the sense of balance; muscular hypotonia, leading to paralysis, temporary blindness, uncontrollable muscle tremors...and many psychotropic effects..."

It was known at the time that some of these effects – such as the impact of bacterial endotoxins like Staphylococcal Enterotoxin B – would certainly have affected the immune system, but the details of the mechanism of action were not known. Moreover, while this agent was incapacitating at low doses, it could certainly be lethal at higher doses.

Later during the closing stages of the Cold War, in the late 1980s and early 1990s, Russian scientists involved in a bioregulator weapons programme constructed a weapon system deliberately mimicking the autoimmune disease Multiple Sclerosis. They successfully infected mammals with a pathogen engineered to produce myelin basic protein. The unusually large amounts of myelin then produced by the bacterium triggered an immune response that attacked the essential myelin coating of nerve fibres and rapidly killed the animals infected.[39] A weapon used in that way would be lethal rather than an incapacitant, but the study of such autoimmune disease, as with the study of highly specific toxins[40] and other natural diseases[41] could reveal neuronal networks and receptors that could be used in various forms of incapacitating agents. For as Perry Robinson[42] noted "if a hitherto unknown molecular pathway serving a process of life comes to be identified, chemical agents capable of interfering with that pathway might also become identifiable and then form the basis for a novel weapon". It is not unreasonable therefore to ask what could a weapons programme look like today given the capabilities and knowledge available three decades later?[43]

4.3.4 Knowledge of Autoimmune Disease Mechanisms Today

We can get an idea of the state of understanding of neuroimmune diseases today by briefly reviewing what is known about the three most prevalent diseases: myasthenia gravis, multiple sclerosis and narcolepsy.

4.3.4.1 Myasthenia Gravis

Myasthenia gravis (MG) is caused by an autoimmune disruption of the acetylcholine receptors at the junction between the motor neurons and the muscles of the body, *i.e.*, it affects the peripheral rather than the central nervous system. This disruption prevents the normal triggering of the muscle contraction by the nerve impulse. The first description of the impact of this disease was given by Thomas Willis in 1672, but it was not until 1960 that it was understood to be caused by an autoimmune reaction.[44] A recent review by Deymeer described the history of attempts to understand and treat the disease in four stages, as shown in Box 4.3.

Box 4.3 Stages in the history of investigations of myasthenia gravis (MG).[44]

Stage 1. 1868–1930

"...all the clinical characteristics of MG were defined.... studies on the transmission at the neuromuscular junction were initiated.... A toxic agent was believed to be the cause of MG..."

Stage 2. 1930–1960

"...acetylcholine was discovered to be the transmitter at the neuromuscular junction.... The autoimmune nature of MG was suspected.... The hallmark of this period was the use of anti-cholinesterases... in the treatment of MG."

Stage 3. 1960–1990

"...a revolutionary era for MG. Important immunological mechanisms (acetylcholine receptor isolation, discovery of anti-acetylcholine receptor antibodies) were clarified and the autoimmune nature of MG was demonstrated..."

Stage 4. 1990–2020

"...more immunological progress .... New drugs emerged and clinical trial performed. Valuable guidelines were published."

MG is a distressing disease that is characterised by weakness in the muscles of the eye, face, and swallowing. It can cause double vision, drooping eyelids and trouble talking and walking. However, it can be treated, for example, with acetylcholinesterase inhibitors that reduce the removal of the acetylcholine transmitter, and with immunosuppressants that reduce the autoimmunity process. The disease is a longterm condition and millions of people are diagnosed with it each year, but with good treatment it is possible for them to lead normal lives and have a normal life expectancy.[45]

4.3.4.2 *Multiple Sclerosis*

It is important to note that multiple sclerosis (MS), unlike MG, affects the central rather than the peripheral nervous system. MS was first described by Charcot in Paris in 1868. It is:[46]

> "...typically manifest by recurrent acute onset of neurological difficulties reflecting damage to multiple areas of the brain and spinal cord, defined clinically as 'attacks' or 'relapses.' Symptoms associated with these events typically remit, but subsequent relapses occur unpredictably.... It is this *dissemination in time and space* that is characteristic of multiple sclerosis and its principal diagnostic feature..." (Original emphasis)

MS is the most common non-traumatic disabling disease in young adults and is found in both developed and developing countries in increasing numbers. It is a complex disease with many genes partly involved and in combinations leading to increased incidence, and a number of clear environmental factors such as infection with Epstein–Barr virus increasing the possibility of MS. The characteristic[47] "pathological hallmark of MS is perivenular inflammatory lesions, leading to demyelinating plaques". The consequent destruction of the myelin sheaths around the neuronal axons leads in turn to loss of function of the nerve fibre.

The pathological process is complex as the damage can occur in different parts of the CNS and the course of the disease can vary greatly. It has been extensively investigated in experimental autoimmune encephalomyelitis (experimentally caused inflammation of the brain) induced in animal models by peripheral immunization with myelin protein components.[48] It is thought that in a genetically susceptible person the disease involves "a breach of self-tolerance toward myelin and other CNS antigens resulting in persistent peripheral activation of autoreactive T cells" triggered, for example, by a viral infection. Then the myelin-reactive T cells can migrate across the blood-brain barrier where they trigger:

> "...an inflammatory cascade leading to release of cytokines and chemokines, recruitment of additional inflammatory cells including T cells, monocytes, and B cells and persistent activation of microglia and macrophages resulting in myelin damage..."

Given this understanding of the autoimmune process there has been significant progress in recent decades in the use of new immunomodulation drugs to slow down the process by broad suppression of the immune response.

4.3.4.3 Narcolepsy

Like multiple sclerosis, narcolepsy affects the central nervous system, but unlike MS the autoimmune malfunction does not affect the nervous system in many different and variable places. In contrast, the autoimmune damage is focussed on one particular set of neurons that produce all of the orexin neurotransmitter within the CNS and the loss of almost all of these neurons has a striking impact – excessive daytime sleepiness and cataplexy – that is now known as Type 1 narcolepsy. As we have shown, in the 1960s scientists discovered that there were small groups of neurons in the lower parts of the mammalian brain that had axons that ramified extensively in the higher reaches of the brain.[49] These groups of neurons, which are thought to have broad neuromodulatory[50] actions as well as functioning as synaptic neurotransmitters, were well described for dopamine, noradrenaline and serotonin at that time. But it was not until just before the turn of the last century that the set of neurons that produce orexin, and had similar extensive ramifications, was discovered. It was soon found that the activity of these neurons was involved in the control of numerous systems and behaviours, including the sleep-wake cycle.

Recently some of the scientists involved in this research provided a comprehensive review[51] of what has been discovered so far about the role of these neurons and the orexin they produce in narcolepsy (Box 4.4). The authors explained that cataplexy – sudden weakness of our voluntary muscles – can have other causes and so can excessive daytime sleepiness, but the two in conjunction are characteristic of this particular disease. It is a rare disease with an estimated incidence of between 25 and 50 per 100 000 people in Europe, Japan and the US.

It is now well known that sleep takes place in two different forms, rapid eye movement (REM) sleep and non-rapid eye movement (NREM) sleep and

Box 4.4 Sections of the review of narcolepsy.[51]

1. Introduction
2. Epidemiology
3. Genetic and non-genetic associations with narcolepsy
4. Inflammatory response
5. Narcolepsy and neuromediator systems
6. Clinical features
7. Pathophysiology and links with other diseases
8. Diagnosis and treatment
9. Conclusions

that we repeatedly move through several rounds of the cycle each night when we sleep.[52] During wakefulness orexin neurons are very active, in NREM sleep they are not very active and during REM sleep they are rarely active at all. Most dreaming takes place in REM sleep periods and the voluntary muscles are quiescent during these periods for obvious protective reasons. As the review authors explain, it takes unaffected people about 90 minutes to enter REM sleep, but "[I]n contrast, narcolepsy patients can enter REM sleep within 15 minutes of starting their sleep cycle during the day. REM sleep causes day-dreaming and muscular paralysis in this way".[53] While there is a clear association between certain genetic human leukocyte antigen (HLA) characteristics of the immune system, it is also apparent that environmental triggers are also involved in the onset of Type 1 narcolepsy, and that the disease involves the progressive destruction of almost all of the orexin producing cells in the CNS.

In regard to the inflammatory mechanism that causes the destruction of the orexin cells, the authors were cautious in stressing that there is much more to be discovered, but added:

> "...The removal of hypocretin neurons is very selective, sparing the melanin-concentrating hormone-releasing neurons found in the lateral hypothalamus, supporting the popular autoimmune hypothesis for narcolepsy disease pathogenesis. *The strong link between the HLADQB1*06:02 allele and polymorphisms in other immune-related genes and the immunological-autoimmune response in narcolepsy patients with cataplexy support the autoimmune hypothesis...*" [54] (Emphasis added)

Fortunately, it is possible to assist people suffering from Type 1 narcolepsy to some extent, but their lives are still difficult and considerable effort is being made to find better treatments. What is of particular interest here is the mechanism by which cataplexy is triggered in this form of narcolepsy.

A 2023 review by Pintwala and Peever focused on the brain circuits underlying narcolepsy.[55] It makes clear that the orexin system must be very important as it is conserved in mammals and therefore it can be studied in model organisms such as dogs and mice. While there are obviously some species differences in the details, it has proved possible to use mice, for example, to construct reasonable models of narcolepsy through the use of modern methods such as genetic engineering and optogenetics. There is good evidence to show that the loss of the orexin neurons that play a role in arousal and maintaining the wake state destabilises the sleep-wake cycle and is the probable cause of the disruption of the REM and NREM periods, prevalent in narcolepsy. However, there are many other neuron and transmitter types involved in the sleep-wake cycle and there is much more to be discovered in this regard. In contrast, the causes of cataplexy are already clearer.

Pintwala and Peever argue that the evidence demonstrates that there is a REM sleep atonia circuit that causes loss of muscle tone needed to protect the body from damage during REM sleep. The circuit is located in the sublaterodorsal tegmental nucleus (SLD) and the ventromedial medulla

(VMM). It consists of excitatory neurons in the SLD that activate inhibitory neurons in the VMM which silence motor neurons during REM sleep, and consequently cause loss of muscle tone. In the normal individual the neurons in the SLD are inhibited first by wake active neurons including the orexin group during wakefulness and therefore the motor neurons are not inhibited. This inhibition of the SLD neurons is substantially increased secondly by an indirect activation of the serotonin neurons of the dorsal raphe (DR) by orexin neurons. These neurons of the DR normally inhibit neurons in the amygdala that inhibit some of the non-orexin neurons of the wake cycle which themselves also inhibit the SLD neurons. What then happens when the orexin neurons are lost is a two-fold loss of inhibition of the SLD neurons in the sleep atonia circuit, thus allowing the neurons of the VMM to inhibit the motor neurons to the muscles and consequently producing loss of muscle tone.

What is important from our perspective is that the neuronal circuit underlying narcolepsy is being progressively uncovered and thus more and more potential targets are being exposed for benign manipulation. This process will continue apace in order to provide better help to the unfortunate people who have Type 1 narcolepsy.[56] However our increasing knowledge of these same neuronal circuits could also be malignly manipulated, potentially facilitating the development of CNS-acting chemical agent weapons disrupting the sleep-wake cycles of those targeted potentially resulting in artificially induced cataplexy and narcolepsy. We turn now to the question of whether the same applies to the unravelling of the mechanisms underlying higher cognitive functions and thus opening up the potential for manipulation of such functions that previously interested the Cold War weapons developers.

4.4 Immunopsychiatry Today

It has become generally recognised that the nervous, endocrine and immune systems are interconnected, and this broad acceptance is exemplified in the use of the term 'Psychoneuroimmunology.' However, the term 'Immunopsychiatry' that has recently been used in the title of books published by Oxford University Press[57] and Cambridge University Press[58] suggests a somewhat different emphasis. As Pariante, the author of the Preface to the Oxford University Press publication, noted:[59]

> "...The term *immunopsychiatry* highlights a new biological and evolutionary framework where our behaviours and emotions are governed by peripheral immune mechanisms, rather than the other way around..." (Original Emphasis)

And the author reiterated this key point by later noting that "central to today's immunopsychiatry concept" is the view that "immune abnormalities *are the cause*, rather than the consequence of psychiatric disorders" (emphasis added). Thus, what is being suggested by this terminology is that a new and very different mechanistic route by which psychiatric behaviours can be caused. Moreover, as the contents' list of these two books indicates such mechanisms could be the basis of many psychiatric disorders.

4.4.1 Autoimmune Encephalitis

One group of such immune-linked disorders is known as autoimmune encephalitis. These diseases are divided into two different types: those in which the autoimmune attack is against intracellular antigens and those that attack the neuronal surface or the synaptic space. A recent review by Zand indicated the very different prospects for the two types:[60]

> "...antibodies which target intracellular antigens are usually associated with paraneoplastic [cancer] syndromes and a poor prognosis.... [In contrast] Antibodies to neuronal surface/synaptic antigens can also be associated with cancer; however, they are more responsive to immunotherapy..."

In this second group the review also stated that:

> "...anti-N-methyl-D-aspartate (NMDA) receptor encephalitis and anti-leucine-rich-glioma-inactivated 1 (LG11) compromise 85% of patients. Anti-NMDA receptor encephalitis has become one of the most frequently recognized autoimmune encephalitides since its discovery..."

The author pointed out that this new information is important because these diseases often strike children and young adults but also often may respond well to early diagnosis and treatment.

4.4.1.1 Anti-NMDA Receptor Encephalitis

This second group of diseases that involve the targeting of synaptic receptors and neuronal cell surface proteins was the subject of a major article by Dalmau *et al.* in *Physiological Reviews* in 2017.[61] At that time the authors stated that there were 16 such diseases identified and that in these diseases:

> "...Studies show that patients' antibodies can impair the surface dynamics of the target receptors eliminating them from synapses (*e.g.*, NMDA receptor), block the function of the antigens without changing their synaptic density (*e.g.*, GABAb receptor), interfere with synaptic protein-protein interactions (LGI1, Caspr2), alter synapse formation (*e.g.*, neurexin-3a), or by unclear mechanisms associate to a new form of tauopathy (IgLON5) ..."

The article set out to trace the recent discovery of these diseases, then to describe the triggers and symptoms of each disease with a special focus on those that had been modelled in experiments with animals.

The authors called all of these 16 diseases autoimmune encephalopathies (AE) and suggested that they were discovered, for example, when patients with unusual symptoms improved after the use of immunotherapy and further investigation led to the discovery of autoantibodies against neuronal cell proteins. For instance, they described a study which:

> "... identified four young women with prominent neuropsychiatric symptoms and neuronal antibodies that produced a characteristic pattern of immunostaining of the neuropil of rat brain and intense immunolabeling of the cell surface of live neurons..."

All four patients had ovarian cancers and three recovered after immunotherapy. Importantly, the authors noted that: *"the target antigen was eventually characterized as the GluN1 subunit of the NMDA receptor, and the disorder is now known as anti-NMDA receptor encephalitis"* (emphasis added). The approach of identifying groups of patients with similar symptoms, demonstration of specific neuronal antibodies and then characterising these antibodies and the associated antigens, then led to the recognition of the other AE diseases known at the time of the review.

It might seem odd that an ovarian cancer could produce an assault on the nervous system, but as the authors noted that in:

> "...ovarian teratoma (which frequently associates with anti-NMDA receptor encephalitis), *the tumor itself contains mature or immature neural tissue, sometimes difficult to differentiate microscopically from normal brain tissue.* As in brain, these neurons contain NMDA receptors that likely contribute in triggering the autoimmune response..." (Emphasis added)

However, anti-NMDA receptor encephalitis can also follow a viral infection, for example as they explain:

> "The discovery that 20% of patients with herpes simplex virus encephalitis (HSE) develop antibodies against cell surface neuronal proteins (mainly NMDA receptor) demonstrates a link between a CNS viral infection and brain autoimmunity..."

And this, they argue, can explain the well-known relapsing neurological symptoms that can occur a few weeks after successful treatment of HSE infection. The review then goes on to detail the mechanism by which the receptor is affected.

When a doctor sees a patient with this disease the cause may not be clear but the resulting impact, that was described in a 2020 report as shown in Box 4.5, is striking.[62]

These behavioural changes are produced by alterations in the functions of the central nervous system with the NMDA receptor (NMDAr) being naturally activated by the major excitatory neurotransmitter glutamate and the

Box 4.5 Behavioural change in NMDA receptor encephalitis.[61]

Initial Stages

> "...Initial stages of this NMDAr [NMDA Receptor] dysfunction can result in loss of executive function and paranoid behaviors manifested as neuropsychiatric symptoms that are often new in onset for the patient, which should trigger clinical concern ..."

Later Stages

> "...If this autoimmune NMDAr antagonism goes untreated or unrecognized, patients often go on to develop more significant cognitive impairment, agitation, and seizures..."

Potential Outcomes

> "...High levels of antagonism manifest as motor stereotypies (dyskinesia, oculogyric crisis, chorea, dystonia), psychosis, mutism, and alteration in consciousness, which are manifestation of pathological frontostriatal syndromes."

receptor having a complex structure with variable components. As the authors point out these effects are similar in outcome to the use of the drugs ketamine and phencyclidine. They state that:

> "In humans, the best studied effects of NMDAr dysfunction and excitotoxicity exist in use of ketamine and phencyclidine (PCP). Both NMDAr antagonists result in reduced NMDAr activity similar to NMDAr antibodies. *However, the antagonists act directly on the receptors to antagonize NMDAr function whereas the pathological antibodies decrease the actual number of receptors on the cell membrane...*" (Emphasis added)

So here we have an impact of an immune dysfunction that has a similar impact to a drug that affects the CNS – although the exact mechanism at the synapse is different. In their 2017 review specifically of anti-NMDA receptor encephalitis Dalmau and his colleagues noted[63] that in cases of such encephalitis *"advanced MRI techniques.... showed a frequently reduced functional connectivity of the hippocampi with the anterior default mode network that correlated with [reduced] memory performance"* (emphasis added). As the hippocampi have a crucial role in memory functions this finding would not have been unexpected.

A 2022 paper by Andrzejak *et al.*, extended these findings by looking at the probable mechanisms involved. It argued that the evidence suggested that in the hippocampus the effect of the loss of the NMDA receptors was to decrease the synaptic activity. As they note:[64]

> "With regard to NMDAR encephalitis, most studies have focused on how autoantibodies cause memory deficits, in particular within hippocampal circuits. A framework for such investigations is based on a long-history of research showing that *NMDARs are ligand-gated ion channels responsible for synaptic integration and plasticity and underlie hippocampal learning and memory processes* ..." (Emphasis added)

Then in the disease condition:

> "...Antibody binding primarily triggers the reversible internalization of NMDARs and thereby a net decrease in surface receptor clusters in hippocampal neurons. Loss of receptors has been shown to decrease synaptic NMDAR-mediated currents and to compromise synaptic plasticity..."

So, the impact of the immune dysfunction here is thought to be direct in reducing the activity of neurons involved in learning processes.

However, they were particularly interested in the impact of the disease on the operations of the *higher cortex* and discovered that what was happening there was rather different. In their work they studied the impact of:

> "...a patient-derived monoclonal antibody targeting the NR1 subunit of NMDAR and...its effects on *in vitro* cultures of rodent cortical neurons, using imaging and electrophysiological techniques...."

So, they were examining in detail what happened in an experimental culture of mammalian cortical neurons when antibodies derived from a person with the disease were used to target the function of the cells. They reported that the:

> "...antibody drives cortical networks to a hyperexcitable state and disrupts mechanisms stabilizing network activity.... Network hyperactivity is in part a result of a reduced synaptic output of inhibitory neurons, as indicated by a decreased inhibitory drive and levels of presynaptic inhibitory proteins, specifically in inhibitory-to-excitatory neuron synapses..."

And they add the following observation:

> "...Importantly, on a single-cell level [the] antibody selectively impairs NMDAR-mediated currents and synaptic transmission of cortical inhibitory neurons, yet has no effect on excitatory neurons, which contrasts with its effects on hippocampal neurons..."

In short, as they stress, the cortex neurons and networks are affected in a very different way from what had previously been discovered in the hippocampal neurons. Moreover, their Significance Statement on the research ends as follows:

> "...*Notably, similar mechanism of NMDAR-mediated inhibitory hypofunction leading to cortical disinhibition has been suggested to underlie pathology of schizophrenia, hence our data provide new evidence for common mechanisms underlying neuropsychiatric disorders...*" (Emphasis added)

So, what we see here is not only the recent discovery of a novel set of disease mechanisms, but also the very rapid determination of the neuronal mechanisms involved in the disruption of the normal behaviour of human beings.

Such findings might leave some to believe that the forecast made by Adolphs in 2015 that the causes of psychiatric and neurological illness might be solved within 50 years[65] as needing to be revised and the date to be brought forward. Certainly, further discoveries in regard to these types of diseases are to be expected. For example, a recent opinion piece by Weickert and Weickert in the journal *Neuropsychopharmacology* argued that the reason drug trials for the treatment of schizophrenia often failed was because this disease is heterogeneous having different underlying biological mechanisms. Thus, even if a drug does work in some cases this effect will be lost in the failure to help in many other cases in the trial. Therefore, progress can only be made when patients can be sorted into those with similar underlying causes by some biomarker of significance. In the view of this, the authors proposed a new approach to treating schizophrenia:[66]

> "A proof of concept for this approach uses peripheral blood or cerebral spinal fluid antibody screening in people with schizophrenia. If anti *N*-methyl-D-aspartate receptor encephalitis is confirmed in patients presenting with psychotic symptoms, they can be successfully treated with plasma electrophoresis, intravenous immunoglobulins, and steroids..."

The authors claim that this treatment approach of "screening for the biomarkers related to underlying pathological mechanisms has been proven to work" and argue that it should be followed in the future. However, it is likely that the role of the immune system will vary in different diseases[67] from a "single-cause direct" (as in anti-NMDA receptor encephalitis) to a "multifactorial and indirect" (much more complex) effect across an "immunopsychiatric continuum". It is therefore necessary to examine the likelihood of much more complex cases being successfully investigated.

4.4.2 Autoimmune Aspects of Neurodegenerative Diseases

Psychiatrists are certainly interested in the autoimmune aspects of diseases where the role of immunity is more complex. This is obvious from the

Box 4.6 Some chapters of immunopsychiatry: a clinician's introduction to the immune basis of mental disorders.[57]

Chapter 10. Immune System Dysregulation and Schizophrenia
Chapter 11. Depression as a Neuroinflammatory Condition
Chapter 12. Immunology of Bipolar Disorder
Chapter 13. Immunology of Post-Traumatic Stress Disorder (PTSD)
Chapter 14. Immunology of Eating Disorders
Chapter 15. Immunology of Late-Life Psychiatric Disorders and Dementias

chapters of the book *Immunopsychiatry: A Clinician's Introduction to the Immune Basis of Mental Disorders* (Box 4.6).

Here we will examine one of the major degenerative diseases as an example of this complex end of the immunopsychiatric continuum.

4.4.2.1 Parkinson's Disease

It is important to understand why much research is devoted to trying to find ways of treating mental illnesses. As a 2017 review by De Haan *et al.* of *Autoimmune Aspects of Neurodegenerative and Psychiatric Diseases* argued:[68]

> "Neurodegenerative and psychiatric diseases (NPDs) are today's most important group of diseases, surpassing both atherosclerotic cardiovascular disease and cancer in morbidity incidence. Although NPDs have a dramatic impact on our society because of their high incidence, mortality, and severe debilitating character, remarkably few effective interventions have become available..."

Moreover, the reviewers add that the long-term use of medications available often has severe adverse side effects. Therefore, they suggest there is an urgent need to find new safe and effective treatments for these diseases. As people lead longer lives, it appears very likely that this search is going to continue apace in order to help reduce these personal and societal costs.

For the purpose of their review, they made the assumption that:

> "...NPD symptoms basically are manifestations of a chronic self-sustaining inflammatory process with detrimental consequences for the patient. Specific inhibition of the destructive immune responses in the brain, leaving the patient's immune system intact, would be the ultimate solution to cure patients from the disease...."

As they noted "[T]o reach this goal, the primary targets, *e.g.*, the primary self-antigens (pSAgs) of the patient's chronic (auto)immune response, need to be identified". They argue that for a few major NPDs the primary self-antigens have been identified (as we have just discussed), but much further research will be needed to discover and define the pSAgs for many other diseases.

The difficulties involved in such research are clear for their view of the potential reasons for the occurrence of these diseases:

> "...Autoimmune diseases are the result of an interplay between environmental stimuli, such as diet, life style, trauma, exposure to microbes, pathogens, or toxic compounds (smoking), and the possession of predisposing gene alleles, which ultimately at a certain moment in time "trigger" a sustained loss of self-tolerance resulting in an immune-mediated damage of autologous tissues..."

However, they felt able to give a list of "Candidate pSAgs of NPDs" set out according to the primary neurotransmitter system and part of the brain affected. Within this list they give a set of diseases that affect the dopamine system which notably includes Parkinson's disease. In this disease they state that the affected parts of the brain are the substantia nigra compactum and ventral tegmentum and the pSAg is the vesicular monoamine transporter-2 (VMAT2).

They note that Parkinson's disease is the most common neurodegenerative disease at younger ages and that it produces characteristic, very well-known movement and cognitive defects. Moreover, these difficulties coincide with chronic inflammation and destruction of the dopamine neurons in the substantia nigra compactum and ventral tegmentum. They describe how these effects can be produced in mice through immunization with dopamine producing neurons or injection with a VMAT2 antagonist which leads to a build-up of high levels of neurotoxic dopamine in the cytoplasm of the neurons and then an immune response and the consequent development of the symptoms similar to Parkinson's disease. Furthermore, mice in which the VMAT2 has been genetically removed (VMAT2-knockout mice) display the symptoms of the disease. Such evidence, they suggest, makes VMAT2 a candidate primary self-antigen (pSAg) for Parkinson's disease.

As we have discussed previously, the dopamine producing neurons of the lower brain and the ascending axonal output tracts to the higher parts of the brain were described in the 1960s and can be summarised as follows:[69]

> "The ascending Dopamine system can be divided into three pathways: The nigrostriatal pathway originates in the substantia nigra cell cluster and innervates the caudate-putamen (striatum). The mesolimbic pathway originates in the ventral tegmental area cell cluster and innervates various limbic structures such as the nucleus accumbens.... The mesocortical pathway also originates in the ventral tegmental area and innervates the cerebral cortex."

Disruption of this system can have impacts on the control of movements through disruption of the nigrostriatal pathway and much else through disruption of the other two pathways. The dopamine neurotransmitter in these neurons is held in vesicles and transporters like VMAT2 which are critically involved in the storage and release of the dopamine from these vesicles. They are therefore of much interest in the study and treatment of many diseases.[70] However, at the present time "it is still unclear if the reduction in VMAT2 expression contributes to or causes PD [Parkinson's disease]". Indeed, the immune system's role in Parkinson's disease is complex and may vary in different subtypes of the disease that will have to be elucidated in further research.[71]

However, novel methods are being applied to investigate Parkinson's disease in more detail in model organisms such as mice. For example, by using modern genetic methods it has been possible to show that the dopamine cluster of neurons in CNS is not homogeneous.[72] And then it has been possible to track the functions of the different genetically typed neurons. One recent report concluded that such work provides new directions for research on Parkinson's disease:[73]

> "...because there is emerging evidence that several of the subtypes studies here [in the model organism] exist in humans. The cell body locations and axonal projections of Aldh1a1^{+} [neuronal subtype] match the pattern of dopamine loss in Parkinson's disease, and these neurons are especially vulnerable in Parkinson's disease..."

Thus, they state that the:

> "... motor deficits may not be due to an absolute dopamine deficiency, but rather, to a loss of dopamine signalling from specific pro-motor subtypes such as Anaxa1^{+} neurons."

Such findings consequently open up the possibility of specific studies of the causes and consequences of the destruction of these neurons.

4.5 The Wider Context of Scientific Advances in Neuroscience

The specific issues discussed in this chapter also need to be considered in the wider context of the revolutionary transformation that is taking place in neuroscience and associated sciences and technologies. An illustration of this change can be seen in the pages of the journal *Neuron*. This journal celebrated its 30th year of publication in 2018 and during that year it provided space for a number of scientists to write short essays on their ideas of what could happen over the next 30 years. All of the

Box 4.7 Some examples of potential neuroscience advances.[74]

Neuroscience: Past and Future

Bridging between Mind and Body
Neurobiology: A Quantitative Science

Visions for the Future of Neuroscience

A Basis for Mental Processes
The Fast-Moving Field of Neural Interfaces
CNS Repair: Look Within
Stress and Psychopathologies

Setting the Stage for the Next Generation of Neuroscience

Understanding Social Behaviour
Cell Biology in the Circuit Age
Neural Dynamics across Timescales
Neuronal Assemblies and Memory
Decoding States and Contexts
Nanoscale Neuroscience

Progress and Future Goals for Neuroscience

Black Box and Cell Assemblies
Human Functional Imaging: Past and Future
Keys to Unlocking Psychiatric Disorders
Bioengineering Advances
Transcriptomes Illuminating Synaptic Network Plasticity

contributions were very positive about future progress in their widely varying different fields of research (Box 4.7), but one of the founding editors of the journal, Lily Jan, put her finger on a different point of great interest here when she noted that[74] "[T]hanks to a wide range of technological breakthroughs, the future of neuroscience is certainly bright. *What will transpire in future neuroscience research is tantalizingly unpredictable*". (Emphasis added).

One example of these perceptions of the future of neuroscience concerned the mammalian ascending neuromodulatory system[75] of cells such as the dopamine neurons just discussed in relation to Parkinson's disease and the orexin producing neurons previously discussed in relation to narcolepsy. The essay contrasted the amazing technical research advances about neuronal circuit functions being reported in the journal with the difficulty of translating the information gained into practical therapeutic medicine.

Another author, Polley, suggested that perhaps a way forward was right in front of our noses stating that:[76]

> "...The small, neuro-chemically diverse clusters of long-range neuromodulatory nuclei buried deep in the brainstem, midbrain, and basal forebrain have evolved the means to modify processing and plasticity in local circuits with a precision that easily surpasses the latest CNS devices and pharmaceutical interventions..."

Yet, the author argues that we know little about how these systems are activated and do not "have satisfactory taxonomies to understand the differences in form, spatial specificity, and temporal endurances of the local circuit changes imparted by such systems", which we would need to translate the research work into medically useful impacts.

This is an instance of a general problem that we have in understanding how the different parts of the brain are integrated across different levels despite our detailed investigations of some of the elements of the system. As noted by Watanabe and colleagues in a recent editorial in *Frontiers in Integrative Neuroscience*:[77]

> "...A multitude of studies, including omics studies and spatial RNASeq are being conducted at the level of components to whole system architecture and begin to characterize the nervous system in certain key aspects..."

But:

> "...There remains a great need to improve our understanding of the nervous system, particularly how different levels of biological processes are interconnected and specifically interacted to generate various brain functions. A better understanding of regional and multi-scale nervous system cooperative or interactive functions would contribute to the development of novel medicines and other related technologies..."

Thus, it might be argued that we are a long way from being able to understand and manipulate these long range neuromodulatory systems. We have a good conceptual apparatus to think about neurons, synapses and neuronal networks, and we understand how these networks function in some cases. We also have some understanding of how such networks can be reset, for example by orexin in the sleep/wake cycle. Nevertheless, we seem to be lacking a concept that could give us an overall view of the role these modulatory neuronal systems play in behavioural change.

However, in the current context it is as well to expect the unexpected. Thus, a news article in *Nature* in late 2023 noted that a novel perspective was developing:[78]

> "The idea that the nervous system passes messages from one nerve cell to another only through synapses – the points at which the cells link up end to end – is changing. Two studies show how messages can pass between cells over longer distances, through a 'wireless' nerve network..."

The work was carried out on the famous nematode worm model organism *Caenorhabditis elegans* where the synaptic connections of all 302 neurons – the connectome – are known. The first of these papers by Ripoll-Sanchez *et al.* described all of the connections between neurons that they could find between neuropeptide producing neurons and their receptors found in other neurons that were anatomically plausible:[79]

> "...we considered two neurons (nodes) to be connected if the first neuron expresses a particular neuropeptide ligand and the second expresses a paired receptor, subject to special constraints on signalling..."

If two neurons produced and had receptors for the same neuropeptide but were never reasonably near in space they would not be considered as connected. The results of the investigation were surprising. As the authors noted:

> "Collectively, the connections mediated by neuropeptide signaling showed only modest overlap with those mediated by synapses. For example, 1,522 *neuron pairs are connected by both chemical synapses and peptidergic interactions; this represents only 5% and 4%, respectively, of total neurons with inferred short- or mid-range peptidergic interactions.*" (Emphasis added)

Significantly, they added:

> "...*A striking consequence of this extensive extrasynaptic peptidergic signaling, for the mid-range network in particular, was its integration of neurons disconnected from the synaptic connectome...*" (Emphasis added)

It should also be noted that in the view of these authors "neuropeptide signaling in nematodes shows surprising conservation and similar diversity to neuropeptide signaling in the human brain, despite vast differences in neuron number and anatomical complexity".

The second paper took the argument a stage further by using optogenetics [the use of light to activate specific neurons] to activate all of the neuron pairs in *C. elegans* and viewing the results using whole-brain calcium imaging.[80] As the results did not fit with what would be expected on the basis of the synaptic connectome (direct connections) they constructed a mutant that enabled them to show that extra-synaptic signalling contributed to the difference. In short there was important 'wireless' signalling by neuronally derived neuropeptides helping to drive the worm's behaviour. So, there are two different, but connected connectomes. As the editors of a set of papers on neuromodulatory systems concluded, such:[81]

> "...Studies demonstrate how different levels of biological activity influence neurophysiological outcomes including pathological phenomena, revealing how properties at one level influence the property of another level..."

And they sensibly suggest that:

> "...Future studies with complementary computational reconstitution of these observed influences in outcome by intervention in other levels may further help our understanding of how the nervous systems work in different levels hand in hand from molecular to cognitive levels..."

And therefore that: "[T]his advance may bring us further versatile discoveries, such as a new type of effective combinatory therapeutic intervention targeting the interconnecting building blocks of several biological levels". Of course, they were thinking of benign outcomes, but there are clearly other possibilities as the detailed mechanisms causing normal and abnormal behaviour are further clarified in future research.

4.6 Conclusions

Benignly-intended research is being carried out in many different fields and aspects of neuroscience. In this chapter we have just explored two fields of work where past research on weapons systems would suggest that there could be the possibility of future misuse. The more general point is that it seems reasonable to assume that much future neuroscience research will continue to reveal increasingly detailed information about the nature, function and actions of neuronal pathways (synaptic and non-synaptic), neurons and receptors that will be helpful in advancing medical treatments for diseases of the CNS. However, this same research may also potentially open up development pathways for novel CNS-acting chemical weapons to malignly manipulate the higher CNS directly or indirectly through their interactions with the endocrine or immune systems. These possibilities are compounded by related developments in converging sciences and technologies, particularly in AI and machine learning which is the subject of the next chapter.

References

1. J. P. Perry Robinson, Bringing the CBW conventions closer together, *CBW Conv. Bull.*, 2008, **80**, 1–4.

2. J. P. Perry Robinson, Difficulties facing the chemical weapons convention, *Int. Aff.*, 2008, **84**(2), 223–239.
3. M. R. Dando, *The Chemical and Biological Nonproliferation Regime after the Covid-19 Pandemic: Dealing with the Scientific Revolution in the Life Sciences*, Springer/Nature, Cham, Switzerland, 2023, See Chapter 3: Neuroscience Developments of Concern, pages 29 to 56, particularly Figures 3.1 The new threat spectrum and Figure 3.2 A plethora of possibilities.
4. D. Nutt and B. Moss, *Psychedelics: The revolutionary drugs that could change your life – a guide from the expert*, Hodder and Stoughton, London, 2023.
5. D. Nutt and B. Moss, *Psychedelics: The revolutionary drugs that could change your life – a guide from the expert*, Hodder and Stoughton, London, 2023, p. 12.
6. S. E. Petersen and O. Sporns, Brain networks and cognitive architectures, *Neuron*, 2015, **88**(1), 207–219.
7. L. Q. Uddin, *et al.*, Towards a universal taxonomy of macroscale functional human brain networks, *Brain Topogr.*, 2019, **32**(6), 926–942.
8. Large-scale brain networks, Wikipedia. Accessed 19, October, 2023.
9. M. E. Raichle, The brain's default mode network, *Annu. Rev. Neurosci.*, 2015, **38**, 433–447.
10. J. J. Gattuso, *et al.*, Default mode network modulation by psychedelics: a systematic review, *Int. J. Neuropsychopharmacol.*, 2023, **26**, 155–188.
11. J. J. Gattuso, *et al.*, Default mode network modulation by psychedelics: a systematic review, *Int. J. Neuropsychopharmacol.*, 2023, **26**, 157.
12. J. J. Gattuso, *et al.*, Default mode network modulation by psychedelics: a systematic review, *Int. J. Neuropsychopharmacol.*, 2023, **26**, 157.
13. D. Nutt and B. Moss, *Psychedelics: The revolutionary drugs that could change your life – a guide from the expert*, Hodder and Stoughton, London, 2023, p. 79.
14. J. J. Gattuso, *et al.*, Default mode network modulation by psychedelics: a systematic review, *Int. J. Neuropsychopharmacol.*, 2023, **26**, 185.
15. L. Johansen, *et al.*, The psychological process of classic psychedelics in the treatment of depression: a systematic review protocol, *Syst. Rev.*, 2022, **11**, 85.
16. See, for example (a) R. L. Carhart- Harris, *et al.*, Neural correlated of the LSD experience revealed by multimodal neuroimaging, *Proc. Natl. Acad. Sci. U. S. A.*, 2016, **113**(17), 4853–4858. (b) R. L. Carhart- Harris, *et al.*, Neural correlates of the psychedelic state as determined by fMRI studies with psilocybin, *Proc. Natl. Acad. Sci. U. S. A.*, 2012, **109**(6), 2138–2143. (c) C. Timmermann, *et al.*, Human brain effects of DMT assessed via EEG-fMRi, *Proc. Natl. Acad. Sci. U. S. A.*, 2022, **120**(13) DOI: 10.1073/pnas.2218949120.
17. J. Bacque-Cazenave, *et al.*, Serotonin in animal cognition and behaviour, *Int. J. Mol. Sci.*, 2020, **21**, 1649.

18. J. S. Meyer and L. F. Quenzer, *Psychopharmacology: Drugs, the Brain and Behaviour*, Sinauer, Sunderland, 2013, pp. 167–184, Mass. Chapter 6 Serotonin, See Figure 6.6 Anatomy of the serotonergic system, p. 174.
19. D. Nutt, *Brain and Mind made Simple*, Waterside Press, Hook, Hampshire, 2021, pp. 35–44, See Chapter 2 Neurotransmitters and their Receptors – The Locks and Keys of Brain Function. Serotonin – The many-faceted neurotransmitter.
20. F. X. Vollenweider and J. W. Smallridge, Classic psychedelic drugs: update on biological mechanisms, *Pharmacopsychiatry*, 2022, **55**, 121–138.
21. J. S. Meyer and L. F. Quenzer, *Psychopharmacology: Drugs, the Brain and Behaviour*, Sinauer, Sunderland, 2013, pp. 440, Mass. Chapter 6 Serotonin, See Figure 6.6 Anatomy of the serotonergic system, p. 174.
22. C. E. Canal, Serotonergic psychedelics: experimental approaches for assessing mechanisms of action, in *Handbook of Experimental Pharmacology*, 2018, vol. 252, pp. 227–260.
23. See M. R. Dando, Table 1.6 Stages in DREADDs technology, in *Neuroscience and the Problem of Dual Use: Neuroethics in the New Brain Research Projects*, Springer/ Nature, Cham, Switzerland, 2020, p. 11.
24. D. S. Grayson, *et al.*, The rhesus monkey connectome predicts disrupted functional networks resulting from pharmacogenetic inactivation of the amygdala, *Neuron*, 2016, **91**(2), 453–466.
25. B. J. He, Integrating Consciousness Science with Cognitive Neuroscience: An introduction to the Special Focus, *J. Cogn. Neurosci.*, 2024, **36**, 1541–1545.
26. S. Sutherland and S. Shafer, The brain and long COVID, *Sci. Am.*, 2023, **2023**, 74–81, Special Edition.
27. S. Sutherland and S. Shafer, The brain and long COVID, *Sci. Am.*, 2023, **2023**, 74, Special Edition.
28. M. Bender, *et al.*, Autoimmune disease, by numbers, pp. 98–101 in new science of the immune system: how the body fights disease and ways we can boost our defenses, *Sci. Am.*, 2022, 100–101, Special Collector's Edition, Spring.
29. R. Jeppesen and M. E. Benros, Autoimmune diseases and psychotic disorders, *Front. Mol. Psychiatry*, 2019, **10**, 131.
30. R. E. Iwama and Y. Moran, Origins and diversification of animal innate immune responses against viral infections, *Nat Ecol Evol*, 2023, 7, 182–193.
31. L. Pavon *et al.*, Overview of the immune system, in *Immunopsychiatry: A Clinician's Introduction to the Immune Basis of Mental Disorders*, ed. A. L. Teixeira and M. E. Bauer, Oxford University Press, Oxford, 2019, ch. 1, pp. 1–20.
32. G. P. Owens, A neuroprimer; principles of central nervous system immunity, *Semin. Pediatr. Neurol.*, 2017, DOI: 10.1016/j.spen.2017.08.004.
33. M. E. Bauer *et al.*, Immune mechanisms affecting the functioning of the central nervous system (CNS), in *Immunopsychiatry: A Clinician's*

Introduction to the Immune Basis of Mental Disorders, ed. A. L. Teixeira and M. E. Bauer, Oxford University Press, Oxford, 2019, ch. 3, pp. 47–64.
34. D. Kwon, Feature: Guardians of the brain, *Nature*, 2022, **606**, 22–24.
35. M. E. Bauer *et al.*, Immune mechanisms affecting the functioning of the central nervous system (CNS), in *Immunopsychiatry: A Clinician's Introduction to the Immune Basis of Mental Disorders*, ed. A. L. Teixeira and M. E. Bauer, Oxford University Press, Oxford, 2019, ch. 3, pp. 49–51.
36. L. Turner and N. Harrison, Basic concepts in immunobiology, in *Immunopsychiatry*, ed. N. Harrison, *et al.*, Cambridge University Press, Cambridge, 2021, ch. 1, pp. 1–24.
37. P. Dettmer, *Immune: A Journey into the Mysterious System That Keeps You Alive*, Hodder and Stoughton, London, 2021, pp. 262–268, See Chapter 40: Autoimmune Disease. pp. 264–266.
38. SIPRI, *The Problem of Chemical and Biological Warfare: Volume II CB Weapons Today*, Almqvist and Wiksell, Stockholm, 1973, pp. 298–300, This book has been made freely available on the internet.
39. M. Crowley and M. R. Dando, *Toxin and Bioregulator Weapons: Preventing the Misuse of the Chemical and Life Sciences*, Springer Nature, Cham, Switzerland, 2022, pp. 111–112, See Section 6.3 Soviet Bioregulator Weapon Research and Development.
40. M. Crowley and M. R. Dando, *Toxin and Bioregulator Weapons: Preventing the Misuse of the Chemical and Life Sciences*, Springer Nature, Cham, Switzerland, 2022, pp. 31–41, See Section 2.3 Soviet Bioregulator Weapons Research and Development, pp. 31–41.
41. (a) S. C. Parija, Neuroparisitology – a new horizon, *Trop. Parasitol.*, 2023, **13**(2), 71–72; (b) F. Libersat, *et al.*, Mind control: how parasites manipulate cognitive functions in their insect hosts, *Front. Psychol.*, 2018, **9**, 572; (c) C. Arnold, Sick as a dog: how understanding canine diseases will save human lives, *Nat. Med.*, 2022, **28**, 1970–1973.
42. J. P. Perry Robinson, Difficulties facing the chemical weapons convention, *Int. Aff.*, 2008, **84**(2), 223–239.
43. C. Davis, Prepared statement, in *Addressing the Biological Weapons Threat: Russia and Beyond. Subcommittee on Europe, Eurasia and Emerging Threats. One Hundred Thirteenth Congress, Second Session, May 7*, Government Printing Office, Washington, D. C., 2014, pp. 31–40, See the eighth questions in the series of 11 questions that Dr Davis suggested that it was not unreasonable for a concerned citizen to ask.
44. F. Deymeer, History of myasthenia gravis revisited, *Arch Neuropsychiatry*, 2021, **58**, 154–162.
45. J. A. Morren and L. Yuebing, Myasthenia gravis: Frequently asked questions, *Cleveland Clin. J. Med.*, 2023, **90**(2), 103–113.
46. A. A. Lizarraga and W. A. Sheremata, Multiple sclerosis: clinical features, immunopathogenesis, and treatment, in *Inflammatory Disorders of the Nervous System*, ed. A. Minagar and J. S. Alexander, Humana Press, Springer Nature, Cham, Switzerland, 2017, ch. 2, pp. 25–76.

47. R. Dobson and G. Giovannoni, Multiple Sclerosis – a review, *Eur. J. Neurol.*, 2019, **26**, 27.
48. N. Garg and T. W. Smith, An update on immunopathogenesis, diagnosis, and treatment of multiple sclerosis, *Brain Behav.*, 2015, **5**(9), e00361.
49. See, for example J. S. Meyer and L. F. Quenzer, *Psychopharmacology: drugs, the brain, and behavior*, Sinauer, Sunderland, 2nd edn, 2013, pp. 167–184, Mass. Chapters 5 Catecholamines (Dopamine and Noradrenaline) pages 143–166, and Chapter 6 Serotonin.
50. For current thinking on how a "wireless" neuropeptide connectome probably operated alongside the directly wired synaptic connectome in nervous systems see L. Ripoll-Sanchez, *et al.*, The neuropeptide connectome of C. elegans, *Neuron*, 2023, **111**, 3570–3589.
51. V. Chavda, *et al.*, Narcolepsy-a neuropathological obscure sleep disorder: a narrative review of current literature, *Brain Sci.*, 2022, **12**, 1473.
52. M. R. Dando, *Neuroscience and the Problem of Dual Use: Neuroethics in the New Brain Projects*, Springer Nature, Cham, Switzerland, 2020, See Section 1.3 Sleep Research, pp. 7–13 and 6.3.2 Brain Research on Circuits: Sleep/Wake pp. 107–108. pp. 2 and 5.
53. V. Chavda, *et al.*, Narcolepsy-a neuropathological obscure sleep disorder: a narrative review of current literature, *Brain Sci.*, 2022, **12**, 2.
54. V. Chavda, *et al.*, Narcolepsy-a neuropathological obscure sleep disorder: a narrative review of current literature, *Brain Sci.*, 2022, **12**, 5.
55. S. C. Pintwala and J. Peever, Brain circuits underlying narcolepsy, *Neuroscientist*, 2023, **29**(6), 751–766, See Figure 4 The REM sleep atonia circuit and cataplexy, p. 760.
56. H. Ito, *et al.*, Deficiency of orexin signalling during sleep is involved in abnormal REM sleep architecture in narcolepsy, *Proc. Natl. Acad. Sci. U. S. A.*, 2023, **120**(41), e2301951120.
57. In *Immunopsychiatry: A Clinician's Introduction to the Immune Basis of Mental Disorders*, ed. A. L. Teixeira and M. E. Bauer, Oxford University Press, Oxford, 2019.
58. In *Textbook of Immunopsychiatry*, ed. G. Khandaker *et al.*, Cambridge University Press, Cambridge, 2021.
59. C. M. Pariante, From Psychoneuroimmunology to Immunopsychiatry: A Brief History. Preface, pages vii – ix, in *Immunopsychiatry: A Clinician's Introduction to the Immune Basis of Mental Disorders*, ed. A. L. Teixeira and M. E. Bauer, Oxford University Press, Oxford, 2019, pp. vii–viii.
60. R. Zand, Autoimmune encephalitis: clinical features, pathophysiology, and treatment, in *Inflammatory Disorders of the Nervous System: Pathogenesis, Immunology, and Clinical Management*, ed. A. Minagar and J. S. Alexander, Humana Press, Springer Nature, Cham, Switzerland, 2nd edn, 2017, ch. 8, pp. 175–186.
61. J. Dalmau, *et al.*, Autoantibodies to synaptic receptors in neuronal cell surface proteins in autoimmune diseases of the central nervous system, *Physiol. Rev.*, 2017, **97**(20), 839–887.

62. B. J. Kolls, *et al.*, Autoimmune encephalitis: NMDA receptor encephalitis as an example of translational neuroscience, *Neurotherapeutics*, 2020, **17**, 404–413.
63. J. Dalmau, *et al.*, Autoantibodies to synaptic receptors in neuronal cell surface proteins in autoimmune diseases of the central nervous system, *Physiol. Rev.*, 2017, **97**(20), 848.
64. E. Andrzejak, *et al.*, Patient-derived anti-NMDAR antibody disinhibits cortical neuronal networks through dysfunction of inhibitory neuron output, *J. Neurosci.*, 2022, **42**(15), 3253–3270.
65. R. Adolphs, The unsolved problems of neuroscience, *Trends Cogn. Sci*, 2015, **19**(4), 173–175.
66. T. W. Weickert and C. S. Weickert, A path to novel effective treatments to schizophrenia, *Neuropsychopharmacology*, 2023, **49**, 321–322.
67. L. J. De Picker, The future of immunopsychiatry: Three milestones to clinical innovation, *Brain, Behav., Immun.:Health*, 2021, **16**, 100314, See Figure 3 on p. 4.
68. P. De Haan, *et al.*, Autoimmune aspects of neurodegenerative and psychiatric diseases: a template for innovative therapy, *Front. Mol. Psychiatry*, 2017, **8**, 46 pp. 1, 2, 4 (Table 1), 5.
69. J. S. Meyer and L. F. Quenzer, *Psychopharmacology: Drugs, the Brain and Behaviour*, Sinauer, Sunderland, 2013, pp. 150, Mass. Chapter 6 Serotonin, See Figure 5.8, p. 150.
70. A. Alwindi and A. Bizanti, Vesicular monoamine transporter (VMAT) regional expression and role in pathological conditions, *Heliyon*, 2023, **9**, e22413.
71. J. Lauritens and M. Romero-Ramos, The systemic immune response in Parkinson's disease: focus on the peripheral immune component, *Trends Neurosci.*, 2023, **48**(10), 863–878.
72. Z. Gaertner, *et al.*, Molecular heterogeneity in the substantia nigra: A roadmap for understanding PD motor pathophysiology, *Neurobiol. Dis.*, 2022, **175**, 105925.
73. M. Azcorra, *et al.*, Unique functional responses differentially map onto genetic subtypes of dopamine neurons, *Nat. Neurosci.*, 2023, **26**, 1762–1774.
74. (a) L. Jan, *et al.*, Neuroscience: Past and future, *Neuron*, 2018, **98**(1), 10–11. (b) For Table 4.8 see also T. Moore *et al.*, Visions for the Future of Neuroscience, *Neuron*, 2018, **98**(3), 464–465. (c) W. Hong *et al.*, Setting the Stage for the Next Generation of Neuroscience, *Neuron*, 2018, **99**(1), 11–12. (d) M-B Moser *et al.*, Progress and Future Goals for Neuroscience, *Neuron*, 2018, **100**(1), 14–15.
75. G. Gambino, *et al.*, Editorial: neuromodulatory ascending systems: Their influence at the microscopic and macroscopic levels, *Front. Neural Circuits*, 2022, DOI: 10.3389/fncir.2022.1028154.
76. D. Polley, CNS repair: look within, *Neuron*, 2018, **98**, 464.

77. Y. Watanabe, *et al.*, Editorial: Bridging the gap: implementing building blocks of the nervous system to simulate neuronal functions at different levels, *Front. Integr. Neurosci.*, 2023, 1–2.
78. C. Lopez Lloreda, Wi-Fi for neurons: first map of wireless nerve signals unveiled, *Nature*, 2023, **623**, 894–895.
79. L. Ripoll-Sanchez, *et al.*, The neuropeptidergic connectome of *C. elegans*, *Neuron*, 2023, **111**, 3570–3589.
80. F. Randi, *et al.*, Neural signal propagation atlas of *Caenorhabditis elegans*, *Nature*, 2023, **623**, 406–413.
81. Y. Watanabe, *et al.*, Editorial: Bridging the gap: implementing building blocks of the nervous system to simulate neuronal functions at different levels, *Front. Integr. Neurosci.*, 2023, 2.

CHAPTER 5

Advances in Other Science and Technology Relevant to Hostile Misuse of CNS-acting Chemicals

5.1 Introduction

Chapters 3 and 4 explored the contemporary advances in neuroscience that could potentially be employed to understand, alter and disrupt circuits in the central nervous system (CNS) for benign or malign purposes. This chapter explores parallel advances in the chemical and life sciences and other relevant disciplines and associated technologies, notably artificial intelligence (AI) and nanotechnology, their growing interaction and convergences, and highlights the potential risk of their being utilised in the development of CNS-acting and other incapacitating chemical agent (ICA) weapons. This chapter will firstly explore how chemical and life science and associated technological advances beyond neuroscience can facilitate the discovery of novel naturally occurring CNS-acting agents or the synthetic development of such new agents and enhance their production.

The chapter will then explore how technological advances could be used to overcome the challenge of ensuring a controlled delivery of CNS-acting or other incapacitating chemical agents to the targets. It will explore State development, since the coming into force of the CWC, of CNS weapons delivering a limited quantity of agent to an individual or small group. It will go on to examine how advances in nanotechnology and encapsulation are being used in a medical context to facilitate delivery, uptake and subsequent movement of CNS-acting chemicals to specific target sites within the body, and how such technologies could be misused in CNS-acting chemical agent

Preventing Weaponization of CNS-acting Chemicals: A Holistic Arms Control Analysis
By Michael Crowley and Malcolm Dando

Published by the Royal Society of Chemistry, www.rsc.org

weaponization. The chapter will then explore how aerosol dispersal technologies and nanotechnology have facilitated more effective agent dissemination. The chapter will finally explore commercial development of "wide area" dual-use aersolisation and agent delivery systems promoted for agricultural and other civilian purposes before examining in detail the development and availability of an extensive range of delivery mechanisms primarily promoted for the "wide area" dispersal of riot control agents (RCAs) which could be utilised or adapted for delivering other toxic chemicals, potentially including weaponised CNS-acting chemicals or other types of ICAs.

5.2 Developments in CNS-acting Chemical Agent Discovery and Synthesis

5.2.1 Screening for Novel Chemical Agents

Trapp has previously argued that "[O]ver an entire century, chemical weapons development programmes have screened thousands of small molecules for their utility as chemical weapons – it is fair to assume that with regard to the 'traditional' types of CW agents, it is unlikely that research would lead to new types of lethal agents that surpass significantly what has already been achieved in the traditional CW programmes".[1] Despite such apparent diminishing returns, it appears that certain States have continued research into potential novel chemical weapons agents with lethal effects, though the intention behind such research is unclear (see below).

In addition, CBW experts have highlighted the continuing State interest in screening, investigating and potentially utilising novel toxic chemicals as 'less lethal' weapons in law enforcement. This has included discovery and investigation of novel CNS-acting chemicals – including pharmaceuticals, toxins and bioregulators. From our Chapter 2 survey of previous State CBW programmes exploring CNS-acting agents, such screening and investigations were resource intensive; initially very broad and scatter-gun in scope; often reliant on industry leads; frequently employed *in vivo* animal testing; sometimes included unethical, human testing; and were informed by then-limited knowledge of CNS regulatory processes. Consequently, the efficiency and effectiveness of these historical programmes was limited. As the Scientific Advisory Board (SAB) of the Organisation for the Prohibition of Chemical Weapons (OPWC) noted in its report to the 2nd CWC Review Conference: "[A]lthough many extremely potent compounds that act on the central nervous system were discovered in period from the 1950s to the 1970s, only two types, anticholinergics and opioids, appear to have been developed into chemical-warfare agents or non-lethal agents for use in law-enforcement".[2] In contrast contemporary State research programmes would potentially benefit from the vastly greater knowledge of the human CNS, its regulatory processes and the nature, roles and modes of action of CNS-acting chemicals in such processes. Such knowledge has been accompanied by advances in screening techniques.

5.2.1.1 *Contemporary State Screening and Investigation of Potential Toxin Weapons Effecting the CNS and Other Core Physiological Systems*

Military and non-military institutions and scientists in certain States have searched out an ever-increasing range of 'novel' toxins that might be weaponised. These include toxins derived, for example, from indigenous poisonous plants, amphibians, reptiles, scorpions and marine animals, some of which affect the human CNS or other core physiological processes. Sometimes the intent of the activities was clearly medical or for other 'protective purposes' permitted under the CWC and BTWC. In other examples the purpose was unstated or unclear. However, in certain cases the intention appears to be for toxin weaponization using a low technology route. In this regard the current authors found that scientists from the Indian National Defence Research and Development Organisation (DRDO) and related entities searched for, identified, and characterised toxins and other substances from a wide range of indigenous Indian stinging and other poisonous plants.[3] Thus, in 2018, DRDO scientists argued that because the BTWC "banned the use or stockpiling of most of the pathogenic bio-threat agents" this would "necessitate... search [for] some novel natural bio-threat agents from stinging plants that may be used as future bio-weapon for self-defence purposes". Consequently, they conducted research to "identify, characterize and screen the potential of... stinging plants on the basis of their secondary metabolite contents that may be used for the formulation of novel future bio-threat agents for self-defence".[4] This DRDO research continued by investigating plants that produce "toxins that have the ability to adversely affect human health in a variety of ways, ranging from relatively mild allergic reactions to serious medical complications, including death". The scientists reported that they had successfully identified "several poisonous plants that can be used for the development of novel multi-system targeted warfare agents for defensive applications". They also identified "poisonous components" from 10 possible candidate plants[5] and investigated their "mode of action", that could "have harmful effect on various biological systems like *nervous*, cardiac, digestive, respiratory, dermal, *etc.* simultaneously". (Emphasis added)[6] A number of the toxins investigated as potential agents are highly toxic and can be lethal; others while not as poisonous might still potentially be employed as 'less lethal' toxin weapons.

Furthermore, screening for novel toxins of dual use concern – but where the intention of such activities was either unclear or stated as being for prophylactic or other protective purposes – has been undertaken by military or military-related entities or researchers in additional States. Chinese military-related researchers have analysed novel toxins including those obtained from sea anemones, jellyfish, salpa, and shellfish,[7] leading the US State Department in 2024 to determine that Chinese "PLA's [Peoples Liberation Army's] research organizations have been conducting and directing military research related to dual-use marine toxins" without adequate Chinese reporting of such activities to BTWC States Parties.[8] Similarly Iranian scientists investigated novel toxins derived from scorpion and box jellyfish venom;[9] and Russian scientists explored novel toxins including those derived from poisonous frogs, scorpion venom and macromycete fungal species.[10]

Although certain contemporary military research into novel toxins that has come to light has employed basic *in vivo* screening techniques, increased *in vitro* screening capabilities that are now common place in pharmaceutical drug discovery could readily be adapted to existing State CBW research and development activities, to more effectively screen novel natural agents and also human synthesised chemicals.

In the last decades screening capabilities have been transformed by *in vitro* high-throughput screening (HTS) which is the use of automated equipment to rapidly test large numbers of samples for biological activity at the model organism, cellular, pathway, or molecular level. In its most common form, HTS is an experimental process in which between a thousand and a million small molecule compounds of known structure are screened. Other substances, such as chemical mixtures, natural product extracts, oligonucleotides, and antibodies, can also be screened. Because HTS typically aims to screen 100 000 or more samples per day, relatively simple and automation-compatible assay designs, robotic-assisted sample handling, and automated data processing are employed in the process. HTS is commonly used in pharmaceutical and biotechnology companies to identify compounds with pharmacological or biological activity. These are used as starting points for medicinal chemical optimization during pharmacological probe or drug discovery and development.[11] HTS is commonly employed in a parallel screening process.

In its 2008 report to the 2nd CWC Review Conference, the SAB warned of the dangers of such increased screening capability. "[N]ew biologically active molecules are being discovered at an unprecedented rate. For example, parallel synthesis and high-throughput screening are producing data on millions of compounds. The tools for such techniques are becoming widely available and could be selectively targeted at toxic molecules. Pharmaceutical and pesticide companies now hold huge databases of biological data that could be a source of prototypes for new chemical-warfare agents".[12] The SAB highlighted one area of such drug research of concern involving "compounds that could be developed legitimately within the constraints of the Convention as non-lethal agents for law-enforcement purposes". The SAB noted that the "accelerated discovery of drugs has resulted in the identification of many new compounds that act very selectively on the central nervous system, both regionally and with regard to receptor subtypes. It remains to be seen what challenge, if any, one or another of these compounds could pose to the Convention".[13]

In its 2012 report to the 3rd CWC Review Conference, the SAB argued that the vastly expanded identification of novel toxic chemicals arising particularly from pharmaceutical drug discovery programmes may not in reality translate into a consequent increase in weaponization risk:

> "There has been some concern that amongst the millions of compounds that have been screened in drug-discovery programmes there will be prototypes for new toxic chemicals. This may be the case, but nearly all of these data are generated using *in vitro* screening and may not reflect toxicity *in vivo*. Comparatively little testing is now undertaken in animals other than for lead compounds. This contrasts with the much smaller number (thousands) of compounds that were screened during the 1950s to the 1970s. For a large percentage of these compounds, toxicity testing was performed in small animal species."[14]

Consequently, the SAB concluded that the existence of large in-house databases in pharmaceutical companies and related institutions posed "no greater risk to the purposes of the Convention than the much smaller databases that existed before entry into force of the Convention". Furthermore, they noted that "all of the centrally acting chemicals known to have been seriously studied as potential incapacitants are products or extensions of 'traditional' drug research programmes of the 1950s to the 1970s".[15]

Whilst the 2012 SAB analysis was a useful corrective to potential over-estimation of the risks in this area at the time, the current authors are less sanguine in their interpretation of the contemporary threats which will in part be exacerbated by more efficient screening capabilities such as virtual high throughput screening, combined with powerful AI enabling greater capacity to identify potential chemical weapon agent types with specific desired characteristics based on the vastly increased knowledge of the CNS and other core human physiological systems.

Already in its 2023 report to the 5th CWC Review Conference the SAB highlighted the growing importance of virtual high-throughput screening which instead of physically testing each potential molecule using *in vitro* high-throughput screening assays, employs developments in machine learning to screen potential candidates from chemical compound libraries. Virtual HTS (vHTS) now sees regular use in the drug development process and allows pharmaceutical companies to screen millions of molecules for bioactivity, toxicity, and other specific characteristics in a fraction of the time needed for traditional methods. Such an application has already demonstrated commercial success, with numerous drugs identified *via* vHTS currently on the market or in clinical trials.[16]

Box 5.1 A short primer on artificial intelligence.

According to the International Standards Organisation (ISO) website artificial intelligence is defined as[17] "a technical and scientific field devoted to the engineered system that generates outputs such as content, forecasts, recommendations or decisions for a given set of human-defined objectives". However, the text quickly adds that this does not make much sense to non-specialists and that AI just "analyses data to extract patterns and make predictions". However, it adds that AI is changing rapidly and that it is moving towards capabilities in generative AI to create new data similar to the data it was trained upon.

AI uses neural networks to analyse large sets of data to recognise patterns in the data. Narrow AI "embodies systems meticulously crafted to excel at specific tasks within well-defined parameters". Generative AI systems, in contrast, "aim to possess a form of general intelligence, allowing them to adapt, learn and apply knowledge across various domains". These systems are in the early stages of development and the text notes that while powerful these newer systems have drawbacks such as the potential for creating convincing misinformation, raising concerns about copyrights and security threats. They also require substantial computing resources that have high costs and environmental impacts.

While the growth of AI has many benefits, it clearly also has potential implications, for example, in regard to chemical and biological threats and regulation of this technology and such social impacts are at a very early stage of development.[18] Currently, the rate of advance in this field is very fast. As the ISO website notes, since 2017 rapid advances in "computer vision, natural language processing, robotics, and autonomous systems have been driven by progress in deep learning and increased computing power". And more recently, since 2023, "the rise of large language models have demonstrated the potential of AI systems to generate human-like text, answer questions and assist in a wide range of tasks". This rapid rate of advance is expected to continue. One particularly important aspect of current developments is deep learning. As explained by Sarker in a 2021 comprehensive review:[19]

> "...In general, AI incorporates human behavior and intelligence to machines or systems, while ML [Machine Learning] is the method to learn from data or experience, which automates analytical model building. DL [Deep Learning] also represents learning methods from data where the computation is done through multi-layer neural networks and processing. The term 'Deep' in the deep learning methodology refers to the concept of multiple levels or stages through which data is processed for building a data-driven model."

The author notes that "[D]eep learning (DL), a branch of machine learning (ML) and artificial intelligence (AI) is nowadays considered as a core technology of today's Fourth Industrial Revolution..." As such it is certain to have increasingly wide-ranging implications.

Perhaps the most important points to be made about the advances in AI are that the new generative AI systems are enormously different from what has gone before, and that they directly affect the potential for misuse in chemical and biological agents. An illustration of this can be found in a recent review[20] of Chemical Language Models "which generate new molecules in the form of strings using deep learning". Chemists use diagrams of the chemical structure of molecules (*e.g.*, the famous benzene ring) but these can easily be translated into strings of symbols that can be used as input for computers. The author notes that chemical molecules can be considered as parts of a 'chemical language' and that:

> "...Like human language, chemical language possesses a *syntax*: finite elements (atoms, like words) can relate (form bonds) to one another only in specific ways to produce 'chemically valid' molecules." (Original Emphasis)

Moreover, the text continues:

> "Molecules also have *semantical properties*: based on which elements are present and how they are connected, different high-level properties (*e.g.*, physicochemical, biological) will emerge." (Original Emphasis)

So, it is possible to train an AI system to learn this language (although we do not necessarily understand how it does this). Thus, the author explains that Chemical Language Models:

> "...borrow and adapt algorithms developed for natural language processing to learning the chemical language. This is allowed by the usage of string notations, such as Simplified Molecular Input Line Entry Systems (SMILES) strings [the input data]."

Once the AI system has been trained on such input in order that it learns the language, it can be used to achieve a variety of objectives, for example, one "in which molecules are generated aiming to optimise one or more goals".

These systems have already been extremely successful (as we illustrate throughout this chapter). The author concludes that:

> "Deep learning models like CLMs [Cumulative Link Models] and beyond are expected to have an increasingly relevant role in drug discovery. Besides improving time- and cost-efficiency, deep learning will accelerate our capacity to explore uncharted regions in the chemical space, as well as to formulate and verify exciting new scientific hypotheses for drug discovery...."

And looking forward she suggests that:

"...In the future, joined forces among AI experts, chemists, and biologists will allow designing innovative algorithms imbued with scientific knowledge and gathering of new scientific insights into human biology driven by AI."

This could radically change the dimensions of the impact that we have already seen from current advances and could clearly be of relevance to novel CNS-acting chemicals.[21]

5.2.2 From Discovery to Rationale Engineering

Over 20 years ago, Petro and his colleagues raised concerns about the danger that a fundamental change in biological weapons development was then, or would, in the future, be taking place. They argued that "[E]merging biotechnologies likely will lead to a paradigm shift in BW [biological weapons] agent development; future biological agents could be *rationally engineered* to target specific human biological systems at the molecular level". (Emphasis added).[22] In this new paradigm, according to Petro *et al.*, instead of focusing only upon readily discoverable agents, and then either employing them directly or attempting to modify them, the biological weapons developer would increasingly be able to focus on the discovery of new biological targets in human physiological systems to attack with bespoke biological agents designed to achieve ever more specific effects. Petro and his colleagues' prescient warnings can be applied just as well to the development of bespoke chemical weapons through the rational engineering of an overlapping range of synthetic toxic chemical agents intended to attack, malignly alter or influence previously identified human physiological systems.

We argue that in the intervening 20 years, this paradigm shift has begun to take place, in part informed by our growing and deepening knowledge of the components, organisation and processes underpinning human physiological systems notably including the CNS as discussed in Chapters 3 and 4. This increased knowledge of the growing range of potential target processes can now be coupled with contemporaneous advances in rational design of novel agents, the establishment of pathways for their retro-synthesis and their subsequent manufacture including through the employment of systems biology and synthetic biology. These advances will also, of course, increase the effectiveness of the established methods of discovery and adaptation of naturally occurring toxic chemicals. Whilst advances in, and convergences of, multiple life science and associated disciplines have fuelled this shift, AI and nanotechnology in particular have had, and will likely continue to have, far reaching effects.

5.2.3 Systems Biology, AI and Rational Drug Design

Systems biology is an approach of combining experimental work with mathematical modelling and the use of information technology, so as to substantially increase the capabilities of researchers to predict the effect of chemicals on the physiological operations of living organisms. This combined approach could now be used to establish how the complex interactions within the human nervous system and linked immune and endocrine systems could be affected by both naturally occurring toxins and bioregulators and also rationally designed synthetic analogues, with attendant malign applications. The power of this approach has been significantly enhanced in recent years through the development and application of AI/machine learning technologies.

The ease with which current AI/machine learning systems normally used to generate new drugs for beneficial purposes could readily be employed in the *de novo* design of novel toxic chemicals for use as chemical weapons, was starkly highlighted by Urbani *et al.* in a proof of concept demonstration in 2021.[23] The authors are civil scientists who build "machine learning models for therapeutic and toxic targets to better assist in the design of new molecules for drug discovery" and normally use "machine learning model predictions of bioactivity for the purpose of finding new therapeutic inhibitors of targets for human disease". For the purpose of their study, however, instead of the computer model penalising toxicity, they altered it so that it rewarded both toxicity and bioactivity. To narrow the range of predictive molecules generated, they chose to drive the model towards compounds like the nerve agent VX. Within 6 hours over 40 000 virtual molecules were generated, and many of these were predicted to be more toxic than publicly known chemical warfare agents. Consequently, they warned that "[B]y inverting the use of our machine learning models, we had transformed our innocuous generative model from a helpful tool of medicine to a generator of likely deadly molecules".[24] They pointed out that while their experiment was focused on VX-like molecules, "it is equally applicable to other toxic small molecules with similar or different mechanisms, with minimal adjustments to our protocol".[25] They also warned that AI generative machine learning tools are equally applicable to larger molecules such as peptides and macrolactones *etc.* "[F]or us, the genie is out of the medicine bottle when it comes to repurposing our machine learning".[26]

However, toxicity is not the only criterion for past State selection of CW agents. The physicochemical properties of the agents are key to whether they can be disseminated to cause their intended effects (*e.g.*, thermal stability). The model therefore cannot yet substitute for a State chemical weapon testing programme.

In addition, recently developed AI modelling software, such as AlphaFold[27] RoseTTAFold,[28] or ESMFold,[29] can now increasingly be employed to predict the 3-D structure of existing amino-acid based molecules, with accuracy that is comparable to that of experimental methods, and consequently give an enhanced insight into such molecules' likely modes of activity. A freely available AlphaFold database now holds the predicted structure of around 220 million protein structures, *i.e.* nearly every protein from known organisms in DNA databases.[30] In addition ESMFold has been used to predict the structures of some 600 million proteins from bacteria, viruses and other microorganisms that haven't yet been characterized.[31] These tools can also be employed to predict structural information on novel amino acid based molecules as long as the amino acid sequence is known. In 2021 Anishchenko *et al.* used machine learning algorithms (or deep neural networks) called the trRosetta network to "dream up" or "hallucinate" new proteins after they had been fed a random sequence of amino acids. The researchers then obtained synthetic genes encoding 129 of the network-'hallucinated' sequences, and physically expressed and purified the proteins in *Escherichia coli;* 27 of the proteins yielded experimental spectra that were consistent with the hallucinated protein structures.[32] They concluded that "deep networks trained to predict *native* protein structures from their sequences can be inverted to design *new* proteins, and such networks and methods should contribute alongside traditional physics-based models to the *de novo design of proteins with new functions*". (Emphasis added).[33] CBW scholars, such as Krin and Jeremias, have highlighted the potential malign application of such modelling tools for the discovery and development of novel biological and chemical weapons.[34] The capabilities of these modelling tools are increasing at a very fast rate. In 2024, AlphaFold3 which employs a machine-learning network called a diffusion model provided scientists with the ability to predict the structures of proteins during interactions with other molecules, including the ability to physically model how well chemicals bind to proteins.[35] Once again whilst this will be beneficial to predict how potential candidate therapeutic drugs could interact with their targets in the patient's body, it would also be useful in exploring the actions of potential chemical weapon agents including CNS-acting agents upon their targets.

5.2.4 Identifying Retro-synthesis Pathways

Identifying a candidate drug (or potential novel chemical weapons agent) is only the start of the process leading to its production. The next step would be to employ retro-synthesis techniques to design the optimal combination of reactions required to synthesise this final product (or target compound). This is accomplished by working backwards from the target compound and using reactions to come up with multiple potential precursors, and back again to establish starting compounds and eventually the full reaction pathways. This is not a straightforward task, in particular for complex multi-step reactions. AI could greatly facilitate such processes, which could if linked with robotics become automated.[36] In addition to concerns over their use in the synthesis of novel chemical weapons agents, these techniques could also be employed to determine alternative pathways for synthesising existing chemical weapons agents that would evade currently scheduled or otherwise regulated precursors.[37]

At present, the key factors limiting AI performance in retrosynthesis are the currently available datasets that must be used for the AI model training. Although a large body of literature has been accumulated over the years on different reaction mechanisms and synthetic pathways which can be used to retrieve information on the synthesis and the components involved and to train the machine learning algorithms, this data is highly heterogeneous often incomplete, and sometimes contradictory.[38] Consequently, no study, has to date, reported a novel synthetic route provided entirely by AI and synthesised in the laboratory or industrial setting.[39] In contrast Urbani *et al.* warned that "With current breakthroughs and research into autonomous synthesis, a complete design-make-test cycle applicable to making not only drugs, but toxins, is within reach. Our proof-of-concept highlights how a non-human autonomous creator of a deadly chemical weapon is entirely feasible".[40]

The potential wide ranging application of AI to many stages in the synthesis of toxic organic chemicals was highlighted in the SAB's 2023 report for the 5th CWC Review Conference. "[M]achine learning is also being applied to organic chemistry, in synthesis route planning, and catalyst and reaction optimisation, with a number of companies employing *in silico* synthetic planning into their overall approach to accessing target molecules".[41]

5.2.5 Biotechnology Techniques Employed for Development and Synthesis of Novel Agents

Biotechnology is a broad term encompassing the application of biological components or processes to advance human purposes. Synthetic biology refers to a set of concepts, approaches, and tools within biotechnology that enable the modification or creation of biological organisms. A hallmark of synthetic biology is the use of concepts and approaches common to engineering disciplines. These can include standardization of components (*e.g.*, well-characterized functions encoded by DNA), the use of software and computational modelling for designing biological systems from those components, and the construction of prototypes based on those designs. Synthetic biologists frequently apply such approaches in iterative Design-Build-Test cycles to accelerate progress.[42]

The fields of synthetic biology and genetic engineering have been revolutionised by the development and proliferation of a range of genome-editing technologies notably CRISPR/Cas9 (or the 'genetic scissors'). Such technology enables precise edits to be made to the genetic material of virtually any living cell and allows those changes to be passed on to subsequent generations. Compared to previous gene editing techniques, CRISPR/Cas9 and further systems such as Cpf1 and C2c2 are simple, accessible and low cost, and have consequently seen rapid global distribution, accelerating the pace of work in gene editing.

Trapp also raises biology-oriented synthesis (BIOS) as a synthetic biology enabling tool worthy of analysis. BIOS uses core structures obtained from natural products as scaffolds of compound collections to access biological diversity around bioactive structures, or deconstructing natural products into fragments and recombining these fragments into biologically active composites.[43] Such genetic engineering and synthetic biology techniques are now being increasingly employed benignly in the development of novel bio-based products including therapeutics, antimicrobials, animal health products and genetically engineered crops, as well as for research in areas such as cell engineering, phenotypic screening and genomics, and industrial biotechnology (microbial fermentation and therapeutics).[44]

However, many observers have warned of the dual-use nature of such technologies.

For example, in its 2018 report *Biodefense in the Age of Synthetic Biology*, the US National Academy of Sciences (NAS) concluded that: "[S]ynthetic biology expands what is possible in creating new weapons. It also expands the range of actors who could undertake such efforts and decreases the time required .... Of the potential capabilities assessed, three currently warrant the most concern: (1) re-creating known pathogenic viruses, (2) making existing bacteria more dangerous, and (3) making harmful biochemicals *via in situ* synthesis..." With regard to "chemicals, biochemicals, and toxins" the report stated that "synthetic biology blurs the line between chemical and biological weapons. High-potency molecules that can be produced through simple genetic pathways are of greatest concern, because they could conceivably be developed with modest resources and organizational footprint".[45]

In its 2023 report to the 5th CWC Review Conference, the SAB stated that "[T]he key implication of modern genetic engineering techniques of relevance to the Convention [CWC] is the ability to transform cells into biofactories for the production of chemicals, either for peaceful or malicious uses".[46] The SAB highlighted the increased accessibility of such dual-use tools and techniques without adequate regulatory control noting that "the genetic sequences of highly pathogenic bacteria and viruses can now be downloaded freely from websites, such as GenBank, EMBL, and DDBJ". And further that "various viral, prokaryotic, and eukaryotic genomes can be synthesised, at low cost, using commercial services". They state that although "many informal and industry-specific controls are in place to screen orders and outputs to prevent dangerous sequences from being widely disseminated...", "few formal regulatory constraints exist to prevent these services from producing these sequences".[47] Consequently the SAB warned that: "[T]he technical barriers for the artificial design and chemical synthesis of dangerous bacteria or a viral genome, *toxin, or bioregulator chemical*, are falling quickly due to advances in synthetic biology technologies that have reduced the cost and increased accessibility". (Emphasis added). They specifically noted that "[S]pecial attention should be paid to the commercialisation of benchtop DNA synthesisers that could be misused to covertly print dangerous sequences that code for the bioproduction of toxic chemicals".[48]

In his *Statement for the Record* in 2016, the Director of National Intelligence highlighted genome editing in the US intelligence community's *Worldwide Threat Assessment* under risks associated with weapons of mass destruction arguing that:

> "Research in genome editing conducted by countries with different regulatory or ethical standards than those of Western countries

> probably increases the risk of the creation of potentially harmful biological agents or products. Given the broad distribution, low cost, and accelerated pace of development of this dual-use technology, its deliberate or unintentional misuse might lead to far-reaching economic and national security implications."[49]

The implications of such technological advances and their widening dissemination (without adequate regulation) were highlighted again by the US State Department in its 2024 report on *Adherence to and Compliance with Arms Control, Non-Proliferation, and Disarmament Agreements and Commitments*. Here the US asserted that North Korea has the "technical capability to produce bacteria, viruses, and toxins that could be used as BW [biological weapons] agents…[it] also has the capability to genetically engineer biological products with technologies such as CRISPR, which have been reported by [North Korea's] State Academy of Sciences and other sources".[50]

Such developments in techniques to manipulate the genetic code are rapidly advancing with corresponding potential for future weaponization. According to Trapp "the development of a new genetic code using non-standard amino acids…is on the horizon…thereby expanding the chemical repertoire available for protein synthesis. This could result in novel compounds with profound differences in structure, stability, activity, and binding as compared to the proteins known from nature".[51] Already, the recoding of genomes by completely replacing a natural codon by an artificial one has been accomplished at the laboratory level. These experiments have been conducted in order to "test the malleability of the genetic code, to expand the chemistry and functionality of proteins, and to search for new materials with properties different from those accessible by using the 20 natural amino acids". The resulting novel substances could have a broad range of benign potential applications including as nanostructures, therapeutics, industrial enzymes, and in drug delivery.[52] However, Trapp warns that "the very same techniques could be used to tailor the properties of designer proteins to the requirements of a novel CW agent that, for example, would selectively act on a specific receptor in the CNS".[53]

Similarly, in the forward-looking section of its 2018 synthetic biology report, the NAS warned:

> "It may be possible to use synthetic biology to *modulate human physiology* in novel ways. These ways include physiological changes that differ from the typical effects of known pathogens and chemical agents. Synthetic biology expands the landscape by potentially allowing the delivery of biochemicals by a biological agent and by potentially allowing the engineering of the microbiome or immune system. Although unlikely today, these types of manipulations may become more feasible as knowledge of complex systems, such as the immune system and microbiome, grows." (Emphasis as in original).[54]

It further concluded that:

> "Some malicious applications of synthetic biology may not seem plausible now but could become achievable if certain barriers are overcome. These barriers include knowledge barriers, as is the case for building a novel pathogen, or technological barriers, as in engineering complex biosynthetic pathways into bacteria or re-creating known bacterial pathogens..."[55]

Clearly such developments need to be carefully watched and regulation considered well in advance of the barriers being surmounted.

5.2.5.1 *Bio-synthesis and Largescale Production of Toxic Chemicals*

In addition to multi-faceted applications in design, construction and adaptation of a range of biologically active substances, advances in biotechnology have significantly increased the effectiveness and efficiency of large-scale production of such substances. Whilst these developments have been occurring in the context of the pharmaceutical and other related commercial industries, there are dual-use implications *i.e.*, that they may facilitate large-scale weaponization of CNS-acting agents and other incapacitating chemical agents.

Biological processes have become more widely used in the manufacturing of chemical products, from high-volume organic products such as alcohols and urea to high-value chemicals and pharmaceuticals. In its 2014 report, the SAB temporary working group on convergence noted that "bulk and fine chemicals are being produced increasingly using biologically mediated processes".[56] Its report highlighted three such processes: fermentation technology and metabolic engineering; enzymes/biocatalysts; and synthetic biology.

In its 2018 report to the 4th CWC Review Conference the SAB noted the changing nature of these industries:

> "The past few years have seen significant investment in the development of biotechnologies for chemical production. Their adoption for commodity chemical production (large volumes, low cost per-unit-volume) has slowed as oil prices have dropped (making traditional chemical production processes more economically viable). Biotechnologies are increasingly being used for producing specialty chemicals (low volumes, high cost per-unit-volume), where the diversity of chemical properties and function available through harnessing metabolic processes as production tools, are highly desirable (in chemical sectors that include pharmaceuticals, flavours, and fragrances). Trends in adoption of biotechnologies across different sectors of chemical production are expected to be dynamic..."[57]

In its 2018 report the SAB also assessed the current capability of large-scale production using biological and biologically-mediated processes of substances of concern:

"Industrial scale chemical production using fermentation processes requires selection of microbial strains capable of reproducibly producing desired chemicals at scales and sufficient quality to be economically viable. This can require significant efforts in strain engineering and selection that requires testing on various production scales.

Industrial processes involve informatics, automation and significant molecular biological characterisation. For these reasons, developing reliable fermentation-based processes with genetically modified organisms may not be as simple as often described in security discussions on biotechnology. However, for malicious purposes, scaled-up and high-quality production may not be needed to obtain material in sufficient quantity to violate the prohibitions of the Convention."[58]

The SAB further noted that:

"Genomic modification capabilities . . .have improved the activity, selectivity, solvent tolerance and general robustness of biocatalysts or enzymes for an increasing range of chemical reactions. In some cases, genes have been introduced into microorganisms to enable them to produce specific chemicals of interest. Synthetic biology is reducing the barrier to entry and time requirement for building customised biological machines. A wide variety of enabling technologies have been used. It is conceivable to use these approaches to synthesise toxic chemicals, toxins, bioregulators, or their precursors, in quantities that could challenge the Convention. . ."[59]

Furthermore, they stated that:

"While the technical capability to chemically synthesise many toxins exists today, there are practical limitations with regard to scale and complexity. The threat of possible misuse of this technology in relation to the Convention is therefore considered low at the present time.. . . Future developments affecting the cost, efficacy and availability of relevant enabling technologies may necessitate reviewing the likelihood of using these approaches to acquire toxins."[60]

In its 2023 report to the 5th CWC Review Conference the SAB highlighted the significant ongoing advances in this field:

"Enabled by innovative metabolic modelling and advances in platform -omics technologies, synthetic biology, and genome editing tools, the

> production of chemicals in microorganisms is no longer restricted to products arising from native metabolic potential, and microbial cell factories may produce value-added chemicals directly. The integration of these analytical developments has offered new knowledge and manifold strategies to unravel biological cellular networks, modify biological systems, and create synthetic bio-devices or biosynthesis platforms."[61]

The SAB recognised that genetic engineering of live cells *in vivo* was still the most widely utilised format for the bioproduction of specific proteins in small quantities, particularly for research purposes. However, they highlighted the "rapid modernisation" in recent years of cell-free (or *in vitro*) protein synthesis (CFPS) – that is the production of protein using biological machinery in a cell-free system without the use of living cells – which has "created many new opportunities for large-scale production of specific proteins, including those that cannot be found in nature". The SAB noted that recently developed CFPS systems boast two major attributes of relevance to the CWC. First, by eliminating the need to sustain life, CFPS can be used to manufacture proteins that are novel, difficult to produce, or would otherwise be highly toxic to the host cell. Second, CFPS systems are scalable, enabling reactions from 10 μL up to 100 L to be performed.[62]

5.2.6 Chemical Synthesis and Production of Biological Agents

In its 2018 report to the 4th CWC Review Conference the SAB stated that:

> "...The technologies that may enable the chemical synthesis of replicating organisms are becoming more accessible, but the skills required remain in relatively few facilities. In most cases, and currently, obtaining bacteria and viruses from their natural sources, or from commercial suppliers, is simpler than through employing chemical synthesis, metabolic engineering or synthetic biology approaches to make them. In regard to ability to produce highly complex genomes, a new initiative, Genome Project-write (GP-write) has been launched to synthesise a complete human genome, it is estimated that such an undertaking (3 billion bases over 23 chromosomes) will require at least 10 years... With regard to producing synthetic genomes, a current limitation is the ability to predict whether or not the incorporation of the genome into a cell will result in a functional organism."[63]

In its 2023 report to the 5th CWC Review Conference the SAB stated that "the number of substances of biological origin that can be produced by chemical synthesis—such as proteins, peptides, and biotoxins—continues to increase".[64] The report then gave illustrative examples of the substances produced through these rapidly developing technologies including:

> Complex proteins... [for example] ... Shiga toxin components and the toxin of the long-chain scorpion, were constructed by chemical synthesis resulting in a toxin material with high biological activity in high yield. In a similar way, the chemical synthesis of functional α-bungarotoxin (a snake venom) has been undertaken, leading to a fully functional toxin.[65]

The report highlighted peptide-based structures – notably, α-amanitin, which is an extremely toxic bicyclic octapeptide isolated from the death-cap mushroom and is highly toxic to eukaryotic cells. A multistep synthesis method for this toxin has been published and can also be used for the construction of endless numbers of variants with different properties.[66] The report also noted that synthetic infectious horsepox virus, a large DNA virus, was constructed starting from chemically synthesised DNA fragments. Noteworthy in this case is the phenomenon of so-called "reverse genetics", in which infectious viruses are generated without the need for a nucleic acid.[67] A bacterial genome of a *Caulobacter* species was constructed through chemical synthesis, highlighting the promise of chemical synthesis for decoding fundamental genome functions and its concurrent utility for the construction of tailor-made organisms for industrial purposes and health benefits.[68]

5.3 Science and Technology Developments in CNS-agent Delivery and Uptake

5.3.1 Development of CNS-acting Agent Weapons Employing Limited Area Delivery Mechanisms

During the Cold War a number of States conducted research and development of CNS-acting chemical agent weapons, including employing mechanisms designed to deliver the agent to a specific individual or small group of individuals.[69] Certain States undertook such activities until shortly prior to, or subsequent to, the coming into force of the CWC in 1997. To date at least three of these States (China, Israel and the US) developed CNS-acting chemical agent weapons that employed individual projectiles or other means to facilitate transdermal delivery of the CNS-acting chemical agent, whilst at least one State (Iran) sought to develop limited aerosol delivery mechanisms.

5.3.1.1 Chinese Development and Use of CNS-acting Weapons for Military and Law Enforcement

From at least the mid-1990s, Chinese State-owned companies developed and promoted a weapon incorporating an unidentified CNS-acting chemical for use in law enforcement operations against individuals. In 1995, marketing materials distributed internationally by China North Industries Corporation (NORINCO) promoted the "BBQ-901 anaesthetic gun system." This weapon discharges a projectile, with an effective range of 40 metres, which on impact injects a liquid incapacitating chemical agent into the target.[70] According to NORINCO "[T]he Model BBQ-901 Anaesthetic system is a fine unlethal [sic] special weapon system for SWAT units and other special usage.... It can be used for reconnaissance and capture of criminals in a concealed place. It is also used as a riot control weapon to subdue the ruffians and maintain public order".[71] The essentially similar if not identical, "BBQ-901 narcosis gun" was promoted by a second Chinese company, State 9616 Plant, at the Asia Pacific China Police Expo held in Beijing in 2004 and again in 2006.[72] Although no further marketing materials from NORINCO or State 9616 Plant have come to light, photographs of this weapon being displayed in May 2011 by the Chinese Peoples' Liberation Army garrison in Hong Kong were subsequently published in 2012.[73]

In June 2013 *Radio Free Asia* broadcast a report alleging that Chinese police had employed "anaesthesia guns" to suppress ethnic Hui villagers resisting a large-scale enforced house demolition operation in Xiangdong village, Hualong County, Qinghai Province. According to a reported eyewitness "[T]he special police beat people and used anaesthesia guns. The villagers who were shot were unable to stand up. After fainting, they were taken away and detained". According to this report, a spokesperson from the Public Security Brigade of the Hualong County Public Security Bureau denied that anaesthetic guns had been used in this operation.[74]

China has provided no public information regarding its stockpiles, if any, of CNS-acting chemical agent weapons nor the specific purposes of their intended employment. To date, China has made no statement clarifying whether any Chinese entity has conducted or is conducting research activities related to the development of CNS-acting chemical agent weapons targeting individuals or groups. In its 2021 *CWC Compliance Report*, the US reported "[S]cientists at a Chinese military institute have expressed interest in military applications of PBAs [Pharmaceutical Based Agents] and are engaged in research involving the synthesis, characterization, and testing of PBAs with potential dual-use applications".[75]

5.3.1.2 Israeli Development and Use of Trans-dermal CNS-acting Weapon

In its 2005 analysis of Israel's biological and chemical programs, the Swedish Defence Research Institute concluded that: "[T]he state previously developed offensive biological and chemical warfare capabilities. It has not been possible to conclude if these offensive programs still remain active today".[76] The report contended that "the focus of the Israeli chemical and biological capacity today is to develop agents for small-scale covert use, *i.e.*, a so-called 'dirty tricks' program".[77] It appears this included Israeli development of a hand-held *trans*-dermal CNS-acting chemical agent weapon intended for use against individuals. In October 1997, the Israeli security services (Mossad) employed this weapon, in either a failed assassination attempt or a kidnapping operation that subsequently went awry.[78] According to a January 1998, *Janes Intelligence Review* article, the chemical agent used was "believed to have been a synthetic opiate called Fentanyl which, absorbed through the skin and quickly metabolised, can kill within 48 hours and leaves no trace".[79] The Israeli intelligence team that conducted the operation reportedly included one physician, who "also carried an antidote known as Narcan or Naloxone in case something went wrong".[80] The target was Mr Khalid Mishal, a then-mid-ranking Hamas leader living in Jordan. The Israeli group reportedly followed Mr Mishal and attempted to deliver the fentanyl *trans*-dermally. Mr Mishal's driver who witnessed the event, described the attack to the *New York Times*: "a man advanced toward Mr. Meshal and then lunged toward the area around his left ear. The Mossad agent's hand... was wrapped in a white bandage, with a small lead-colored protuberance in the palm".[81] The paper also interviewed Mr Mishal who stated: "[I] felt a loud noise in my ear... It was like a boom, like an electric shock. Then I had [a] shivering sensation in my body like an electric shock".[82] Mr Mishal was able to escape the attack, but following the event, was reportedly seriously affected by the drug, and required significant medical attention. King Hussein of Jordan reportedly demanded that Israel provide an antidote, which they did, and Mr Mishal subsequently made a full recovery.[83] On 5th October 1997, the Israeli Government publicly admitted responsibility for the attack.[84]

5.3.1.3 US Development of Fentanyl Felt Pad Projectiles

In 1992 the National Institute of Justice, part of the US Justice Department, funded a project at Lawrence Livermore National Laboratory to research the feasibility of "utilising anaesthetic compounds in combination with antidotes to enhance the dose safety of chemical incapacitants" for both civilian law enforcement and "special military operations, and low intensity conflict".[85] The project explored a variety of delivery mechanisms. Researchers designed and tested prototype 22-calibre projectiles, fired from air rifles, incorporating small felt pads soaked with a mixture of a CNS-acting chemical agent together with a skin penetrating solvent dimethyl sulfoxide (DMSO) for rapid transdermal delivery. They subsequently developed a 38-calibre cartridge to deliver the felt pad holding the CNS-acting chemical agent/DMSO cocktail. The researchers recommended developing frangible paintball-type projectiles containing this cocktail.[86] In December 1999, the U.S. Army's Edgewood Chemical Biological Center (ECBC) solicited research proposals for a three phase project – CBD00-108 *Chemical Immobilizing Agents for Nonlethal Applications* – the objective of which was to "demonstrate the feasibility of a safe, reliable chemical immobilizing agent(s) for non-lethal (NL) applications in appropriate military missions and law enforcement situations".[87] Under Phase 1 researchers would "conduct an analysis of promising new chemical immobilizing agents or combinations of agents". Under Phase 2, input would be gathered from potential military and law enforcement users on the desired performance and operational characteristics. Subsequently, an "appropriate delivery technique", for example, "an aerosol generator for dissemination for the inhalation route of entry, or a dart for injection in the intra-muscular route of entry" would be designed and demonstrated. Under Phase 3, potential law enforcement and military applications of the technology were to be analysed.[88] In June 2000, ECBC awarded a contract for Phase 1 of this project to OptiMetrics, Inc.[89] In November 2002, it was reported that Phase 1 had been completed.[90] It is not known whether Phase 2 or 3 were undertaken and if so, when or by whom. A 2003 study by the National Research Council on non-lethal weapons science and technology stated that military research on CNS-acting weapons was ongoing, including research by ECBC on a sponge projectile soaked with a CNS-acting chemical agent and antidote.[91] No subsequent information is available on such CNS-acting projectile weapons.

5.3.1.4 Iranian Military University Development of CNS-agent Grenades

In its 2024 *CWC Compliance Report*, the US accused Iran of "maintain[ing] a chemical weapons program that includes pursuit of certain PBAs [pharmaceutical-based agents] as part of a broader category of 'incapacitating agents' for offensive purposes".[92] In addition to allegations that since 2005, some of Iran's military-controlled facilities, namely Imam Hossein University and Malek Ashtar University of Technology have researched chemicals that had a "wide range of sedation, dissociation, and amnestic incapacitating effects", the *US Compliance Report* also includes allegations of development of related means of delivery. "[M]ost recently on September 23, 2023, an X (formerly Twitter) account representing Lab Dookhtegan – whose members claim to be hacker organization in opposition to Iranian state-sponsored cyber actors – posted confidential documents detailing an Iranian military university's development of grenades meant to disseminate medetomidine, an anesthetic that is a central nervous system-acting chemical. According to the documents, this development included information on the production and testing of prototype weapons".[93]

While the original Iranian documents cited by the US are not currently publicly accessible, the US based NGO, the Institute for Science and International Security claims to have acquired them and details their contents in its November 2024 report: *The Islamic Republic's Work on Pharmaceutical Based Agents*.[94] According to the authors:

> "The hacked documents describe a project titled "Deterrence." They detail the development of incapacitating ammunition containing

> medetomidine by the IHU Department of Chemistry. The documents imply a goal of the project as altering the incapacitating agent in the existing ammunition to include medetomidine. The 23 hacked pages appear to be part of a much larger set of documents, presumably containing records of IHU researchers evaluating the military application of different incapacitating technologies. This document reports on advanced stages of testing medetomidine-based ABC-M7A2 and ABC-M7A3 hand grenades and 38 mm MK 2 anti-riot munition."[95]

The authors summarise information translated from the hacked documents including the formulation of various 18 g chemical fillers for use in a medetomidine hand grenade subsequently tested by the IRGC. This grenade seems to be an adaptation of a standard riot control grenade that normally contains the RCA, CS, but where this is replaced with a chemical filler now containing medetomidine instead. Consequently "different pyrotechnic mixtures have been tested to ensure reliability and effectiveness of the grenade in parameters such as burn rate, smoke volume, duration, and dispersion". The various fillers include the "M6-40% formulation [that] introduces medetomidine hydrochloride as the incapacitating agent which composes 40 percent of the filler mixture".[96]

According to the authors: "In a different set of experiments, three different formulations of medetomidine grenades were tested in open-air conditions. According to the report, among all of the different variations tested, the 40% medetomidine formulation showed satisfactory results in terms of smoke emission, smoke dispersion, and uniform burning".[97] They also state that "Although there are no detailed test results available about it, it is reasonable to assume from the introductory section at the end of the available pages about the MK2 CS cartridge and its launcher that the same test have been performed using medetomidine-filled MK2 cartridges".[98]

5.3.2 Developments in CNS-agent Uptake and Targeting the Brain and CNS

5.3.2.1 Nanotechnology and Drug Delivery and Uptake

Nanotechnology and nanoscience refer to the investigation and use of functional components and materials that are smaller in size than 100 nanometres in at least one dimension. Most nanomaterials have physical, chemical, or biological properties that differ significantly from those at a larger scale. Nano-sized particles have a high surface area to mass ratio, which may impart higher chemical reactivity, increased strength, and modified electrical properties. Nanoparticles may also exhibit altered optical and magnetic properties. Nanotechnology is a rapidly expanding technology having many diverse applications some of which raise dual use concerns particularly with regard to the delivery and update of toxic chemicals.[99]

In 2008 the International Union of Pure and Applied Chemistry (IUPAC) noted the wide-ranging applications of this technology yet also highlighted its potential dangers:

> "...Nanotechnology may allow the development of capsules for safe enclosure and targeted delivery of biologically active chemicals. It may enhance the use of active groups for bonding to specific targets in organs or cells. It may provide new means of facilitating entry into the body or cells, *in particular in the brain*, for selective reaction with specific gene patterns or proteins, or for overcoming the immune reaction of the target organism. These developments promise significant benefits in the form of new medicines but at the same time could be used in a program aimed at developing more effective CW..."[100] (Emphasis added).

Nanoparticle-based formulations (nanocarriers) are being explored in medicine for enhanced or 'smart' drug delivery. Examples include controlled drug release, enhanced penetration of the blood-brain barrier for delivery of therapeutic peptides, and the targeting of specific organs or cells with the majority of such drugs being developed for cancer cells. The inter-related utilisation of nanotechnology to facilitate improved aerosol delivery of therapeutic drugs will be discussed in Section 5.3.3.1.

Examples of the types of nanomaterials that have been explored as potential drug delivery systems include polymeric nanoparticles such as polysaccharide chitosan nanoparticles which function due to their ability to facilitate both protein and drug conjugation; poly-(lactic-*co*-glycolic acid) as a carrier across the BBB; inorganic ceramic nanoparticles such as silica, titania and alumina that can be used as drug carriers due to their porous nature; metallic nanoparticles including iron oxide, gold shell and titanium oxide nanoparticles; and carbon nanoparticles (fullerenes and nanotubes) which are widely used for drug delivery purposes given the multiple attachment points of the fullerenes and the high electrical conductivity and strength of the nanotubes.[101] Further drug delivery nanocarrier structures include those in the form of nanocrystalline drugs, liposomes, polymer-drug conjugates, polymeric micelles, protein nanoparticles, and lipid colloidal dispersions.[102] Solid lipid nanoparticles, also, have been developed for targeted delivery and controlled release of a wide range of drugs.[103] Further types used in therapeutic drug formulations include: imprinted polymers, dendrimers and nano-emulsions.[104] There are essentially two types of nano-polymers that have been evaluated as drug delivery systems: nano-capsules (a core of encapsulated drug surrounded by a polymeric membrane or shell) and nanospheres (where the drug is distributed or absorbed throughout the polymeric matrix).[105] Additional bio-based nanocarriers are being researched including DNA- and virus-based systems.[106]

Nanocarrier-based delivery systems present several advantages over traditional drug delivery approaches, in particular overcoming solubility problems, protecting the drug from the environment (temperature, photo damage, pH), and controlling its release profile. These delivery systems permit more precise and controlled targeting at the site of action, while reducing the time of exposure of non-targeted tissues. This can potentially increase efficacy, and reduce toxicity and side-effects.[107]

An important property of nanoparticles intended for drug delivery is biodegradability. Biodegradable nano-systems may be effective tools to transport drug payloads across biological barriers and could be used to deliver chemicals to targets they normally would not reach. Nanomaterials used as drug delivery vehicles also can be designed to protect their chemical payload from biological factors (enzymes in particular) that would otherwise metabolise the compounds, thereby altering (normally reducing) their toxicity and/or promoting their elimination from the body.[108]

Recent years have seen advances in the engineering of 'smart' nanoparticles which are designed to respond to environmental stimuli such as changes in temperature, pressure, electric or magnetic fields, light, or chemical parameters such as pH or ionic strength. Such adapted nanoparticle capabilities potentially allow the combining of their carrier properties with drug release mechanisms that respond to external factors, enabling targeted delivery of drugs in precise locations in the body when combined with, for example, imaging techniques. While such developments have significant medical applications, they could instead potentially be malignly employed for highly targeted delivery and timed release of chemical warfare agents for example in certain special forces operations (see below).[109] In contrast, Trapp has previously voiced his scepticism of the utility of such smart nanoparticles in scenarios where much cheaper, more easily produced traditional chemical warfare agents have previously been used over wide areas for mass effects on target groups and populations without the need to control the environment.[110]

5.3.2.2 *Transporting CNS-acting Chemicals to the Brain*

The potential dual use implications of developments in particle engineering and nanotechnology that could allow the delivery and uptake of biologically active chemicals to specific target organs or receptors have been highlighted by several scientific bodies[111] and arms control experts. Of relevance to these discussions has been their potential utility in transporting CNS-acting chemicals across the blood-brain barrier (BBB) that restricts the passage of many types of molecules as well as microorganisms from the bloodstream into the neural tissue. While the BBB protects the brain from many common bacterial infections, it presents a significant challenge in delivering benign therapeutic agents to the brain for the treatment of brain disorders. It comprises an endothelial cell monolayer associated with pericytes and astrocytes. The BBB separates the blood from the cerebral parenchyma and prevents the penetration of drugs into the CNS. The BBB prevents the uptake of all large-molecule drugs and more than 98 percent of pharmaceutical small-molecule drugs. Only very small (<5 KDa), lipid-soluble, electrically neutral molecules and weak bases are able to diffuse passively across the BBB.[112]

A range of strategies have been developed to carry CNS active drugs across the BBB. These include bypass through intranasal delivery, invasive transport across the BBB, the use of cell-penetrating peptides, passive targeting of diseased tissue, receptor-mediated endocytosis and combinations of these strategies. Nanoparticles have been engineered to transport drug payloads, including small molecules, therapeutic DNA and peptides, through biological barriers, including the BBB.[113,114] Nanoparticle brain targeting drug delivery remains, however, still in an experimental stage.[115]

In 2008 the US National Research Council warned that nanotechnologies could be used to overcome the BBB and thereby "enable unparalleled access to the brain. Nanotechnologies can also exploit existing transport mechanisms to transmit substances into the brain in analogy with the Trojan horse".[116] In 2012 the UK Royal Society highlighted how the continuing advances in nanotechnology-based drug delivery systems were demonstrating the potential to deliver peptides to the brain by preventing peptide degradation within the blood and promoting transport across the brain endothelial capillaries.[117] For example research in 2011 by Lalasta had already demonstrated the oral and intravenous delivery to the brain of two peptides, both analgesic opioid receptor agonists, using nanoparticle technology.[118] While only a small percentage of the administered peptide penetrated the blood-brain barrier in this research, the Royal Society noted that such advances had great potential in the development of therapeutics but also potential applications for the delivery of incapacitating chemical agents. However, the Royal Society recognised that the current state of development, delivery of agents with nanocarriers by oral or intravenous route would have very limited application in the development of incapacitating chemical agents.[119]

However, as nanotechnology advances the nature and level of threat may change. For example, in 2018 Trapp warned that engineered nanoparticles may be of potential utility in "the targeted delivery of chemical agents that act on the CNS. This type of delivery becomes particularly relevant when one considers the development of incapacitating chemical agents based on toxins or bioregulators which are otherwise difficult to deliver and are rapidly metabolised or eliminated. Such types of novel chemical agents and delivery systems would be rather expensive and technically challenging to use, but they might be of interest for special-operations types of scenarios".[120]

5.3.3 Dual Use Technologies that Could be Employed for Aerosolised Delivery of CNS-acting Agents

5.3.3.1 Aerosolised Delivery of Therapeutic Drugs and Vaccines to Individuals

Aerosols are particles in the form of a liquid or a powder that are suspended in air and can be inhaled. Aerosolision of particles whether of natural origin, such as infectious micro-organisms, or synthesised including therapeutic drugs and vaccines facilitates their entry into the body through the mucous membranes lining the nasal and respiratory tracts as well as the intestinal tract. The size of the particles to a great extent determines where they will be deposited in the airway after inhalation. Particles up to 5 micrometers in diameter can reach deep lung areas (alveoli or air sacs); larger particles will be deposited in more anterior parts of the respiratory tract.[121]

Delivery of therapeutic drugs and vaccines *via* the aerosol route is attractive for a number of reasons. The surface area of the lung is between 80 and 140 square meters. Also, the alveolar (air sac) epithelium (cell lining) in most pulmonary regions is only about 0.1–0.2 micrometers thick, and the distance between epithelial surface and the blood is much less than it is in the bronchial system, which should facilitate drug uptake.[122] Aerosol delivery of vaccines has the added advantage that direct delivery to the mucous membranes of the nasal and lung surfaces is achieved, where many infections begin. Among the various mucosal sites, nasal delivery is especially attractive for vaccine administration because this is a site of relatively high tissue permeability, low activity of destructive enzymes and the presence of a considerable number of immune response cells.[123] There are, however, a number of absorption barriers in the human lung including the mucus layer, the alveolar lining fluid layer, and competing uptake pathways such as particle engulfment by macrophages (white blood cells that can take up and destroy foreign particles). In the last two decades aerosol technology has been advancing rapidly and is already in use to deliver effective inhaled drug therapy for the treatment of disease. Propellant metered-dose inhalers, dry powder inhalers, and nebulisers are used to deliver drugs directly to the lungs promoting rapid absorption into the blood.

The nasal route has emerged as being particularly advantageous for the delivery of drugs in part because this is a site of relatively high tissue permeability and low activity of destructive enzymes.[124] For CNS-acting chemicals, this route has the added potential of providing direct access of drugs to the brain, and many peptides and proteins (among others: Orexin-A, insulin, leptin, erythropoietin) can be detected in the central nervous system after intranasal delivery.[125] Uptake after nasal delivery is achieved by various methods: absorption into olfactory blood vessels and entry into the general circulation; absorption into olfactory lymphatic vessels draining to the deep cervical lymph nodes of the neck; and extracellular diffusion or convection in compartments associated with olfactory nerve bundles and entry into the cranial compartment.[126]

An example of successful aerosol delivery of a commercial pharmaceutical drug with potential utility as a CNS-acting chemical agent weapon is the neuropeptide oxytocin, which was reported to increase trusting behaviour in humans given a single dose by nasal spray.[127] It has subsequently been marketed by Vero Labs as a liquid spray, "formulated to emphasize its key role in human bonding to improve confidence, enhance relationships and strengthen bonds".[128] A growing number of studies have investigated the effects of oxytocin after nasal administration on human behaviour, cognition and brain activation, particularly in connection with specific disorders such as anxiety autism and schizophrenia.[129] In addition, Shoyele and Slowey[130] have produced a list of 15 peptides/proteins that are effective when administered over the aerosol route. This list includes among others several cytokines, erythropoietin, calcitonin, insulin, amylin and the growth hormone.

5.3.3.2 *Concerns About Malign Employment of Aerosol Drug Delivery*

Concerns about potential malign application of advancing aerosol drug delivery technologies have been raised by a number of arms control and scientific bodies including IUPAC,[131] the UK Royal Society and the US NAS. For example, in its 2008 report, IUPAC stated:

> "Advances in nanotechnology and aerosol drug delivery, driven by the needs of such sectors as the pharmaceutical or pesticide industries, are also significant. They may be exploited for more effective and targeted delivery methods for toxic chemicals. Many of the considerations that promote the design of particles for effective and targeted drug delivery *via* the respiratory system would equally be applicable in a program to improve the dispersion of a CW agent in the atmosphere. An example is the use of large porous aerosol particles that allow the delivery of drugs into the deep regions of the lungs to promote absorption in the alveolar region—if used for CW purposes this could increase the systemic toxicity of a given agent."[132]

These concerns have also been taken up by the OPCW SAB and repeatedly highlighted in their reports to CWC States Parties. In its 2008 report to the 2nd CWC Review Conference the SAB noted that:

> "Pharmaceutical companies are showing an increasing interest in administering drugs by inhalation as an alternative non-invasive method of delivery. In the case of drugs with a low molecular mass, many of which are readily absorbed by the lungs, this route circumvents the extensive metabolism that occurs in the intestinal tract following ingestion. Larger molecules, such as proteins, are also absorbed through the lung membrane, albeit at a slower rate.... Some drug companies are also focussing on the lungs as a point of entry for the rapid administration of drugs to the central nervous system. The physical properties that promote rapid absorption through the lungs are similar to those that promote the penetration of the blood-brain barrier, and a number of devices have been developed for administering such drugs as opioids, anti-migraine drugs, and anti-convulsants as aerosols."[133]

They further noted that:

> "Many of the considerations that promote the design of particles for the effective and targeted delivery of drugs *via* the respiratory system would be applicable to the dissemination of a chemical-warfare agent as an aerosol. The efficiency of absorption has been improved, for instance, through the use of large porous aerosol particles that allow the delivery of drugs into the deep alveolar regions of the lungs and promote their absorption there. The spray-drying equipment needed to create such particles is relatively inexpensive and widely available, although the optimisation of a well-engineered particle requires expertise and considerable effort. This type of technology can be combined with nanotechnology to deliver nanoparticulate aggregates that will, once absorbed, disperse in the body, where their design (*e.g.*, multifunctional polymeric design) could facilitate improved and more-selective delivery and targeting of drugs."[134]

These concerns were again explored by the SAB in its 2012 report to the 3rd CWC Review Conference. They highlighted continuing interest within the pharmaceutical industry in "administering drugs by inhalation...that...can deliver drugs with suitable physical properties, or in appropriate formulations, rapidly to the central nervous system".[135] The SAB again raised the conjunction of advances in aerosol delivery and nanotechnology, highlighting "use of porous nanoparticles as carriers composed, for example, of silica or L-lactide that allow delivery of drugs into the deep alveolar regions of the lungs".[136] The SAB stated that "although the optimisation of a well-engineered particle requires expertise and considerable effort...the equipment needed to create such particles is relatively inexpensive...[and]... *the technology could be exploited in the design of incapacitants*".[137] (Emphasis added).

In 2012 the Royal Society also noted that advances in research into inhalation-based methods of drug and vaccine delivery "may offer potential applications in the delivery of agents for incapacitation". They noted that the use of an aerosolised fentanyl derivative in the Moscow theatre siege was predated by investigations into the administration of fentanyl as a nebulised aerosol for clinical purposes.[138] Research had been conducted some years earlier demonstrating the comparable effect of aerosolised fentanyl to intravenous administration at the same dosage.

It is however important to underline that the advances in technology facilitating targeted aerosol delivery of therapeutic drugs intended to treat disease and other illnesses of individuals will not necessarily be directly applicable to the development of wide area aerosolised CNS-acting weapons. As the SAB noted in its 2008 report "the sophisticated engineering of a high-value drug for targeted delivery might not be appropriate for a chemical weapons delivery system, where as much as 99.9% of the toxic chemical disseminated will not reach its target organ within the human body".[139]

The Royal Society 2012 report also highlighted the differences between small scale therapeutic aerosol delivery to individual patients and wide area weaponised dispersal:

> "... However, as the Moscow example demonstrated, a uniform dose is difficult to achieve in non-clinical, wide area dissemination. Air currents may disperse the aerosol unequally through the room and vicinity of the target to the ventilation system would also influence individual exposure dose considerably."[140]

The Royal Society report also noted that wide area dissemination of an agent is particularly problematic since effective dissemination of an agent is not necessarily compatible with effective uptake. In addition, and crucially, they noted that "the problem of controlling the dose would remain". However, whilst recognising these differences, it is worth recalling that they may be overcome by continuing technological advances and targeted application. And they may be of less importance if the employers of such wide area aerosol CNS-acting weapons place a higher value on ensuring incapacitation of the targeted groups over safety and healthcare considerations of those targeted.

Perhaps the starkest warning for the potential dangers of the development and convergence of aerosol delivery techniques and nanotechnology has come from the US National Academy of Sciences (NAS) 2008 report. In the section reviewing the then-current state of nanotechnologies and gas-phase technologies, the NAS stated that:

> "Pharmacological agents are not used as weapons of mass effect, because their large-scale deployment is impractical; it is currently impossible to get an effective dose to a combatant. However, technologies that could be available in the next 20 years would allow dispersal of agents in delivery vehicles that would be analogous to a pharmacological cluster bomb or a land mine."[141]

The NAS subsequently recommended the arms control community to watch for potential triggers of concern, notably:

> "...development of such [nanotechnology or gas-phase] technologies for agricultural or pesticide purposes. Precedents exist for using toxins as weapons delivered by ingestion, transdermally, or by inhalation. Analysts should watch for development of standardized delivery systems that can distribute small-molecule 'payloads' over large areas, such as crop fields. Those delivery systems would protect agents from meteorological conditions and then release agents on contact with a soldier. Of particular concern would be a single delivery system that could be easily loaded with different agents, as warheads are switched in an artillery round. This type of system would allow easy crossover to nefarious purposes."[142]

The NAS warned that the consequence of the development of such a system would be an "off-the-shelf small-molecule delivery system that could be leveraged against U.S. forces. Highly potent compounds could cause troop incapacitation or performance degradation. Coupling off-the-shelf agents with an off-the-shelf delivery system would create new threat opportunities against U.S. forces. Spreading agents that would be released on contact would deny U.S. military operations some strategic infrastructure or geographic positions".[143]

The NAS findings have informed subsequent analysis in this area, including the 2012 UK Royal Society report. Given these concerns the following subsections of this chapter attempt to survey the contemporary development and promotion of commercially available off-the-shelf aerosol delivery systems, as well as more specialised delivery mechanisms developed by States and commercial companies for law enforcement and military delivery of riot control agents and examine the threats of their employment to deliver CNS-acting agents.

5.3.4 Commercial Dual Use Wide Area Delivery Aerosol Devices that Could be Employed for Delivery of CNS-acting Agents

There is a range of commercially available devices which generate and disperse aerosols including particulate smoke or dense vapour fogs. Common types of foggers employ inert gas or electric pumps to propel mineral oil, glycol, or a glycerine and water mixture into a heat exchanger, where the solution is vaporized, the vapour then mixing with the cooler external air at a controlled rate to condense to a fog or mist with a controlled droplet size. Such devices are commonly and legitimately employed in the entertainment industry, emergency services training and by the military for creating smokescreens to obscure troop movement. However, such devices could potentially be employed for the weaponised dispersal of toxic chemicals including CNS-acting agents. Indeed, certain commercial manufacturers of multipurpose smoke devices highlight their potential use as RCA delivery mechanisms (see Section 5.3.6.2.1).

Other spray devices are widely employed in agriculture and livestock farming. The increasing capacity and range of commercially available devices, together with their ability to be employed on diverse ground, marine or aerial vehicles, raise concerns of potential misuse by both State and non-State actors. In its 2012 report to the 3rd CWC Review Conference, the OPCW SAB highlighted the potential misuse of equipment such as "spray and fogging devices developed by the pesticide industry or developed for veterinary treatment of large-scale animal farms" by non-State actors for the dissemination of chemical weapons and biological weapons agents.[144] In its 2023 report to the 5th CWC Review Conference the SAB noted that since the 4th CWC Review Conference, "there has been a surge in the use of electrostatic sprayers and foggers for the disinfection of surfaces contaminated by COVID-19. Although this is not a new technology and is commonly used to apply pesticides to crops, devices are more widely and readily available than ever before".[145]

Of particular concern have been the implications of the dramatic increase in the last two decades of research, development and possession of UAVs[146] by State military and security forces. Similarly, in the civilian sphere, there has been contemporary development and increased online accessibility of a growing range of small hobbyist and mid-size commercially available UAVs capable of delivering several kilogram payloads. In addition to multipurpose UAVs, there has been associated development of a variety of specialized UAVs (notably agricultural and firefighting drones) intended to carry (sometimes significant) liquid payloads. In its 2023 report to the 5th CWC Review Conference, the SAB noted that whilst COVID 19 decontamination foggers and sprayers could "be person-portable, cordless, and compact" they could also be integrated onto a "UAV or an unmanned ground vehicle (UGV) to decontaminate large areas such as offices, schools, airports, and arenas".[147] They also highlighted "the use and availability of UAVs to spray crops has also increased, with better battery life and payload capacity". And noted that "this legitimate use requires monitoring, or at least certain codes of conduct applied to the sale of these commercially available products". [148]

It is predictable that future commercially available UAVs will be able to carry heavier payloads, fly and loiter longer, venture farther from their controller, survive in more difficult weather and be able to do so *via* more secure communications links that are less susceptible to disruption.[149] In addition, further dual-use technological developments have resulted in increased autonomous capability and inter-operability between UAVs. Such advances provide so-called 'swarming' capability, whereby numerous autonomous aircraft, networked together, can be employed towards common goals. Whilst such swarming technology has benign applications in agriculture or for search and rescue missions, it also has potential offensive applications, whereby a UAV swarm could be tasked with overwhelming a target or targets through weight of numbers and avenues of attack.[150] In certain posited scenarios, large swarms of relatively small UAVs could be armed with chemical or biological agent payloads, collectively delivering significantly large amounts of agent onto or dispersing such agents over the target populations.[151] The use of drones to deliver chemical agents in the war in the Ukraine is discussed in Section 5.3.6.3.2.

In addition, once again the convergence of AI with UAV technological advances has been highlighted as a concern. For example, in its 2023 study for the 5th CWC Review Conference, the SAB noted that:

> "Incorporation of AI into agricultural sprayers, both manned and unmanned, is now gaining traction and is leading to targeted application of chemicals. Facial recognition technology integrated within UGVs or UAVs could be exploited by non-State actors to direct a targeted release

of toxic chemicals at specific people. Although access to facial recognition technology was very limited a decade ago, it is now widely available through various online platforms."[152]

5.3.5 Wide Area Aerosolised CNS-agent Weapons in Enclosed Spaces

5.3.5.1 Russian Use of Wide Area CNS-acting Weapon in Enclosed Spaces

Whilst the full details of the Russian Special Forces' employment of a CNS-acting chemical agent weapon to end the Dobrovka Theatre siege in October 2002 are unlikely now to come to light, certain elements of the operation are in the public domain. On the evening of 23rd October 2002, approximately 40 Chechen separatists took approximately 900 theatre-goers hostage, demanding Russia withdraw its military from Chechnya. The Chechens were a mixed group of men and women; the men armed with assault rifles, the women armed with handguns and controlling detonators for numerous explosive devices, some worn on their bodies, others placed nearby. The hostage-takers threatened to detonate all their devices, destroying the theatre and killing all the hostages unless their demands were met. Following failure in negotiations, the Russian Special Forces attempted to use the CNS-acting chemical agent weapon to incapacitate all the theatre occupants – hostages and hostage takers alike – before they forcibly entered the building. The Special Forces waited 30 minutes before they entered the building by which time all the occupants were unconscious. The hostage takers were all executed whilst unconscious and the hostages were taken to nearby hospitals. Unfortunately, approximately 130 hostages died from asphyxiation or the toxic effects of a CNS-acting weapon.

It is clear from subsequent research and analysis that the Russian Special Forces had access to and employed a previously developed weaponised CNS-acting chemical agent of the opiate class, carfentanil. They are thought to have mixed this highly toxic chemical agent with the shorter-acting remifentanil to make a less potent and toxic opioid cocktail for the purpose of the operation. However, it appears that they did not have access to, or did not choose to employ, a corresponding mechanism specifically designed for effectively and swiftly dispersing the CNS-acting agent in enclosed spaces. Instead, the security forces employed an *ad hoc*, and what turned out to be only partially satisfactory, existing dispersal mechanism, *i.e.*, the theatre's ventilation system. After the Russian Special Forces pumped the CNS-acting agent into this system and the aerosol spread through the chutes to the main auditorium, the hostage takers noticed its presence and reportedly had time to react before any incapacitating effects were fully manifested – indeed hostages describe making calls to loved ones saying they were in the process of being "gassed". Whilst the male hostage-takers exited into surrounding corridors to fight the Russian security forces they expected to be soon storming the building, the female hostage-takers remained with the hostages and the explosives in the main auditorium and had plenty of time if they had wished to detonate all their explosive devices.

5.3.5.2 *Russian Computer Modeling of Calmative Agents in Enclosed Spaces*

Despite the clear shortcomings of this operation, there are indications that following the Moscow theatre incident, Russian scientists continued work related to the development and future employment of CNS-acting chemical agent weapons. This included research specifically intended to overcome the difficulties of delivering "calmative" agents in enclosed spaces. A paper by Klochikhin *et al.* presented at the 3rd Ettlingen Symposium on Non-Lethal Weapons in May 2005, described their computer modelling of a scenario in which an airborne chemical "calmative" agent was introduced into a building where hostages were held captive. The paper stated that "[I]f the level of 95% efficiency is absolutely required to neutralize terrorists and to prevent mass destruction, there is no chance to eliminate hard consequences and fatalities. Calculations show that the majority of hostages can get serious poisoning and part of them – fatality. This is the cost of releasing if no other solutions [are] left".[153]

The authors reported that "[O]ne possible solution under discussion is to apply gaseous calmative agent and antidote together in the same composition or consequently after some delay. This is the way to control the value of impact and to decrease collateral damage".[154]

The researchers noted that "the real problem of chemical NLW is rather difficult. It requires serious efforts to develop reliable techniques and mathematic instruments for calculation of various scenarios…the full solution for such challenge demands the big intensive work of many scientific teams within several years".[155]

It appears that Russian researchers continued work to develop computer models for the application of what they describe as "calmatives" against groups of individuals in enclosed spaces. In November 2009, Klochikhin and Selivanov presented a "[R]eport on the 1st phase of the *Project 'Gas Flow'*" to a meeting in London.[156] In the presentation, the authors described their work to develop computer code to generate 3-D simulations of "an effective scenario of calmative application" utilising existing medical data on "calmatives" and physical data describing the nature of "gas" movement in enclosed spaces. The authors stated that the resultant computer code "draws the gas; simulates gas transfer with air between rooms; calculates its concentrations in rooms; evaluates the calmative effects; shows the realistic simulation to define characters' status and gas concentration field to optimize the scenario gas effects".[157] No further information about *Project "Gas Flow"* is currently in the public domain.

5.3.6 Wide Area RCA Dissemination Devices

5.3.6.1 Overview

Investigations by Crowley and colleagues at the Omega Research Foundation uncovered the development, production, possession or promotion by State and commercial entities in at least 16 countries of "wide area" RCA means of delivery of potential concern since the CWC came into force in 1997.[158] These systems are capable of delivering significantly greater amounts of RCA over wider areas or more extended distances than currently possible with standard law enforcement RCA dispersal mechanisms, such as hand-held sprays. In addition to their potential misuse for collective ill treatment or punishment of crowds, such wide-area RCA delivery mechanisms could be employed as 'force multipliers' in conjunction with firearms, making lethal force deadlier on a large scale. Although nominally developed or acquired for law enforcement, they may also be incorporated into military arsenals, potentially resulting in subsequent use in armed conflict.

Certain RCA delivery mechanisms, though designated as RCA munitions, could also be used to disperse alternative toxic chemicals such as CNS-acting chemical agents and other incapacitating chemical agents. In addition, current weak trade controls in this area could result in acquisition of wide area RCA delivery mechanisms and their misuse by non-State actors, including armed opposition forces, unregulated private military and security companies, criminal groups, or terrorist organizations.

5.3.6.2 Review of Wide Area RCA Delivery Mechanisms of Concern

5.3.6.2.1 Indoor and Perimeter RCA Dispersion System. Companies have developed and promoted RCA delivery systems designed to disperse RCAs in an enclosed space, some smaller models are designed for personal household protection whilst others are intended for law enforcement or military applications. Whilst these systems are clearly designed and marketed for RCA dispersal, they or similar systems could potentially be repurposed to disperse CNS-acting chemicals to pacify or otherwise incapacitate prisoners and other targets within enclosed spaces. The risks of such systems being employed for the malign dispersal of CNS-acting agents are underlined by the previous weaponised employment of such agents in enclosed spaces by Russia and the subsequent computer modelling of such activities by Russian scientists.

5.3.6.2.1.1 TG Guard System [US]. Two US companies—Non Lethal Technologies, Inc. and Mace Security International, Inc.— have promoted a fixed installation RCA dispersion device called the TG Guard System which can employ either [oleoresin capsicum] OC or CS.[159] According to Non Lethal Technologies, Inc., the system is designed to provide "critical force protection for personnel and facilities" and has been promoted for use inside prisons[160] and for perimeter protection of "embassies, military bases, courts, and other government installations".[161] When remotely initiated by an operator, the system can "deploy OC irritant dust from strategically positioned squib fired dispensers within, or around perimeters of critical value buildings...to deny access to hostile individuals or crowds attempting to enter such government buildings to destroy or disrupt government operations".[162] The basic TG Guard Control Unit can "operate and discharge up to 25 tear gas Dispensers selectively ... [whilst]... larger facilities can use multiple Control Units and Dispensers to accommodate their security needs".[163] Each 'Dispenser' can hold and instantaneously dispense 110 grammes of either "CS powder" or "OC irritant dust".[164]

5.3.6.2.2 Foggers, Sprayers and Water Cannon

5.3.6.2.2.1 Fog Maker System [Iran]. The Iranian company, Shahid Meisami Group (SMG) which markets riot control agents, has also been reported to have promoted a means of delivery of potential concern. In its 2020–2022 *CWC Compliance Reports*, the US asserted that SMG had "provided fact sheets to interested users on a 'Fog Maker System' that can be used to make smoke and fog at high volume in a short time".[165] The Compliance Reports include images from an undated "[I]ranian advertisement" for a 'Fog Maker' mounted on a speed boat and a truck. According to the US, "this is noteworthy because it can disseminate debilitating chemicals, like [the RCA] CR, over a large area quickly".[166]

5.3.6.2.2.2 US Smoke and Agent Dispenser [US]. The US company MSI Delivery Systems Inc. developed and until at least 2016 promoted the Afterburner 2000 system (AB2K).[167] Company marketing materials described the product as a "robust multi-mission, multi-purpose smoke generator capable of rapidly blanketing large areas with dense smoke".[168] The standard non-toxic camouflage/training smoke "can be *mixed with irritants such as OC, CS, Pepper* [that] upgrades mission capabilities to include: *Crowd Control and Civil Unrest*, SWAT Teams and Tactical Incursions, *Prison Systems for Riot Control/Prisoner* Extraction, Non-lethal Terrorist Suppression/ Repel Pirates, *Military Operations (MOUT/COIN)* [Military Operations on Urban Terrain/Counter-Insurgency]".[169] (Emphases added). According to the company, the Afterburner 2000 could release over 1500 cubic feet of smoke with a range greater than 100 feet (30 metres) in one second.[170] The marketing material stated that "[the] standalone version" of the Afterburner 2000 "expels 50 000 cubic feet (1,416 cubic meters) of smoke on a single charge"; whilst the "dependent version with high-capacity backpack expels 320 000 cubic feet (9,061 cubic meters) of smoke on a single charge."[171]

5.3.6.2.2.3 Water Cannon [China, Germany]. In 2018, the Hong Kong police acquired three, reportedly German-manufactured, water cannons.[172]

These vehicles each contained two water spray devices capable of employing large quantities of water that can be mixed with an RCA and/or a dye (to identify protestors for subsequent arrest), and can target individuals or crowds, up to 50 metres away.[173] From August 2019, there were repeated reports of Hong Kong police employing water cannon firing water mixed with blue dye and/or the RCA against both violent and non-violent protestors, in disputed circumstances.[174] The Hong Kong police reportedly stated that the blue dye was "harmless" and confirmed that the RCA employed was PAVA (pelargonic acid vanillylamide, a synthetic capsaicinoid) which they stated would cause only temporary discomfort.[175] In December 2021, the Hong Kong newspaper, *The Standard*, reported that Hong Kong police had acquired three additional water cannon trucks, manufactured in China. Although smaller, they were similar to the previous vehicles and were capable of spraying water mixed with PAVA.[176]

5.3.6.2.3 **Multiple Barrel and Automatic Launchers.** Companies including in China, France, India, Israel, South Korea, and the United States have previously developed a variety of multi-barrel launchers capable of firing large salvoes of RCA projectiles, whilst Chinese and Russian companies marketed automatic grenade launchers that could fire large numbers of RCA projectiles from a single barrel. Both launcher types potentially enable users to indiscriminately deluge substantial numbers of people with RCA in a very short period.[177]

5.3.6.2.3.1 Vehicle Mounted 38 mm and 64 mm Automatic Grenade Launchers [China]. In 2012, China Ordnance Equipment Research Institute began to promote a vehicle mounted 38 mm automatic grenade launcher and a 64 mm automatic riot grenade launcher, both of which could be remotely controlled by an operator inside the vehicle and were capable of firing a variety of munitions including tear gas grenades, smoke grenades and stun grenades.[178] These two weapon systems are designed to "cope with (large-scale) mass events quickly and effectively".[179] The 38 mm launcher has a maximum firing rate of 200 rounds per minute and a maximum effective range of at least 300 metres. The 68 mm launcher has a maximum firing rate of 60 rounds per minute and a maximum effective range of 600 metres.[180]

5.3.6.2.3.2 38 mm and 40 mm Iron Fist [United States]. US manufacturer, Non Lethal Technologies Inc., developed and as of August 2025 markets the IronFist "less lethal 38 mm weapon system with up to 36 barrels ... [intended] to rapidly deploy a blanket of less lethal munitions into, or over, a hostile crowd".[181] When the IronFist 36-barrel system is loaded with high capacity 10-inch CS rounds, each containing 10 mini-grenades, up to 360 CS mini-grenades can be fired from a single vehicle station.[182] The IronFist system is designed to be hard-mounted on a wide range of land vehicles and marine vessels, or to permanent structures "such as prisons, government buildings, military base perimeters, or embassy compounds".[183]

5.3.6.2.3.3 37 mm and 40 mm VENOM Systems [United States]. The VENOM launcher is manufactured and, as of August 2025, marketed by the US company Combined Systems, Incorporated (CSI), for use by military security and police forces in a variety of scenarios. It was originally developed for the US Marine Corps.[184] According to CSI, the standard VENOM is a "30-tube, lightweight, high capacity 38 mm grenade launcher [intended for] mounted use on vehicles, tripod mounts, and static installations".[185] It can be "deployed remotely or on land and marine platforms". It "delivers non-lethal flash and sound, smoke obscuration, irritant [RCA], and blunt trauma effects. These capabilities support escalation of force, determination of intent, crowd dispersal and area denial objectives".[186] CSI also produce the VENOM Multi-Calibre Tube Launched Munitions System which is modular and is available in single, double and triple bank configurations. Each bank can launch ten 38 mm or 40 mm grenades, and five 66 mm grenades. VENOM can launch a range of 37/38 mm, 40 mm, and 66 mm projectiles either individually or in salvoes of up to 30 projectiles, depending on the system and configuration.[187]

5.3.6.2.4 Drones, Unmanned Ground Vehicles and Unmanned Weapons Systems. A growing range of unmanned ground vehicles (UGVs), unmanned aerial vehicles and unmanned weapons systems – some incorporating semi-autonomous capabilities – have been developed intended to perform a variety of military, security or policing operations. Companies including in Israel, Spain, the United Arab Emirates, and the US have previously developed UGVs incorporating RCA sprayers, foggers or projectile launchers.[188]

5.3.6.2.4.1 Modular Advanced Armed Robotic System (MAARS) [United States]. QinetiQ North America, Inc. (QNA) developed the Modular Advanced Armed Robotic System (MAARS) through "partnership with various agencies in the [US] Department of Defense".[189] MAARS was designed specifically for "reconnaissance, surveillance and target acquisition (RSTA) missions", and is intended to "keep soldiers at a safe distance from enemy fire while effectively executing security missions such as ambushes, hostage rescue, forced entry, booby-trapped areas, detainee riots, site security and IED [Improvised Explosive Device]".[190] Consequently, it is remotely controlled by an operator who can be over 1 kilometre away. MAARS incorporates a four-barrelled 40 mm grenade launcher that can fire either 40 mm high explosive rounds or a range of 'less-lethal' ammunition, including 40 mm tear gas grenades.[191] According to a November 2010 *New York Times* article, US Army Special Forces bought six MAARS "for classified missions", and the National Guard requested "dozens more to serve as sentries on bases in Iraq and Afghanistan".[192] Although there have been no reports of its field deployment by the US military to date, MAARS was trialled for example in July 2016 by the US Marine Corps Warfighting Laboratory;[193] and as of August 2025 is promoted by QNA on its website.

Companies, law enforcement bodies or other State entities including in Austria, Brazil, China, France, India, Israel, Russia, South Africa, and the US have previously explored or are currently exploring development and application of a range of UAVs as platforms for RCA delivery mechanisms.[194] Certain small UAVs and associated 'less lethal' weapons appear to be designed to deliver limited quantities of RCA in a highly focussed manner, against individuals or small groups. Other UAVs appear capable of delivering significantly larger quantities of RCAs over wide areas affecting large numbers of people, potentially in an excessive manner, such concerns increasing if multiple drones are employed.

5.3.6.2.4.2 Drone Tear Smoke Launcher System [India]. The Tear Smoke Unit of the Indian Border Security Force (BSF) developed, in September 2022 unveiled, and now promotes its Drone Tear Smoke Launcher.[195] A BSF spokesman stated that it "will be a potential force multiplier for the security forces working in the law-and-order management domain".[196] The launcher carries 6 "Multi Pellet Tear Smoke Grenades" each comprising 4 "Tear Smoke Pellets", and each pellet emitting "smoke" containing CS for about 20 seconds.[197] The grenades can be launched individually or together in a volley from the drone. While the associated drone has a range of 2–3 km and can be flown at altitude of upto 120 metres, the "ideal launching height" is stated as between 35–50 metres to prevent injuries to the targets from direct impact of tear gas projectiles and the "ideal deployment range" is 1km.[198]

5.3.6.2.4.3 Cyclone Riot Control Drone System [Israel]. The Israeli company, ISPRA, has developed and, as of August 2025, promotes the Cyclone riot control drone system on its website and at trade fairs.[199] According to the company's marketing materials, the Cyclone "provides police forces with the capability to disperse less lethal munitions from drones – allowing maximum accuracy, real time control of riot situations and minimum injuries to civilians while maintaining distance between police forces and rioters".[200] The company promotes the use of this drone system "when a barrier or obstacle is separating between the parties" and states that "with several drones, a continuous presence over the rioters can be maintained, dispersing nonlethal ammunition and providing commanders with real-time video image of the situation on the ground".[201] The Cyclone system incorporates 12 'less lethal' sub-munitions and the total weight of the "pyrotechnic mixture" contained is 400+/− 20 grammes.[202]

5.3.6.2.4.4 Skunk Riot Control Copter [South Africa]. The South African company Desert Wolf, as of August 2025, promotes the Skunk riot control drone. Capable of holding up to 4000 pepper balls, this drone was equipped with four high-capacity paintball barrels that can cumulatively release "80 Pepper balls per second, stopping any crowd in its tracks".[203] In October 2015, the company director announced development of a modified Skunk copter capable of carrying up to 48 tear gas grenades, supplied by an unnamed South American company, which could be dropped over crowds.[204] Multiple UAVs can be "operated in formation by a single operator [employing] the Desert Wolf Pangolin ground control station".[205]

5.3.6.2.5 Large RCA Projectiles. Prior to the coming into force of the CWC in 1997, certain States including Iraq, South Africa, the United States, the USSR and Yugoslavia undertook research and development for military purposes of a range of large calibre munitions and projectiles or other means of delivering RCAs over long distances and/or wide areas.[206] A number of such large calibre RCA munitions, projectiles and 'wide area' means of RCA delivery were reportedly utilised in armed conflicts, most notably by the US in Vietnam and by Iraq in its conflict with Iran.[207] Following the coming into force of the CWC certain States reportedly developed a variety of large RCA projectiles, apparently for use by their military.

5.3.6.2.5.1 Malodorant Cluster Munition [US]. From 2004, the US company General Dynamics Ordnance and Tactical Systems worked under the direction of the US Army's Armament Research, Development and Engineering Center (ARDEC) to develop a 155 mm artillery projectile.[208] The XM1063 was designed to carry out three interrelated functions, to: "separate combatants from non-combatants; suppress, disperse or engage personnel [and] deny personnel access to, use of, or movement through a particular area, point or facility".[209] The XM1063 was intended to have a range of at least 20 kilometres, and potentially up to 28 kilometres.[210] The multiple submunitions would be released above the target and then fall to the ground, dispersing their payloads,[211] with an estimated area coverage of at least 5000 square metres.[212] The proposed payload was described as a "liquid payload"[213] and a "non-lethal personnel suppression agent".[214] Payload agent effectiveness was apparently tested at the US Army's Edgewood Chemical Biological Center (ECBC),[215] suggesting a chemical fill. A Joint Non-Lethal Weapons Directorate (JNLWD) reference book on "non-lethal" weapons included a reference to a legal review conducted in 2007 of the "XM1063 Malodorant155 mm Artillery Round"[216] that indicated that such malodorant agents were considered for this munition. The project was reportedly suspended in 2008.[217] In a 2012 *New Scientist* article, Hambling noted that although "the project is on hold, [it] has been developed by General Dynamics…to the stage of test firings and could be reactivated".[218]

5.3.6.2.5.2 Large Calibre RCA Projectiles [Russia]. At least one Russian company reportedly developed a range of large calibre munitions and other devices intended to disseminate "irritant-action pyrotechnic composition" over wide areas or extended distances. According to the 2009 English language version of the 2006 *Ordnance and Munitions* volume of *Russia's Arms and Technologies*, the 'wide area' RCA means of delivery included 105 mm rocket-propelled grenades, 82 mm mortar shells weighing 3.5 kilograms and capable of reaching distances of more than 2.6 kilometres, 120 mm munitions weighing 16 kilograms for use in mortars or artillery that could reach up to 6.8 kilometres, munition dispensers carried by helicopters, and even 500-kilogram cluster bombs packed with chemical irritant submunitions designed to be dropped from low-altitude helicopters or aircraft flying up to 12 000 metres above the target. Information on these RCA means of delivery was last made publicly available in 2009.[219] It is not known whether production of these munitions and devices continues, or whether stocks remain.[220] Whilst the nature of the "irritant-action pyrotechnic composition" employed was not identified, these delivery mechanisms appeared capable of employing a range of RCAs.

5.3.6.3 Misuse of Wide Area RCA Delivery Systems

Although the deployment of wide area RCA means of delivery for law enforcement is not yet a common practice, there are indications that we may now be at a tipping point, with development, acquisition, use and misuse of these systems reported in an increasing number of States. For example, water cannon firing water laced with RCAs or malodorants were reported to have been used in Hong Kong in 2019–2020,[221] in Israel from 2013–2021,[222] and Thailand in 2020.[223] In 2019, Lebanese security forces inappropriately employed French multi-barrel launcher to fire salvoes of tear gas projectiles into protesting crowds.[224] In 2021 Colombian security forces inappropriately used US multi-barrel projectile launcher, capable of firing 30 tear gas projectile salvoes, against protests in several cities.[225] In 2018, the Israeli security forces used a range of tear gas drones in response to mass protests, in some cases against peaceful protestors, bystanders, journalists and field medical facilities (as discussed below). In February 2024, Indian police deployed drones dropping tear gas projectiles to disperse large-scale protests of farmers in Haryana province in Northern India,[226] in a manner considered inappropriate and excessive.[227] And from late 2022 to at least April 2025 Russian military forces repeatedly employed drones to drop RCA grenades upon Ukrainian forces in embedded field positions (as discussed below).

5.3.6.3.1 Israel/Occupied Territories: Use of Wide Area RCA Delivery Systems. Although Israeli security forces (ISF) employment of certain multiple barrel launchers was previously reported, a significant escalation in the scale of 'remote control' RCA delivery systems use was witnessed from March to June 2018,[228] most notably during the ISF response to widespread Palestinian protests marking the 'Great March of Return', and particularly on the border between Gaza and Israel. A UN mandated Independent International Commission of Inquiry into the protests stated that the Israeli security forces "fired heavy amounts of tear gas at demonstrators from vehicle-mounted multiple launchers and released tear gas from drones".[229] According to media and NGO sources, the ISF employed a range of commercially available drones against the protesters – apparently including some that had been originally designed and marketed for RCA delivery. The majority, however, appear to have been general purpose or surveillance drones that were adapted in an *ad hoc* manner for the delivery of either large single RCA projectiles or multiple RCA grenades and canisters.[230] Human rights organisations and the media documented the widespread and repeated inappropriate use of RCA dispersing drones against large numbers of protestors over several days, this included the insufficiently-targeted large scale release of tear gas projectiles onto crowds, only some of whom were violent; among those injured or affected were peaceful protesters and bystanders.[231] In certain cases RCA drones appear to have been used to deliberately target non-violent individuals or groups, including journalists,[232] medical workers and their patients. The UN Commission of Inquiry documented allegations of an apparent attack upon a health centre in Khan Yunis (which was clearly marked with medical insignia) on 13 April 2018 by one or more ISF drones.[233] An eye-witness stated "The drones, there were many of them coming one after another, hovering over us, over the tents, and then they would release tear gas canisters which would unload the canisters [sic] on and around the big medical tent, the field hospital. Tear gas canisters were being dropped at the door of the tent and around the tent and then they exploded.... Many health workers fainted".[234] In 2021 Israel authorised further employment of such RCA drones for law enforcement purposes.[235]

5.3.6.3.2 Russia: Use of Drones to Disperse RCAs in Armed Conflict in Ukraine. There have been repeated media reports that from late 2022 to at least April 2025, Russian military forces have used RCAs against Ukrainian military forces during their ongoing invasion of Ukraine, as part of what they call a 'Special Military Operation'. This has included the repeated reported use of drones to drop K-51, VOH, RH-VO and KS-23 Cherymukha ('Bird Cherry') grenades filled with the RCAs CS[236] and CN,[237] onto Ukrainian troops embedded in fortified positions, in order to dislodge them and enable subsequent military engagement.[238] In a video broadcast on 2 May 2023, on the Russian State-controlled *Channel One*, soldiers assigned to Russia's 810th Naval Infantry Brigade openly admitted to dropping RCA grenades into Ukrainian trenches "in order to smoke them out from fortified positions".[239] In December 2023, Russia's 810th Naval Infantry Brigade stated on its Telegram channel that the brigade is using a "radical change in tactics" by dropping K-514 grenades on Ukrainian forces.[240] In April 2024, Colonel Serhii Pakhomov, acting head of the

Ukrainian military's Atomic, Biological and Chemical Defence Forces, claimed that Kyiv had recorded "around 900 uses of riot control agents by Russia in the past six months out of over 1400 since the Feb[ruary] 2022 invasion".[241] In a July 2024 *Note Verbale*, Ukraine formally requested technical assistance from the OPCW in "countering the threat of chemical weapons use by the Russian Federation". In its *Note Verbale* Ukraine stated that between 15 February 2023–25 April 2024, 2968 cases of the enemy's use of riot control agents were recorded and documented. It also stated that from 25 April 2024 to 24 May 2024 there were an additional 715 cases of RCA dropped from UAVs and that 48% of these used K-51 and RG-VO gas grenades.[242] At the request of Ukraine, the OPCW Technical Secretariat has to date undertaken three separate Technical Assistance Visits (TAV) to the country following alleged incidents of use of toxic chemicals as a weapon. The three OPCW TAV reports issued in November 2024,[243] February 2025[244] and June 2025[245] collectively recorded evidence of the repeated use of drones to drop grenades (including RG-Vo-type grenades) containing CS against Ukraine armed forces positions. The OPCW reports did not identify the perpetrator of these attacks.

5.3.6.4 OPCW SAB Concerns About Wide Area RCA Delivery Mechanisms

Over the last decade, the OPCW SAB has highlighted its growing concern at the potential misuse of 'wide area' RCA means of delivery. In its 2012 report to States Parties on developments in science in technology, in preparation for the Third CWC Review Conference, the SAB "note[d] with concern isolated reports of the commercial availability of munitions apparently designed to deliver large amounts of riot control agents over long distances".[246] In its 2018 report to the 4th Review Conference, the SAB warned that the "continued development, testing, production and promotion of diverse munition systems capable of disseminating RCAs by remote control remains of concern, as these systems could also enable the delivery of toxic chemicals". The SAB identified certain delivery mechanisms of concern including "fixed indoor devices, external area clearing or area denial devices, grenade and multiple munition launchers, and delivery systems on unmanned ground and aerial vehicles (UGVs and UAVs)". The SAB stated that the "availability of such systems opens up the possibility that they could be filled intentionally with alternate types of chemicals including CWAs or *CNS-acting compounds*".[247] (Emphasis added).

In its 2023 report to the 5th Review Conference, the SAB again highlighted its concerns regarding the "continued development, testing, production, and promotion of diverse munition systems capable of disseminating RCAs on a large scale or over a long distance". They noted that a review of developments and specifications of commercial products showed that a large variety of delivery systems and ammunition carrying chemical payloads are now available. They noted that the "capabilities being developed *increasingly resemble military equipment*". (Emphasis added). In addition, building on the concerns highlighted in their report to the 4th Review Conference, the SAB stated that "these systems could be repurposed and filled with other chemicals including CWAs, CNS-acting chemicals, *and bioregulators*".[248] (Emphasis added). However, despite the repeated and increasingly stark warnings of the SAB, this issue has not been addressed collectively by the CWC States Parties.

5.4 Conclusions

The chemical and life sciences are undergoing a revolution in capabilities that is allowing scientists to understand and manipulate living systems in unprecedented ways. This revolution is capable of producing immense health and broader societal benefits, but these capabilities can also be abused for malign purposes. Beyond the neurosciences, discussed in Chapters 3 and 4, of particular potential relevance are developments and convergences in biotechnology, as well as their interactions with nanoscience, and artificial intelligence and computer science (including machine learning). Such overarching processes have significantly impacted upon research of CNS-acting chemical agents and broader incapacitating chemical agents. Important advances have been made in the discovery and screening of novel naturally occurring chemicals including toxins and bioregulators and in the synthesis of their chemical analogues. These advances have been paralleled by advances in analysis notably with regard to modes of activity, interactions with corresponding cell receptor and sub-receptor sites, and their roles in and effect upon the wider operation of the brain, CNS and other core regulatory and physiological systems in humans. These developments have happened at the same time as our capacity to manufacture an ever-wider range of biological substances, notably proteins and peptides including bioregulators, has advanced. This has been matched by our increasing capability to rationally design and synthesise analogues of such biologically active chemicals incorporating modifications resulting in the chemically synthesised substances possessing different physiological properties from their natural forebears.

Significant technological advances have also been made that can potentially be utilised to overcome the challenge of ensuring controlled delivery of CNS-acting or other types of incapacitating chemical agents to targeted individuals, or alternatively facilitating wide area dispersal potentially affecting large numbers of people. Developments in nanotechnology and encapsulation currently being used in a medical context to facilitate delivery, uptake and subsequent movement of CNS acting chemicals to active sites within the patient could instead be employed in CNS-weaponization. Developments in aerosol dispersal technologies and nanotechnology could also facilitate more effective particulate agent dissemination on a widescale for both benign and malign applications. Such advances could now be coupled with the contemporary commercial development of 'wide area' dual-use aerosolisation and agent delivery systems promoted for agricultural and other civilian purposes. A more direct concern could be their employment with the extensive range of delivery mechanisms primarily promoted for the 'wide area' dispersal of RCAs, resulting in the delivery of weaponised CNS-acting chemicals or other ICAs.

Undeniably, these multiple strands of research and the associated technological advances will yield major health and wider societal benefits. However, because of their intrinsic dual or rather multi-faceted applicability, they could instead be exploited for the weaponization of CNS-acting chemical agents, or for other malign manipulation of core human physiological systems. As the SAB noted in its 2023 report to the 5th CWC Review Conference:

> "...[T]here is limited available information about the analysis, generation, synthesis, and safe handling of biotoxins, bioregulators, and other chemicals of biological origin or biological inspiration. The ability to rapidly synthesise, modify, and produce large quantities of these materials may exacerbate safety and security related to this problem. Advances in technologies at the convergence of biology and chemistry increase the risk of a novel toxin or bioregulator that evades traditional defensive countermeasures (including medical and protective countermeasures and detection technologies) that are rapidly emerging and creating a proliferation concern..."[249]

References

1. R. Trapp, Convergence of chemistry and biology, and nanotechnology, in *Preventing chemical weapons: arms control and disarmament as the sciences converge*, ed. M. Crowley, M. Dando and L. Shang, Royal Society of Chemistry, 2018, p. 201.
2. OPCW, *Report of the Scientific Advisory Board on developments in science and technology for the Second Special Session of the Conference of the States Parties to review the operation of the Chemical Weapons Convention*, RC-2/DG.1. OPCW, The Hague, 2008, paragraph 2.4.
3. See: M. Crowley and M. R. Dando, *Toxin and Bioregulator Weapons: Preventing the Misuse of the Chemical and Life Sciences*, Palgrave Macmillan, Springer Nature, Cham, Switzerland, 2022, pp. 79–85, for a detailed discussion together with associated references to relevant Indian scientific research papers.
4. S. M. Gupta, *et al.*, Phytochemical Analysis of Indian Stinging Plants: An Initiative Towards Development of Future Novel Biothreat Agents for Self-defence, *Proc. Natl. Acad. Sci., India, Sect. B*, 2018, **88**(2), 819–825.
5. *Heracleum canescens, Delphinium brunonianum, Silybum marianum, Aconitum chasmanthum, Melia azedarach, Convallaria majalis, Solanum*

xanthocarpum, Hyoscyamus niger, Arisaema triphyllum, and Cannabis sativa.

6. S. M. Gupta, *et al.*, Himalayan Toxic Plants of Defense Importance, *ACTA Sci. Med. Sci.*, 2018, **2**(3), 44–48.
7. For further discussion together with associated references to relevant Chinese scientific research papers see: M. Crowley and M. R. Dando, *Toxin and Bioregulator Weapons: Preventing the Misuse of the Chemical and Life Sciences*, Palgrave Macmillan, Springer Nature, Cham, Switzerland, 2022, pp. 50–56.
8. US State Department, *Adherence to and compliance with arms control, non-proliferation, and disarmament agreements and commitments*, Washington, D.C., April 2024, page 23.
9. For further discussion together with associated references to relevant Chinese scientific research papers see: M. Crowley and M. R. Dando, *Toxin and Bioregulator Weapons: Preventing the Misuse of the Chemical and Life Sciences*, Palgrave Macmillan, Springer Nature, Cham, Switzerland, 2022, pp. 94–100.
10. For further discussion together with associated references to relevant Russian scientific research papers see: M. Crowley and M. R. Dando, *Toxin and Bioregulator Weapons: Preventing the Misuse of the Chemical and Life Sciences*, Palgrave Macmillan, Springer Nature, Cham, Switzerland, 2022, pp. 119–124.
11. M. Attene-Ramos *et al.*, High Throughput Screening, in *Encyclopedia of Toxicology*, 3rd edn, 2014, pp. 916–917.
12. OPCW, *Report of the Scientific Advisory Board on developments in science and technology for the Second Special Session of the Conference of the States Parties to review the operation of the Chemical Weapons Convention*, RC-2/DG.1. OPCW, The Hague, 2008, paragraphs 2.3.
13. OPCW, *Report of the Scientific Advisory Board on developments in science and technology for the Second Special Session of the Conference of the States Parties to review the operation of the Chemical Weapons Convention*, RC-2/DG.1. OPCW, The Hague, 2008, paragraphs 2.4.
14. OPCW, *Report of the Scientific Advisory Board on developments in science and technology for the Third Special Session of the Conference of the States Parties to review the operation of the Chemical Weapons Convention*, RC-3/DG.1. OPCW, The Hague, 2012, paragraph 47.
15. OPCW, *Report of the Scientific Advisory Board on developments in science and technology for the Third Special Session of the Conference of the States Parties to review the operation of the Chemical Weapons Convention*, RC-3/DG.1. OPCW, The Hague, 2012, paragraph 47.
16. OPCW, *Report by the Director-General: Report of the Scientific Advisory Board on Developments in Science and Technology to the Fifth Special Session of the Conference of the States Parties to Review the Operation of the Chemical Weapons Convention*, RC-5/DG.1. OPCW, The Hague, 22 February, 2023, paragraphs 37 & 38.

17. International Standards Organisation, *Information Technology – Artificial Intelligence – Artificial Intelligence Concepts and Terminology*, ISO/IEC 22989:2022, 2022.
18. Countering Weapons of Mass Destruction Office, *Department of Homeland Security Report on Reducing the Risks at the Intersection of Artificial Intelligence and Chemical, Biological, Radiological, and Nuclear Threats*, Department of Homeland Security, Washington, D.C, April 26, 2024.
19. I. H. Sarker, Deep Learning; A Comprehensive Overview on Techniques, Taxonomy, Applications and Research Directions, *SN Comput. Sci.*, 2021, **2**, 420.
20. F. Grisoni, Chemical language models for *de novo* drug design: Challenges and opportunities, *Curr. Opin. Struct. Biol.*, 2023, **79**, 102527.
21. (a) See, for example, Director-General, Response to the Report of the Thirty-Eighth Session of the Scientific Advisory Board. EC-106/DG.19, 4 July, 2024, paragraph 19 referring to the work using chemical language models to predict structures and properties of chemical compounds; (b) M. A. Skinnider, *et al.*, A deep generative model enables automated structure elucidation of novel psychoactive substances, *Nat. Mach. Intell.*, 2021, **3**, 973–984. (c) B. Taflaj, *et al.*, New Psychoactive Substances Toxicity: A Systematic Review of Acute and Chronic Psychiatric Effects, *Int. J. Mol. Sci.*, 2024, **25**, 9484.
22. J. B. Petro, *et al.*, Biotechnology: Impact on biological warfare and biodefense, *Biosecur. Bioterrorism*, 2003, **1**, 161–168.
23. F. Urbina, *et al.*, Dual use of artificial-intelligence-powered drug discovery, *Nat. Mach. Intell.*, 2021, **4**, 189–191.
24. F. Urbina, *et al.*, Dual use of artificial-intelligence-powered drug discovery, *Nat. Mach. Intell.*, 2021, **4**, 189–191.
25. F. Urbina, *et al.*, Dual use of artificial-intelligence-powered drug discovery, *Nat. Mach. Intell.*, 2021, **4**, 189–191.
26. F. Urbina, *et al.*, Dual use of artificial-intelligence-powered drug discovery, *Nat. Mach. Intell.*, 2021, **4**, 189–191.
27. J. Jumper, *et al.*, Highly accurate protein structure prediction with AlphaFold, *Nature*, 2021, **596**, 583–589.
28. M. Baek, *et al.*, Accurate prediction of protein structures and interactions using a three-track neural network, *Science*, 2021, **373**, 871–876.
29. Z. Lin *et al.*, Evolutionary-scale prediction of atomic level protein structure with a language model, Preprint at bioRxiv, 2022, https://www.biorxiv.org/content/10.1101/2022.07.20.500902v2.
30. E. Callaway, 'The entire protein universe': AI predicts shape of nearly every known protein, *Nature*, 2022, **608**, 15–16.
31. (a) E. Callaway, AlphaFold's new rival? Meta AI predicts shape of 600 million proteins, *Nature*, 2022, **611**, 211–212; (b) Z. Lin *et al.*, Evolutionary-scale prediction of atomic level protein structure with a language model, Preprint at bioRxiv, 2022, https://www.biorxiv.org/content/10.1101/2022.07.20.500902v2.

32. I. Anishchenko, *et al.*, De novo protein design by deep network hallucination, *Nature*, 2021, **600**, 547–552.
33. I. Anishchenko, *et al.*, De novo protein design by deep network hallucination, *Nature*, 2021, **600**, 547.
34. (a) A. Krin and G. Jeremias, Artificial Intelligence: possible risks and benefits for BWC and CWC, CBWNet, Working Paper, 5 July, 2023, p. 7; (b) See also S. R. Carter *et al.*, The Convergence of Artificial Intelligence and the Life Sciences: Safeguarding Technology, Rethinking Governance, and Preventing Catastrophe, Nuclear Threat Initiative, Washington, D.C. October, 2023; (c) Department of Homeland Security, Department of Homeland Security Report on Reducing the Risks at the Intersection of Artificial Intelligence and Chemical, Biological, Radiological, and Nuclear Threats, Department of Homeland Security, Washington, D.C. April 26, 2024.
35. E. Callaway, Major AlphaFold upgrade offers boost for drug discovery, *Nature*, 2024, DOI: 10.1038/d41586-024-01383-zNews.
36. A. Krin and G. Jeremias, Artificial Intelligence: possible risks and benefits for BWC and CWC, *CBWNet, Working Paper*, 5 July, 2023, pp. 9–11.
37. F. Urbina, *et al.*, Dual use of artificial-intelligence-powered drug discovery, *Nat. Mach. Intell.*, 2021, **4**, 190.
38. A. Krin and G. Jeremias, Artificial Intelligence: possible risks and benefits for BWC and CWC, *CBWNet, Working Paper*, 5 July, 2023, pp. 10–11.
39. Z. Zhong *et al.*, Recent advances in artificial intelligence for retrosynthesis, 2023. This paper was submitted to Cornell University's Annual Global Survey of Machine Learning on 14 January 2023.
40. F. Urbina, *et al.*, Dual use of artificial-intelligence-powered drug discovery, *Nat. Mach. Intell.*, 2021, **4**, 189–191.
41. OPCW, *Report by the Director-General: Report of the Scientific Advisory Board on Developments in Science and Technology to the Fifth Special Session of the Conference of the States Parties to Review the Operation of the Chemical Weapons Convention*, RC-5/DG.1. OPCW, The Hague, 22 February, 2023, p. 39.
42. National Academies of Sciences, Engineering, and Medicine, *Biodefense in the Age of Synthetic Biology*, The National Academies Press, Washington, DC, 2018, p. 15.
43. A. Adibekian and P. Stallforth, Cutting Edger Chemical Biology: Report from the 2016 International Symposium on Chemical Biology, January, 13-15 2016, Geneva, Switzerland, *ACS Chem. Biology*, 2016, **11**(4), 816–820.
44. Spiez Laboratory and CSS ETH Zurich, Spiez Convergence – Report on the First Workshop, 6-9 October 2016, as cited in R. Trapp, Convergence of chemistry and biology, and nanotechnology, in *Preventing chemical weapons: arms control and disarmament as the sciences converge*, ed. M. Crowley, M. Dando and L. Shang, Royal Society of Chemistry, 2018, p. 203.

45. National Academies of Sciences, Engineering, and Medicine, *Biodefense in the Age of Synthetic Biology*, The National Academies Press, Washington, DC, 2018, p. 3–4.
46. OPCW, *Report by the Director-General: Report of the Scientific Advisory Board on Developments in Science and Technology to the Fifth Special Session of the Conference of the States Parties to Review the Operation of the Chemical Weapons Convention*, RC-5/DG.1. OPCW, The Hague, 22 February, 2023, p. 67.
47. OPCW, *Report by the Director-General: Report of the Scientific Advisory Board on Developments in Science and Technology to the Fifth Special Session of the Conference of the States Parties to Review the Operation of the Chemical Weapons Convention*, RC-5/DG.1. OPCW, The Hague, 22 February, 2023, p. 68.
48. OPCW, *Report by the Director-General: Report of the Scientific Advisory Board on Developments in Science and Technology to the Fifth Special Session of the Conference of the States Parties to Review the Operation of the Chemical Weapons Convention*, RC-5/DG.1. OPCW, The Hague, 22 February, 2023, p. 68.
49. J. Clapper, *Director of National* Intelligence, Statement for the Record: Worldwide Threat Assessment of the US Intelligence Community, Senate Armed Services Committee, 9 February 2016, p. 9.
50. US State Department, *Adherence to and compliance with arms control, non-proliferation, and disarmament agreements and commitments*, State Department, Washington, D.C., April 2024, pp. 25–26.
51. R. Trapp, Convergence of chemistry and biology, and nanotechnology, in *Preventing chemical weapons: arms control and disarmament as the sciences converge*, ed. M. Crowley, M. Dando and L. Shang, Royal Society of Chemistry, 2018, p. 203.
52. Spiez Laboratory and CSS ETH Zurich, Spiez Convergence – Report on the First Workshop, 6-9 October 2016, as cited in R. Trapp, Convergence of chemistry and biology, and nanotechnology, in *Preventing chemical weapons: arms control and disarmament as the sciences converge*, ed. M. Crowley, M. Dando and L. Shang, Royal Society of Chemistry, 2018, p. 203.
53. R. Trapp, Convergence of chemistry and biology, and nanotechnology, in *Preventing chemical weapons: arms control and disarmament as the sciences converge*, ed. M. Crowley, M. Dando and L. Shang, Royal Society of Chemistry, 2018, p. 204.
54. National Academies of Sciences, Engineering, and Medicine, *Biodefense in the Age of Synthetic Biology*, The National Academies Press, Washington, DC, 2018, p. 4.
55. National Academies of Sciences, Engineering, and Medicine, *Biodefense in the Age of Synthetic Biology*, The National Academies Press, Washington, DC, 2018, p. 4.
56. OPCW, Convergence of Chemistry and Biology Report of the Scientific Advisory Board's Temporary Working Group Adopted by the Scientific

Advisory Board at its Twenty-First Session, 27 June 2014, SAB/REP/1/14, p. 3.
57. OPCW, *Report of the Scientific Advisory Board on Developments in Science and Technology for the Fourth Special Session of the Conference of the States Parties to Review the Operation of the Chemical Weapons Convention*, RC-4/DG, 2018, paragraph 4.
58. OPCW, *Report of the Scientific Advisory Board on Developments in Science and Technology for the Fourth Special Session of the Conference of the States Parties to Review the Operation of the Chemical Weapons Convention*, RC-4/DG, 2018, paragraph 6.
59. OPCW, *Report of the Scientific Advisory Board on Developments in Science and Technology for the Fourth Special Session of the Conference of the States Parties to Review the Operation of the Chemical Weapons Convention*, RC-4/DG, 2018, paragraph 7.
60. OPCW, *Report of the Scientific Advisory Board on Developments in Science and Technology for the Fourth Special Session of the Conference of the States Parties to Review the Operation of the Chemical Weapons Convention*, RC-4/DG, 2018, paragraph 9.
61. OPCW, *Report by the Director-General: Report of the Scientific Advisory Board on Developments in Science and Technology to the Fifth Special Session of the Conference of the States Parties to Review the Operation of the Chemical Weapons Convention*, RC-5/DG.1. OPCW, The Hague, 22 February, 2023, paragraph 21.
62. OPCW, *Report by the Director-General: Report of the Scientific Advisory Board on Developments in Science and Technology to the Fifth Special Session of the Conference of the States Parties to Review the Operation of the Chemical Weapons Convention*, RC-5/DG.1. OPCW, The Hague, 22 February, 2023, paragraph 69.
63. OPCW, *Report of the Scientific Advisory Board on Developments in Science and Technology for the Fourth Special Session of the Conference of the States Parties to Review the Operation of the Chemical Weapons Convention*, RC-4/DG, 2018, paragraphs 10 & 11.
64. OPCW, *Report by the Director-General: Report of the Scientific Advisory Board on Developments in Science and Technology to the Fifth Special Session of the Conference of the States Parties to Review the Operation of the Chemical Weapons Convention*, RC-5/DG.1. OPCW, The Hague, 22 February, 2023, paragraph 15.
65. OPCW, *Report by the Director-General: Report of the Scientific Advisory Board on Developments in Science and Technology to the Fifth Special Session of the Conference of the States Parties to Review the Operation of the Chemical Weapons Convention*, RC-5/DG.1. OPCW, The Hague, 22 February, 2023, paragraph 16.
66. OPCW, *Report by the Director-General: Report of the Scientific Advisory Board on Developments in Science and Technology to the Fifth Special Session of the Conference of the States Parties to Review the Operation of*

the Chemical Weapons Convention, RC-5/DG.1. OPCW, The Hague, 22 February, 2023, paragraph 17.

67. OPCW, *Report by the Director-General: Report of the Scientific Advisory Board on Developments in Science and Technology to the Fifth Special Session of the Conference of the States Parties to Review the Operation of the Chemical Weapons Convention*, RC-5/DG.1. OPCW, The Hague, 22 February, 2023, paragraph 18.
68. OPCW, *Report by the Director-General: Report of the Scientific Advisory Board on Developments in Science and Technology to the Fifth Special Session of the Conference of the States Parties to Review the Operation of the Chemical Weapons Convention*, RC-5/DG.1. OPCW, The Hague, 22 February, 2023, paragraph 18.
69. (a) M. Crowley and M. Dando, Down the Slippery Slope? A study of contemporary dual-use chemical and life science research potentially applicable to incapacitating chemical agent weapons, Biochemical Security 2030 Project, University of Bath, October 2014; (b) M. Dando and M. Furmanski, Midspectrum Incapacitant Programs, in *Deadly Cultures: Biological Weapons Since 1945*, ed. M. Wheelis, L. Rózsa and M. Dando, Harvard University Press, Harvard, 2006; (c) N. Davison, *Nonlethal weapons*, Palgrave Macmillan, Basingstoke, 2009.
70. China North Industries Corporation (NORINCO), undated, Security, Anti-Riot Weapons and Ammunition brochure, distributed at MILIPOL security exhibition, Paris, France, 1995, p. 11.
71. China North Industries Corporation (NORINCO), undated, Security, Anti-Riot Weapons and Ammunition brochure, distributed at MILIPOL security exhibition, Paris, France, 1995, p. 11.
72. State 9619 Plant company brochures, distributed during Asia Pacific China Police Expo 2004 (23rd–26th June 2004), and Asia Pacific China Police Expo 2006 (24th–27th May 2006) held at Beijing Exhibition Centre, Beijing, China, undated.
73. G. Arthur, New Equipment in Hong Kong, Defence Review Asia, 19th March 2012, p. 35. http://www.defencereviewasia.com/articles/153/NEW-EQUIPMENT-IN-HONG-KONG.
74. Radio Free Asia, Thousands of people forcibly demolished a Hui village, police used anesthesia guns against Hui people, 7 June 2013. https://www.rfa.org/cantonese/news/demolition-06072013094102.html (accessed 17 September 2020) [original in Cantonese, translation using google translate].
75. United States, Compliance with the Convention on the Prohibition of the Development, Production, Stockpiling, and Use of Chemical Weapons and on Their Destruction [*2021 CWC Compliance Report*], April 2022, p. 11.
76. M. Normack *et al.*, *Israel and WMD: Incentives and capabilities*, Swedish Defence Research Agency (FOI), December, 2005, p. 41.
77. M. Normack *et al.*, *Israel and WMD: Incentives and capabilities*, Swedish Defence Research Agency (FOI), December, 2005, p. 41.

78. (a) See: M. Crowley, The Man Who Haunts Israel, Time Magazine, 29 July 2014; (b) Y. A. Hilalah, A. Jazeera, Kill Him Silently: Mossad vs Khaled Meshaal: The story behind Mossad's bungled bid to assassinate Hamas leader Khaled Meshaal; (c) Harvard Sussex Programme, News Chronology, CBW Conventions Bulletin, Issue no. 38, December 1997, p. 29; (d) Physician Member of Hit Team, Paper Says, *Can. Med. Assoc. J.*, 1997, **157**(11), 1504; (e) L. Beyer, Don't Try This at Home–Or in Aman, *Time*, 1997, **150**(17), 27 – as cited in The Moscow Theater Hostage Crisis: Incapacitants and Chemical Warfare, Chemical and Biological Weapons Nonproliferation Program, CNS, James Martin Center for Nonproliferation Studies, Monterrey Institute of International Studies; (f) M. Ginsburg, 'Should there be a need': The inside story of Israel's chemical and biological arsenal, Times of Israel, 17th September 2013; (g) A. Cowell, The Daring Attack That Blew Up in Israel's Face, New York Times, 15th October 1997.
79. Israeli Intelligence Agenices Come Under Fire, *Janes Intelligence Review*, 1st January 1998.
80. Israeli Intelligence Agenices Come Under Fire, *Janes Intelligence Review*, 1st January 1998.
81. A. Cowell, The Daring Attack That Blew Up in Israel's Face, *New York Times*, 15th October 1997. http://www.nytimes.com/1997/10/15/world/the-daring-attack-that-blew-up-in-israel-sface.html?pagewanted=all&src=pm (accessed 16th April 2014).
82. A. Cowell, The Daring Attack That Blew Up in Israel's Face, *New York Times*, 15th October 1997. http://www.nytimes.com/1997/10/15/world/the-daring-attack-that-blew-up-in-israel-sface.html?pagewanted=all&src=pm (accessed 16th April 2014).
83. (a) M. Crowley, The Man Who Haunts Israel, Time Magazine, 29 July 2014; (b) Harvard Sussex Programme, News Chronology, CBW Conventions Bulletin, Issue no. 38, December 1997, p. 29.
84. (a) Amnesty International, Attempt to kill Hamas leader follows a pattern of extrajudicial killings, News Service 168/97, 8th October, AI Index: MDE 15/89/97, 1997; (b) Harvard Sussex Programme, News Chronology, CBW Conventions Bulletin, Issue no. 38, December 1997, p. 29.
85. (a) B. Anderson and P. Grant, Dose Safety Margin for Enhancement for Chemical Incapacitation and Less-than-Lethal Targeting, NIJ Final Report and Recommendations. Lawrence Livermore National Laboratory, Forensic Science Centre. For further discussion see: Governments researched fentanyls as weapons for decades, Associated Press, 8 October 2016, 1997; (b) N. Davidson, *'Non-Lethal' Weapons*, Palgrave Macmillan, Basingstoke, 2009; (c) A. M. Pearson, Late and Post Cold War Research and Development of Incapacitating Biochemical Weapons, in *Promise of Peril: Incapacitating Biochemical Weapons*, ed. A. M. Pearson and M. I. Chevrier and M. Wheelis, Lexington Books, Lanham, MD, 2007, ch. 4, pp. 67–101.

86. (a) B. Anderson and P. Grant, Dose Safety Margin for Enhancement for Chemical Incapacitation and Less-than-Lethal Targeting, NIJ Final Report and Recommendations. Lawrence Livermore National Laboratory, Forensic Science Centre. For further discussion see: Governments researched fentanyls as weapons for decades, Associated Press, 8 October 2016, 1997; (b) N. Davidson, *'Non-Lethal' Weapons*, Palgrave Macmillan, Basingstoke, 2009; (c) A. M. Pearson, Late and Post Cold War Research and Development of Incapacitating Biochemical Weapons, in *Promise of Peril: Incapacitating Biochemical Weapons*, ed. A. M. Pearson and M. I. Chevrier and M. Wheelis, Lexington Books, Lanham, MD, 2007, ch. 4, pp. 67–101.
87. United States Army, Topic CBD00-108, Chemical Immobilizing Agents for Non-lethal Applications, Small Business Innovation Research Solicitation, CBD 00.1, December 1999.
88. United States Army, Topic CBD00-108, Chemical Immobilizing Agents for Non-lethal Applications, Small Business Innovation Research Solicitation, CBD 00.1, December 1999.
89. United States Army, CBD,26 Phase I Selections from the 00.1 Solicitation. Summary available from http://web.archive.org/web/20101123094323/, http://dodsbir.net/selections/abs001CBD.htm (accessed 2 February 2025).
90. D. Ruppe, United States: U.S. Military Studying Nonlethal Chemicals, *Global Security Newswire*, 4th November 2002.
91. (a) National Research Council, *An Assessment of 'Non-Lethal' Weapons Science and Technology*. National Academies Press, Washington, D.C., 2003, pp. 124–125; (b) N. Davidson, *'Non-Lethal' Weapons*, Palgrave Macmillan, Basingstoke, 2009.
92. United States, *Annual Report on Compliance with the Chemical Weapons Convention (CWC)*, Condition (10)(C), Bureau of Arms Control, Deterrence, and Stability, State Department, Washington, D.C., 4 April 2024.
93. United States, *Annual Report on Compliance with the Chemical Weapons Convention (CWC)*, Condition (10)(C), Bureau of Arms Control, Deterrence, and Stability, State Department, Washington, D.C., 4 April 2024.
94. M. Giveh and the Good ISIS Team, The Islamic Republic's Work on Pharmaceutical Based Agents, Institute for Science and International Security, 25 November 2024.
95. M. Giveh and the Good ISIS Team, The Islamic Republic's Work on Pharmaceutical Based Agents, Institute for Science and International Security, 25 November 2024, p. 3.
96. M. Giveh and the Good ISIS Team, The Islamic Republic's Work on Pharmaceutical Based Agents, Institute for Science and International Security, 25 November 2024, p. 4.
97. M. Giveh and the Good ISIS Team, The Islamic Republic's Work on Pharmaceutical Based Agents, Institute for Science and International Security, 25 November 2024, p. 5.

98. M. Giveh and the Good ISIS Team, The Islamic Republic's Work on Pharmaceutical Based Agents, Institute for Science and International Security, 25 November 2024, p. 5.
99. OPCW, Convergence of Chemistry and Biology, Report of the SAB's Temporary Working Group, SAB/REP/1/14, June 2014. OPCW, The Hague.
100. M. Balali-Mood, *et al.*, Impact of scientific developments on the Chemical Weapons Convention/ IUPAC Report, *Pure Appl. Chem.*, 2008, **80**, 175–200, paragraph 12.
101. Khanna *et al.* (2015) As cited in R. Trapp, Convergence of chemistry and biology, and nanotechnology, in *Preventing chemical weapons: arms control and disarmament as the sciences converge*, ed. M. Crowley, M. Dando and L. Shang, Royal Society of Chemistry, 2018, p. 208.
102. S. Bamrungsap, *et al.* (2012), *Nanomedicine*, 2012, 7(8), 1253. As cited in R. Trapp, Convergence of chemistry and biology, and nanotechnology, in *Preventing chemical weapons: arms control and disarmament as the sciences converge*, ed. M. Crowley, M. Dando and L. Shang, Royal Society of Chemistry, 2018, p. 208.
103. N. Yadav, *et al.*, *Int. J. Appl. Pharm.*, 2013, 5(2), 8. As cited in R. Trapp, Convergence of chemistry and biology, and nanotechnology, in *Preventing chemical weapons: arms control and disarmament as the sciences converge*, ed. M. Crowley, M. Dando and L. Shang, Royal Society of Chemistry, 2018, p. 208.
104. OPCW, *Report of the Scientific Advisory Board on Developments in Science and Technology for the Fourth Special Session of the Conference of the States Parties to Review the Operation of the Chemical Weapons Convention*, RC-4/DG, 2018, paragraph 157.
105. R. Trapp, Convergence of chemistry and biology, and nanotechnology, in *Preventing chemical weapons: arms control and disarmament as the sciences converge*, ed. M. Crowley, M. Dando and L. Shang, Royal Society of Chemistry, 2018, p. 209.
106. OPCW, *Report of the Scientific Advisory Board on Developments in Science and Technology for the Fourth Special Session of the Conference of the States Parties to Review the Operation of the Chemical Weapons Convention*, RC-4/DG, 2018, paragraph 157.
107. OPCW, *Report of the Scientific Advisory Board on Developments in Science and Technology for the Fourth Special Session of the Conference of the States Parties to Review the Operation of the Chemical Weapons Convention*, RC-4/DG, 2018, paragraph 158.
108. R. Trapp, Convergence of chemistry and biology, and nanotechnology, in *Preventing chemical weapons: arms control and disarmament as the sciences converge*, ed. M. Crowley, M. Dando and L. Shang, Royal Society of Chemistry, 2018, p. 209.
109. R. Trapp, Convergence of chemistry and biology, and nanotechnology, in *Preventing chemical weapons: arms control and disarmament as the sciences converge*, ed. M. Crowley, M. Dando and L. Shang, Royal Society of Chemistry, 2018, pp. 209–210.

110. R. Trapp, Convergence of chemistry and biology, and nanotechnology, in *Preventing chemical weapons: arms control and disarmament as the sciences converge*, ed. M. Crowley, M. Dando and L. Shang, Royal Society of Chemistry, 2018, pp. 209–210.
111. (a) See also: M. Balali-Mood, *et al.*, Impact of scientific developments on the Chemical Weapons Convention/ IUPAC Report, *Pure Appl. Chem.*, 2008, **80**, 175–200, paragraphs 11 and 12; (b) OPCW, *Report of the Scientific Advisory Board on developments in science and technology for the Second Special Session of the Conference of the States Parties to review the operation of the Chemical Weapons Convention*, RC-2/DG.1. OPCW, The Hague, 2008, paragraphs 2.5–2.8; (c) OPCW, *Report of the Scientific Advisory Board on developments in science and technology for the Third Special Session of the Conference of the States Parties to review the operation of the Chemical Weapons Convention*, RC-3/DG.1. OPCW, The Hague, 2012, paragraphs 50–53, 55, 57 and 58.
112. National Research Council, *Emerging Cognitive Neuroscience and Related Technologies*, National Academies Press, Washington, DC, 2008, p. 51.
113. M. S. Kalaiarasi, *et al.*, Development of biocompatible nanogel for sustained drug release by overcoming the blood brain barrier in zebrafish model, *J. Appl. Biomed.*, 2016, **14**(2), 157.
114. Y. Liu, *et al.*, Brain-targeted co-delivery of therapeutic gene and peptide by multifunctional nanoparticles in Alzheimer's disease mice, *Biomaterials*, 2016, **80**, 33.
115. T.-T. Zhang, *et al.*, Strategies for transporting nanoparticles across the blood–brain barrier, *Biomater. Sci.*, 2016, **4**, 219.
116. National Research Council, *Emerging Cognitive Neuroscience and Related Technologies*, National Academies Press, Washington, DC, 2008, p. 135.
117. S. S. Suri, Nanotechnology-based drug delivery systems, *J. Occup. Med. Toxicol.*, 2007, **2**, 16, as cited in Royal Society, *Neuroscience, conflict and security*, Royal Society, London, 2012, p. 51.
118. A. Lalatsa *et al.* (2011) A prodrug nanoparticle approach for the delivery of peptides to the brain *via* the oral and intravenous routes. Submitted to L. Martin-Banderas, M. A. Holgado, J. L. Venero, J. Alvarez-Fuentes and M. Fernández-Arévalo, Nanostructures for Drug Delivery to the Brain. *Curr. Med. Chem.*, 2011, **18**(34), 5303–5321, as cited in Royal Society, *Neuroscience, conflict and security*, Royal Society, London, 2012, p. 52.
119. Royal Society, *Neuroscience, conflict and security*, Royal Society, London, 2012, p. 52.
120. R. Trapp, Convergence of chemistry and biology, and nanotechnology, in *Preventing chemical weapons: arms control and disarmament as the sciences converge*, ed. M. Crowley, M. Dando and L. Shang, Royal Society of Chemistry, 2018, p. 210.
121. G. Scheuch, *et al.*, Clinical perspectives on pulmonary systematic and macromolecular delivery, *Adv. Drug Delivery Rev.*, 2006, **58**, 996–1008.
122. G. Scheuch, *et al.*, Clinical perspectives on pulmonary systematic and macromolecular delivery, *Adv. Drug Delivery Rev.*, 2006, **58**, 996–1008.

123. N. Csaba, Nanoparticles for Nasal Vaccination, *Adv. Drug Delivery Rev.*, 2009, **61**, 140–157.
124. N. Csaba, Nanoparticles for Nasal Vaccination, *Adv. Drug Delivery Rev.*, 2009, **61**, 140–157.
125. J. J. Lochhead and R. G. Thorne, Intranasal delivery of biologics to the central nervous system, *Adv. Drug Delivery Rev.*, 2012, **64**, 614–628.
126. J. J. Lochhead and R. G. Thorne, Intranasal delivery of biologics to the central nervous system, *Adv. Drug Delivery Rev.*, 2012, **64**, 614–628.
127. M. Kosfeld, *et al.*, Oxytocin increases trust in humans, *Nature*, 2005, **435**, 673–676.
128. Vero Labs, http://www.verolabs.com (last accessed September 12, 2016).
129. O. Weissman, *et al.*, Oxytocin administration to parent increases infant physiological and behavioral readiness for social engagement, *Biol. Psychiatry*, 2012, **72**(12), 982–989.
130. S. Shoyele and A. Slowey, Prospects for formulating proteins/peptides as aerosols for pulmonary delivery, *Int. J. Pharm.*, 2006, **314**(1), 1–8.
131. M. Balali-Mood, *et al.*, Impact of scientific developments on the Chemical Weapons Convention/ IUPAC Report, *Pure Appl. Chem.*, 2008, **80**, 175–200.
132. M. Balali-Mood, *et al.*, Impact of scientific developments on the Chemical Weapons Convention/ IUPAC Report, *Pure Appl. Chem.*, 2008, **80**, 175–200, paragraph 11.
133. OPCW, *Report of the Scientific Advisory Board on developments in science and technology for the Second Special Session of the Conference of the States Parties to review the operation of the Chemical Weapons Convention*, RC-2/DG.1. OPCW, The Hague, 2008, paragraph 2.7.
134. OPCW, *Report of the Scientific Advisory Board on developments in science and technology for the Second Special Session of the Conference of the States Parties to review the operation of the Chemical Weapons Convention*, RC-2/DG.1. OPCW, The Hague, 2008, paragraph 2.8.
135. OPCW, *Report of the Scientific Advisory Board on developments in science and technology for the Third Special Session of the Conference of the States Parties to review the operation of the Chemical Weapons Convention*, RC-3/DG.1. OPCW, The Hague, 2012, paragraph 57.
136. OPCW, *Report of the Scientific Advisory Board on developments in science and technology for the Third Special Session of the Conference of the States Parties to review the operation of the Chemical Weapons Convention*, RC-3/DG.1. OPCW, The Hague, 2012, paragraph 58.
137. OPCW, *Report of the Scientific Advisory Board on developments in science and technology for the Third Special Session of the Conference of the States Parties to review the operation of the Chemical Weapons Convention*, RC-3/DG.1. OPCW, The Hague, 2012, paragraph 58.
138. P. M. Wax, *et al.*, Unexpected 'Gas' Casualties in Moscow: A Medical Toxicology Perspective, *Ann. Emerg. Med.*, 2003, **41**(5), 700–705.
139. OPCW, *Report of the Scientific Advisory Board on developments in science and technology for the Second Special Session of the Conference of the*

States Parties to review the operation of the Chemical Weapons Convention, RC-2/DG.1. OPCW, The Hague, 2008, paragraph 2.6.
140. P. M. Wax, *et al.*, Unexpected 'Gas' Casualties in Moscow: A Medical Toxicology Perspective, *Ann. Emerg. Med.*, 2003, **41**(5), 700–705.
141. National Research Council, *Emerging Cognitive Neuroscience and Related Technologies*, National Academies Press, Washington, DC, 2008, p. 137.
142. National Research Council, *Emerging Cognitive Neuroscience and Related Technologies*, National Academies Press, Washington, DC, 2008, p. 137.
143. National Research Council, *Emerging Cognitive Neuroscience and Related Technologies*, National Academies Press, Washington, DC, 2008, p. 137
144. OPCW, *Report of the Scientific Advisory Board on developments in science and technology for the Third Special Session of the Conference of the States Parties to review the operation of the Chemical Weapons Convention*, RC-3/DG.1. OPCW, The Hague, 2012, paragraph 56.
145. OPCW, *Report by the Director-General: Report of the Scientific Advisory Board on Developments in Science and Technology to the Fifth Special Session of the Conference of the States Parties to Review the Operation of the Chemical Weapons Convention*, RC-5/DG.1. OPCW, The Hague, 22 February, 2023, paragraph 81.
146. Unmanned aerial vehicles (UAVs) have been defined as "powered aerial vehicles that do not carry a human operator, use aerodynamic forces to provide vehicle lift, can fly autonomously or be piloted remotely, can be expendable or recoverable, and can carry a lethal or nonlethal payload" [See: E. Bone and C. Bolkcom (2003) Unmanned aerial vehicles: background and issues for Congress, Report for Congress, RL31872, 25 April].
147. OPCW, *Report by the Director-General: Report of the Scientific Advisory Board on Developments in Science and Technology to the Fifth Special Session of the Conference of the States Parties to Review the Operation of the Chemical Weapons Convention*, RC-5/DG.1. OPCW, The Hague, 22 February, 2023, paragraph 81.
148. OPCW, *Report by the Director-General: Report of the Scientific Advisory Board on Developments in Science and Technology to the Fifth Special Session of the Conference of the States Parties to Review the Operation of the Chemical Weapons Convention*, RC-5/DG.1. OPCW, The Hague, 22 February, 2023, paragraph 83.
149. See for example: D. Rassler, Remotely piloted innovation, terrorism, drones and supportive technology, *Combatting Terrorism Center at West Point*, US Military Academy, October 2016, p. 60.
150. (a) L. Friese *et al.*, Emerging unmanned threats: The use of commercially-available UAVs by armed non-state actors, Armament Research Services, 2016, p. 35; (b) See also: D. Rassler, Remotely piloted innovation, terrorism, drones and supportive technology, Combatting Terrorism Center at West Point, US Military Academy, October 2016, p. 60–61.
151. (a) L. Friese *et al.*, Emerging unmanned threats: The use of commercially-available UAVs by armed non-state actors, Armament Research Services,

2016, p. 35; (b) See also: D. Rassler, Remotely piloted innovation, terrorism, drones and supportive technology, Combatting Terrorism Center at West Point, US Military Academy, October 2016, p. 60–61.
152. OPCW, *Report by the Director-General: Report of the Scientific Advisory Board on Developments in Science and Technology to the Fifth Special Session of the Conference of the States Parties to Review the Operation of the Chemical Weapons Convention*, RC-5/DG.1. OPCW, The Hague, 22 February, 2023, paragraph 82.
153. V. Klochikhin *et al.*, Principles of Modelling of the Scenario of Calmative Application in a Building with Deterred Hostages, *Proceedings of the 3rd European Symposium on Non-Lethal Weapons*, Ettlingen, Germany, 10–12th May, V17, Pfinztal: Fraunhofer ICT, 2005, p. 3.
154. V. Klochikhin *et al.*, Principles of Modelling of the Scenario of Calmative Application in a Building with Deterred Hostages, *Proceedings of the 3rd European Symposium on Non-Lethal Weapons*, Ettlingen, Germany, 10–12th May, V17, Pfinztal: Fraunhofer ICT, 2005, p. 3.
155. V. Klochikhin, Principles of Modelling of the Scenario of Calmative Application in a Building with Deterred Hostages, *Proceedings of the 3rd European Symposium on Non-Lethal Weapons*, Ettlingen, Germany, 10–12th May, V17, Pfinztal: Fraunhofer ICT, 2005, pp. 3–4.
156. V. Klochikhin and V. Selivanov, Report on the 1st phase of the Project "Gas Flow", *Presentation in MBDA, 24th–27th November*, London, 2009. [Copy of presentation held by authors]. Further details of this London meeting are not available.
157. V. Klochikhin and V. Selivanov, Report on the 1st phase of the Project "Gas Flow", *Presentation in MBDA, 24th–27th November*, London, 2009. [Copy of presentation held by authors]. Further details of this London meeting are not available., slide 77.
158. (a) See: M. Crowley and M. R. Dando, *Toxin and Bioregulator Weapons: Preventing the Misuse of the Chemical and Life Sciences*, Palgrave Macmillan, Springer Nature, Cham, Switzerland, 2022; (b) M. Crowley, Development and Hostile Use of Toxic Chemical Means of Delivery and Dispersal, in *Preventing Chemical Weapons: Arms Control and Disarmament as the Sciences Converge*, ed. M. Crowley *et al.*, Royal Society of Chemistry, 2018, ch. 12, pp. 332–380; (c) M. Crowley, *Chemical Control: Regulation of Incapacitating Chemical Agent Weapons, Riot Control Agents and their Means of Delivery*, Palgrave Macmillan, Basingstoke, 2016, pp. 88–106; (d) M. Crowley, Tear gassing by remote control: The development and promotion of remotely operated means of delivering or dispersing riot control agents, Remote Control Project, December 2015; (e) M. Crowley, Chinese "wide area" riot control agent means of delivery: implications for the Chemical Weapons Convention, November; Bradford University & Omega Research Foundation, 2015; (f) M. Crowley, Drawing the line: Regulation of "wide area" riot control agent delivery mechanisms under the Chemical Weapons Convention, Omega Research Foundation/University of Bradford, April 2013.

159. (a) NonLethal Technologies, Inc., Less Lethal, Tactical Munitions and Weapons Systems for Military, Corrections and Law Enforcement, Product catalogue, 2017. Available on https://www.nonlethaltechnologies.com/pdf/NLT_Brochure.pdf. Accessed 1 August 2025; (b) Mace Security International, Inc., TG Guard Security Protection System, 2019. https://web.archive.org/web/20231215041105/https://tgguard.com/index.html (last captured 15 December 2023).
160. NonLethal Technologies, Inc., *Less Lethal, Tactical Munitions and Weapons Systems for Military, Corrections and Law Enforcement, Product catalogue*, 2017, p. 11. Available on https://www.nonlethaltechnologies.com/pdf/NLT_Brochure.pdf. Accessed 2 February 2025.
161. NonLethal Technologies, Inc., *Product Data Sheets, TG-100-OC – Irritant Powder Dispenser OC Powder*, 2018. https://www.nonlethaltechnologies.com/pdf/DS/TG-OC.pdf. Accessed 1 August 2025.
162. NonLethal Technologies, Inc., *Product Data Sheets, TG-100-OC – Irritant Powder Dispenser OC Powder*, 2018. https://www.nonlethaltechnologies.com/pdf/DS/TG-OC.pdf. Accessed 6 September 2022.
163. NonLethal Technologies, Inc., *Less Lethal, Tactical Munitions and Weapons Systems for Military, Corrections and Law Enforcement, Product catalogue*, 2017, p. 11. Available on https://www.nonlethaltechnologies.com/pdf/NLT_Brochure.pdf. Accessed 2 February 2025.
164. NonLethal Technologies, Inc., *Product Data Sheets, TG-100-OC – Irritant Powder Dispenser OC Powder*, 2018. https://www.nonlethaltechnologies.com/pdf/DS/TG-OC.pdf. Accessed 6 September 2022.
165. (a) United States, Compliance with the Convention on the Prohibition of the Development, Production, Stockpiling, and Use of Chemical Weapons and on Their Destruction, 2020, p. 10; (b) United States, Compliance with the Convention on the Prohibition of the Development, Production, Stockpiling, and Use of Chemical Weapons and on Their Destruction, 2021, p. 14. (c) United States, Compliance with the Convention on the Prohibition of the Development, Production, Stockpiling, and Use of Chemical Weapons and on Their Destruction, 2022, p. 13.
166. (a) United States, Compliance with the Convention on the Prohibition of the Development, Production, Stockpiling, and Use of Chemical Weapons and on Their Destruction, 2020, pp. 10–11; (b) United States, Compliance with the Convention on the Prohibition of the Development, Production, Stockpiling, and Use of Chemical Weapons and on Their Destruction, 2021, p. 14; (c) United States, Compliance with the Convention on the Prohibition of the Development, Production, Stockpiling, and Use of Chemical Weapons and on Their Destruction, 2022, pp. 13–14.
167. MSI Delivery Systems Inc., *AB2K MMADS User and Safety Manual*, 3rd edn, February 1 2016. Available at: https://dl-manual.com/doc/ab2k-mmads-user-safety-manual-third-edition-0vejljpejnop. (Accessed 1 August 2025).

At the time of writing the company website was no longer operational.

168. (a) MSI Delivery Systems Inc., AB2K MMADS User and Safety Manual, 3rd edn, February 1 2016. Available at: https://dl-manual.com/doc/ab2k-mmads-user-safety-manual-third-edition-0vejljpejnop. (Accessed 18 February 2022). At the time of writing the company website was no longer operational; (b) MSI Delivery Systems Inc., Afterburner 2000 Smoke Delivery System, White Paper, 25 January 2010.
169. (a) MSI Delivery Systems Inc., Afterburner 2000 aerosol delivery system, slideshow presentation, slide 6. The company further stated that it only provides the non-toxic training smoke. "Irritants" are purchased [by the customer] from local suppliers, undated; (b) MSI Delivery Systems Inc., Afterburner 2000 Smoke Delivery System, White Paper, 25 January 2010, p. 4.
170. MSI Delivery Systems Inc., *Afterburner 2000 Smoke Delivery System, White Paper,* 25 January 2010, p. 3.
171. MSI Delivery Systems Inc., *Afterburner 2000 Smoke Delivery System, White Paper,* 25 January 2010, p. 3.
172. (a) K. Cheng, Hong Kong police to receive 3 water cannon vehicles, worth HK$27m, this year, Hong Kong Free Press, 19 January 2018, 2018, updated 31 March 2020, https://hongkongfp.com/2018/01/19/hong-kong-police-receive-3-water-cannon-vehicles-worth-hk27m-year/; (b) P. Oltermann, Hong Kong activist to Germany: stop selling riot control kit to city, The Guardian, 11 September 2019. https://www.theguardian.com/world/2019/sep/11/hong-kong-activist-to-germany-stop-selling-riot-control-kit-to-city.
173. Cops buy three more water cannon trucks at HK$9.9 million, The Standard *(Hong Kong)*, 22 December 2021. https://www.thestandard.com.hk/breaking-news/section/4/184952/Cops-buy-three-more-water-cannon-trucks-at-HK$9.9-million.
174. (a) For media coverage see for example: J. Creery, Hong Kong police water cannon truck fires blue-coloured water as protesters besiege gov't HQ, Hong Kong Free Press, 31 August 2019, updated 31 March 2020. https://hongkongfp.com/2019/08/31/breaking-police-water-cannon-truck-fires-blue-coloured-water-hong-kong-protesters-besiege-govt-hq/; (b) BBC, Hong Kong: Blue-dyed water fired at protesters defying ban. BBC, 31 August 2019. https://www.bbc.co.uk/news/av/world-asia 49536000; (c) Agence France-Presse/The Guardian, Hong Kong protests: police fire water cannon with blue dye as crowds defy ban, 21 October 2019. https://www.theguardian.com/world/2019/oct/21/hong-kong-protests-police-fire-water-cannon-with-blue-dye-as-crowds-defy-ban.
175. H. Chan, Greenpeace questions Hong Kong police claim that blue dye from water cannon is 'harmless', Hong Kong Free Press, 25 October 2019, updated 31 March 2020.
176. Cops buy three more water cannon trucks at HK$9.9 million, The Standard *(Hong Kong)*, 22 December 2021. https://www.thestandard.

com.hk/breaking-news/section/4/184952/Cops-buy-three-more-water-cannon-trucks-at-HK$9.9-million.

177. (a) M. Crowley, Tear gassing by remote control: The development and promotion of remotely operated means of delivering or dispersing riot control agents, Remote Control Project, December 2015, pp.10–15; (b) M. Crowley and M. R. Dando, Toxin and Bioregulator Weapons: Preventing the Misuse of the Chemical and Life Sciences, Palgrave Macmillan, Springer Nature, Cham, Switzerland, 2022.
178. China Ordnance Equipment Research Institute, Undated Catalogue, Distributed at Asia Pacific China Police Expo 2012, (copy held by Omega Research Foundation).
179. China Ordnance Equipment Research Institute, Undated Catalogue, Distributed at Asia Pacific China Police Expo 2012, (copy held by Omega Research Foundation), pp. 8 and 13.
180. China Ordnance Equipment Research Institute, Undated Catalogue, Distributed at Asia Pacific China Police Expo 2012, (copy held by Omega Research Foundation), p. 13.
181. (a) NonLethal Technologies, The IronFist, 2016; (b) See also: NonLethal Technologies, Product Catalogue, 2017. Both available from company website: http://nonlethaltechnologies.com/. Accessed 1 August 2025.
182. NonLethal Technologies, The IronFist, 2016.
183. NonLethal Technologies, Iron Fist, undated brochure, distributed at IDEX 2013, Abu Dabhai, United Arab Emirates, 17–23 February 2013, 2013.
184. (a) Combined Systems, Inc., The Less-Lethal Product Source, Product catalogue (undated) p. 17; (b) See also: Combined Systems, Inc. Defense Solutions, Military product catalogue (undated), p. 13. both available at https://combinedsystems.com/downloads?page=1.
185. Combined Systems, Inc., The Less-Lethal Product Source, Product catalogue. Available at: https://www.combinedsystems.com/download-our-catalog/.
186. Combined Systems, Inc., VENOM Specifications Sheets (undated), p. 1. https://web.archive.org/web/20231208024312/https://www.combinedsystems.com/wp-content/uploads/2020/04/V3037-_-VENOM-FLIER.pdf.
187. Combined Systems, Inc., Defense Solutions, Military product catalogue (undated), pp. 12–15, https://www.combinedsystems.com/wp-content/uploads/2020/05/CSI-MilitaryBrochure-FINAL.pdf.
188. (a) M. Crowley, Tear gassing by remote control: The development and promotion of remotely operated means of delivering or dispersing riot control agents, Remote Control Project, December 2015, pp. 15–21; (b) M. Crowley and M. R. Dando, Toxin and Bioregulator Weapons: Preventing the Misuse of the Chemical and Life Sciences, Palgrave Macmillan, Springer Nature, Cham, Switzerland, 2022.
189. QinetiQ North America, MAARS Advanced armed roboticsystem, http://www.qinetiq-na.com/products/unmanned-systems/maars/ (accessed 28 October 2019, now removed).

190. QinetiQ North America, MAARS Datasheet, available from https://www.qinetiq.com/en/what-we-do/services-and-products/maars-weaponized-robot (accessed 1 August 2025).
191. QinetiQ North America, MAARS Datasheet, available at https://www.qinetiq.com/en/what-we-do/services-and-products/maars-weaponized-robot (accessed 1 August 2025).
192. J. Markoff, War Machines: Recruiting Robots for Combat, *The New York Times*, 27th November, 2010.
193. See: Marine Corps Warfighting Laboratory MAGTAF Integrated Experiment (MCWL), Defense Imagery Distribution System, 9 July 2016, https://www.dvidshub.net/image/2729263/marine-corps-warfighting-laboratory-magtaf-integrated-experiment-mcwl#.V8mPkBmM0o8.facebook. Accessed 1 August 2025.
194. (a) M. Crowley, Tear gassing by remote control: The development and promotion of remotely operated means of delivering or dispersing riot control agents, Remote Control Project, December 2015, pp. 22–29; (b) M. Crowley and M. R. Dando, Toxin and Bioregulator Weapons: Preventing the Misuse of the Chemical and Life Sciences, Palgrave Macmillan, Springer Nature, Cham, Switzerland, 2022; (c) NDTV, Border Security Force Develops Tear Gas Shell-Dropping Drone System, 2 September 2022. https://www.ndtv.com/india-news/border-security-force-develops-tear-gas-shell-dropping-drone-system-3311348 (accessed 2 August 2025); (d) Condor Non-Lethal Technologies, Condor Drop, https://condornaoletal.com.br/product/condor-drop (accessed 1 August 2025).
195. Tear Smoke Unit, Drone Tear Smoke Launcher, https://tearsmoke.org/products/drone-tear-smoke-launcher/ (accessed 2 August 2025).
196. NDTV, Border Security Force Develops Tear Gas Shell-Dropping Drone System, 2 September 2022.
197. Tear Smoke Unit, Multi Pellet TS Grenade, https://tearsmoke.org/grenade/multi-pellet-ts-grenade/.
198. Tear Smoke Unit, Drone Tear Smoke Launcher, https://tearsmoke.org/products/drone-detailed-information/ (accessed 2 August 2025).
199. (a) Ispra, products, riot control drone system, https://www.ispraltd.com/Riot-Control-Drone-System.html (accessed 1 August 2025); (b) Cyclone riot control drone system, ISPRA product leaflet, January 2015, http://www.ispraltd.com/image/users/423329/ftp/my_files/Riot%20Control%20Drone%20-%20Cyclone-1.pdf?id=30746915 (accessed 1 August 2025); (c) for image of ISPRA promotion of Cyclone at ADEX 2018, held in Baku, Azerbaijan, in September 2018 see: https://www.armyrecognition.com/adex_2018_news_official_show_daily/adex_2018_ispra_promotes_cyclone_riot_control_drone_system.html (accessed 1 August 2025).
200. Cyclone riot control drone system, ISPRA product leaflet, January 2015, http://www.ispraltd.com/image/users/423329/ftp/my_files/Riot%20Control%20Drone%20-%20Cyclone-1.pdf?id=30746915 (accessed 1 August 2025).

201. ISPRA Smart solutions for crowd control, Product Catalogue, 2019, p. 22, https://sfilev2.f-static.com/image/users/423329/ftp/my_files/Ispra_Catalog_(2019).pdf?id=32113724 (accessed 1 August 2025).
202. Cyclone riot control drone system, ISPRA product leaflet, January 2015, http://www.ispraltd.com/image/users/423329/ftp/my_files/Riot%20Control%20Drone%20-%20Cyclone-1.pdf?id=30746915 (accessed 1 August 2025).
203. Skunk Riot Control Copter, Desert Wolf website. Available at http://www.desert-wolf.com/dw/products/unmanned-aerial-systems/skunk-riot-control-copter.html (accessed 1 August 2025).
204. G. Martin, Desert Wolf Adding Grenades to Skunk Riot Control UAV, Defenceweb, 7 October 2015.
205. Skunk Riot Control Copter, Desert Wolf website, available at http://www.desert-wolf.com/dw/products/unmanned-aerial-systems/skunk-riot-control-copter.html. Accessed 1 August 2025.
206. (a) J. Perry Robinson and M. Leitenberg, The problem of chemical and biological warfare, in *The rise of CB weapons*, Stockholm, Sweden, 1971, vol. 1; (b) J. Perry Robinson, The problem of chemical and biological warfare, in *CB Weapons Today*, SIPRI, Stockholm, Sweden, 1973, vol. II.
207. M. Crowley, *Drawing the line: regulation of 'wide area' riot control agent delivery mechanisms under the Chemical Weapons Convention*, University of Bradford/Omega Research Foundation, April 2013, pp. 6–8.
208. (a) United States Army, Picatinny Centre, Non-Lethal Artillery Structural Firing (FYO4) Purchase Order Contract in Support of the FY04 155mm Non-Lethal Artillery Projectile Program. Contract Number W15QKN-04-M-0328, 14th September 2004. Available at www.sunshine-project.org/incapacitants/jnlwdpdf/XM1063.pdf (accessed 14th February 2013); (b) J. McCormick and R. Lee, Presentation on 155mm XM1063 Non-Lethal Personnel Suppression Projectile, General Dynamics OTS, National Defense Industrial Association, 42nd Annual Armament Systems: Gun and Missile Systems Conference and Exhibition, Charlotte, North Carolina, USA, 2007, p. 4. Available at http://www.dtic.mil/ndia/2007gun_missile/GMTuePM2/McCormickPresentation.pdf (accessed 28 June 2022).
209. J. McCormick and R. Lee, *Presentation on 155mm XM1063 Non-Lethal Personnel Suppression Projectile*, General Dynamics OTS, National Defense Industrial Association, 42nd Annual Armament Systems: Gun and Missile Systems Conference and Exhibition, Charlotte, North Carolina, USA, 2007, p. 4. Available at http://www.dtic.mil/ndia/2007gun_missile/GMTuePM2/McCormickPresentation.pdf (accessed 28 June 2022).
210. J. McCormick and R. Lee, *155mm XM1063 Non-Lethal Personnel Suppression Projectile*, Presentation to the 41st Annual Armament Systems: Guns and Missile Systems, Conference & Exhibition, National Defense Industrial Association (NDIA), Sacramento, US, 27th–30th March 2006, p. 12. Available at http://www.dtic.mil/ndia/2006garm/tuesday/mccormick.pdf (accessed 1 August 2025).

211. J. McCormick and R. Lee, *155mm XM1063 Non-Lethal Personnel Suppression Projectile*, Presentation to the 41st Annual Armament Systems: Guns and Missile Systems, Conference & Exhibition, National Defense Industrial Association (NDIA), Sacramento, US, 27th–30th March 2006, p. 5. Available at http://www.dtic.mil/ndia/2006garm/tuesday/mccormick.pdf (accessed 28 June 2022).
212. NLOS-C Non-Lethal Personnel Suppression, US Army ARDEC brochure, 2005, As cited by N. Davison *"Off the Rocker" and "On the Floor": The Continued Development of Biochemical Incapacitating Weapons*. Bradford Disarmament Research Centre, Report, No. 8, August 2007, 2007, p. 34.
213. US Army AEDEC, Solicitation, R—155mm XM1063 Non-lethal Artillery Engineering Support Contract, Solicitation number: W15QKN-04-X-0819, 30th September 2004. as cited in N. Davison, *"Off the Rocker" and "On the Floor": The Continued Development of Biochemical Incapacitating Weapons*. Bradford Disarmament Research Centre, Report, No. 8, August 2007, 2007, p. 34.
214. J. McCormick and R. Lee, *155mm XM1063 Non-Lethal Personnel Suppression Projectile*, Presentation to the 41st Annual Armament Systems: Guns and Missile Systems, Conference & Exhibition, National Defense Industrial Association (NDIA), Sacramento, US, 27th–30th March 2006, p. 7. Available at http://www.dtic.mil/ndia/2006garm/tuesday/mccormick.pdf (accessed 28 June 2022).
215. US Army ARDEC, Solicitation (Modification) R-155mm XM1063 Non-lethal Artillery Engineering Support Contract (Ref: W15QKN-04-X-0819), *FBO Daily*, 30th September 2004. As cited in N. Davison, *"Off the Rocker" and "On the Floor": The Continued Development of Biochemical Incapacitating Weapons*. Bradford Disarmament Research Centre, Report, No. 8, August 2007, 2007, p. 34.
216. Joint Non-Lethal Weapons Directorate, *Non-Lethal Weapons (NLW) Reference Book*, 30th June 2011, p. viii. Copy of report available at http://publicintelligence.net/dod-non-lethal-weapons-2011/ (accessed 1 August 2025).
217. D. Hambling, US military malodorant missiles kick up a stink, *New Scientist*, Issue 2867, 4th June 2012.
218. D. Hambling, US military malodorant missiles kick up a stink, *New Scientist*, Issue 2867, 4th June 2012.
219. Non-lethal munitions section, Volume 12 "Ordnance and Munitions" (English language version), Russia's Arms and Technologies, *The XXI Century Encyclopedia*, (version 2006.1eng), Arms and Technologies Publishing House, 5th May 2009.
220. (a) M. Crowley, *Dangerous Ambiguities: Regulation of Riot Control Agents and Incapacitants under the Chemical Weapons Convention*, Bradford Non-lethal Weapons Research Project, 2009, pp. 108–110; (b) M. Crowley, Drawing the line: regulation of 'wide area' riot control agent delivery mechanisms under the Chemical Weapons Convention,

University of Bradford/Omega Research Foundation, April 2013, pp. 25–38; (c) M. Crowley, *Chemical Control: Regulation of Incapacitating Chemical Agent Weapons, Riot Control Agents and their Means of Delivery*, Palgrave Macmillan, Basingstoke, 2016, pp. 99–104.

221. (a) J. Creery, Hong Kong police water cannon truck fires blue-coloured water as protesters besiege gov't HQ, Hong Kong Free Press, 31 August 2019, updated 31 March 2020. https://hongkongfp.com/2019/08/31/breaking-police-water-cannon-truck-fires-blue-coloured-water-hong-kong-protesters-besiege-govt-hq/; (b) BBC, Hong Kong: Blue-dyed water fired at protesters defying ban. BBC, 31 August 2019. https://www.bbc.co.uk/news/av/world-asia 49536000; (c) Agence France-Presse/The Guardian, Hong Kong protests: police fire water cannon with blue dye as crowds defy ban, 21 October 2019. https://www.theguardian.com/world/2019/oct/21/hong-kong-protests-police-fire-water-cannon-with-blue-dye-as-crowds-defy-ban.

222. (a) S. Michaeli, Crowd Control Israel's Use of Crowd Control Weapons in the West Bank, B'Tselem [The Israeli Information Center for Human Rights in the Occupied Territories], January 2013; (b) Association for Civil Rights in Israel, Spraying "skunk" at homes—violation of the right to dignity and freedom of demonstration. ACRI, 23rd January 2013 [original in Hebrew]; (c) Amnesty International, Israel and Occupied Palestinian Territories: Trigger-happy: Israel's Use of Excessive Force in the West Bank. MDE 15/002/2014, 27th February 2014; (d) The Economist, A whiff from hell: Skunk, a high-tec Israeli weapon against stone-throwers, 6th June 2015; (e) Physicians for Human Rights/International Network of Civil Liberties Organisations, Lethal in Disguise: The health consequences of crowd control weapons, 2016; (f) Physicians for Human Rights/International Network of Civil Liberties Organisations, Lethal in Disguise 2: How Crowd-Control Weapons Impact Health and Human Rights, 2023.

223. Human Rights Watch, *Thailand: Water Cannon Used Against Peaceful Activists*, 17 October 2020.

224. (a) Amnesty International, Suppressing Protests: French Less-Lethal Weapons Used In Lebanon, January 2021; (b) Amnesty International, Lebanon: New evidence reveals French law enforcement equipment unlawfully used to crush protests, 28 January 2021.

225. (a) Human Rights Watch, Colombia: Egregious Police Abuses Against Protesters, 9 June 2021; (b) Amnesty International, Cali: In the epicentre of repression: human rights violations during the 2021 national strike in Cali, Colombia, AMR 23/1405/2021, 30 July 2021.

226. BBC News, India farmers' protest: Tear gas fired as thousands march on fortress Delhi, 13 February 2024, https://www.bbc.co.uk/news/world-asia-india-68271356 (accessed 1 August 2025).

227. (a) Amnesty International, India: Right to peaceful protest under threat due to mounting restrictions and escalating crackdown on farmers' march, 14 February 2024; (b) Human Rights Watch, India: Protect Farmers' Right to Peaceful Protest, 29 February 2024, https://www.hrw.

Section III: Holistic Arms Control Analysis

org/news/2024/02/29/india-protect-farmers-right-peaceful-protest (accessed 1 August 2025).
228. In first, Israel uses drones to drop tear gas on Gaza protesters — report, *Times of Israel*, 12 March 2018.
229. United Nations, *Report of the detailed findings of the Independent International Commission of Inquiry on the protests in the Occupied Palestinian Territory*, Human Rights Council, Fortieth Session, 25 February–22 March 2019, A/HRC/40/CRP.2, 18 March 2019, paragraph 478.
230. (a) New drone technology drops tear gas on Gaza protesters, 30 March 2018, AFP/Ynetnews.com, https://www.ynetnews.com/articles/0,7340,L-5211564,00.html (accessed 1 August 2025); (b) D. Hilton, Drones over Gaza: How Israel tested its latest technology on protesters, Middle East Eye, 16 May 2018, https://www.middleeasteye.net/news/drones-over-gaza-how-israel-tested-its-latest-technology-protesters (accessed 1 August 2025); (c) N. Waters, First ISIS, then Iraq, now Israel: IDF Use of Commercial Drones, Belingcat, 18 June 2018, https://www.bellingcat.com/news/mena/2018/06/18/first-isis-iraq-now-israel-idf-use-commercial-drones/ (accessed 1 August 2025).
231. (a) Israel deploys drones to drop tear gas on Gaza protesters, Times of Israel, 31 March 2018; (b) Amnesty International, Israel/OPT: Israel/OPT: Israeli forces must end the use of excessive force in response to "Great March of Return" protests, Amnesty International, 13 April 2018; (c) Drones rain tear gas down on protesters as 35,000 Palestinians clash with Israeli security forces on the Gaza border, Daily Mail, 14 May 2018; (d) Gaza protests: All the latest updates, Al Jazeera, 12 November 2018; (e) F. Greenwood and O. Zaqqout, Drones Don't Wear Uniforms, They Should, Foreign Policy, 22 May 2018; (f) N. Waters, First ISIS, then Iraq, now Israel: IDF Use of Commercial Drones, Belingcat, 18 June 2018, https://www.bellingcat.com/news/mena/2018/06/18/first-isis-iraq-now-israel-idf-use-commercial-drones/ (accessed 19 February 2019).
232. A. Jazeera, Israeli drone targets journalists, 12 November 2018, https://www.aljazeera.com/news/2018/11/12/gaza-protests-all-the-latest-updates (accessed 1 August 2025).
233. United Nations, *Report of the detailed findings of the Independent International Commission of Inquiry on the protests in the Occupied Palestinian Territory*, Human Rights Council, Fortieth Session, 25 February–22 March 2019, A/HRC/40/CRP.2, 18 March 2019, paragraph 478.
234. United Nations, *Report of the detailed findings of the Independent International Commission of Inquiry on the protests in the Occupied Palestinian Territory*, Human Rights Council, Fortieth Session, 25 February–22 March 2019, A/HRC/40/CRP.2, 18 March 2019, paragraph 563.
235. J. Breiner, Israel Using Drones to Tear Gas Palestinian Demonstrators in West Bank, *Haaretz*, 28 April 2021. https://www.haaretz.com/israel-news/2021-04-28/ty-article/.premium/israeli-police-using-drones-to-drop-tear-gas-on-palestinian-demonstrators/0000017f-f49d-d887-a7ff-fcfd91cd0000 (accessed 1 August 2025).

236. 2-Chlorobenzylidenemalonitrile.
237. Chloroacetophenone or phenacyl chloride.
238. (a) L. Phillips and D. Crouch, Have Chemical Weapons been Used in Ukraine? Royal United Services Institute, 20 June 2023; (b) D. Hambling, Russia's Tear Gas Bombings In Ukraine May Be First Step In Dangerous Chemical Escalation, Forbes, 1 November 2022; (c) F. Hoske *et al.*, Ukraine says Russia steps up illegal use of tear gas to clear trenches, Associated Press, 17 April 2024; (d) for alleged CS drone use against non-military targets see: L. Harding, 'I haven't slept for two days': Kharkiv residents reel from Russian attacks, The Guardian, 26 April 2025.
239. Russian TV, *Channel 1, "Time" programme*, 2 May 2023, https://www.1tv.ru/news/issue/2023-05-02/21:00#6 (accessed 1 August 2025).
240. See United States, Condition, *(10)(C) Annual Report on Compliance with the Chemical Weapons Convention (CWC)*, Bureau of Arms Control, Deterrence, and Stability, 4 April 2024.
241. F. Hoske *et al.*, Ukraine says Russia steps up illegal use of tear gas to clear trenches, *Associated Press*, 17 April 2024.
242. Ukraine, *Note Verbale from Permanent Representation of Ukraine to the OPCW, to the Technical Secretariat of the OPCW*, No.61219/35-196/50-81468, 13 June 2024, p. 1.
243. OPCW Technical Secretariat, Report of the OPCW Technical Assistance Visit on the activities carried out in support of a request by Ukraine (TAV/04/24) S/2338/2024, 18 November 2024, 2025.
244. OPCW Technical Secretariat, Report of the OPCW Technical Assistance Visit on the activities carried out in support of a request by Ukraine (TAV/05/24 AND TAV/01/25) S/2370/2025, 14 February 2025.
245. OPCW Technical Secretariat, Report of the OPCW Technical Assistance Visit on the activities carried out in support of a request by Ukraine (TAV/02/25 AND TAV/03/25), S/2415/2025, 20 June 2025.
246. OPCW, *Report of the Scientific Advisory Board on Developments in Science and Technology for the Third Special Session of the Conference of the States Parties to Review the Operation of the Chemical Weapons Convention*, RC-3/DG.1, Third Review Conference, 8th–19th April 2013, 29th October 2012, paragraph 56.
247. OPCW, *Report of the Scientific Advisory Board on Developments in Science and Technology for the Fourth Special Session of the Conference of the States Parties to Review the Operation of the Chemical Weapons Convention*, RC-4/DG.1, 30 April 2018, paragraph 154.
248. OPCW, *Report of the Scientific Advisory Board on Developments in Science and Technology to the Fifth Special Session of the Conference of the States Parties to Review the Operation of the Chemical Weapons Convention*, RC-5/DG.1, February 2023, paragraph 80.
249. OPCW, *Report by the Director-General: Report of the Scientific Advisory Board on Developments in Science and Technology to the Fifth Special Session of the Conference of the States Parties to Review the Operation of the Chemical Weapons Convention*, RC-5/DG.1. OPCW, The Hague, 22 February, 2023, paragraph 72.

CHAPTER 6

The Biological and Toxin Weapons Convention and Preventing the Weaponization of CNS-acting Chemicals

6.1 Introduction

The Biological and Toxin Weapons Convention (BTWC) entered into force on 26 March 1975.

As of February 2025 the BTWC has 187 States Parties, with Micronesia being the most recent.[1] Four States have signed but not yet ratified the Convention: Egypt, Haiti, Somalia, and Syria – by signing they have given political support to the objectives and principles of the BTWC and committed themselves to not undermining its objectives. Seven additional States have neither signed nor acceded: Chad, Comoros, Djibouti, Eritrea, Israel, and Kiribati.

The object and purpose of the BTWC is set out in the final two paragraphs of the Convention's Preamble:

> "Determined for the sake of all mankind, to *exclude completely* the possibility of bacteriological (biological) agents *and toxins being used as weapons*.
>
> Convinced that such use would be repugnant to the conscience of mankind and that no effort should be spared to minimize this risk."[2] (Emphases added).

Preventing Weaponization of CNS-acting Chemicals: A Holistic Arms Control Analysis
By Michael Crowley and Malcolm Dando

Published by the Royal Society of Chemistry, www.rsc.org

As will be explored in the chapter, the BTWC has a number of mechanisms intended to ensure that it remains relevant in the light of, and responds appropriately to, the opportunities and challenges arising from the continuous developments in the life sciences and related technologies. One important mechanism built into the Convention was the establishment of five yearly Review Conferences intended to allow States Parties the opportunity to collectively review and improve the Convention's implementation, and which were specifically tasked to "take into account any new scientific and technological developments relevant to the Convention".[3]

However, due to the restrictions imposed to deal with the COVID-19 pandemic, the preparations for and the actual holding of the 9th BTWC Review Conference in 2022 were much delayed. Unfortunately, these delays meant that BTWC State Parties met against the background of the war in Ukraine. Subsequently, due primarily to the diplomatic consequences of the ongoing international conflict, only a largely *status quo* outcome was achieved at the Review Conference with the agreement of a bare-bones *Final Declaration* and a commitment to continue the development of the existing annual Meetings of States Parties. In the light of the current very difficult international environment, this chapter explores how the BTWC and attendant control regime can be strengthened and more effectively implemented by States Parties to meet its objectives, and specifically to prevent and address the development and use of CNS-acting weapons employing toxins, bioregulators and other naturally occurring substances as well as their synthetic analogues.

6.2 General Obligations Under the BTWC

Article I of the BTWC establishes the principal obligations and prohibitions of the Convention, declaring that:

> "Each State Party to the Convention undertakes never in any circumstances to develop, produce, stockpile or otherwise acquire or retain:
>
> (1) Microbial or other biological agents, or toxins, whatever their origin or method of production, of types and in quantities that have no justification for prophylactic, protective or other peaceful purposes.
> (2) Weapons, equipment or means of delivery designed to use such agents or toxins for hostile purposes or in armed conflict."[4]

Under Article II, the BTWC requires the destruction or diversion for peaceful purposes of all biological and toxin weapons.[5] Under Article III, the BTWC also prohibits any direct or indirect transfer of biological and toxin weapons to anyone, as well as any assistance, encouragement or inducement of any State, group of States or international organisation to manufacture or otherwise acquire them.[6]

The BTWC does not *explicitly* establish a prohibition on the *use* of biological and toxin weapons. However, such prohibition is *implicit* through bans on the development, production, stockpiling or acquisition or retention by other means of these agents. It would only be after one or other of these prohibitions had been breached that use could occur. Furthermore, under Article VIII, the BTWC invokes the provisions of the 1925 Geneva Protocol which explicitly contain such a prohibition.[7] The BTWC States Parties have clarified their collective interpretation that any use of a biological or toxin weapon would imply a violation of the BTWC, and this has been repeatedly affirmed by the BTWC State Parties in Extended Understandings enunciated in the *Final Declarations* of, for example the Fourth,[8] Sixth,[9] and Seventh[10] and Eighth Review[11] Conferences.

6.3 CNS-acting Chemicals and the Scope of Coverage

Since a number of CNS-acting chemicals and a broader range of other incapacitating chemical agents (ICAs) are either components or synthetic biologically active analogues of such substances, they appear to fall within the scope of Article I of the BTWC, particularly if one considers the terms 'toxins' and 'other biological agents' broadly, as explored below. This interpretation is reinforced as a result of the expansive scope of Article I coverage affirmed in successive Extended Understandings of BTWC Review Conferences.[12]

Unfortunately, the term 'toxins' was not explicitly defined under the Convention nor has a definition been collectively agreed subsequently by BTWC States Parties in their Extended Understandings. However, scientific and medical organisations have explored the issue[13] and certain States have explicitly defined toxins in their national measures.[14] Consequently, toxins are commonly taken to denote chemical compounds produced by living organisms that are toxic or harmful to another living organism; and although this has not been formally recognised by the BTWC States Parties, there appears to be a general acceptance of such wide scope by the States Parties when addressing toxins.

The importance of the BTWC's expansive scope of coverage for 'toxins' has been underlined by the World Health Organisation (WHO), specifically highlighting its implications for the application of the Convention to bioregulators.[15]

> "In the sense of the Biological and Toxin Weapons Convention, 'toxin' includes substances to which scientists would not normally apply the term. For example, there are chemicals that occur naturally in the human body that would have toxic effects if administered in large enough quantity. Where a scientist might see a bioregulator, say, the treaty would see a poisonous substance produced by a living organism, in other words a toxin..."

And the text continued:

> "...nor is this unreasonable. Wasp venom, for example, is clearly a toxin, yet its active principle is histamine, which is also a human bioregulator. Although histamine might not itself be made into an effective weapon, the same cannot necessarily be said for other bioregulators."

Similarly, the term 'other biological agents' has not been defined by BTWC States Parties, but can once again be considered extensive in the range of substances encompassed. As Chevrier and Leonard have noted: "[T]he ordinary meaning of 'other biological agents', unmodified by any restricting adjectives or clauses, implies a very broad scope, provided that broad scope is consistent with the object and purpose of the Convention..."[16]

Consequently, taken collectively 'toxins' and 'biological agents' would appear to encompass a very wide range of potential agents, including those that affect the CNS and other core physiological systems. Furthermore, the breadth of substances covered by the Convention as described in Article 1 is not limited on the grounds of lethality of the agent. Consequently, the development and use of purportedly 'less lethal' CNS-acting weapons or other incapacitating chemical agent weapons acting on the core physiological systems, employing toxins or other biological agents, would come under the scope of the BTWC.

A then-senior U.S. State Department official interviewed by Crowley in 2010 stated that:

> "If an incapacitant is a microbial or other biological agent or is a toxin then it would be covered. For example, as part of the U.S. BW programme there were incapacitating agents – SEB [Staphylococcal enterotoxin B] – certainly we would interpret the Convention to cover agents whose purpose was to incapacitate."[17]

Similarly, another Western Government expert previously interviewed by Crowley in 2011 stated that:

> "The BTWC applies to incapacitating biological and toxin agents. SEB would be [an] example of an incapacitating agent... weaponised in past offensive programmes."[18]

However, the full extent of the range of incapacitating agents covered by the Convention is contentious. The expert has noted that although certain States believe the term "other biological agents" extends to bioregulators, this "has never been tested fully in a BTWC context such as a Review Conference".[19]

In addition, certain States, at least at the national level, have previously explicitly recognised the applicability of both the BTWC and the CWC to incapacitating chemical agent weapons. For example, in August 2009 as part of its response to the *Fourth Report* from the U.K. House of Commons Foreign Affairs Committee, the U.K. Government stated that: "[D]evelopment, production, retention, acquisition or use of 'Incapacitating biochemical weapons' are prohibited by both [BTWC and CWC] Conventions. Use of the word 'weapons' here is crucial".[20]

When considering the range of CNS-acting agents and other incapacitating chemical agents that may be covered under the BTWC it is important to take account of General Purpose Criterion, under Article I (1) that links the Convention's prohibitions to "types and quantities" that have "no justification for prophylactic, protective or other peaceful purposes". This allows the Convention to be 'open' in scope and thereby cover the full range of biological agents and toxins, even those that have not been discovered or synthesized yet.

The open-ended nature of this coverage has been repeatedly emphasised and clarified through the Extended Understandings agreed at successive BTWC Review Conferences, which make it clear that the Convention is intended to be fully inclusive with regard to the range of substances which it covers and also with regard to the potential targets (which include plants, as well as animals and humans). For example, the *Final Declaration* of the Eighth BTWC Review Conference of 2016 reaffirmed that:

> "...*all naturally or artificially created or altered microbial and other biological agents and toxins, as well as their components, regardless of their origin and method of production* and whether they affect humans, animals or plants, of types and in quantities that have no justification for prophylactic, protective or other peaceful purposes, *are unequivocally covered by Article I*.... [and that] ... *Article I applies to all scientific and technological developments in the life sciences and in other fields of science relevant to the Convention*..."[21] (Emphases added).

Furthermore, consideration should also be given to the background papers prepared by States Parties for Review Conferences, a number of which have highlighted the relevance of peptides, bioregulators and their analogues to the BTWC, and have highlighted the potential dangers that misuse of such agents could pose. For example, of the ten scientific papers submitted by States Parties as contributions to the BTWC Conference Secretariat's 2006 *Background Information Document on New Scientific and Technological Developments Relevant to the Convention* for the Sixth Review Conference,[22] the Netherlands, Russian Federation, Sweden, UK and US highlighted the potential dangers to the BTWC of the misuse of bioregulators and peptides (or other biologically active agents that could be utilized as CNS-acting or other incapacitating chemical agents). In its paper Sweden, for example, stated:

> "The growing understanding of how bioregulators and other compounds affect and control central biological processes of mammalian

> systems, including human, is... judged as an increased dual-use risk.... The techniques to deliver biologically active organisms or compounds to their target structures have evolved as a key technology for pharmaceutical industry as well as for biotechnology in general.... Advances in this field, including aerosol delivery, is one of the reasons behind the increasing concern regarding the use of bioregulators for nefarious purposes."[23]

In its paper the United States warned:

> "While naturally-occurring threat agents, such as anthrax, and "conventionally" genetically engineered pathogenic organisms are the near-term threats we must confront today; *the emerging threat spectrum will become much wider and will include biologically active agents such as bioregulators."*[24] (Emphasis added).

Similar concerns about such agents were included by certain States in their submissions to the Implementation Support Unit (ISU) review of *New Scientific and Technological Developments Relevant to the Convention*,[25] conducted in advance of the Seventh Review Conference. For example, in its discussion of the "rapidly developing field" of neuroscience Australia noted that "the utility of neuropeptides, bioregulators and other agents of biological (BWC) as so-called non-lethal weapons has been a popular point of discussion".[26] Likewise in its discussion of neuroscience, the UK stated that:

> "...Technologies for the discovery and development of compounds that act on the central nervous system (CNS) are of particular relevance to the BTWC. These include advances in molecular and genetic neuroscience, drug discovery technology, bioregulators and drug delivery to the brain."[27]

And furthermore that:

> ...Developments in [drug development and delivery] ... could also result in the identification of compounds with potential for misuse as biological *or toxin weapons agents* since drugs acting on the brain to produce toxic or incapacitating effects could also have utility in a BW programme. *Methods to facilitate delivery of such agents could also be exploited for harmful purposes, for example, to facilitate the entry of peptide neurotoxins across the BBB [blood– brain barrier].* However, *the prohibitions of Article I of the BTWC fully cover all such biological agents and toxins, whether naturally occurring or artificially created, regardless of their origin or method of production..."*[28] (Emphases added).

Subsequently, concerns were highlighted by the ISU in its 2011 synthesis report of scientific and technological advances prepared for the Seventh

Review Conference. Considering neurobiology amongst the range of "advances with potential for weapon applications", the ISU stated that: "[S]ince the last review conference, there have been advances in: understanding the role of neuroregulators; how to influence psychological states and alter physical performance".[29] This inclusion of neuroregulators by the ISU in its analysis of *New Scientific and Technological Developments Relevant to the Convention* underlines the general acceptance of bioregulators as being within the BTWC scope.

The constructive discussions of science and technology at the 2022 Ninth BTWC Review Conference were mainly concerned with the development of a science and technology review mechanism, but some States did submit Working Papers on advances in science and technology. The United States, for example, opened its paper on the subject by stating that while no new developments have fundamentally changed the nature of science and technology or its relationship to the Convention:[30]

> "...the pace of developments identified in 2011 has continued to accelerate, expanding the possibilities not only for benefits to public health and medicine, agriculture, the environment, and other peaceful applications, but also for misuse for purposes banned by the Convention. Individual scientific discoveries and technological tools continue to grow more accessible, to develop more rapidly, and to converge with other disciplines such that the threat from biological weapons and our ability to respond effectively is evolving rapidly...."

And the paper went on to discuss, amongst several other issues, advances in protein folding analysis and the new possibilities of the synthesis of biological molecules. The United Kingdom in its contribution to the Science and Technology discussions paid particular attention to peptide synthesis, stating that:[31]

> "A scientific advancement that has seen much progress since the eighth Review Conference in 2016 is peptide synthesis technology. This technology is rapidly advancing, and the emerging fields of peptide-based drugs and biomaterials are increasing accessibility of custom peptide synthesis and driving down its cost. Peptide synthesis equipment is available at all scales, although currently from a limited number of manufacturers globally. However, companies providing custom peptide synthesis services are numerous and globally widespread..."

Clearly, such advances could have significance in relation to concerns about toxins and bioregulators.

In their Extended Understandings, the BTWC States Parties have clarified that the range of substances covered by the BTWC Article 1 prohibitions further extends to include "all ...*artificially created or altered* microbial and *other biological agents and toxins*, as well as their components". (Emphasis added). This position has subsequently been reaffirmed, and its importance highlighted, by certain individual State Parties. For example, during discussions at the September 2021 BTWC Meeting of Experts (MX-2) to Review of Developments in the Field of Science and Technology Related to the Convention, the US stated:[32]

> "Toxins are covered by the Chemical Weapons Convention, but they are also covered by the Biological and Toxin Weapons Convention…. So there is a clear area of overlap there, in fact the overlap goes a bit further because the interpretive statements that have been adopted over the years by the Review Conferences have made clear that the BWC also…*captures synthetic analogues of toxins, regardless of the method of production…. So there are a whole series of chemical compounds that may not be natural in origin, but have similar modes of action, that would be a violation of the Biological Weapons Convention as well as the Chemical Weapons Convention*. We at least see this overlap as a good thing, not a bad one, and it certainly ensures that there are no gaps…" (Emphasis added).

6.4 Scope of Activities Regulated

The text of Article I prohibits a range of activities involving biological agents and toxins, except those that can be justified for "prophylactic, protective or other peaceful purposes". During the *travaux préparatoires* States Parties clarified the term 'prophylactic' as including medical activities, such as diagnosis, therapy and immunization, and also agreed that 'protective' activities would cover development of protective masks and clothing, air and water filtration systems, detection and warning devices and decontamination equipment, and that these applications must not be interpreted as to permit possession of biological agents and toxins for defence, retaliation or deterrence.[33] There has, however, been no such definition agreed for 'other peaceful purposes',[34] though Goldblat believed that "one can assume that it includes scientific experimentation".[35] Although these phrases have been used in the BTWC Review Conference *Final Declarations*, no Extended Understandings as to their meaning have been agreed at these Review Conferences. These ambiguities have contributed to concerns by some commentators that biodefence programmes may potentially be employed to mask prohibited activities. This has notably become apparent in the discussions around certain previous US 'threat assessment' development activities which came close to and some believe did cross the line violating the BTWC.[36]

A further closely-related issue not given adequate clarity in Article 1 or elsewhere in the Convention concerns research directly involving biological agents and toxins or with regard to potential weapons, equipment or means of delivery; and the dividing line between such research and actual weapons development. No explicit guidance is given in the Convention or in subsequent Extended Understandings to enable States Parties to determine if and when such research should be prohibited. Similarly, whilst recognising the attempts the BTWC States Parties have made to create additional transparency in certain research domains, notably through CBMs, such mechanisms have not yet been adapted to take account of the specific risks associated with dual-use activities involving CNS-acting agents. Consequently, there are no adequate regulatory, transparency or other measures in the Convention or developed subsequently that provide States with sufficient confidence that dual-use research conducted in another State is not being employed in the development of CNS-acting or other incapacitating chemical agent weapons that employ bioregulators, toxins and other biologically active substances, or that such research is not being misused for other malign purposes. The authors have highlighted previous dual use research of potential applicability to CNS-acting weapons development in a range of States including China, Czech Republic, India, Iran, Russia, Syria, the UK and the US.[37]

Further ambiguity surrounds the application of the Article I prohibition on "weapons, equipment or means of delivery designed to use such agents or toxins for hostile purposes or in armed conflict". Although "armed conflict" is a well-defined and understood term in international humanitarian law[38] the application of Article I with regard to "hostile purposes" not amounting to "armed conflict" has not been clarified. Specifically, it is unclear how the use of CNS-acting or other incapacitating chemical agent weapons employing toxins and bioregulators for military operations other than war (MOOTW) including counter-insurgency operations, peace-enforcing and peace-keeping operations would be regulated by the BTWC. Guidance on the application of the BTWC in these areas would be beneficial, including clarity on the scope of any permitted activities, the range of toxins and bioregulators that could be employed, and the constraints upon use in such circumstances.

An additional area where greater clarity from the BTWC States Parties would be beneficial concerns the development and use of weapons employing toxins and bioregulators for law enforcement purposes – in particular with regard to CNS-acting or other incapacitating chemical agent weapons. Whilst 'law enforcement' would appear to fall within the ambit of "other peaceful purposes", clarity is needed on the scope of such permitted activities, the range of toxins and bioregulators that could be employed and the constraints upon such use. Although these issues have not been formally addressed at BTWC Review Conferences, the situation with regard to the development and use of certain toxin-based RCAs (as opposed to CNS-acting or other incapacitating chemical agent weapons) for law enforcement activities seems to be established, given the substantive State Practice in this area. The employment of capsacinoids and chemically synthesized analogues is widespread amongst law enforcement agencies around the world and does not appear to have been raised publicly as an issue of concern *per se* in the context of the BTWC by any State Party. The acceptability of toxin-based RCA use in a specific instance would depend on the type and quantity of agent employed, as well as the context. In contrast, the permissibility under the BTWC of research, development, stockpiling and use of CNS-acting or other incapacitating chemical agent weapons employing toxins, bioregulators, other biologically active peptides, whether of natural origin or their synthetic analogues, is currently unclear and is likely to be highly contested by BTWC States Parties. Whether, when and how this issue is raised and addressed in the context of the BTWC is likely to be heavily influenced by the speed and manner in which CWC States Parties implement the 2021 CSP Decision that recorded an "understanding" that use of aerosolised CNS-acting chemicals would be inconsistent with the purpose of law enforcement under the CWC (as explored in Chapter 7).

6.5 Measures to Facilitate Implementation, Verification and Compliance

Since its establishment, important long-standing limitations on the value of the BTWC (and its control regime) as a tool to prevent and respond to development and use of toxin and bioregulator weapons (including CNS-acting and other ICA weapons) became increasingly apparent. This has been notably due to its lack of effective verification and compliance mechanisms, and the absence of an international organisation comparable to the OPCW to coordinate such activities and facilitate implementation by States Parties.

As the current revolution in biotechnology got underway in the latter part of the last century these limitations in the BTWC began to be ever more clearly exposed, and efforts were made by the BTWC States Parties during the 1990s to address these deficiencies and provide a more assured means of keeping the Convention up-to-date in application. In 1995 BTWC States Parties initiated a process to develop a legally binding Protocol to the BTWC that would have established an implementation body and created a verification system and other institutional structures to manage the implementation of the Convention at the international level. However, in July 2001, just as the negotiations were nearing conclusion, the US unilaterally withdrew its support and the negotiations collapsed.[39] Subsequently, for over two decades, States Parties tried, not very successfully, to find other less far-reaching means of strengthening the Convention. This has been the situation up to the present time.[40] Despite the lack of comprehensive measures, important, albeit more limited and piecemeal, advances have been made to improve confidence building measures, review of science and technology, and support State implementation. In addition, the BTWC and attendant control regime do provide the opportunity for States Parties to employ certain consultation, fact-finding, investigation and compliance measures if they are willing to do so.

6.5.1 Confidence Building Measures

In 1991, the States Parties updated a range of politically-binding Confidence-Building Measures (CBMs) to facilitate transparency and information exchange concerning certain activities and facilities that might raise compliance questions that had originally been agreed in 1987. Improvements in the mechanism of submission and some changes were agreed at subsequent review conferences in 2006 and 2011 as summarised in Box 6.1. The CBMs include a "[D]eclaration of past activities in offensive and/or defensive biological research and development programs" and also an "[E]xchange of information on national biological defense research and development programmes, including declarations on facilities where biological defense research and development programs are conducted".[41]

The CBMs potentially provide an important mechanism for States to alleviate concerns regarding their toxin and bioregulator research and address any misconceptions into the nature and purpose of specific research programmes. However, their effectiveness has to date been limited given their voluntary nature; although the number of States submitting CBMs has steadily grown from 17 in 1987, only 111 of the 187 States Parties submitted CBMs in 2024.[42] Furthermore, without a process of external peer review, the veracity of such information cannot be confirmed. Finally, given the extremely low level of public transparency in this area – with for example only 32 States openly publishing their CBM reports in 2024[43] (and some of these being redacted versions of the full CBM report circulated to States) – public knowledge and confidence in such State reporting is extremely limited.[44] In 2021, during the third strand of the Meetings of Experts (MX3) dealing with Strengthening National Implementation, there were discussions on

Box 6.1 Summary of current CBM measures.[41]

As agreed at the Third and Seventh Review Conferences, the CBMs consist of six measures or forms, A to G (Form D was deleted by the Seventh Review Conference):

CBM A

Part 1: Exchange of data on research centres and laboratories

Part 2: Exchange of information on national biological defence research and development programmes

CBM B Exchange of information on outbreaks of infectious diseases and similar occurrences caused by toxins

CBM C Encouragement of publication of results and promotion of use of knowledge

CBM E Declaration of legislation, regulations and other measures

CBM F Declaration of past activities in offensive and/or defensive biological research and development programmes

CBM G Declaration of vaccine production facilities

improving the CBMs and on the use of voluntary visits and peer review mechanisms to increase confidence in compliance. While confidence building and transparency measures were amongst the issues agreed for further work at the 2022 9th Review Conference, discussion of these issues, like everything else, ground to a halt at the 2023 Meeting of States Parties (see below).[45]

6.5.2 Review of Science and Technology (S&T)

Following the collapse of the BTWC Verification Protocol negotiations in 2001, the BTWC States Parties started an Intersessional Process initially between the Fifth and Sixth Review Conferences and then repeated between subsequent Review Conferences.[46] These processes which comprise an annual Meeting of Experts followed by a Meeting of States Parties, provide an important mechanism which could be utilized to monitor advances in science and technology that could be applicable to the development of new forms of toxin and bioregulator weapons including CNS-acting or other ICA weapons.

For example, the Seventh BTWC Review Conference decided to include, as part of its 2012–2015 intersessional programme, a standing agenda item to review developments in the field of science and technology related to the Convention.[47] Consequently, in 2014: "advances in the understanding of pathogenicity, virulence, toxicology, immunology and related issues" were reviewed; whilst in 2015 "advances in production, dispersal and delivery technologies of biological agents and toxins" were considered.[48] Such structured reviews by the BTWC States Parties present important opportunities for individual States Parties and civil society to highlight research and development trajectories of potential concern, including those relating to the development of CNS-acting toxin and bioregulator weapons. For example, in 2016 at the Eighth Review Conference the UK submitted a paper[49] that attempted to summarise the discussions on science and technology since the Seventh Review Conference and concluded that one of the seven issues requiring further consideration in an enhanced review mechanism was:

> "(d) **developments derived from the convergence of biology, chemistry and other scientific disciplines**, for example: advances in neuroscience, including design and delivery of peptides and other neurotransmitters; trends in production processes, including for toxins and bioregulators; nanotechnology, including its application to defensive countermeasures, such as detection and diagnosis, protective clothing and equipment, medical countermeasures and decontamination. This could include exploring appropriate ways and means to promote greater collaboration with the CWC to analyse potential benefits, risks and threats resulting from relevant advances in science and technology." (Original emphasis)

The importance of the BTWC establishing effective mechanisms to review relevant science and technology developments, including those relating to toxins and bioregulators, was underscored by growing concerns that continuing biotechnological development will lower the technological barriers to toxin weaponization. For example, in its November 2021 analysis of science and technology trends of relevance to the BTWC, the US stated:

> "...Rapid advances in science and technology have expanded the ability to manipulate biological agents and toxins, potentially enabling production and delivery or allowing for more facile engineering of agents with enhanced characteristics. For instance, some toxins that exist in nature have traditionally been considered to have a low likelihood of weaponization because of a limited ability to isolate the toxin from its host microbe. Today, several of those toxins can be produced in higher quantities, using genetic engineering or chemical synthesis techniques, thus making them more accessible as a potential weapon."[50]

Momentum appeared to be building towards establishing a formal mechanism to review relevant science and technology and its implications for the BTWC in a more structured and in-depth manner than had previously occurred, with hopes that the 9th BTWC Review Conference would be the forum where this mechanism could be agreed. However, as noted above, due primarily to the diplomatic consequences of the ongoing war in Ukraine, only a largely *status quo* outcome was achieved with the agreement of a *Final Declaration* and continuation of the existing Meetings of States Parties. However, there were two positive inter-related decisions adopted which provide potential mechanisms for future strengthening of the BTWC regime, notably including more effective review of science and technology.

Firstly, the *Final Report* of the 9th Review Conference included a Decision to establish a new Working Group on strengthening the Convention stating that:

> "8. Determined to strengthen the effectiveness and to improve the implementation of the Convention in all its aspects, the Conference decides to establish a Working Group open to all States Parties.
>
> The aim of the Working Group is to identify, examine and develop specific and effective measures, including possible legally-binding measures, and to make recommendations to strengthen and institutionalise the Convention in all its aspects, to be submitted to States Parties for consideration and any further action. These measures should be formulated and designed in a manner that their implementation supports international cooperation, scientific research and economic and technological development, avoiding any negative impacts".

In this context, the Working Group will address the following:

(a) Measures on international cooperation and assistance under Article X;
(b) *Measures on scientific and technological developments relevant to the Convention;*
(c) Measures on confidence-building and transparency;
(d) Measures on compliance and verification;
(e) Measures on national implementation of the Convention;
(f) Measures on assistance, response and preparedness under Article VII;
(g) Measures on organizational, institutional and financial arrangements

(Emphasis added)[51]

Secondly, the Final Document included the creation of a specific new arrangement for reviewing science and technology developments:

"19. The Conference decides to develop with a view to establishing a mechanism to review and assess scientific and technological developments relevant to the Convention and to provide States Parties with relevant advice. In order for this mechanism to be established, the Working Group on the strengthening of the Convention will make appropriate recommendations."[52]

In early 2023 the dates and topics for the Working Group and annual Meetings of States Parties (which had responsibility for overseeing the whole Intersessional Process) were agreed. Encouragingly, discussions related to scientific and technological developments and the scientific and technological review mechanism were prominent topics in the second Meeting of the Working Group, which took place in August 2023. Since the topic of a science and technology review mechanism had been discussed in some detail before and during the 9th Review Conference a paper[53] by the Friends of the Chair was produced in order to summarise the state of these discussions (Box 6.2).

Intensive discussions continued on this topic during this second Working Group meeting, and an observer's report by the civil society Biological Weapons Prevention Project (BWPP) suggested a potential agreement on the key issue of the structure, pointing out that:[54]

"...while there are still proponents for the models of either a limited-membership committee/board or of a body open to all states parties, a considerable majority of interventions during this session that expressed a preference spoke in favour of a hybrid model containing elements of the two. Some outlined how elements within a hybrid

Box 6.2 "Food for thought" non-paper by the Friends of the Chair.[53]

Draft Elements of a BWC Science and Technology Advisory Board

1. Role and Functions
2. Structure and composition
   Open-Ended Science and Technology Advisory Group
   Science and Technology Reporting Committee
   Temporary Working Groups
3. Independence
4. Resources
5. Relationship with the Implementation Support Unit (ISU)
6. Rules of Procedure

Draft Procedural Elements

1. Mode of Operation
2. Temporary working groups
3. Observers
4. Administrative and technical support
5. Languages
6. Reports

> system might interact with each other, and with any temporary working groups (TWGs) that might be asked to closely examine particular issues…"

As a result of initial progress, there seemed to be reason to believe that during future Working Group meetings an agreement on the structure and function of the science and technology review mechanism would be possible.

The third Meeting of the Working Group was completed in December 2023 and the Chair of the Working Group reported on progress to the December 2023 Meeting of States Parties (MSP). However, this MSP was characterised by complete deadlock over Russian proposals concerning limitations on statements that international organisations (notably NATO) and non-governmental organisations should be allowed to make to these meetings, and only a minimal procedural report of the MSP was possible. Given this serious and currently unresolved deadlock, the future of the Intersessional Process and a process for establishing a mechanism to review science and technology were both uncertain. Meetings continued to address this issue in 2024 and an assessment prior to the December 2024 MSP was optimistic there would be agreement at that meeting to recommend the establishment of an S&T review mechanism in 2025.[55] However, due to procedural objections by

Russia it was not possible to reach agreement at the December 2024 MSP and only the dates for the meetings in 2025 were agreed.[56]

6.5.3 Implementation Support Unit

A significant limitation of the BTWC is the current lack of sufficient institutional support to encourage and facilitate effective implementation of the Convention by all State Parties. Instead, there is a minimal BTWC Implementation Support Unit (ISU) within the UNODA of the United Nations Office at Geneva. The ISU was established after the Sixth Review Conference in 2006 to provide administrative support to meetings agreed by the Review Conference as well as comprehensive implementation and universalization of the Convention and the exchange of confidence-building measures.[57] Its mandate has been repeatedly renewed and extended by successive Review Conferences and in 2022, the 9th Review Conference renewed the mandate and funding of the ISU which will now run until 2027.[58] The 9th Review Conference also agreed to increase the ISU's staffing by one person from three to four, and it was hoped, by some observers, that this extra staff member would be, at least in part, working on science and technology issues.[59]

6.5.4 Non-compliance Measures

Although lacking formal verification measures, the BTWC does contain two Articles that were intended to address non-compliance concerns by other means. Article V encouraged States Parties to "consult one another and to cooperate in solving any problems which may arise in relation to the objective of, or in the application of the provisions of, the Convention".[60] Although these provisions have been employed by States Parties, such bilateral consultations are necessarily confidential in nature, and unless the participants agree to wider information dissemination, can result in a lack of transparency amongst States Parties as a whole. Such considerations are exacerbated for civil society organisations, making it even more difficult for them to ascertain how frequently such consultations have been utilized and how effective they have been. As well as such bilateral consultations, Article V also allows consultations to be conducted on a multilateral basis "through appropriate international procedures within the framework of the United Nations and in accordance with its Charter".[61] Prior to 2022, this provision has been utilized once, by Cuba, resulting in a formal consultative meeting in 1997.[62]

In addition, Article VI of the BTWC allows any State Party to lodge a complaint with the United Nations Security Council (UNSC) about non-compliance by another State Party. The UNSC may then launch an investigation, with which all parties are enjoined to cooperate.[63] Such investigations would be conducted by the UN Secretary-General utilising the resources and capabilities available to them. This may include making use of the resources of the UN

Secretary General's Mechanism (USGM), and in this case would cover suspected use of CNS-acting or other ICA toxin or bioregulator weapons in all forms of armed conflict, and potentially also the use of such weapons to commit human rights abuses in the absence of such armed conflict.[64]

6.5.4.1 Russian Disinformation Campaign and Invoking of Non-compliance Measures

In recent years Russia, supported by China, appears to have conducted active and concerted biological weapons disinformation campaigns. Previously, these apparent disinformation campaigns focused on concerns around US biodefence as well as US support for similar activities taking place in former Soviet Union States, notably in Georgia. Recently, such Russian apparent disinformation has been given a further dangerous twist following its invasion, on 24 February 2022, of Ukraine. On 11 March 2022, Russia initiated a debate by the UN Security Council (UNSC), of its allegations that Ukraine has "a network of at least 30 biological laboratories that host extremely dangerous biological experiments, aimed at *enhancing the pathogen properties* of... *lethal diseases with the help of synthetic biology [and that] This work is funded and directly supervised by ... the United States*". (Emphasis added).[65] The UN High Representative for Disarmament Affairs, Izimu Nakamitsu, commented on the Russian allegations by stating that the "United Nations is not aware of any biological weapons programmes".[66] The Russian allegations were forcefully refuted by a number of States during the UNSC meeting[67] and also in an emergency session of the UN General Assembly.[68] Ukraine, along with the US and other States that provide support to biological research in Ukraine, has previously been transparent about its activities. They annually declare such activities under the BTWC's Confidence-Building Measures, and they voluntarily share this information publicly. In contrast, Russia has not publicly presented substantive, credible evidence to support its allegations.

On 29 June 2022, the Russian Federation submitted a request for the convening of a Formal Consultative Meeting under Article V of the BTWC. This meeting was held from 5–9 September 2022 and was chaired by Ambassador György Molnár of Hungary.[69] The meeting ended without agreement on whether Russia's claims were valid. However, an overwhelming majority of the participating States Parties rejected Russia's allegations as being baseless. Consequently, the Final Document of the meeting concluded that "No consensus was reached regarding the outcome of the Formal Consultative Meeting".[70]

On 24 October 2022, Russia addressed a letter to the President of the UNSC, which summarised Russia's allegations against Ukraine and the US, reiterated the outstanding questions after the Formal Consultative Meeting, and lodged a formal complaint with the UNSC under Article VI of the BTWC.[71] Russia presented a draft UN resolution proposing the establishment of a commission consisting of all UNSC members to investigate its claims against Ukraine and the US.[72] On 2 November 2022, this draft resolution was put to a vote of the UNSC but did not secure approval: whilst China and Russia voted in favour, France, the UK and the US voted against and all non-permanent UNSC members abstained.[73]

6.5.5 Education and Awareness Raising

When BTWC State Party efforts to negotiate a Verification Protocol for the BTWC failed in the first years of this century, the UK Foreign and Commonwealth Office produced a Green Paper that discussed a range of possible options to strengthen the Convention.[74] Amongst the potential topics for exploration that the paper highlighted were codes of conduct for scientific professional bodies. It stated that "such codes would be developed by academic and professional bodies to lay out standards internationally for work relevant to the prohibitions of the Convention". These proposals were consistent with discussions and conclusions that had been reached previously by State Parties at the Fourth BTWC Review Conference in 1996. The *Final Declaration* of that Review Conference stating under Article IV:

> "The Conference notes the importance of....
>
> Inclusion in textbooks and in medical, scientific and military education programmes of information dealing with the prohibitions and provisions contained in the Biological and Toxin Weapons Convention and the Geneva Protocol of 1925.
>
> The Conference believes that such measures which States Parties might undertake in accordance with their constitutional processes would strengthen the effectiveness of the Convention, as requested by the Second and Third Review Conferences."[75]

This possible means of strengthening the Convention was taken up by the BTWC States Parties, and formed part of the subsequent Intersessional Processes that followed (Box 6.3).

More recently, in good part because of efforts by China, the Tianjin Guidelines for life scientists were developed and endorsed by the Inter-Academy Panel of National Scientific Academies.[76] The development and refinement of these guidelines involved a joint effort funded by different countries and reported to meetings of BTWC State Parties in Geneva. In general, the presentations of the Guidelines have been well received, but as yet there has not been a formal endorsement of the Guidelines by States Parties.[77] The Tianjin Guidelines and the broader utility of codes of conduct will be explored in more depth in Chapter 11.

Box 6.3 Codes of conduct topics in the Intersessional Processes (from Review Conference Final Documents).

The First Intersessional Process

> "... in 2003-2005 States Parties to the BTWC considered, *inter alia,* the topic of 'content, promulgation, and adoption of codes of conduct for scientists...'"

The Second Intersessional Process

> "... in 2007-2010 States Parties considered the topic of 'adoption and/or development of codes of conduct with the aim of preventing misuse in the context of advances in bio-science and bio-technology research with the potential of use for purposes prohibited by the Convention...'"

The Third Intersessional

> ".... in 2012-2015 the topic of 'voluntary codes of conduct and other measures to encourage responsible conduct by scientists, academia and industry' was considered by States Parties under the Standing Agenda Item on the 'Review of developments in the field of science and technology related to the Convention...'"

The Fourth Intersessional Process

> "... During the ... Intersessional Process 2018–2020, States Parties ... agreed to consider the topic of the 'development of a voluntary model code of conduct for biological scientists and all relevant personnel...'"

Moreover, in 2005 when this subject was first discussed, a Working Paper by Australia made the important point that it could not be assumed that such biosecurity codes of conduct would be quickly and easily assimilated by the life and associated scientific communities. The paper began by stating that:[78]

> "Amongst the Australian scientific community, there is a low level of awareness of the risk of misuse of the biological sciences to assist in the development of biological or chemical weapons. Many scientists working in 'dual-use' areas simply do not consider the possibility that their work could inadvertently assist in a biological or chemical weapons programme..."

The paper therefore argued that:

> "...For most of these researchers, biological weapons issues may seem irrelevant and therefore strong advocacy is required to overcome natural resistance or ignorance. Introducing Codes of Conduct that highlight these issues is an important step in raising awareness. However, it is not enough simply to put such Codes in place. *Without effective measures to educate scientists about the existence and importance of such Codes, attitudes and awareness will remain largely unchanged.*" (Emphasis added)

The available evidence at that time strongly suggested that such a low level of awareness of the security implications of their work was not confined to the Australian scientific community.[79] Unfortunately, the situation has not improved a great deal over the last 20 years as the World Health Organisation concluded in its *Global Guidance Framework for the Responsible Use of the Life Sciences*:[80] Of course, some State Parties have supported biosecurity education projects and reported on best practice to other States Parties in Geneva.[81] However, to date there has not been a concerted effort formally agreed to follow up the obvious need for such education identified as long ago as the Second Review Conference in 1986 when the biotechnology revolution began to be recognised as having significant implications for the Convention.

6.6 Conclusions

The BTWC under its Article I General Purpose Criterion, and through Extended Understandings agreed at successive Review Conferences, makes clear that the Convention is comprehensive in scope and covers both naturally occurring and synthetic toxins and bioregulators. Furthermore, the development, production, acquisition, stockpiling and use of toxins and bioregulators other than for prophylactic, protective or other peaceful purposes is prohibited. The BTWC prohibitions cover weaponised agents intended to attack plants, as well as animals and humans. Furthermore, such prohibitions make no distinction with regard to the lethality of the intended weapons. Consequently, the development and use of CNS-acting or other ICA weapons employing toxins and bioregulators or their synthetic analogues are covered (and prohibited) by the BTWC.

However, there are a number of long-standing ambiguities and limitations in the Convention that its States Parties have failed to address. These include failing to: define toxins and clarify the range of substances covered by this term; delineate whether there are any specific toxin and bioregulator weapons that can be developed and employed for use in law enforcement and if so identify which toxins and bioregulators can be so employed, in what circumstances and under what constraints; and clarify how dual use research is constrained under the Convention to ensure it does not facilitate or is not employed in the development of toxin and bioregulator weapons. In addition, it must also be recognised that there are important long-standing and systemic weaknesses that severely limit the value of the BTWC as a tool to prevent and address the development of toxin and bioregulator weapons. These include the current lack of effective transparency, verification and compliance mechanisms; inadequate monitoring and review of advances in relevant science and technology; and the absence of a sufficiently resourced international organisation empowered to coordinate such activities and to facilitate effective implementation by States Parties.

Despite the modest advances that were agreed at the 9th BTWC Review Conference and that may begin to strengthen BTWC State Parties' abilities to review science and technology in the longer term, the 9th Review Conference did not substantively address issues relating to toxins, bioregulators or other substances of biological origin. More generally it is hard to avoid the conclusion that the devastating effects of the COVID-19 pandemic did not galvanise States Parties nor create a sufficiently collegiate environment in which they felt it possible and necessary to significantly improve the Convention and its operation. Given the continuation of the war in Ukraine for the foreseeable future, progress on these issues and more broadly on strengthening the BTWC is likely to remain slow and difficult.

References

1. ICRC, *Treaties, States Parties and Commentaries: Convention on the Prohibition of the Development, Production and Stockpiling of Bacteriological (Biological) and Toxin Weapons and on Their Destruction (BTWC)*, 1972. Available at https://ihl-databases.icrc.org/applic/ihl/ihl.nsf/INTRO/450?OpenDocument.
2. *Convention on the Prohibition of the Development, Production and Stockpiling of Bacteriological (Biological) and Toxin Weapons and on Their Destruction (BTWC)*, 1972. Available at https://ihl-databases.icrc.org/applic/ihl/ihl.nsf/INTRO/450?OpenDocument; Preamble.
3. *Convention on the Prohibition of the Development, Production and Stockpiling of Bacteriological (Biological) and Toxin Weapons and on Their Destruction (BTWC)*, 1972. Available at https://ihl-databases.icrc.org/applic/ihl/ihl.nsf/INTRO/450?OpenDocument, Article XII.
4. *Convention on the Prohibition of the Development, Production and Stockpiling of Bacteriological (Biological) and Toxin Weapons and on Their Destruction (BTWC)*, 1972. Available at https://ihl-databases.icrc.org/applic/ihl/ihl.nsf/INTRO/450?OpenDocument, Article I.
5. *Convention on the Prohibition of the Development, Production and Stockpiling of Bacteriological (Biological) and Toxin Weapons and on Their Destruction (BTWC)*, 1972. Available at https://ihl-databases.icrc.org/applic/ihl/ihl.nsf/INTRO/450?OpenDocument, Article II.

6. *Convention on the Prohibition of the Development, Production and Stockpiling of Bacteriological (Biological) and Toxin Weapons and on Their Destruction (BTWC)*, 1972. Available at https://ihl-databases.icrc.org/applic/ihl/ihl.nsf/INTRO/450?OpenDocument, Article III.
7. *Geneva Protocol for the Prohibition in War of Asphyxiating, Poisonous or other Gases, and Bacteriological Methods of Warfare*, 17 June 1925.
8. United Nations, *Final Document of the Fourth BTWC Review Conference. BWC/CONF. IV/9, 25th November – 6th December. United Nations, Geneva*, (1996), Article I, paragraph 3.
9. United Nations, *Final Document of the Sixth BTWC Review Conference. BWC/CONF.VI/6, 20th November – 8th December. United Nations*, Geneva, 2006, Article I, paragraph 3.
10. United Nations, *Final Document of the Seventh Review Conference, BWC/CONF.VII/7, 5th–22nd December, United Nations*, Geneva, 2011, Article I, paragraph 3.
11. United Nations, *Final Document of the Eighth Review Conference*. BWC/CONF.VIII/CRP.2, 7th–25th November. United Nations, Geneva, 2016, Article I, paragraph 3.
12. See for example, United Nations, Final Document of the Seventh Review Conference, BWC/CONF.VIII/CRP.2, 7th–25th, United Nations, Geneva, November 2016. See Article I, paragraphs 1 & 2, p. 10.
13. World Health Organisation, *Public Health Response to Biological and Chemical Weapons*, WHO, Geneva, 2nd edn, 2004, p. 215.
14. United States Code, *Title 18, Crimes and Criminal Procedure*, Chapter 10, Biological Weapons, Section 178, Definitions.
15. World Health Organisation, *Public health response to biological and chemical weapons: WHO Guidance*, WHO, Geneva, 2nd edn, 2004, p. 216.
16. M. Chevrier and J. Leonard, Incapacitating Biochemicals and the Biological Weapons Convention, pp. 209–224, in *Incapacitating Biochemical Weapons*, ed. A. Pearson, M. Chevrier and M. Wheelis, Lexington Books, Lanham, MD, 2007, p. 211.
17. Interview by Crowley with then-senior US State Department official, 27th May 2010. [Interviewee was speaking in his private capacity].
18. Crowley's interview with Western Government expert, 28th April 2011.
19. Crowley's interview with Western Government expert, 20th April 2011.
20. United Kingdom, *Response of the Secretary of State for Foreign and Commonwealth Affairs, Fourth Report from the Foreign Affairs Committee, Session 2008-09, Global Security: Non-Proliferation*, London, 2009, August 2009, p. 22.
21. United Nations, *Final Document of the Seventh Review Conference*, BWC/CONF.VIII/CRP.2, 7th–25th, United Nations, Geneva, November 2016, See Article I, paragraphs 1 & 2, p. 10.
22. See relevant *National Background Papers on Scientific and Technological Developments Relevant to the Biological Weapons Convention*, prepared for the Sixth Review Conference, Available from https://web.archive.org/web/20060411211122/http://www.opbw.org/ (accessed 3 February 2025).

23. Sweden, *Scientific and Technological Developments Relevant to the Biological Weapons Convention*, submitted by Sweden, 2006. https://web.archive.org/web/20081201092231/http://www.opbw.org/rev_cons/6rc/docs/adv/BWC.Conf.VI_S&T_sweden_en.pdf (accessed 3 February 2025).
24. United States, *Scientific and Technological Developments Relevant to the Biological Weapons Convention*, submitted by United States of America, paragraph 30, 2006. http://www.opbw.org/rev_cons/6rc/docs/adv/BWC.Conf.VI_S&T_usa_en.pdf (accessed 3 February 2025).
25. United Nations, Seventh Review Conference of the States Parties to the Convention on the Prohibition of the Development, Production and Stockpiling of Bacteriological (Biological) and Toxin Weapons and on Their Destruction, *New scientific and technological developments relevant to the Convention, Background Information Document Submitted by the Implementation Support Unit*. BWC/CONF.VII/INF.3/Add.1, United Nations, Geneva, 23 November 2011, https://docs.un.org/en/BWC/CONF.VII/INF.3/Add.1 (accessed 3 February 2025).
26. United Nations, Seventh Review Conference of the States Parties to the Convention on the Prohibition of the Development, Production and Stockpiling of Bacteriological (Biological) and Toxin Weapons and on Their Destruction, *New scientific and technological developments relevant to the Convention*, Background Information Document Submitted by the Implementation Support Unit. BWC/CONF.VII/INF.3/Add.1, United Nations, Geneva, 23 November 2011, https://docs.un.org/en/BWC/CONF.VII/INF.3/Add.1 (accessed 3 February 2025), 2011, paragraph 68.
27. United Nations, Seventh Review Conference of the States Parties to the Convention on the Prohibition of the Development, Production and Stockpiling of Bacteriological (Biological) and Toxin Weapons and on Their Destruction, *New scientific and technological developments relevant to the Convention*, Background Information Document Submitted by the Implementation Support Unit. BWC/CONF.VII/INF.3/Add.1, United Nations, Geneva, 23 November 2011, https://docs.un.org/en/BWC/CONF.VII/INF.3/Add.1 (accessed 3 February 2025), 2011, Paragraph 120.
28. United Nations, Seventh Review Conference of the States Parties to the Convention on the Prohibition of the Development, Production and Stockpiling of Bacteriological (Biological) and Toxin Weapons and on Their Destruction, *New scientific and technological developments relevant to the Convention*, Background Information Document Submitted by the Implementation Support Unit. BWC/CONF.VII/INF.3/Add.1, United Nations, Geneva, 23 November 2011, https://docs.un.org/en/BWC/CONF.VII/INF.3/Add.1 (accessed 3 February 2025), 2011, Paragraph 125.
29. United Nations, *New scientific and technological developments relevant to the Convention, Background information document submitted by the Implementation Support Unit*, 7th BTWC Review Conference, BWC/CONF.-VII/INF.3, 10 October 2011, Paragraph 13.
30. United States, *S&T Developments of Relevance to the Biological Weapons Convention*, *9th Review Conference of the Biological and Toxin Weapons*

Convention, United Nations, Geneva, 12 September 2022. Available at the ODA Meeting Place, p. 1.

31. United Kingdom, *National Contribution from the United Kingdom of Great Britain and Northern Ireland to the Background Information Document on New Scientific and Technological Developments Relevant to the Biological and Toxin Weapons Convention (BTWC)*, United Nations, Geneva. September. Available at the ODA Meeting Place, 2022, p. 9.
32. *Intervention by United States*, BTWC Meeting of Experts MX2 – Review of Developments in the Field of Science and Technology Related to the Convention, 2 September 2021, Text taken from an automatically generated transcription provided by the UNOG, as cited in M. Crowley and M. Dando, *Toxin and Bioregulator Weapons: Preventing the misuse of the chemical and life sciences*, Palgrave, 2022, pp. 176–186.
33. (a) J. Goldblat, The Biological Weapons Convention: An Overview, *Int. Rev. Red Cross*, 1997, **79**, 318; (b) M. Sossai, Drugs as weapons: disarmament treaties facing the advances in biochemistry and non-lethal weapons technology, *J. Conflict Secur. Law*, 2010, **15**(5–24), 5.
34. (a) J. Goldblat, The Biological Weapons Convention: An Overview, *Int. Rev. Red Cross*, 1997, **79**, 318; (b) M. Sossai, Drugs as weapons: disarmament treaties facing the advances in biochemistry and non-lethal weapons technology, *J. Conflict Secur. Law*, 2010, **15**(5–24), 5.
35. J. Goldblat, The Biological Weapons Convention: An Overview, *Int. Rev. Red Cross*, 1997, **79**, 318.
36. M. Crowley and M. Dando, *Toxin and Bioregulator Weapons: Preventing the misuse of the chemical and life sciences*, Palgrave, 2022, pp. 176–186.
37. (a) For relevant country case studies see: M. Crowley and M. Dando, Down the Slippery Slope? A study of contemporary dual-use chemical and life science research potentially applicable to incapacitating chemical agent weapons, Biochemical Security 2030 policy paper 8, University of Bath and University of Bradford, 2014, pp. 14–73; (b) M. Crowley and M. Dando, *Toxin and Bioregulator Weapons: Preventing the misuse of the chemical and life sciences*, Palgrave, 2022, pp. 47–196.
38. In an authoritative review of international humanitarian law, Henckaerts and Doswald-Beck, consider that "[S]tate practice establishes. . .the use of biological weapons meant to affect humans is prohibited. . .as a norm of customary international law applicable in both international and non-international armed conflicts." See: Henckaerts, L. Doswald-Beck, *Customary International Humanitarian Law, volume 1: Rules*, Cambridge University Press, Cambridge, 2005, p. 256.
39. (a) See, for example R. Whitehair and S. Brugger, BWC Protocol Talks in Geneva Collapse Following U.S. Rejection, Arms Control Today, September, 2001; (b) M. R. Dando, *Preventing Biological Warfare: The Failure of American Leadership*, Palgrave, Basingstoke, 2002; (c) J. Littlewood, *The Biological Weapons Convention: A Failed Revolution*, Ashgate, Aldershot, 2005.

40. M. R. Dando, *The Chemical and Biological Nonproliferation Regime after the Covid-19 Pandesmic: Dealing with the Scientific Revolution in the Life Sciences*, Palgrave Macmillan, Springer Nature, Cham, Switzerland, 2023.
41. United Nations Office for Disarmament Affairs, *Guide to Participating in the Confidence- Building Measures of the Biological Weapons Convention*, United Nations, Geneva, 2015, p. 2.
42. United Nations, *BWC Confidence Building Measures*, Overall rate of CBM Report submissions, https://bwc-ecbm.unog.ch/ (accessed 3 February 2025).
43. The 32 States publishing their 2024 CBMs were: Australia, Austria, Belgium, Bulgaria, Cameroon, Canada, Cyprus, Czech Republic, Denmark, Estonia, Finland, Germany, Hungary, Ireland, Japan, Kenya, Latvia, Lithuania, Luxembourg, Mexico, Netherlands, New Zealand, Norway, Portugal, Moldova, Slovakia, Slovenia, Sweden, Switzerland, Uganda, UK and US.
44. United Nations, *BWC Confidence Building Measures*, Submissions made by States Parties by year: 2024. Top of Form; https://bwc-cbm.un.org/ (accessed 3 February 2025).
45. BWPP, *The closing day of the 2023 BWC MSP – a meeting of lost opportunities*, BioWeapons Prevention Project, 16th December 2023. Available at https://www.bwpp.org/reports.html.
46. United Nations, Fifth Review Conference of the States Parties to the Convention on the Prohibition of the Development, Production and Stockpiling of Bacteriological (Biological) and Toxin Weapons and on their Destruction, *Final Report*, BWC/CONF.V/17. United Nations, Geneva, December 2002.
47. United Nations, *Final Document of the Seventh Review Conference*, BWC/CONF.VII/7, 5th–22nd December, United Nations, Geneva, 2011, Article 1, paragraph 2. p. 10.
48. United Nations, *Final Document of the Seventh Review Conference*, BWC/CONF.VII/7, 5th–22nd December, United Nations, Geneva, *Decisions and Recommendations*, 2011, paragraph 23. pp. 23, 24.
49. United Kingdom, *Review of Developments in Science and Technology: Key Points from the 2012–2015 Biological and Toxin Weapons Convention Intersessional Programme*, BWC/CONF.VCIII/WP.17, United Nations, Geneva, 1 November 2016, p. 7.
50. United States, *Current Trends and their Implications to the Biological Weapons Convention*, 2020 Meeting of the States Parties to the BTWC, Geneva, 22–25 November 2021, BWC/MSP/2020/WP.12, Paragraph 2. p. 1.
51. United Nations, *Final Document of the Ninth Review Conference*, BWC/CONF.IX.9, United Nations, Geneva, 21 December 2022, pp. 9–10.
52. United Nations, *Final Document of the Ninth Review Conference*, BWC/CONF.IX.9, United Nations, Geneva, 21 December 2022, p. 11.
53. Working Group on the Strengthening of the Convention on the Prohibition of the Development, Production and Stockpiling of Bacteriological (Biological) and Toxin Weapons and on Their Destruction, *Food*

for Thought Non-Paper by the Friends of the Chair for the development of a Mechanism to Assess Scientific and Technological Developments Relevant to the BWC and Provide States Parties with Relevant Advice, BWC/WG/2/CRP.5, United Nations, Geneva, 18 August 2023, pp. 3–8.
54. Bioweapons Prevention Project, *Discussions on the possible S&T developments review mechanism*, Report 2023-12, 17 August 2023a, p. 1.
55. J. Littlewood and F. Lentzos, Strengthening the Biological Weapons Convention, *Arms Control Today*, 2024, 1–7.
56. BWPP, *A minimal MSP and some reflections on the 2024 BWC* meetings, Report 2024-17, 2024, https://www.cbw-events.org.uk/BWC24-17.pdf.
57. For further details of the key tasks of the ISU see: *United Nations, Implementation Support Unit, UN Office for Disarmament Affairs*, https://www.un.org/disarmament/biological-weapons/implementation-support-unit (accessed 8 November 2021).
58. United Nations, *Final Document of the Ninth Review Conference*, BWC/CONF.IX.9, United Nations, Geneva, 21 December 2022, paragraph 24, p. 12.
59. United Nations, *Final Document of the Ninth Review Conference*, BWC/CONF.IX.9, United Nations, Geneva, 21 December 2022, paragraph 25, p. 12.
60. *Convention on the Prohibition of the Development, Production and Stockpiling of Bacteriological (Biological) and Toxin Weapons and on Their Destruction (BTWC)*, 1972. Available at https://ihl-databases.icrc.org/applic/ihl/ihl.nsf/INTRO/450?OpenDocument, Article V.
61. *Convention on the Prohibition of the Development, Production and Stockpiling of Bacteriological (Biological) and Toxin Weapons and on Their Destruction (BTWC)*, 1972. Available at https://ihl-databases.icrc.org/applic/ihl/ihl.nsf/INTRO/450?OpenDocument, Article V.
62. For further discussion see: J. Tucker, *Strengthening Consultative Mechansims Under Article V to Address BWC Compliance Concerns, Harvard Sussex Program*, May 2011, pp. 8–9. Available at http://hsp.sussex.ac.uk/new/_uploads/hspop/HSPOP_1.pdf.
63. *Convention on the Prohibition of the Development, Production and Stockpiling of Bacteriological (Biological) and Toxin Weapons and on Their Destruction (BTWC)*, 1972. Available at https://ihl-databases.icrc.org/applic/ihl/ihl.nsf/INTRO/450?OpenDocument, Article VI.
64. If the Security Council decides it wants an investigation and tasks the Secretary- General to conduct such an investigation, the SG would have the freedom to use the resources and capabilities available to them, and that would include the UNSGM. However, investigation of human rights abuses would not be part of a BTWC process.
65. United Nations Security Council, *Statement by Permanent Representative Vassily Nebenzia at UNSC briefing on biological laboratories in Ukraine*, 11 March 2022.
66. United Nations Security Council, *High Representative's briefing to the Security Council on Threats to International Peace and Security, Statement*

by Ms. Izumi Nakamitsu, High Representative for Disarmament Affairs, 11 March 2022, p. 2.

67. United Nations Security Council 8991ST Meeting (AM), Meeting coverage, United Nations Not Aware of Any Biological Weapons Programmes, Disarmament Chief Affirms as Security Council Meets to Address Related Concerns in Ukraine, SC/14827, 11 March 2022.
68. United Nations, *11th Emergency Special Session of the UN General Assembly on Ukraine*, Convened by S/RES/2623, 2022.
69. For details of the meeting and statements and associated background documents see: UNODA, *Biological Weapons Convention – Formal Consultative Meeting 2022*, https://meetings.unoda.org/meeting/65052.
70. United Nations, *Formal Consultative Meeting of the States Parties to the Convention on the Prohibition of the Development, Production and Stockpiling of Bacteriological (Biological) and Toxin Weapons and on Their Destruction*, BWC/CONS/2022/3, 19 September 2022, p. 10.
71. Russian Federation, *Letter from the Permanent Representative of the Russian Federation to the United Nations addressed to the President of the Security Council, UNSC, S/2022/796, 24 October 2022. Available on The Trench website*, chrome-extension://efaidnbmnnnibpcajpcglclefindmkaj/https://www.the-trench.org/wp-content/uploads/2022/11/UNSC-20221024-Russia-letter-Article-VI-Documentation.pdf (accessed 4 February 2025).
72. For analysis of this process see: J. Zanders, Russia's apoplexy over biological research – Implications for the BTWC and its Articles V and VI, *Arms Control Law*, 15 November 2022. https://armscontrollaw.com/2022/11/15/russias-apoplexy-over-biological-research-implications-for-the-btwc-and-its-articles-v-and-vi/ (accessed 12 July 2024).
73. For full details of the vote and associated statements by UNSC States see: J. Zanders, *Russia's apoplexy over biological research – Implications for the BTWC and its Articles V and VI, Arms Control Law*, 15 November 2022. https://armscontrollaw.com/2022/11/15/russias-apoplexy-over-biological-research-implications-for-the-btwc-and-its-articles-v-and-vi/ (accessed 12 July 2024).
74. United Kingdom, *Strengthening the Biological and Toxin Weapons Convention: Countering the Threat from Biological Weapons, Cm 54484*, The Stationary Office, London, 2002.
75. United Nations, Fourth Review Conference of the States Parties to the Convention on the Prohibition of the Development, Production and Stockpiling of Bacteriological (Biological) and Toxin Weapons and on Their Destruction, *Final Document*, BWC/CONF.1V/9, United Nations, Geneva, p. 18.
76. L. Wang, *et al.*, Tianjin Biosecurity Guidelines for codes of conduct for scientists: Promoting responsible sciences and strengthening biosecurity, *J. Biosaf. Biosecur.*, 2021, **3**, 82–83.
77. M. R. Dando, *The Chemical and Biological Nonproliferation Regime after the Covid-19 Pandemic: Dealing with the Scientific Revolution in the Life*

Sciences, Palgrave Macmillan, Springer/Nature, Cham, Switzerland, 2023, pp. 91–99.
78. Australia, *Raising Awareness: Approaches and Opportunities for Outreach*. BWC/MSP/MX/WP.29, United Nations, Geneva, 21 June 2005, p. 1.
79. M. R. Dando and B. Rappert, *Codes of Conduct for the Life Sciences: Some Insights from UK Academia*, University of Bradford, Briefing Paper No 16 (Second Series), 2005.
80. World Health Organisation, *Global Guidance Framework for the Responsible Use of the Life Sciences: Mitigating biorisks and governing dual-use research*, WHO, Geneva, 2022, p. 28.
81. (a) Australia *et al.*, Possible approaches to education and awareness-raising among life scientists, BWC/CONF. VII/WP.20, https://digitallibrary.un.org/record/717988, 2011; (b) Ukraine and United Kingdom of Great Britain and Northern Ireland, Awareness-raising, education and outreach: An example of best practice, BWC/CONF.VIII/WP.10, 2016; (c) Ukraine, Japan and the United Kingdom of Great Britain and Northern Ireland, Awareness-raising, education and outreach: recent developments, BWC/MSP/2017/WP.222, 2017.

CHAPTER 7

The Chemical Weapons Convention and Preventing the Weaponization of CNS-acting Chemicals

7.1 Introduction

The Chemical Weapons Convention (CWC) entered into force on 29th April 1997 and, as of 1 February 2025, comprises 193 States Parties that have ratified, acceded or succeeded to it – the highest number of any comparable arms control or disarmament treaty. The CWC States Parties represent about 98% of the global population and landmass.[1] An additional Signatory State (Israel) has signed the CWC, thus rendering political support to the objectives and principles of the Convention and committing itself to not undermining the Convention's objectives. Only three Non-Signatory States – DPRK (North Korea), Egypt and South Sudan – have not taken relevant action on the Convention. The Treaty is of unlimited duration and was intended by the negotiating States to be far more ambitious and comprehensive in scope and application than any prior international agreement on chemical weapons.

In the Convention's preambular paragraphs, the States Parties voice their "determine[ation] for the sake of all mankind, to exclude completely the possibility of the use of chemical weapons" whilst ensuring "that achievements in the field of chemistry should be used exclusively for the benefit of mankind". And consequently, they voice their conviction "that the complete and effective prohibition of the development, production, acquisition, stockpiling, retention, transfer and use of chemical weapons, and their destruction, represent a necessary step towards the achievement of these common objectives".[2]

Preventing Weaponization of CNS-acting Chemicals: A Holistic Arms Control Analysis
By Michael Crowley and Malcolm Dando

Published by the Royal Society of Chemistry, www.rsc.org

To date, most inter-governmental discourse regarding measures to prevent weaponization of CNS-acting chemicals and other types of ICAs has taken place within the context of the CWC and its attendant control regime and through its implementing body, the Organisation for the Prohibition of Chemical Weapons (OPCW). This chapter will examine how these chemicals and the associated science and technologies are addressed by the Convention, exploring the strengths, weaknesses and ambiguities in this instrument and associated measures developed by CWC States Parties to reinforce and implement it; and assessing the real-world consequences for preventing the development and use of these weapons.

7.2 General Obligations Under the Chemical Weapons Convention

Article 1 of the Chemical Weapons Convention (CWC) prohibits the development, production, stockpiling, transfer and use of chemical weapons "under any circumstances".[3] States Parties are also prohibited from engaging in "any military preparations to use chemical weapons"[4] or to "assist, encourage or induce, in any way, anyone to engage in any activity prohibited to a State Party under this Convention".[5] In addition, Article I also requires that all existing stocks of chemical weapons[6] and chemical weapons production facilities[7] be destroyed. Although the Convention prohibits chemical weapons, it allows for the controlled peaceful use of toxic chemicals.[8]

7.3 Scope of Coverage

The comprehensive nature of the CWC in terms of substances covered is established under Article II(2) which defines 'toxic chemicals' as:

> "Any chemical which through its chemical action on life processes, can cause death, temporary incapacitation or permanent harm to humans or animals.[9] *This includes all such chemicals, regardless of their origin or of their method of production,* and regardless of whether they are produced in facilities, in munitions or elsewhere"[10] (Emphasis added).

Given the emphasised text, it is clear that the Convention is not restricted to pharmaceutical chemicals or other chemicals manufactured by humans but would also cover those produced naturally, including certain poisons produced by living organisms, *i.e.*, toxins.

The importance of the Convention's comprehensive coverage has been widely recognised, not least by the World Health Organisation (WHO) in its 2004 report on the public health response to biological and chemical weapons, including CNS-acting or other ICA weapons.[11] The WHO report highlighted the dangers of the weaponization of a spectrum of chemical and toxin agents: "now that large-scale production processes for *biologically active peptides* and similar substances are undergoing rapid commercial development, *bioregulators and other toxins* constitute a field rich in potential weapons as well as pharmaceuticals, and in particular weapons of intense disabling or incapacitating power" (Emphasis added).[12] It subsequently noted that "[I]t is fortunate, therefore, that this advance in biotechnology should have coincided with the adoption of the *Chemical Weapons Convention*, since it *places its States Parties under the express obligation to ensure that bioregulators and other toxins,* like all other toxic chemicals, *are used only for the purposes that the Convention does not prohibit.*" (Emphases added).[13]

7.4 The Definition of Chemical Weapons

Whilst the Convention includes, in one of its annexes, three Schedules of toxic chemicals and precursors, these are specifically "identified for the application of verification measures"[14] and are not intended in any way to delineate or restrict the range of chemicals that may be considered to be "chemical weapons" under the Convention. Instead the Convention defines such weapons under Article II(1) *inter alia* as:

> "(a) Toxic chemicals and their precursors, *except where intended for purposes not prohibited under this Convention, as long as the types and quantities are consistent with such purposes*;" (Emphasis added).
>
> (b) Munitions and devices specifically designed to cause death or other harm through the toxic properties of those toxic chemicals specified in subparagraph (a), which would be released as a result of the employment of such munitions and devices;
>
> (c) Any equipment specifically designed for use directly in connection with the employment of the munitions and devices referred to in (b).[15]

A critical aspect of this definition – allowing it to be comprehensive, open-ended and flexible – is the emphasised text in Article II(1)(a) above. This has become known as the General Purpose Criterion (GPC). Krutszch and Trapp emphasised the implications of this formulation:

> "...Under this concept, *all* toxic or precursor chemicals are regarded as chemical weapons *unless* they have been developed, produced, stockpiled, or used for purposes *not* prohibited. The definition thus covers any toxic or precursor chemical if intended for chemical weapons purposes, irrespective of whether it has been listed in one of the Schedules." (Original emphasis).[16]

Indeed, the use of the toxic properties of *any* chemical (including natural toxins and bioregulators, as well as their synthetic analogues) as a weapon against either humans or animals – even those chemicals not originally developed or intended for such purposes – is captured by the GPC. Furthermore, because the GPC establishes a prohibition based on intent rather than on a limited list of toxic chemical agents, it allows the Convention to accommodate and reflect developments in science and technology; consequently, as Meselson and Perry Robinson have highlighted, "even toxic chemicals whose existence is not yet known are covered" by its provisions.[17]

To determine whether the use of a pharmaceutical chemical, toxin, bioregulator or other toxic chemical would be in conformity with the CWC, the intention or purpose for its use needs to be established. The range of "purposes not prohibited" under the CWC has been defined under Article II.9 as:

"(a) Industrial, agricultural, research, medical, pharmaceutical or other peaceful purposes;

(b) Protective purposes, namely those purposes directly related to protection against toxic chemicals and to protection against chemical weapons;

(c) Military purposes not connected with the use of chemical weapons and not dependent on the use of the toxic properties of chemicals as a method of warfare;

(d) Law enforcement including domestic riot control purposes."[18]

In summary, the use of any toxic chemical (whether naturally occurring or artificially synthesised) other than for "purposes not prohibited", as defined in Article II.9, would be considered use of a chemical weapon and prohibited under the CWC. Furthermore, the use of toxic chemicals for "purposes not prohibited" would be acceptable only "as long as the types and quantities [of these pharmaceutical chemicals, toxins and bioregulators, *etc*] are consistent with such purposes".[19]

7.5 Definitional Lacunae and Their Potential Consequences

Although many terms utilised in the CWC are defined in the text, several important concepts were left undefined by the negotiators, including notably "law enforcement" and "domestic riot control purposes". Certain commentators have argued that such ambiguity was intentional, accommodating the interests of certain States. For example, the March 1994 editorial of the *Chemical Weapons Convention Bulletin* argued that: "[S]ome, by no means a majority, of the negotiating States wished to protect possible applications of disabling chemicals that would either go beyond, or might be criticized as going beyond, applications hitherto customary in the hands of domestic police forces".[20]

It can be argued that a certain degree of constructive ambiguity was useful and perhaps was indispensable in developing a Convention text that all negotiating States could sign up to, given the opposing positions of some States on certain issues. Such textual inexactitude is by no means unusual in international agreements. Where such ambiguity exists in international agreements, States have recourse to legal tools and guidelines to aid them in their interpretations of the text, most importantly, the UN Vienna Convention on the Law of Treaties.[21] However, even utilising such interpretive tools, the absence of definitional clarity and the consequent textual ambiguity that can result, if left unresolved over a lengthy period of time, can become dangerous, leading to differing interpretations of the CWC by States Parties and divergent State practice. This in turn could potentially lead to breaches of the Convention by some States Parties, and also to an erosion of the stability and key prohibitions of the regime as a whole.

Despite these concerns, analysis by the authors of all publicly available OPCW documents shows that, as of February 2025, no policy making body of the OPCW has issued guidance on the interpretation of the meaning of "law enforcement including domestic riot control purposes" and the nature and scope of activities covered under this term, nor established specific dedicated processes for doing so.[22] To date, it has been left to States Parties to interpret the meaning of this term and the appropriate implementation of the Convention in this area. The implications of this lack of definitional clarity for the employment of CNS-acting chemical agents and other types of incapacitating chemical agents in law enforcement are explored below.

7.6 Regulating CNS-acting Chemical Research, Development and Use

The CWC does not explicitly prohibit research relating to chemical weapons (for example, if it is for "protective purposes"), but instead prohibits the development and production of such weapons under Article I. However, where research is an intrinsic part of a weapons development programme it clearly will fall within the scope of the Article I prohibition. There appear to be four potential scenarios where research into CNS-acting chemicals that could be employed as weapons may fall within the scope of the CWC, triggering different obligations upon CWC States Parties:

7.6.1 Scenario 1: State Research and Development of CNS-acting Weapons Intended for Use in Armed Conflict

The use in armed conflict of the toxic properties of chemical agents – including CNS-acting chemical agents or other types of ICAs – is absolutely prohibited,[23] as is their development (and by implication, associated research), production, acquisition, stockpiling, retention or transfer when intended for such purposes, under Articles I and II. If States have undertaken programmes to research and develop CNS-acting chemicals or other types of ICAs, and/or associated means of delivery for such purposes, they are required to halt such activities, declare any chemical weapons and chemical weapons production facilities (CWPFs) they possess (under Article III[24]) and ensure they are verifiably destroyed (under Article I, and in accordance with Articles IV and V respectively[25]). It should, however, be noted that in practice certain facilities working with CNS-acting/ICA toxins and bioregulators may not currently be considered as falling under the definition of a CWPF. This would be the case if such facilities: (a) did not produce a Schedule 1 chemical, (b) did not cross the annual production capacity threshold of one tonne per year for non-Schedule 1 chemicals, or (c) if they did produce more than one tonne per year but this chemical had a legitimate use (*e.g.* as a pharmaceutical). In such circumstances, these facilities would not currently trigger corresponding declaration and destruction requirements under the CWC.[26]

7.6.2 Scenario 2: State Research and Development of CNS-acting Weapons Intended for Use for Law Enforcement Purposes

Ambiguities in the CWC surrounding application of Article II 1(a) and Article II.9(d) have previously led to differing interpretations by States as to whether certain toxic chemicals beyond RCAs can legitimately be employed for law enforcement purposes. Following the Russian use of CNS-acting chemicals to end the Moscow theatre siege in 2002, several CWC States Parties and civil society organisations raised concerns regarding the consequences of further development, proliferation and use of weapons employing such chemicals in law enforcement.

Since 2018, discussions within the OPCW have focused on whether the use of aerosolised CNS-acting toxic chemicals for law enforcement purposes was permissible under the CWC. In recent years, a growing group of CWC States Parties pressed for this issue to be addressed by the OPCW. Russia has previously opposed such measures. In a November 2018 paper issued during the Fourth CWC Review Conference, Russia stated that the use of aerosolised CNS-acting chemicals for law enforcement purposes "is not regulated under the [Chemical Weapons] Convention".[27]

On 1 December 2021, the 26th Conference of States Parties to the CWC (CSP-26) adopted an "[U]nderstanding" – on the aerosolized use of CNS-acting chemical agents for law enforcement purposes, which "Decides that the aerosolised use of CNS-acting chemicals is understood to be inconsistent with law enforcement purposes as a 'purpose not prohibited' under the Convention".[28] This CSP Decision did not introduce a new prohibition but rather clarified the correct application of the existing prohibitions and rules of the CWC. The Decision also noted that "munitions and devices specifically designed to cause death or other harm" through the release of aerosolised CNS-acting chemicals would "constitute a 'chemical weapon'",[29] and consequently should be declared and verifiably destroyed. Whilst 85 States supported the Decision, 10 States – including China, Iran and Russia – voted against it. On 29 November 2021, China, Iran, Russia and Syria issued a joint statement declaring their rejection of the Decision which they considered to be "an *ultra vires* act" that "went beyond the powers and functions of the Policy-Making Organs of the OPCW", and "so could not have any legal effect(s) on the States Parties' rights and obligations under the Convention".[30]

In addition to the contested nature of its adoption and legal status, aspects of the Decision are ambiguous. Although it addresses "CNS-acting chemicals", there is no definition of this phrase, nor an indication of the range of chemicals that would be covered by it. It is also limited in its scope – its prohibitions are specifically restricted to *CNS-acting* chemicals. Consequently, any existing or future law enforcement weapons that use toxic chemicals that act on other core human physiological processes (*i.e.*, other types of incapacitating chemical agents) would not be covered by this prohibition. In this regard, additional concern arises given the inter-relationship between the CNS and other physiological systems, notably the endocrine system and the immune system, as described in Chapter 4 of this publication. Consequently, if such systems are not also protected from nefarious attack, then an ICA weapon that modifies the immune system to indirectly modify the nervous system may not be prohibited under this Decision.

The Decision is further limited in the scope of the means of delivery addressed. It explicitly prohibits only *aerosolized* CNS-acting weapons, excluding, for example, gaseous CNS-acting agents or CNS-acting agents delivered by non-aerosolized mechanisms such as the Chinese law enforcement CNS weapon highlighted in Chapter 5 of this publication. The Decision further restricts application to "munitions and devices *specifically designed* to cause death or other harm" and therefore the use of general purpose munitions and delivery devices such as air blowers and generic aerosol delivery systems may not be covered. In summary, although this Decision is a significant advance in constraining weaponised use of CNS-acting chemicals, its full implications will only become apparent as State Parties attempt to implement it. To facilitate consistent implementation, a States Party process to clarify outstanding areas of ambiguity in the text should be established. If this is not agreed, alternatively, State practice (where it remains unchallenged) could be left to shape the norm for implementation of this Decision. The current lack of clarity with regard to the scope and nature of this Decision, exacerbated by the lack of a clear dedicated process to facilitate its application by CWC States Parties and the OPCW as a whole, risks its implementation becoming patchy. Consequently, the permissibility under the CWC of research, development and use of law enforcement weapons employing pharmaceutical chemicals, toxins and bioregulators, is likely to remain unclear and contested for some time.[31]

7.6.3 Scenario 3: State Research and Development of CNS-acting Chemical Agent Weapons for Protective Purposes

Under Article X of the CWC, States are permitted to "conduct research into, develop, produce, acquire, transfer or use means of protection against chemical weapons, for purposes not prohibited under this Convention"[32] even where this involves production in appropriate quantities of potential chemical weapons agents, including CNS-acting or other types of ICAs. The range of permissible activities established under Article X potentially allows States to mask illicit weapons research, development and production under the guise of "protective purposes". One area of particular potential concern, or where misconceptions can readily arise, is with regard to threat (or risk) assessment studies. As Krutzsch and Trapp have noted: "[C]hemical defence programmes sometimes include threat assessment studies related to new, hitherto unknown, agents (new lethal agents, new incapacitants). When such assessments include research work that borders on what would be undertaken in chemical weapons development, they are no longer consistent with the object and purpose of the Convention".[33]

Consequently, in order to address these concerns and to increase "transparency of national programmes related to protective purposes", Article X obliges each State Party to "provide annually to the Technical Secretariat information on its programme".[34] In 2004, the 9th Conference of States Parties (CSP) adopted a template form for States Parties to utilise when submitting their Article X.4 declarations,[35] to facilitate consistent implementation by a greater number of States. In its 2008 report, the SAB stated that the "potential risks" to the CWC associated with advances in science and technology would "increase significantly, should dedicated chemical weapons programmes exist and should they take advantage of new toxic chemicals".[36] Consequently, the SAB argued that "there is therefore good reason to call for transparency in chemical-defence programmes"[37] and recommended that the "[R]eview Conference may wish to take this up when it addresses issues related to the annual submission by States Parties of information on their national protective programmes".[38]

Whilst publicly available OPCW reporting and analysis of this issue has been limited and infrequent, on 31 January 2024, the OPCW did provide a substantive report on implementation of Article X.[39] It stated that: "[B]etween entry into force of the Convention and 31 December 2023, 160 out of 193 States Parties submitted information on their national programmes for protection against chemical weapons".[40] Whilst this at first sight appears to be a substantive proportion of State responses, it is not known how many States have submitted information regularly throughout this period. Indeed

Table 7.1 CWC State submission of Article X information on National Protection Programmes, covering 2013–2023. Data from ref. 41.

Year	2013	2014	2015	2016	2017	2018	2019	2020	2021	2022	2023
Number of submissions	96	83	75	65	75	57	71	78	78	71	53

an annex to the January 2024 report quantifying the numbers of Article X submissions by States detailing their national protection programmes, broken down by State and by year, between 31 December 2013 and 31 December 2023, indicates that respondent States were likely to be in the minority.[41] A summary of total Article X submissions by year (Table 7.1) shows that at its highest point in 2013, 96 States provided Article X submissions whilst the lowest number of submissions received was 53 in 2023.

Although a small number of these States have unilaterally made their annual Article X declarations public, most have not. Consequently, it is not possible to determine whether the vast majority of States Parties have provided information to the OPCW on research for "protective purposes" related to CNS-acting agents and ICAs more broadly, and if so whether such information is accurate, complete, and includes any activities of concern.

7.6.4 Scenario 4: Research and Development of CNS-acting Chemicals and Delivery Mechanisms Purportedly for Medical or Other Non-prohibited Purposes, but Which Can Be Employed as Weapons

From a combined reading of Article I and Article II, development, production, acquisition, stockpiling, retention or transfer of toxic chemicals (including those that could be employed as CNS-acting agents or other types of ICAs) for "[I]ndustrial, agricultural, research, medical, pharmaceutical or other peaceful purposes" would be considered as "[P]urposes Not Prohibited" under the CWC,[42] provided such activities conformed to the "types and quantities" restriction.[43] All States Parties are, however, required under Article VII to "adopt the necessary measures to implement [their] obligations under this Convention" and shall "[P]rohibit natural and legal persons anywhere...under [their] jurisdiction... from undertaking any activity prohibited to a State Party under this Convention, including enacting penal legislation with respect to such activity".[44]

Consequently, States Parties need to adopt and enforce the appropriate national implementation measures to ensure that no research and development activities related to toxic chemicals (including pharmaceutical chemicals, toxins, bioregulators or other biologically active peptides), ostensibly for peaceful purposes, are being utilised for activities contrary to the object and purposes of the CWC or in violation of international law, such as in the development of CNS-acting or other types of ICA weapons intended for armed conflict or human rights abuses, or (following the 2021 CSP-26 Decision) for aerosolised CNS-acting chemical agent law enforcement weapons. As well as constraining the activities of State entities, these provisions include obligations upon all States to actively prevent the development, acquisition, transfer and use of all weapons employing toxic chemicals including CNS-acting or other types of ICA weapons by any non-State actors including criminals, terrorist organisations, private military or security companies, and militias.

7.7 OPCW 'Culture' and Working Practices

The potentially deleterious consequences following from the ambiguities in the CWC text have been played out through, and further exacerbated by, the limitations in the structure, working practices and 'culture' of the attendant control regime established to facilitate implementation of the Convention, *i.e.*, OPCW. Indeed a number of commentators have previously highlighted the long-standing limited ability of the OPCW to address certain important or pressing issues, even if delays in action could potentially seriously weaken the effectiveness of the Convention.[45] Factors considered to have contributed to this situation have included the Organisation's culture of decision-making by consensus and the consequent avoidance of difficult or controversial issues;[46] the wide disparity in resources and scientific and technical expertise available to State Party delegations and National Authorities; the limitations on the autonomy of the OPCW's Technical Secretariat including its restricted ability to receive and act on open source information;[47] limited transparency and accountability of the OPCW to civil society; and a reticence by the Organisation to receive information from, and interact with, relevant civil society organisations in a systematic manner.[48]

In recent years, the OPCW has on certain significant occasions found innovative ways to work around organisational obstacles – notably in its ground-breaking work facilitating destruction of all declared chemical weapons facilities and stockpiles in Syria in the midst of the ongoing conflict – and has been attempting to address some of its systemic failings. However, as illustrated later in this chapter, the 'consensus culture' and avoidance of contentious issues have clearly influenced how the Organisation has reviewed and addressed the application of the Convention to CNS-acting weapons.

7.8 Measures to Facilitate Implementation, Verification and Compliance

The CWC is implemented by its States Parties with the assistance of the Organisation for the Prohibition of Chemical Weapons (OPCW).[49] The OPCW comprises all CWC States Parties together with a Technical Secretariat with approximately 500 staff, headed by the Director General. The Technical Secretariat carries out the daily work of monitoring, verifying and facilitating the implementation of the CWC.[50] It receives State Parties' declarations, detailing chemical weapons-related activities or materials and relevant industrial activities. After receiving declarations, the Technical Secretariat inspects and monitors States Parties' facilities and activities that are relevant to the CWC, aiming to ensure compliance.[51]

The CWC categorises certain individual toxic chemicals and families (groups of related chemicals) and precursors into three Schedules, and activities and facilities that involve these chemicals above certain thresholds are subject to national implementation measures as well as international routine reporting and verification measures by the Technical Secretariat, to ensure they are not misused for chemical weapons purposes. Schedule 1 chemicals and precursors pose a 'high risk' to the Convention; many have been developed, produced, stockpiled and used as chemical weapons in the past, and they have few, if any peaceful uses. Schedule 2 chemicals have a variety of legitimate uses but also pose a 'significant risk' to the Convention because they can be directly used as chemical weapons or used as precursors in the production of Schedule 1 chemicals or other Schedule 2 chemicals. Schedule 3 chemicals are usually produced in large quantities for 'purposes not prohibited' by the CWC but still pose a risk to the Convention because they are precursors to Schedule 1 or 2 chemicals, or have been stockpiled as chemical weapons. Although the Convention has provided for the Schedules to be updated under a simplified technical change procedure that can be initiated by a State Party,[52] the process has appeared to be heavily politicized and the Schedules have not been revised on a regular basis. To date the only such revision took place in November 2019 with the addition of several groups of chemicals associated with the notion of Novichok nerve agents and some carbamates to Schedule 1.[53]

Currently, one CNS-acting chemical weapons agent, 3-Quinuclidinyl benzilate (BZ),[54] is considered to be a Schedule 2.A. chemical, and two of its immediate precursors, Quinuclidin-3-ol and 2,2-Diphenyl-2-hydroxyacetic acid (Benzilic Acid) are considered to be Schedule 2.B. chemicals.[55] Consequently, any facility that produces, processes or consumes more than 1 kg of BZ or 1 tonne of its two listed precursors must be declared.[56] In addition, it must be opened to routine on-site inspection if it produces 10 kg of BZ or 10 tonnes of precursors in a 3 year period.[57] Furthermore, transfers of any Schedule 2 chemical including BZ and its two listed precursors are only permitted between CWC States Parties.[58] Additional CNS-acting chemicals or other types of ICAs, including those that have previously been explored as potential CNS-acting or ICA weapons, are not currently included in any of the CWC Schedules and consequently lie outside the OPCW's routine reporting, control and verification procedures.

In addition to routine verification measures, the CWC incorporates a range of consultation, clarification and fact-finding mechanisms that a State Party can initiate when it is concerned about the possible misuse of any toxic chemicals (including non-Scheduled chemicals) or other possible noncompliance by another State Party.[59] These mechanisms range from informal bilateral consultations to clarification procedures involving the Executive Council and full-fledged challenge inspections and investigations of alleged use of chemical weapons. On-site challenge inspections may be invoked by a State Party at any facility or location in the territory or any other place under the jurisdiction or control of another State Party.[60] The results of such challenge inspections and investigations, including any evidence that may help to clarify the situation or evidence that may suggest non-compliance, as well as information on the level of cooperation of the inspected State Party, must be passed to the Executive Council (EC), and as the case may be a Special Session of the Conference of States Parties (CSP) for resolution.[61] Cases of grave noncompliance can be brought to the attention of the UN General Assembly as well as the UN Security Council.

7.9 Assessing and Responding to Science and Technology Developments

The CWC has a number of mechanisms intended to ensure that it remains relevant in the light of, and responds to the challenges arising from the continuous developments in the chemical and life sciences, and related technologies, including those related to the weaponization of CNS-acting chemicals. One important mechanism built into the Convention under Article VIII was the establishment of five-yearly Review Conferences intended to allow States Parties the opportunity to collectively review and improve the Convention's implementation, and which were specifically tasked to "take into account any relevant scientific and technological developments".[62]

The Convention also makes provision under Article VIII for the establishment of a Scientific Advisory Board (SAB).[63] This is an OPCW subsidiary body that enables the Director-General to render specialised advice in science and technology to the Review Conference, Conference of States Parties, Executive Council, or States Parties to the Convention. The role of the SAB is to assess and report to the Director-General developments in scientific and technological fields that are relevant to the Convention. On request, the SAB provides advice on technical matters related to the implementation of the Convention, including on co-operation and assistance to the Technical Secretariat.[64] The SAB is normally not asked and does not express its views on matters of policy but restricts its recommendations to technical issues related to the Convention's implementation. SAB recommendations can support the scientific and technical work of the Technical Secretariat and of States Parties, but they also can become the basis for regulatory steps or policy decisions by the OPCW. In the case of CNS-acting chemicals, as will be illustrated below, the SAB's analysis and recommendations had clear policy implications.

The SAB reports directly to the Director-General, who regularly submits its reports, alongside his own response, to the Executive Council. Every 5 years, the SAB prepares a wide-ranging comprehensive analysis of developments in science and technology of relevance to the implementation of the Convention. This substantial report is submitted to the States Parties prior to and is discussed during, the Review Conference, becoming an intrinsic part of States' systematic process at such Conferences of collectively reviewing and improving the Convention's implementation, taking "into account any relevant scientific and technological developments".[65]

The SAB is made up of 25 independent experts from OPCW Member States. The members serve in a personal capacity (not as representatives of their respective governments) for up to two consecutive 3-year terms. The SAB chair and vice-chair are elected annually.[66] The OPCW Technical Secretariat has facilitated the work of the SAB since its inception, and since 2013 a dedicated position of Science Policy Advisor has been included in the Technical Secretariat staff to render dedicated support to the SAB.

At the request of the Director General, the SAB can establish and co-ordinate dedicated temporary working groups (TWGs) to draw upon extended expertise for assessment and reporting on specific issues of relevance to the Convention. A number of such SAB TWGs have analysed areas of relevance to the weaponization of CNS-acting chemicals and ICAs more broadly, see notably the 2014 report of the TWG on convergence of biology and chemistry,[67] and the 2023 report of the TWG on the analysis of biotoxins.[68]

A concerted consideration by the SAB of CNS-acting agents and the more broadly defined ICAs (or incapacitants) was initiated in April 2010, at its 15th meeting:

> "The SAB recognised the complexities presented by riot control agents and incapacitating chemical agents, and their treatment under the Convention. It recalled that both the SAB itself and the Director-General had made reference to the matter on several occasions. The SAB further recognised that it could be of assistance to the Director-General in categorising toxic chemicals that fall within the general definitions of riot control agents or incapacitants for law enforcement".[69]

In addition, "the SAB recommended that it start deliberations on riot control agents and incapacitating chemical agents by receiving briefings on the different technical, legal, law enforcement, military, and political aspects surrounding the subject in order to identify the technical areas in which it can be of most assistance".[70]

Consequently, the SAB considered the technical issues relating to "incapacitants" at four of its subsequent meetings.[71] Former SAB Chair, Mr Stefan Mogl, in a 2013 interview with Crowley, explained that: "[F]ollowing a review and discussion of literature data on the potency and toxicity of potential candidate chemicals for ICAs, the SAB concluded that incapacitants should be considered just as toxic chemicals, most which would be non-Scheduled".[72] In addition, Mogl noted that "utilizing ICA in the field poses great operational challenges most notably that you cannot properly control the dose in the area of agent release and that you don't know the susceptibility of individuals exposed to the agent, and therefore cannot exclude that the agent may cause serious harm: incapacitants should not be called "non-lethal" agents".[73]

7.9.1 The SAB and the Third Review Conference

The SAB findings and recommendations with regard to "incapacitants *i.e.*, ICAs" were incorporated into the SAB report on "Developments in Science and Technology" for the Third CWC Review Conference. The SAB report highlighted the extreme limitations in operational employment of ICA weapons and clearly implied that current ICAs were not suitable for use in law enforcement:

> "...The Board considers the term "non-lethal" as inappropriate when referring to chemicals intended for use as incapacitants, because for all chemicals toxicity is a matter of dosage. The Board noted that chemicals considered having high safety margins in the context of controlled pharmaceutical use can have very low safety margins in the context of incapacitants when factors such as uneven dissemination, variability in human response, and the possible need for a rapid onset are required. It was also emphasised that the issue is not just what incapacitating chemical is used for law enforcement purposes, but how it is used, and the consequences such a use may have. The types of chemicals and pharmaceuticals, known to have been considered as incapacitants from open literature sources, were discussed. Most are centrally acting compounds that target specific neuronal pathways in the brain. All of them emerged from drug programmes undertaken from the 1960s to the 1980s, as far as can be judged by the research that has been published".[74]

In its previous meetings, the SAB received briefings of the findings of workshops held by Spiez laboratory/VERIFIN[75] and also the ICRC[76] on ICAs, from SAB members who participated in these events. The SAB consequently stated that:

> "In the view of the SAB the *technical* discussion on the potential use of toxic chemicals for law enforcement purposes has been exhaustive. It may continue its discussions once technical information about specific candidate chemicals and/or dissemination systems is made available..."[77] (Original emphasis)

The SAB was not tasked and consequently did not make a determination of whether any specific existing agents could be used as ICAs without significant risk to the health or life of the target population. However according to Mogl, "[T]he SAB stand ready to respond if they are asked by the DG [Director General], in future, to assess the effects of a specific chemical agent that has been discovered (or developed) and proposed (or deployed) for use as an incapcitant".[78]

In terms of its proposals, the SAB recommended that:

> "...the Secretariat start preparations for verification activities, relevant to incapacitating chemicals, that could be required in an investigation of alleged use (IAU). Such preparations should include developing analytical methods and procedures, as well as collecting analytical reference data for the analysis of such chemicals. The Secretariat should invite laboratories in Member States to contribute to this effort".[79]

7.9.2 SAB and the Fourth Review Conference

In its Science and Technology (S&T) Report to the 4th Review Conference, the SAB largely summarised its findings from its S&T Report to the 3rd Review Conference and in its recommendations, stated:

> "The SAB refers to its advice from the report to the Third Review Conference.... and maintains the view that the technical discussion on the potential use of toxic chemicals for law enforcement purposes has been exhaustive; .... and that the Secretariat should commence preparations for verification activities that could be required during an investigation of alleged use (IAU), including sample collection and the use of analytical data".[80]

And it continued:

> "...This issue is important to the Convention and is now in the policy domain. The SAB sees no value in revisiting this topic as the scientific facts remain unchanged since the SAB first considered the issue, in the SAB's Report to the Third Review Conference.... The SAB recognises that CNS-acting acting compounds are not RCAs as they do not meet criteria specified in Article II paragraph 7. Furthermore, the SAB notes that there have been examples of the use of CNS-acting chemicals in law enforcement that have resulted in permanent harm and death due to an irreversible action on life processes."[81]

7.9.3 SAB and the Fifth Review Conference

In its S&T Report to the 5th Review Conference the SAB first restated and underlined its position as established in its reports to the 3rd and 4th Review Conferences. It noted that:

> "When CNS-acting chemicals are disseminated in an airborne form, the delivered dose cannot be controlled. There have been examples of the use of CNS-acting chemicals in law enforcement that have resulted in permanent harm and death due to an irreversible action on life processes. *No CNS-acting chemicals with a sufficient margin of safety adequate for use for riot control or law enforcement purposes have been identified.*" (Emphasis added).[82]

The SAB then made a significant technical intervention with regard to the "Understanding" on CNS-acting chemicals adopted by the States Parties at CSP-26 "deciding that, *inter alia*, the aerosolised use of CNS-acting chemicals is understood to be inconsistent with law enforcement purposes as a "purpose not prohibited" under the Convention".[83] The SAB pointedly "note[d] that the term "aerosolised" specifically refers to small solid particles or liquid droplets dispersed in a gas. The term does not, however, cover gases or vapours" and furthermore that the "wording would need to be revised to extend the understanding to cover all possible airborne formulations".[84]

This of course begs a number of questions as to why the original draft "Understanding" was worded in this limited manner. Was this the original text proposed by the US, Swiss and Australian delegations spearheading the initiative to address CNS-acting chemicals? Or was the final "Understanding" text the result of subsequent negotiation and change to secure approval by a wider grouping of States? Was this limited scope intentional or was it an oversight? If it was an oversight then this raises questions as to a potential disconnect between diplomats and technical experts in relevant State Party delegations. There are broader procedural questions for the OPCW as well, as to whether the draft text of resolutions proposed by States is currently circulated in advance to the Technical Secretariat, and specifically the SAB, for advice/comment on technical grounds. And if this is not the case, whether such advice should be routinely sought in future.

For this specific case, however, we are now left with the situation where the SAB has pointed out the limited scope of the "Understanding" on "aerosolised" CNS-acting chemicals, and indicated that it could be revised to "cover all possible airborne formulations", but it is currently unclear whether and how this will be addressed collectively by the CWC States Parties.

The SAB also noted that:

> "...no compounds were added to the schedules with the recent adoption of [the Understanding] decision C-26/DEC.10.... at the Twenty-Sixth Session of the Conference. While no compounds were added,

> the SAB encourages States Parties to share technical information on CNS-acting chemicals of concern, particularly information for inclusion in the Validation Group Working Database and the OCAD, as appropriate. Information sharing is vital for the successful implementation of the Convention..."[85]

The SAB also noted that:

> "The decision adopted by the Conference requested that the Director-General task the SAB with continuing its review of relevant developments in science and technology related to CNS-acting chemicals, and provide updates to the Conference as appropriate. The SAB therefore remains watchful of the rapidly evolving pharmaceutical and chemical industries, with the incorporation of new technologies such as artificial intelligence (AI) and machine learning, and will remain vigilant for any consequent impact on this topic."[86]

Furthermore, the SAB noted that: "[A]lthough CNS-acting chemicals are unscheduled, an understanding of the analysis, synthesis, and safe handling of these chemicals would aid in the prevention of CNS-acting chemicals emerging as chemical warfare agents (CWAs)".[87]

Consequently, the SAB recommended that:

> "... the Director-General consider establishing a TWG to study current developments concerning CNS-acting chemicals, and it is further recommended that this TWG include the participation and expertise of other relevant international institutions (*e.g.*, the United Nations Commission on Narcotic Drugs)..."[88]

This SAB recommendation was subsequently supported by the US in its statement to the 106th meeting of the Executive Council:

> "...We look forward to its continued work on emerging threats, such as the establishment of a SAB Temporary Working Group on central nervous system (CNS)-acting chemicals, as well as new opportunities for efficient, relevant, and enhanced verification. As the Director-General has rightly identified, it is our responsibility as States Parties to ensure the Organization remains fit for purpose in the face of tomorrow's threats."[89]

Prioritisation of possible subjects for future SAB TWGs was discussed at the 38th Session of the SAB in May 2024. During his introduction to the meeting the Director General "described how the threat spectrum is expanding, facilitated by the rapid developments in science and technology, and noted that there is an increased risk of misuse of non-traditional agents, such as biotoxins and CNS-acting chemicals".[90]

However, given its limited human and time resources, and "based on deliberations at [this] 38th Session", the SAB recommended to the Director-General that "a TWG on the impact of artificial intelligence (AI) and related technologies relevant to the Convention be [first] established".[91] The SAB also recommended that "scientific and technical considerations regarding chemicals acting on the central nervous system (CNS)...[be one of two] future TWGs [that should] continue to be considered".[92] Consequently, although CNS-acting chemicals are acknowledged as a key focus for the SAB, a TWG on this issue has not yet been established or scheduled.

7.10 Allegations Made by CWC States of Development, Transfer and Use of CNS-acting Weapons Undertaken by Another CWC State Party

Since the CWC came into force in 1997, certain CWC States Parties have alleged that other CWC States Parties had undertaken research and development, transfer and/or use of CNS-acting or other types of ICA weapons. These State allegations can be differentiated into two broad categories: Firstly, there are allegations, principally by the US, that China, Iran and Russia have conducted such activities. As summarised below, such allegations are normally documented in annual US assessment reports of State compliance with the CWC, and have also been included in high-level US statements to the UN and/or to meetings of the OPCW. In addition to the allegations themselves, the *US CWC Compliance Reports* document actions the US has taken to address its concerns, including attempted bilateral engagement with the States concerned. The second series of allegations has been made by Russia against the US and other 'Western' countries. The allegations are normally broad in nature and contain little supporting evidence, that if present, is of a dubious nature.

7.10.1 Allegations by the US

7.10.1.1 Alleged Chinese Research of CNS-acting Chemical Agent Weapons

In its 2019–2022 *US CWC Compliance Reports*, the US stated that: "Scientists at a Chinese military institute have expressed interest in military applications of PBAs [Pharmaceutical Based Agents] and are engaged in research involving the synthesis, characterization, and testing of PBAs with potential dual-use applications".[93]

In its *2022 CWC Compliance Report* the US further stated:

"Reinforcing these concerns, during the November 2021 CSP, the [People's Republic of China] PRC submitted a joint statement with

> Iran, Russia, and Syria making clear its objections to the decision adopted by the Conference on the aerosolized use of CNS-acting chemicals. In the statement, the co-signatories stated that, 'Article II, paragraphs 1(a) and 9(d) of the Convention permits the use of toxic chemicals for law enforcement including domestic riot control purposes' as long as types and quantities are consistent with such purposes. Therefore, the Convention allows States Parties to choose their own method of law enforcement in full conformity with the provisions of the Convention including the General Purpose Criterion (GPC)."[94]

In its *2024 CWC Compliance Report* the US stated that:

> "During the November 2023 CSP, in bilateral discussions with the PRC, the United States reiterated its concerns about the PRC's potential development of aerosolized fentanyl. In response to U.S. concerns, the PRC reiterated that it does not use fentanyl for law enforcement purposes. . ."[95]

To date, it appears that the US has not attempted to initiate the appropriate formal multilateral consultation, cooperation, and fact-finding mechanisms under the CWC to raise and resolve these issues.

7.10.1.2 Alleged Iranian Research and Development of CNS-acting Weapons

In its *2019–2023 CWC Compliance Reports*, the US stated that since 2005, some of Iran's military-controlled facilities, notably Imam Hossein University (IHU) and Malek Ashtar University (MUT), have researched chemicals that have a wide range of sedation, dissociation, and amnestic incapacitating effects. According to the US, published Iranian papers cited the potential weapons applications of the PBAs; one specifically referenced the use of fentanyl during the 2002 Dubrovka theatre hostage crisis. According to the US, in 2014, the IHU chemistry department sought "kilogram quantities of medetomidine—*a sedative it has researched as an incapacitant*—from Chinese exporters". (Emphasis added).[96] According to the US, the IHU chemistry department has "little history of veterinary or even medical research", and the quantities sought (over 10 000 effective doses) were "inconsistent with the reported end use of research".[97]

In its *2024 CWC Compliance Report,* the US stated:

> ". . .Most recently on September 23, 2023, an X (formerly Twitter) account representing Lab Dookhtegan – whose members claim to be hacker organization in opposition to Iranian state-sponsored cyber actors — posted confidential documents detailing an Iranian military

university's development of grenades meant to disseminate medetomidine, an anesthetic that is a central nervous system-acting chemical. According to the documents, this development included information on the production and testing of prototype weapons."[98]

On 3 December 2020, the US introduced sanctions against the Iranian company, Shahid Meisami Group (SMG) and its director, under Executive Order 13382 [Blocking Property of Weapons of Mass Destruction Proliferators and Their Supporters].[99] The US accused SMG of responsibility for projects "involving the testing and production of chemical agents for use as so-called incapacitation agents".[100] The US further highlighted that it was "concerned about the [Iranian] regime's true intent with regard to the testing and production of these so-called chemical incapacitation agents, which could be used either to further oppress Iranian citizens or for offensive purposes". And stated that it "will continue to work to shed light on Iran's activities, including by highlighting the recently adopted CSP decision [on aerosolization of CNS-acting chemicals]". It noted that "[T]he last CWC compliance-related bilateral exchanges occurred in 2001 and 2004, on the margins of OPCW EC [Executive Council] meetings, but discussions did not resolve any of the issues raised by the United States".

The US has raised these concerns during meetings of the OPCW, including at the 105th Session of the Executive Council in March 2024:

"...Mr. Chairperson, it is important to raise awareness to deter and dissuade further development and future use of chemical weapons .... The United States ... assesses that since acceding to the CWC, Iran has developed and filled weapons with pharmaceutical-based agents in violation of its obligations to the Convention. On September 28, 2023, a hacker organization published documents online, which they identified as confidential reports from the Iranian regime. The documents explain in detail how to weaponize medetomidine, a pharmaceutical-based agent. These documents confirm our prior concerns about Iranian development of such agents...We call on the Iranian government to end its CW program."[101]

To date, it appears that the US, has not attempted to initiate the appropriate formal multilateral consultation, investigation, and fact-finding mechanisms under the CWC to raise and resolve these issues.

7.10.1.3 Russia: Use of CNS-acting Weapons and Alleged Subsequent Research and Development

Extensive Russian and international media documented the Russian security force's use of a CNS-acting chemical agent weapon to end the Moscow theatre siege. On 30th October 2002 the Russian Health Minister, Yuri

Shevchenko, identified the incapacitating agent as "a mixture of derivative substances of the fast action opiate Fentanyl".[102] Mr Shevchenko further stated that: "[I] officially declare: chemical substances which might have fallen under the jurisdiction of the international convention on banning chemical weapons were not used during the special operation".[103] Although Russia refused to fully identify the CNS-acting chemicals employed, analysis by the UK's Defence Science and Technology Laboratory (DSTL) indicated that the weapon comprised two analgesic-anaesthetics, carfentanil and remifentanil.[104] The UK DSTL authors also highlighted contemporary Russian research asserting that: "[S]cientific papers published by Russian military officers indicate an interest in fentanyls extending back 12 years: opioid receptor studies,[105,106] fentanyl analysis[107] and synthesis of fentanyl precursors".[108,109]

To date, Russia has provided no public information as to whether stockpiles of weaponised CNS-acting chemical agents are currently held by Russian authorities, nor whether research into the development and employment of such weapons is taking place, and if so, for what purposes.

The US has repeatedly raised its concerns with Russian CNS activities including in its *CWC Compliance Reports* from 2018–2024. In its *2022 CWC Compliance Report*, the US stated:

> "The United States is...concerned that Russia has a pharmaceutical-based agent (PBA) program intended for purposes inconsistent with the CWC. Russian Health Minister Shevchenko acknowledged to the press that Russian special operations employed "derivatives of fentanyl" to resolve the October 2002 Dubrovka theater hostage crisis. The United States is concerned that Russia is pursuing these types of agents for offensive purposes, which would be a violation of Article I. According to a non-paper Russia published at the OPCW in November 2018, entitled, *Aerosolisation of Central Nervous System-Acting Chemicals for Law Enforcement Purposes*, Russia believes use of these agents 'is not regulated under the [Chemical Weapons] Convention.'"[110]

In addition, in its 2022 *CWC Compliance Report*, the US described its attempts to address these issues including its leadership of:

> "...a successful effort at the OPCW that in 2021 resulted in the adoption of a CSP decision affirming the understanding that "the aerosolized use of CNS-acting chemicals is understood to be inconsistent with law enforcement purposes as a 'purpose not prohibited' under the Convention." The United States has long had concerns about Russia's activities related to PBAs and will continue to work to shed light on these activities, including through highlighting the CSP decision."[111]

To date, it appears that neither the US, the UK or other States that have publicly voiced concerns regarding Russian CNS weapons, have attempted to initiate the appropriate formal multilateral consultation, investigation, and fact-finding mechanisms under the CWC to raise and resolve these issues.

7.10.2 Russian Allegations

7.10.2.1 Russian Allegation of Use of a CNS Acting Agent Rather than Novichok in an Assassination Attempt in UK

On 4 March 2018, a former Russian military intelligence officer (in fact working as a double-agent for UK intelligence services [MI6]) Sergey Skripal, his daughter Yulia and a police officer Nick Bailey, were exposed to a toxic chemical in Salisbury, UK. On 30 June 2018 Dawn Sturgess and Charlie Rowley were also exposed to the same toxic chemical in nearby Amesbury. Bailey, Rowley and the Skripals recovered from their exposure, Sturgess died on 8 July 2018.[112] Samples from the victims of the Salisbury attack were tested at the UK's Defence Science and Technology Laboratory at Porton Down. On 12 March 2018, Theresa May, UK Prime Minister, stated that the nerve agent used in the Salisbury attack was a Novichok nerve agent and stated that the UK Government believed that it was "highly likely that Russia was responsible" for the attack.[113]

On 16 March 2018, the UK formally requested technical assistance from the OPCW Technical Secretariat, under Article VIII, paragraph 38e of the CWC. In its summary report of activities carried out in support of the UK request for technical assistance, the OPCW Technical Secretariat stated that the results of the analysis by OPCW designated laboratories of environmental and biomedical samples collected by the OPCW team "confirm the findings of the United Kingdom relating to the identity of the toxic chemical that was used in Salisbury and severely injured three people".[114]

Russia attempted to place in doubt the credibility of the OPCW analysis of the samples obtained, questioning whether a Novichok agent was the cause of the poisonings. Russia instead alleged that one of the OPCW designated laboratories undertaking the sample testing – the Spiez Laboratory in Switzerland – had found the presence of the CNS-acting agent BZ. On 14 April 2018 Russian Foreign Minister, Sergey Lavrov, addressing the 26th Council for Foreign and Defence Policy in Moscow claimed to be quoting sections of what he described as the Spiez Laboratory report to the OPCW, which had been "obtained on condition of confidentiality".[115] Lavrov claimed that the samples "indicate traces of the toxic chemical BZ and its precursor which are second category chemical weapons". He further claimed that: "[T]his composition was in operational service in the armies of the US, the UK and other NATO countries". And that "[T]he Soviet Union and Russia neither designed nor stored such chemical agents".[116]

On the same day, 14 April 2018, the Russian allegation was refuted on *Twitter/X* by Spiez Laboratory:

> "Only OPCW can comment [on] this assertion. But we can repeat what we stated 10 days ago: We have no doubt that Porton Down has

identified Novichok. PD [Porton Down] - like Spiez - is a designated lab of the OPCW. The standards in verification are so rigid that one can trust the findings."[117]

On 18 April 2018, the Russian allegation was pointedly refuted by the OPCW Director General at a meeting of the Executive Council:

"As it was clearly shown in the detailed and technical presentation, we should not have an iota of doubt on the reliability of the system of the OPCW Designated Laboratories. The Labs were able to confirm the identity of the chemical by applying existing, well-established procedures. There was no other chemical that was identified by the Labs. The precursor of BZ that is referred to in the public statements, commonly known as 3Q, was contained in the control sample prepared by the OPCW Lab in accordance with the existing quality control procedures. Otherwise, it has nothing to do with the samples collected by the OPCW Team in Salisbury. This chemical was reported back to the OPCW by the two designated labs and the findings are duly reflected in the report."[118]

In its statement at the Executive Council meeting, Switzerland also refuted Russia's attempts to undermine the OPCW analysis of the Salisbury samples: "[W]hether or not Spiez Laboratory was one of the designated laboratories involved in the analysis of the Salisbury samples, an analysis report of our designated laboratory would not have been drafted in the way and contained the type of language alleged to be a quote from a Spiez Laboratory report". Switzerland further noted that: "[S]uch actions weaken the credibility and integrity of this Organisation and are as such unacceptable".[119]

In addition, a number of States condemned Russia's disinformation campaign. Canada stated:

"We also call on the Russian Federation to refrain from spinning any more of its fanciful theories about the source of the agent, the means of delivery, or the involvement of third States. We heard new such theories today, including that the chemical agent was administered after the Skripals fell into a coma. Canada finds it deeply regrettable that, as part of this disinformation campaign, the Russian Federation's Foreign Minister this past weekend publicly named one of the OPCW's designated laboratories it believes was involved in the Secretariat's analysis. If the minister's statement is true, it represents a breach of the treaty's Confidentiality Annex, and prompts serious questions about the means by which the Russian Federation obtained this information. If the statement was not true, it shows a craven intent to mislead the international community by attacking the integrity of the Technical Secretariat."[120]

7.10.2.2 Alleged "Western" Transfer of CNS-acting Agent to Ukrainian Forces

Russian Defence Ministry officials have repeatedly alleged that Ukraine has been supplied with CNS acting weapons by the US or other Western States and that Ukraine had used such weapons.

According to a 4th March 2024 report by *TASS*, the Russian Security Service FSB's public relations centre told the Russian News Agency that "a terrorist attack plotted by Ukrainian security services involving the use of an analogue of the BZ chemical warfare agent in Zaporozhye Region had been prevented. Criminal proceedings were launched and three Ukrainian citizens detained. The seized chemicals are used to create weapons of mass destruction. All were developed in the United States".[121]

The 4th March 2024 *TASS* report cites and clips an interview by Lieutenant General Igor Kirillov the chief of Russia's radiation, chemical and biological protection forces for Russian TV channel *Rossiya-24*, in which he claimed that: "What has now been discovered is a substance close to BZ, a poisonous blacklisted substance. Some precursors that are used in its manufacture were seized, too. This poison was created at Edgewood Arsenal in the United States".[122]

According to *TASS*, on 19 February 2024, Kirillov said that "a toxic chemical similar to BZ, included in Schedule 2 of the Chemical Weapons Convention, had been used against Russian servicemen on August 19, 2022. A similar substance, he said, was found on January 28, 2024 during a detection and search operation in a cache in Melitopol".[123]

On 8 July 2024 *TASS* further reported that Lieutenant-General Igor Kirillov claimed: "With the connivance of Western countries, Ukraine does not limit itself to using non-lethal chemical agents, and actively employs listed chemicals such as BZ, prussic acid [hydrogen cyanide], and chlorine cyanide [cyanogen chloride]".[124]

On 8 July 2024, the Russian Delegation to the OPCW requested that the OPCW Technical Secretariat "circulate among the [CWC] States Parties the attached alarming information from Ukrainian and Western countries' open sources regarding the use by Ukrainian army and foreign mercenaries of chemicals".[125] The package of materials included a translation of a "*Briefing by Lieutenant General Igor Kirillov, Chief NBC Protection Troops of the RF Armed Forces, on large-scale provocations using toxic chemicals in Ukraine prepared by the West*" dated 23 February 2023. Although Lieutenant General Kirillov described the use of RCAs, he did not allege the use of CNS-acting chemicals. He did however allege the clandestine transfer of BZ, to Ukrainian forces. Although the alleged perpetrator is not directly stated, the implication is that it was the US. Lieutenant General Kirilov stated that:

> "...The Ministry of Defence of the Russian Federation was informed that on 10 February 2023, a train arrived in Ukraine (Kramatorsk), with a cargo of chemicals in one of its cars, accompanied by a group of

> foreign nationals. The car was detached and towed to the territory of the Kuibyshev metallurgical plant in Kramatorsk, where it was unloaded under the control of Security Service of Ukraine and AFU command representatives. The cargo included 16 sealed metal containers, eight of which had a chemical hazard label, BZ inscription and markings with two red bands, corresponding to the category of incapacitating agents.... The cargo was placed on US-manufactured armoured vehicles moving towards the line of contact in a convoy."[126]

According to Kirilov: "The facts of the simultaneous delivery of toxic chemicals and means of protection against them testify to an attempt to carry out large-scale provocations using the BZ military incapacitating agent during the conflict".[127]

Lieutenant General Kirilov continued:

> "I recall that under the Chemical Weapons Convention, BZ agent is a controlled and a Schedule 2 chemical, and its use is prohibited under Article 1 of the CWC. This chemical causes acute phrenoplegia, disorientation, hallucinations, and memory impairment. BZ is a standard toxic agent of the US Army and was widely used during the Vietnam War.... The United States declared the complete elimination of stocks of BZ back in 1990, when about 50 tons of the substance were disposed of, while samples were kept. In addition, the synthesis of BZ precursors at pharmaceutical production facilities in the amount of up to several tens of tons per year is still possible."[128]

Despite the circulation of information to all CWC States Parties, Russia has not sought to initiate formal CWC consultation, clarification and fact-finding mechanisms to resolve such matters. Considering the nature of the Russian allegations and the lack of credible publicly available information that Russia has provided to justify their claims, the Russian activities appear to be part of a disinformation campaign against the US, 'Western States' and Ukraine rather than an attempt at raising legitimate concerns over CNS-acting chemical agent weapons.

7.11 Conclusions

The CWC through its definitions of "toxic chemicals" and "chemical weapons" incorporating a General Purpose Criterion, ensures a comprehensive and essentially open coverage of substances within its scope including all existing and yet to be discovered or synthesised CNS-acting chemicals – including pharmaceutical chemicals, toxins and bioregulators. While the CWC prohibits the development and use of all chemical weapons against animals and humans, it acknowledges the legitimacy of the use of toxic chemicals for a range of "purposes not prohibited", which includes their use in law enforcement, but only "as long as the types and quantities are consistent with such purposes". The CWC specifically lists one CNS-acting chemical BZ and two of its immediate precursors (3-Quinuclidinol and Benzilic Acid), in its Schedules of chemicals requiring specific regulation and that are subject to routine verification measures. However, the application of the Convention in certain relevant areas is contested, notably regarding the legitimacy of research, development, and use of law enforcement CNS-acting chemical agent weapons, despite the welcome CSP-26 Decision clarifying that aerosolised use of CNS-acting chemicals was inconsistent with law enforcement as a 'purpose not prohibited' under the Convention. The application of the Convention in other relevant areas, including appropriate regulation of dual-use research for 'protective purposes', and notably with regard to wide area RCA means of delivery which could alternatively be employed for CNS-acting weapon delivery systems, has not been formally raised by States Parties in any OPCW policy-making organ.

One longstanding systemic concern has been the frequent failure of CWC States Parties to employ the existing formal multilateral consultation, co-operation and fact-finding mechanisms under the CWC to raise and address activities of other CWC States that may breach the Convention. The US, for example, has publicly alleged that China, Iran and Russia have been undertaking research, development and/or use of CNS-acting chemical agent weapons. Despite such allegations, neither the US nor any other CWC States Party has attempted to resolve these issues through the CWC multilateral processes. Continued failure to utilise the formal collective measures risks allowing States engaged in CNS-acting chemical agent weapons development to more easily ignore or dismiss well-founded concerns as being groundless politically motivated attempts to undermine or tarnish the international reputation of the accused States. Failure to use the formal CWC mechanisms has the additional broad corrosive effect of undermining confidence in the control regime. And more importantly, if there are strong indications that these State Party activities are (or may be) violations of the CWC, then the failure of CWC States Parties to urgently address them undermines the CWC itself and the wider norm against development, possession and use of chemical weapons. Indeed, it can be argued that breaching the norm against chemical weapons does not by itself weaken that norm, but failure to respond to such a breach does.

Russia has likewise made allegations that the US and 'Western States' have developed and/or transferred CNS-acting chemical agent weapons to Ukraine. Despite the circulation of these allegations and attendant information to all CWC States Parties, Russia has not sought to initiate formal CWC mechanisms to clarify and resolve such matters. Considering the nature of the Russian allegations and the lack of credible publicly available information that Russia has provided to justify their claims, the Russian activities appear to be part of a disinformation campaign against the US, 'Western States' and Ukraine, rather than an attempt at raising legitimate concerns over CNS-acting chemical agent weapons. In a separate case regarding OPCW analysis of the toxic chemical used to attack the Skripals in the UK,

Russia alleged that BZ (a CNS-acting chemical agent weapon) had been present in the sample studied, and implied the US had been the originator of this weapon; Russian allegations also attempted to undermine the credibility, working methods and independence of the OPCW Technical Secretariat itself. If such activities are not formally addressed through the appropriate OPCW mechanisms they could lead to an incredibly damaging deterioration of trust in the Organisation itself.

From this analysis, it is clear that the CWC text suffers from significant ambiguities and limitations (including those affecting the regulation of CNS-acting chemicals). These were deliberately introduced, for certain contentious issues, during the Convention's negotiation as part of the process of compromise necessary to achieve a treaty agreed by all. What matters in the long run is how States subsequently have interpreted and implemented the Convention in such contested areas. For CNS-acting chemicals there has clearly been divergence in interpretation and State practice, which has only been partially addressed by the CWC States collectively to date.

However, it is evident that despite its limitations, the CWC does contain a range of existing mechanisms that could be strengthened and effectively applied to regulate CNS-acting chemicals and prevent their weaponization, if States Parties were willing. Consequently, it appears that the OPCW and the regular cycle of CWC State Party meetings provide significant opportunities in the short to medium term for addressing CNS-acting chemical dual-use research and associated activities of concern to ensure that they are not, and are not perceived to be, exploited for malign purposes.

Unfortunately, the 5th CWC Review Conference held in May 2023 took place in a challenging global context notably involving the implications of the war in Ukraine, and the unresolved questions around Syria's chemical weapon possession and use. In addition, the OPCW was grappling with the organisational consequences of the incredibly welcome and significant approaching end of the verified destruction of all declared chemical weapons stockpiles, as well as the relentless revolution in the chemical, life and associated sciences and related technologies. Unfortunately, it appears that the worsening geopolitical environment notably as a consequence of the Ukraine war, had a significant negative impact on the Review Conference. Consequently, the States Parties were unable to adopt by consensus a substantive, strategically oriented *Final Report* of the 5th Review Conference, and instead could only adopt a factual procedural report of the Conference proceedings.[129] Whilst this failure may increase the sense of long-term strategic 'drift' for the Organisation and its future priorities, in the short to medium term the OPCW structures, consultation mechanisms and, vitally importantly, the day-to-day work of the Technical Secretariat will not be significantly affected. In the light of the current very difficult international environment, the final chapter of this publication will explore how the CWC and attendant control regime can be strengthened and more effectively implemented by States Parties to meet their objectives, and specifically to prevent and address the development and use of CNS-acting chemical agent weapons.

References

1. Organisation for the Prohibition of Chemical Weapons (OPCW), *Convention on the Prohibition of the Development, Production, Stockpiling and Use of Chemical Weapons and on their Destruction* (Chemical Weapons Convention [CWC]), 1993. The Convention CWC text, details of States Parties and all OPCW documents are available from the OPCW, website: http://www.opcw.org.
2. Organisation for the Prohibition of Chemical Weapons (OPCW), *Convention on the Prohibition of the Development, Production, Stockpiling and Use of Chemical Weapons and on their Destruction* (Chemical Weapons Convention [CWC]). 1993, Preamble.
3. Organisation for the Prohibition of Chemical Weapons (OPCW), *Convention on the Prohibition of the Development, Production, Stockpiling and Use of Chemical Weapons and on their Destruction* (Chemical Weapons Convention [CWC]). 1993, Article I.
4. Organisation for the Prohibition of Chemical Weapons (OPCW), *Convention on the Prohibition of the Development, Production, Stockpiling and Use of Chemical Weapons and on their Destruction* (Chemical Weapons Convention [CWC]). 1993, Article I.1.c.
5. Organisation for the Prohibition of Chemical Weapons (OPCW), *Convention on the Prohibition of the Development, Production, Stockpiling and Use of Chemical Weapons and on their Destruction* (Chemical Weapons Convention [CWC]). 1993, Article I.1.d.
6. Organisation for the Prohibition of Chemical Weapons (OPCW), *Convention on the Prohibition of the Development, Production, Stockpiling and Use of Chemical Weapons and on their Destruction* (Chemical Weapons Convention [CWC]). 1993, Article I.3.
7. Organisation for the Prohibition of Chemical Weapons (OPCW), *Convention on the Prohibition of the Development, Production, Stockpiling and Use of Chemical Weapons and on their Destruction* (Chemical Weapons Convention [CWC]). 1993, Article I.4.
8. Organisation for the Prohibition of Chemical Weapons (OPCW), *Convention on the Prohibition of the Development, Production, Stockpiling and Use of Chemical Weapons and on their Destruction* (Chemical Weapons Convention [CWC]). 1993, Articles II(9) (a), V(14), and XI(2) (c).
9. The CWC's definition of "toxic chemical" does not include toxicity against plants. Instead the CWC makes reference to herbicides only in the Preamble, where it recognises the prohibition of their use as a method of warfare.
10. Organisation for the Prohibition of Chemical Weapons (OPCW), *Convention on the Prohibition of the Development, Production, Stockpiling and*

Use of Chemical Weapons and on their Destruction (Chemical Weapons Convention [CWC]). 1993, Article II.2.

11. World Health Organisation, *Public health response to biological and chemical weapons: WHO Guidance*, 2nd edn, WHO, Geneva, Switzerland, 2004.
12. World Health Organisation, *Public health response to biological and chemical weapons: WHO Guidance*, 2nd edn, WHO, Geneva, Switzerland, 2004, p. 216.
13. World Health Organisation, *Public health response to biological and chemical weapons: WHO Guidance*, 2nd edn, WHO, Geneva, Switzerland, 2004, p. 216; Chemical Weapons Convention, Article VI(1).
14. Organisation for the Prohibition of Chemical Weapons (OPCW), *Convention on the Prohibition of the Development, Production, Stockpiling and Use of Chemical Weapons and on their Destruction* (Chemical Weapons Convention [CWC]). 1993, Article II.(2).
15. Organisation for the Prohibition of Chemical Weapons (OPCW), *Convention on the Prohibition of the Development, Production, Stockpiling and Use of Chemical Weapons and on their Destruction* (Chemical Weapons Convention [CWC]). 1993, Article II.1.
16. W. Krutzsch, E. Meyjer and R. Trapp, *The Chemical Weapons Convention: A Commentary*, Oxford University Press, 2014, p. 77.
17. M. Meselson and J. P. Robinson, New technologies and the loophole, editorial, *Chemical Weapons Convention Bulletin*, 1994, vol. 23, pp. 1–2.
18. Organisation for the Prohibition of Chemical Weapons (OPCW), *Convention on the Prohibition of the Development, Production, Stockpiling and Use of Chemical Weapons and on their Destruction* (Chemical Weapons Convention [CWC]). 1993, Article II.9.a-d.
19. Organisation for the Prohibition of Chemical Weapons (OPCW), *Convention on the Prohibition of the Development, Production, Stockpiling and Use of Chemical Weapons and on their Destruction* (Chemical Weapons Convention [CWC]). 1993, Article II.1.a.
20. M. Meselson and J. P. Robinson, New technologies and the loophole, editorial, *Chemical Weapons Convention Bulletin*, 1994, vol. 23, p. 2.
21. UN, *Vienna Convention on the Law of Treaties*, Vienna, 1155 UN Treaty Series (UNTS) 331, 23rd May 1969.
22. Analysis was undertaken of all OPCW documents publicly available on the OPCW website http://www.opcw.org. since the Organisation's inception up until 4 February 2025.
23. In addition to the Chemical Weapons Convention, the use of toxins as weapons in armed conflict is prohibited under the 1925 Geneva Protocol, the BTWC (and Extended Understandings of the BTWC States Parties), and customary international humanitarian law.
24. Organisation for the Prohibition of Chemical Weapons (OPCW), *Convention on the Prohibition of the Development, Production, Stockpiling and Use of Chemical Weapons and on their Destruction* (Chemical Weapons Convention [CWC]). 1993, Article III (1)a-c.
25. Organisation for the Prohibition of Chemical Weapons (OPCW), *Convention on the Prohibition of the Development, Production, Stockpiling and*

Use of Chemical Weapons and on their Destruction (Chemical Weapons Convention [CWC]). 1993, Article I (2) and (4); Article IV and Article V.
26. Organisation for the Prohibition of Chemical Weapons (OPCW), *Convention on the Prohibition of the Development, Production, Stockpiling and Use of Chemical Weapons and on their Destruction* (Chemical Weapons Convention [CWC]). 1993, Article IV. See in particular Article IV, 8 (a) (i) and (b)(i).
27. Russian Federation, *Aerosolisation of Central Nervous System-Acting Chemicals for Law Enforcement Purposes*, RC-4/NAT.9, OPCW, 21 November 2018, p. 3.
28. Conference of States Parties, Decision: *Understanding regarding the aersolised use of central nervous system-acting-chemicals for law enforcement purposes*, C-26/DEC.10. OPCW, The Hague, 1 December 2021.
29. Conference of States Parties, Decision: *Understanding regarding the aersolised use of central nervous system-acting-chemicals for law enforcement purposes*, C-26/DEC.10. OPCW, The Hague, 1 December 2021.
30. Iran, *Joint Statement on behalf of 4 Delegations*. Delivered by the Delegation of the Islamic Republic of Iran at the 26th Session of the Conference of States Parties of the OPCW under the Subitem 26.1 "Any Other Business" on the Draft Decision entitled "Understanding Regarding the Aerosolized Use of Central Nervous System-Acting Chemicals for Law Enforcement Purposes". OPCW, The Hague, 29 November 2021.
31. M. Crowley and M. Dando, Central nervous system weapons dealt a blow, *Science*, 2022, **375**(6577), 153–154.
32. Organisation for the Prohibition of Chemical Weapons (OPCW), *Convention on the Prohibition of the Development, Production, Stockpiling and Use of Chemical Weapons and on their Destruction* (Chemical Weapons Convention [CWC]). 1993, Article X, paragraph 2.
33. W. Krutzsch and R. Trapp, Article II: Definitions and Criteria, in *The Chemical Weapons Convention: A Commentary*, ed. W. Krutzsch, E. Myjer and R. Trapp, Oxford University Press, Oxford, 2014, p. 93.
34. Organisation for the Prohibition of Chemical Weapons (OPCW), *Convention on the Prohibition of the Development, Production, Stockpiling and Use of Chemical Weapons and on their Destruction* (Chemical Weapons Convention [CWC]). 1993, Article X, paragraph 4.
35. Conference of the States Parties, *Decision, Submission of information regarding national programmes related to protective purposes pursuant to Article X, Paragraph 4 of the Convention*, C-9/DEC.10. OPCW, The Hague, 30th November 2004.
36. Conference of the States Parties, *SAB Report*, Second Review Conference RC-2/DG.1. OPCW, The Hague. 28 February 2008, paragraph 3.14.
37. Conference of the States Parties, *SAB Report*, Second Review Conference RC-2/DG.1. OPCW, The Hague. 28 February 2008, paragraph 3.14.
38. Conference of the States Parties, *SAB Report*, Second Review Conference RC-2/DG.1. OPCW, The Hague. 28 February 2008, paragraph 3.7.

39. OPCW, *Status of implementation of Article X of the Chemical Weapons Convention as at 31 December 2023*, EC-105/DG.8, Report of the Director General, Executive Council, 105th Session, 5–8 March 2024, 31 January 2024.
40. OPCW, *Status of implementation of Article X of the Chemical Weapons Convention as at 31 December 2023*, EC-105/DG.8, Report of the Director General, Executive Council, 105th Session, 5–8 March 2024, 31 January 2024, Article 2.3.
41. OPCW, *Status of implementation of Article X of the Chemical Weapons Convention as at 31 December 2023*, EC-105/DG.8, Report of the Director General, Executive Council, 105th Session, 5–8 March 2024, 31 January 2024, Annex I, Submissions of information on National Protection Programmes, by year, in accordance with Paragraph 4 of Article X of the Convention, between 31 December 2013 and 31 December 2023.
42. Organisation for the Prohibition of Chemical Weapons (OPCW), *Convention on the Prohibition of the Development, Production, Stockpiling and Use of Chemical Weapons and on their Destruction* (Chemical Weapons Convention [CWC]). 1993, Article II(9)a.
43. Organisation for the Prohibition of Chemical Weapons (OPCW), *Convention on the Prohibition of the Development, Production, Stockpiling and Use of Chemical Weapons and on their Destruction* (Chemical Weapons Convention [CWC]). 1993, Article II(1)a.
44. Organisation for the Prohibition of Chemical Weapons (OPCW), *Convention on the Prohibition of the Development, Production, Stockpiling and Use of Chemical Weapons and on their Destruction* (Chemical Weapons Convention [CWC]). 1993, Article VII (1).a.
45. (a) For example, see: A. Kelle, The CWC After Its First Review Conference: Is the Glass Half Full or Half Empty? Disarmament Diplomacy, issue 71, 2003 June–July; (b) J. Perry Robinson, Non lethal warfare and the Chemical Weapons Convention, Further HSP submission to the OPCW Open-Ended Working Group on Preparations for the Second CWC Review Conference, 24th October 2007.
46. W. Krutzsch and T. Dunworth, Article VIII: The Organization, in *The Chemical Weapons Convention: A Commentary*, ed. W. Krutzsch, E. Myjer and R. Trapp, Oxford University Press, Oxford, 2014, pp. 258–260.
47. OPCW, Director General, *Report of the [Ekeus] advisory panel on future priorities of the Organisation for the Prohibition of Chemical Weapons*, S/951/2011, 25th July 2011, Paragraph 48.
48. (a) D. Feakes, Workshop Report: Pugwash Study Group on the Implementation of the Chemical and Biological Weapons Conventions CWC Implementation: Balancing Confidentiality and Transparency, 15th–17th May 1998; (b) J. Tucker, Introduction, in *The Chemical Weapons Convention: Implementation challenges and solutions*, ed. J. Tucker, Monterey Institute of International Studies, Washington DC, 2001; (c) M. Moodie, Issues for the First CWC Review Conference, J. Tucker, Introduction, in *The Chemical Weapons Convention: Implementation*

challenges and solutions, ed. J. Tucker, Monterey Institute of International Studies, Washington DC, 2001; (d) A. Sands and J. Pate, CWC Compliance Issues, in: J. Tucker, Introduction, in *The Chemical Weapons Convention: Implementation challenges and solutions*, ed. J. Tucker, Monterey Institute of International Studies, Washington DC, 2001; (e) A. Kelle, The CWC After Its First Review Conference: Is the Glass Half Full or Half Empty? *Disarmament Diplomacy*, issue 71, 2002; (f) W. Krutzsch, Never Under Any Circumstances: The CWC in the Third Year After Its First Review Conference, *CBW Conv. Bull.*, 2005, **68**, 6–12; (g) W. Krutzsch, Ensuring True Implementation of the CWC, *CBW Conv. Bull.*, 2007, **76/77**, 16–18.

49. Organisation for the Prohibition of Chemical Weapons (OPCW), *Convention on the Prohibition of the Development, Production, Stockpiling and Use of Chemical Weapons and on their Destruction* (Chemical Weapons Convention [CWC]). 1993, Articles VIII.1 to VIII.8
50. Organisation for the Prohibition of Chemical Weapons (OPCW), *Convention on the Prohibition of the Development, Production, Stockpiling and Use of Chemical Weapons and on their Destruction* (Chemical Weapons Convention [CWC]). 1993, Articles VIII.37 to VIII.47
51. Organisation for the Prohibition of Chemical Weapons (OPCW), *Convention on the Prohibition of the Development, Production, Stockpiling and Use of Chemical Weapons and on their Destruction* (Chemical Weapons Convention [CWC]). 1993, Articles VIII.37 to VIII.47
52. Organisation for the Prohibition of Chemical Weapons (OPCW), *Convention on the Prohibition of the Development, Production, Stockpiling and Use of Chemical Weapons and on their Destruction* (Chemical Weapons Convention [CWC]). 1993, Article XV (4–5)
53. (a) OPCW, Decision: Changes to Schedule 1 of the Annex on Chemicals to the Chemical Weapons Convention. C-24/DEC.5, 27 November 2019; (b) OPCW, Decision: Technical Change to Schedule 1(A) of the Annex on Chemicals to the Chemical Weapons Convention, 2019, C-24/DEC.4, 27.
54. Organisation for the Prohibition of Chemical Weapons (OPCW), *Convention on the Prohibition of the Development, Production, Stockpiling and Use of Chemical Weapons and on their Destruction* (Chemical Weapons Convention [CWC]). 1993, Annex on Chemicals, B. Schedule of Chemicals, Schedule 2.A. Toxic Chemicals.
55. Organisation for the Prohibition of Chemical Weapons (OPCW), *Convention on the Prohibition of the Development, Production, Stockpiling and Use of Chemical Weapons and on their Destruction* (Chemical Weapons Convention [CWC]). 1993, Annex on Chemicals, B. Schedule of Chemicals, Schedule 2.B. Precursors.
56. Organisation for the Prohibition of Chemical Weapons (OPCW), *Convention on the Prohibition of the Development, Production, Stockpiling and Use of Chemical Weapons and on their Destruction* (Chemical Weapons Convention [CWC]). 1993, Article VI and Verification Annex, Part VII,

Regime for Schedule 2 Chemicals and Facilities Related to Such Chemicals, Section A, Article 3 (a) & (c).

57. Organisation for the Prohibition of Chemical Weapons (OPCW), *Convention on the Prohibition of the Development, Production, Stockpiling and Use of Chemical Weapons and on their Destruction* (Chemical Weapons Convention [CWC]). 1993, Article VI and Verification Annex, Part VII, Regime for Schedule 2 Chemicals and Facilities Related to Such Chemicals, Section B, Article 12 (a) & (c).
58. Organisation for the Prohibition of Chemical Weapons (OPCW), *Convention on the Prohibition of the Development, Production, Stockpiling and Use of Chemical Weapons and on their Destruction* (Chemical Weapons Convention [CWC]). 1993, Article VI and Verification Annex, Part VII, Regime for Schedule 2 Chemicals and Facilities Related to Such Chemicals, Section C, Article 31.
59. Organisation for the Prohibition of Chemical Weapons (OPCW), *Convention on the Prohibition of the Development, Production, Stockpiling and Use of Chemical Weapons and on their Destruction* (Chemical Weapons Convention [CWC]). 1993, Article IX.
60. Organisation for the Prohibition of Chemical Weapons (OPCW), *Convention on the Prohibition of the Development, Production, Stockpiling and Use of Chemical Weapons and on their Destruction* (Chemical Weapons Convention [CWC]). 1993, Articles IX.8 to IX.25.
61. Organisation for the Prohibition of Chemical Weapons (OPCW), *Convention on the Prohibition of the Development, Production, Stockpiling and Use of Chemical Weapons and on their Destruction* (Chemical Weapons Convention [CWC]). 1993, Articles IX.3 to IX.7, IX.23.
62. Organisation for the Prohibition of Chemical Weapons (OPCW), *Convention on the Prohibition of the Development, Production, Stockpiling and Use of Chemical Weapons and on their Destruction* (Chemical Weapons Convention [CWC]). 1993, Article VIII, paragraph 22.
63. Organisation for the Prohibition of Chemical Weapons (OPCW), *Convention on the Prohibition of the Development, Production, Stockpiling and Use of Chemical Weapons and on their Destruction* (Chemical Weapons Convention [CWC]). 1993, Article VIII, paragraph 21(h) and 45.
64. OPCW (undated), *Scientific Advisory Board, Keeping pace with scientific and technological change*, https://www.opcw.org/about/subsidiary-bodies/scientific-advisory-board.
65. Organisation for the Prohibition of Chemical Weapons (OPCW), *Convention on the Prohibition of the Development, Production, Stockpiling and Use of Chemical Weapons and on their Destruction* (Chemical Weapons Convention [CWC]). 1993, Article VIII, paragraph 22.
66. OPCW, *Terms of Reference of the Scientific Advisory Board/ Rules of Procedure for the Scientific Advisory Board and Temporary Working Groups of Scientific Experts*, June 2011.
67. OPCW SAB, *Convergence of Chemistry and Biology: Report of the Scientific Advisory Board's Temporary Working Group*, SAB/REP/1/14, 26 June 2014.

68. OPCW, SAB, *Analysis of Biotoxins Report of the Scientific Advisory Board's Temporary Working Group*, SAB/REP/1/23, 20 April 2023.
69. OPCW, SAB, *Report of the Fifteenth Session of the Scientific Advisory Board*, SAB-15/1 14 April 2010, paragraph 13.2.
70. OPCW, SAB, *Report of the Fifteenth Session of the Scientific Advisory Board*, SAB-15/1 14 April 2010, paragraph 13.3
71. (a) OPCW, Report of the Nineteenth Session of the Scientific Advisory Board, SAB-19/1, 12 September 2012; (b) OPCW, Report of the Eighteenth Session of the Scientific Advisory Board, SAB-18/1, 19 April 2012; (c) OPCW, Report of the Seventeenth Session of the Scientific Advisory Board, SAB-17/1, 23 November 2011; (d) OPCW, Report of the Sixteenth Session of the Scientific Advisory Board, SAB-16/1, 6 April 2011.
72. M. Crowley, Application of the OPCW mechanisms for reviewing and addressing S&T developments: The case of ICAs, University of Bath, [author's interview with Mr Stefan Mogl, former Chair of the SAB], November 2013, p. 23.
73. M. Crowley, Application of the OPCW mechanisms for reviewing and addressing S&T developments: The case of ICAs, University of Bath, [author's interview with Mr Stefan Mogl, former Chair of the SAB], November 2013, p. 23.
74. OPCW, *Report of the Scientific Advisory Board on Developments in Science and Technology to the Third Special Session of the Conference of the States Parties to Review the Operations of the Chemical Weapons Convention, RC-3/D.G.1*. Annex 1, paragraph 12. OPCW, The Hague. 29 October 2012.
75. *Technical Workshop on Incapacitating Agents*, held in Spiez, Switzerland, 8 and 9 September 2011, cited in OPCW, *Report of the Scientific Advisory Board on Developments in Science and Technology to the Third Special Session of the Conference of the States Parties to Review the Operations of the Chemical Weapons Convention, RC-3/D.G.1*. Annex 1, paragraph 12. OPCW, The Hague. 29 October 2012
76. (a) International Committee of the Red Cross, Incapacitating Chemical Agents: Implications for International Law, held in Montreux, Switzerland, 24 to 26 March 2010; (b) Expert meeting entitled: Incapacitating Chemical Agents: Law Enforcement, Human Rights Law and Policy Perspectives, from 24 to 26 April 2012, Montreux, Switzerland, as cited in: OPCW, Report of the Scientific Advisory Board on Developments in Science and Technology to the Third Special Session of the Conference of the States Parties to Review the Operations of the Chemical Weapons Convention, RC-3/D.G.1. Annex 1, paragraph 12. OPCW, The Hague. 29 October 2012
77. OPCW, *Report of the Scientific Advisory Board on Developments in Science and Technology to the Third Special Session of the Conference of the States Parties to Review the Operations of the Chemical Weapons Convention, RC-3/D.G.1.*, 2012, paragraph 13.
78. M. Crowley, Application of the OPCW mechanisms for reviewing and addressing S&T developments: The case of ICAs, University of Bath,

[author's interview with Mr Stefan Mogl, former Chair of the SAB], November 2013, p. 24.
79. OPCW, *Report of the Scientific Advisory Board on Developments in Science and Technology to the Third Special Session of the Conference of the States Parties to Review the Operations of the Chemical Weapons Convention, RC-3/D.G.1.,* 2012, paragraph 13.
80. OPCW, *Report of the Scientific Advisory Board on Developments in Science and Technology to the Fourth Special Session of the Conference of the States Parties to Review the Operations of the Chemical Weapons Convention*, RC-4, DG-1. OPCW, The Hague, 30 April 2018, paragraph 100.
81. OPCW, *Report of the Scientific Advisory Board on Developments in Science and Technology to the Fourth Special Session of the Conference of the States Parties to Review the Operations of the Chemical Weapons Convention*, RC-4, DG-1. OPCW, The Hague, 30 April 2018, paragraph 101.
82. OPCW, *Report of the Scientific Advisory Board on Developments in Science and Technology to the Fifth Special Session of the Conference of the States Parties to Review the Operations of the Chemical Weapons Convention.* RC-5/DG.1, OPCW, The Hague, 22 February, 2023, Annex 1, paragraph 8.
83. OPCW, R*eport of the Scientific Advisory Board on Developments in Science and Technology to the Fifth Special Session of the Conference of the States Parties to Review the Operations of the Chemical Weapons Convention.* RC-5/DG.1, OPCW, The Hague, 22 February, 2023, Annex 1 paragraph 11.
84. OPCW, *Report of the Scientific Advisory Board on Developments in Science and Technology to the Fifth Special Session of the Conference of the States Parties to Review the Operations of the Chemical Weapons Convention.* RC-5/DG.1, OPCW, The Hague, 22 February, 2023, Annex 1, paragraph 12.
85. OPCW, *Report of the Scientific Advisory Board on Developments in Science and Technology to the Fifth Special Session of the Conference of the States Parties to Review the Operations of the Chemical Weapons Convention.* RC-5/DG.1, OPCW, The Hague, 22 February, 2023, Annex 1, paragraph 30
86. OPCW, *Report of the Scientific Advisory Board on Developments in Science and Technology to the Fifth Special Session of the Conference of the States Parties to Review the Operations of the Chemical Weapons Convention.* RC-5/DG.1, OPCW, The Hague, 22 February, 2023, Annex 1, Annex 1, paragraph 13.
87. OPCW, *Report of the Scientific Advisory Board on Developments in Science and Technology to the Fifth Special Session of the Conference of the States Parties to Review the Operations of the Chemical Weapons Convention.* RC-5/DG.1, OPCW, The Hague, 22 February, 2023, Annex 1, paragraph 14.
88. OPCW, *Report of the Scientific Advisory Board on Developments in Science and Technology to the Fifth Special Session of the Conference of the States Parties to Review the Operations of the Chemical Weapons Convention.* RC-5/DG.1, OPCW, The Hague, 22 February, 2023, Annex 1, paragraph 31.
89. United States, *Statement by Ambassador Nicole Shampaine, Permanent Representative of the US to the OPCW.* 106th Session of the Executive Council OPCW, The Hague. 8 July 2024.

90. OPCW, SAB, *Report of the Scientific Advisory Board at its Thirty-Eighth Session*, SAB-38/1. OPCW, The Hague, 31 May 2024, paragraph 3.2.
91. OPCW, SAB, *Report of the Scientific Advisory Board at its Thirty-Eighth Session*, SAB-38/1. OPCW, The Hague, 31 May 2024, paragraph 1.5.(b).
92. OPCW, SAB, *Report of the Scientific Advisory Board at its Thirty-Eighth Session*, SAB-38/1. OPCW, The Hague, 31 May 2024, paragraph 1.5.(c).
93. See for example, United States, *Compliance with the Convention on the Prohibition of the Development, Production, Stockpiling, and Use of Chemical Weapons and on Their Destruction*, April 2022, pp. 10–11.
94. United States, Compliance with the Convention on the Prohibition of the Development, Production, Stockpiling, and Use of Chemical Weapons and on Their Destruction, April 2022, p. 11.
95. United States, *Compliance with the Convention on the Prohibition of the Development, Production, Stockpiling, and Use of Chemical Weapons and on Their Destruction*, April 2024), [Digitial version, no page number given].
96. See for example: US, *CWC Compliance Report*, April 2022, p. 15.
97. US, CWC Compliance Report, April 2022, p. 15.
98. United States, *Compliance with the Convention on the Prohibition of the Development, Production, Stockpiling, and Use of Chemical Weapons and on Their Destruction* [2024 *CWC Compliance Report*], April 2024, [Digitial version, no page number given].
99. (a) United States, Compliance with the Convention on the Prohibition of the Development, Production, Stockpiling, and Use of Chemical Weapons and on Their Destruction, 2022, p. 15; (b) See also: G. Bush, US Executive Order 13382, Blocking Property of Weapons of Mass Destruction Proliferators and Their Supporters, 28 June 2005.
100. United States, *Sanctions on Iranian Entities Involved in Chemical Weapons Activities of Concern, press statement*, US Department of State, 3 December 2020.
101. United States, *Statement by Ms. Laura Gross, US Representative to the OPCW*. 105th Session of the Executive Council, OPCW, The Hague, 1 March 2024.
102. ITAR-TASS, from Moscow in English, 2112 hrs GMT 30th October 2002, 2002, as in FBIS-SOV-2002-1030, 'Russian experts discuss use of Fentanyl in hostage crisis,' as cited by J. Perry Robinson (2007) *Non lethal warfare and the Chemical Weapons Convention, Further HSP submission to the OPCW Open-Ended Working Group on Preparations for the Second CWC Review Conference*, 24th October.
103. S. Alison, [from Moscow for Reuters], 1257 hrs ET 30th October 2002, 'Russian confirms siege gas based on opiate fentanyl', as cited in J. Perry Robinson (2007) *Disabling Chemical Weapons, A Documented Chronology of Events 1945-2011, 20th* November 2012 (copy circulated to author).

104. J. Riches, *et al.*, Analysis of Clothing and Urine from Moscow Theatre Siege Casualties Reveals Carfentanil and Remifentanil Use', *J. Anal. Toxicol.*, 2012, **36**(9), 647–656.
105. N. E. Kuzmina and V. S. Kuzmin, Development of concepts on the interaction of drugs with opioid receptors, *Russ. Chem. Rev.*, 2011, **80**, 145–169.
106. F. S. Dukhovich, *et al.*, The agonist paradox: Agonists and antagonists of acetylcholine receptors and opioid receptors, *Chem. Biodiversity*, 2005, **2**, 354–366.
107. V. A. Zlobin, *et al.*, Analysis of opiates by HPLC with indirect spectrophotometric detection, *Pharm. Chem. J.*, 2000, **34**, 279–280.
108. A. V. Panfilov, *et al.*, Sodium borohydride in reductive amination reactions, *Pharm. Chem. J.*, 2000, **34**, 76–78.
109. A. V. Panfilov, Reactions of sodium borohydride in acetic acid: Reductive amination of carbonyl compounds, *Pharm. Chem. J.*, 2000, **34**, 371–373.
110. United States, *Compliance with the Convention on the Prohibition of the Development, Production, Stockpiling, and Use of Chemical Weapons and on Their Destruction*, 2022, p. 17.
111. US, *CWC Compliance Report*, April 2022, p. 22.
112. For overview of this incident see: C. Mcleish, *The Skripal case: Assassination attempt in the United Kingdom using a toxic chemical*, Stokholm International Peace Research Institute (SIPRI) Yearbook, 2019, pp. 408–417.
113. Prime Minister Theresa May, Statement, *House of Commons, Hansard*, volume, 637, columns 620–21, 12 March 2018.
114. Technical Secretariat, Summary of the report on activities carried out in support of a request for technical assistance by the United Kingdom of Great Britain and Northern Ireland (technical assistance visit TAV/02/18), Note by the *Technical Secretariat*, S/1612/2018, paragraph 47. OPCW, The Hague, 12 April 2018.
115. Russian Ministry of Foreign Affairs, *Foreign Minister Sergey Lavrov's address at the 26th Assembly of the Council for Foreign and Defence Policy, Moscow, April 14, 2018*, Press release, 14 April 2018.
116. Russian Ministry of Foreign Affairs, *Foreign Minister Sergey Lavrov's address at the 26th Assembly of the Council for Foreign and Defence Policy, Moscow, April 14, 2018*, Press release, 14 April 2018.
117. Spiez Laborartory Twitter/X account, 14 April 2018. https://x.com/SpiezLab/status/985243574123057152
118. Director General, Opening statement by the Director General to the Executive Council, at its 59th meeting, EC-M-59/DG.1. OPCW, The Hague, 18 April 2018.
119. Executive Council, *Statement by Nadine Olivier Lozano, Deputy Permanent Representative of Switzerland to the OPCW at the Fifty-ninth Meeting of the Executive Council*, Under Agenda item 3, Report on the activities carried out in support of a request for technical assistance by the United Kingdom of Great Britain and Northern Ireland (TAV/02/18), EC-M-59/NAT.2. OPCW, The Hague, 18 April 2018, p. 2.

120. Executive Council, Statement by Ambassador Sabine Nolke, Permanent Representative of Canada to the EC Fifty-Ninth Meeting, EC-M-59/NAT.17. OPCW, The Hague, 18 April 2018.
121. TASS, *Components of chemical agent close to BZ found in Zaporozhye Region — Russian military*, 4 March 2024, https://tass.com/politics/1755117 (accessed 5 February 2025).
122. TASS, *Components of chemical agent close to BZ found in Zaporozhye Region — Russian military*, 4 March 2024, https://tass.com/politics/1755117 (accessed 5 February 2025).
123. TASS, *Components of chemical agent close to BZ found in Zaporozhye Region — Russian military*, 4 March 2024, https://tass.com/politics/1755117 (accessed 5 February 2025).
124. TASS, Russian Defense Ministry: First Ukrainian chemical lab found during special operation, 8 July 2024. https://tass.com/defense/1814199 (accessed 6 February 2025).
125. Russian Fededration, *Correspondence to the OPCW Technical Secretariat from Permanent Representation of the Russian Federation to the OPCW*, 8 July 2024, https://www.opcw.org/sites/default/files/documents/2022/Compendium%20of%20correspondence%20shared%20by%20States%20Parties%20on%20Ukraine.pdf.
126. Lieutenant General Igor Kirillov, Chief NBC Protection Troops of the RF Armed Forces, *Briefing on large-scale provocations using toxic chemicals in Ukraine prepared by the West*, 23 February 2023, pp. 1–2.
127. Lieutenant General Igor Kirillov, Chief NBC Protection Troops of the RF Armed Forces, *Briefing on large-scale provocations using toxic chemicals in Ukraine prepared by the West*, 23 February 2023, p. 2.
128. Lieutenant General Igor Kirillov, Chief NBC Protection Troops of the RF Armed Forces, *Briefing on large-scale provocations using toxic chemicals in Ukraine prepared by the West*, 23 February 2023, pp. 2–3.
129. OPCW, *Report of the Fifth Review Conference of the Chemical Weapons Convention, 15 – 19 May 2023*, RC-5/3. OPCW, The Hague, 7 June 2023.

CHAPTER 8

International Human Rights Law Constraining the Weaponization of CNS-acting Chemical Agents[†]

8.1 Introduction[1]

Law enforcement can be considered to be those activities intended to maintain or restore public security, law and order.[2] While law enforcement officials may potentially use certain types of force if necessary to carry out legitimate law enforcement functions, the nature and degree of such force are constrained and demarcated by international human rights law (IHRL). Consequently, although IHRL does not explicitly focus on CNS-acting chemical agent weapons, it has significant relevance as to whether and if so when, how, and by whom such weapons could be employed for law enforcement. An important strength of IHRL is its applicability in a wide range of circumstances where use of weapons employing CNS-acting chemical agents, or the broader category of incapacitating chemical agents (ICAs), might be considered. The International Court of Justice has affirmed that human rights law continues to apply in situations to which international humanitarian law (IHL) is applicable,[3] whilst the UN Human Rights Committee has stated that, in situations of armed conflict, "both spheres of law are complementary, not mutually exclusive".[4] A recent overview of relevant international law by Hoffberger-Pippan has concluded "it is nowadays an

[†]This chapter draws upon, updates and expands the authors' previous research and publications, notably M. Crowley, *Chemical Control*, Palgrave MacMillan, 2016, pp. 166–189, which is used with permission from Palgrave MacMillan, Copyright 2016.

Preventing Weaponization of CNS-acting Chemicals: A Holistic Arms Control Analysis
By Michael Crowley and Malcolm Dando

Published by the Royal Society of Chemistry, www.rsc.org

almost undisputed fact that human rights also apply in armed conflict and may inform or supplement norms of humanitarian law, including cases where concrete answers of humanitarian law are entirely or partially missing".[5]

Consequently, IHRL is applicable to, and constrains, the use of force by a State in domestic policing situations, in a non-international armed conflict, whether or not that State has acknowledged it as being such, and to those aspects of an international conflict that take place in the national territory of that State. The internationally renowned legal scholar, Francoise Hampson, has highlighted the significance of this extensive IHRL coverage: "[S]tates frequently refuse to characterize an internal armed conflict as such, preferring to call it criminal or terrorist activity. In such a situation, they can hardly challenge the applicability of human rights law".[6]

While a number of human rights norms may constrain or prohibit the weaponization or malign application of CNS-acting chemical agents, the most important and pertinent are the right to life; to freedom from torture and other cruel, inhuman or degrading treatment or punishment (other ill-treatment); to freedom of association and assembly; to liberty and security; and to health. These rights and the associated obligations upon States to regulate and restrict the use of force will be examined, as will the consequent implications for the employment of CNS-acting chemical agents.

8.2 Restrictions on the Use of Force

The United Nations (UN) and certain regional organisations have developed a range of human rights standards and guidance documents that attempt to clarify the circumstances when the use of force may be appropriate, by whom it may legitimately be employed, and under what constraints. Although these standards are not themselves legally-binding upon States, they do distil existing international norms and reflect contemporary good practice. Of particular relevance are two international instruments, developed under the auspices of the UN, which codify the overarching rules by which law enforcement officials should carry out their duties: the UN Code of Conduct for Law Enforcement Officials (UNCoC)[7] and the UN Basic Principles on the Use of Force and Firearms by Law Enforcement Officials (UNBP).[8] In addition, the UN Standard Minimum Rules for the Treatment of Prisoners, which provides protection for those in custodial settings, includes limitations on the use of force by custodial staff.[9] There are also a range of documents developed by regional bodies including the Guidelines for the Policing of Assemblies by Law Enforcement Officials in Africa,[10] the Principles and Guidelines on Human and Peoples' Rights while Countering Terrorism in Africa,[11] the Inter-American Commission's Report on Citizen Security and Human Rights[12] and the European Code of Police Ethics.[13]

These normative standards incorporate certain essential elements which have been labelled by the acronym 'PLAN' referring to the principles of proportionality, legality, accountability, and necessity.[14] Under the principle of necessity, any force used by law enforcement officials must be only the minimum necessary in the circumstances.[15] Under the principle of proportionality, the degree of force used and the potential harm it can cause must be strictly proportionate to the seriousness of the offence and the legitimate objective to be achieved.[16] Under the principle of legality, the use of force would only be permitted if a sufficiently detailed legal basis allows for it in a specific law enforcement situation. The consequent requirement to enact national legislation dealing with the use of force would imply that the relevant domestic laws would "need to be sufficiently detailed, precise, and clear as well as unambiguous in order to prevent abuse and arbitrariness".[17] Under the principle of accountability any use of force that causes death or injury must be reported promptly. Any excessive or otherwise arbitrary use of force has to be properly investigated and, when required, punished as a criminal offence and/or incur disciplinary measures depending on the seriousness of the case. Both superior officers and State officials under their command can be held responsible.[18] A further essential element of many normative standards has been the principle of precaution requiring law enforcement officials and relevant State bodies to take appropriate action beforehand in the planning and control phase of operations. Law enforcement operations have to be carefully planned to avoid, as much as possible, the use of force and to minimize the risk to bystanders.[19]

These normative standards attempt to constrain the use of force including the use of lethal and 'less lethal' weapons[20] and are thus of direct relevance to establishing the potential legitimacy of any weapons employing CNS-acting chemical agents. Of particular importance in this regard is Principle 4 of the UNBP which states that: "Law enforcement officials, in carrying out their duty, shall, as far as possible, apply non-violent means before resorting to the use of force and firearms. They may use force and firearms only if other means remain ineffective or without any promise of achieving the intended result".[21] In addition, Article 5 determines that

> "Whenever the lawful use of force and firearms is unavoidable, law enforcement officials shall:
>
> (a) Exercise restraint in such use and act in proportion to the seriousness of the offence and the legitimate objective to be achieved;
> (b) Minimize damage and injury, and respect and preserve human life;
> (c) Ensure that assistance and medical aid are rendered to any injured or affected persons at the earliest possible moment ..."[22]

The UNBP specifically addresses the development, deployment and use of so-called "non-lethal incapacitating weapons". Article 2 states:

> "Governments and law enforcement agencies should develop a range of means as broad as possible and equip law enforcement officials with

> various types of weapons and ammunition that would allow for a differentiated use of force and firearms. These should include the development of *non-lethal incapacitating weapons*[23] for use in appropriate situations, with a view to increasingly restraining the application of means capable of causing death or injury to persons ..." (Emphasis added)[24]

In addition, Article 3 States: "The development and deployment of non-lethal incapacitating weapons should be carefully evaluated in order to minimize the risk of endangering uninvolved persons, and the use of such weapons should be carefully controlled".[25] While these Articles set broad constraints on "non-lethal incapacitating weapons", no detailed operational guidance on their application in practice is provided. Further ambiguity arises as the term "non-lethal incapacitating weapons" is not defined, nor is the range of weapons coming under its coverage identified.

Further specialised advice on the application of international human rights law to 'less lethal' weapons was developed through an expert group process conducted under the auspices of the Office of the High Commissioner for Human Rights (OHCHR), and its findings issued by OHCHR in 2020 as the *UN Human Rights Guidance on Less-Lethal Weapons in Law Enforcement*.[26] Whilst this Guidance identified a non-exhaustive list of weapons and related equipment that violated international human rights law, and identified a contrasting additional range of 'less lethal' weapons employed in law enforcement; CNS-acting chemical agent weapons were not contained in either category and were not specifically addressed in this document. However, its "[G]eneral considerations pertaining to 'less lethal' weapons and related equipment" – covering design and production, legal review, testing and procurement, monitoring, transparency, training, medical assistance, transfer and international cooperation and assistance – are applicable and will be highlighted where appropriate.[27]

8.3 Protection of the Right to Life

A number of international,[28] and regional, human rights instruments[29] establish and protect the innate right to life. Of particular importance is the International Covenant on Civil and Political Rights (ICCPR) which states "every human being has the inherent right to life. This right shall be protected by law. No one shall be arbitrarily deprived of his life".[30] Furthermore, the UN Human Rights Committee (HRC), the body responsible for monitoring the ICCPR's implementation, has stated that the right to life is "the supreme right from which no derogation is permitted even in time of public emergency which threatens the life of the nation".[31] In addition, the HRC has stated that:

> "States Parties should take measures not only to prevent and punish deprivation of life by criminal acts, but also to prevent arbitrary killing

by their own security forces. The deprivation of life by the authorities of the State is a matter of the utmost gravity. Therefore, the law must strictly control and limit the circumstances in which a person may be deprived of his life by such authorities."[32]

The HRC has further argued that the right to life, at least under the ICCPR, is a right "which should not be interpreted narrowly".[33] However, the right to life is not absolutely guaranteed in all circumstances. A number of human rights instruments permit States to intentionally take human life through judicially sanctioned executions (as explored below) or during the conduct of armed conflicts to halt and nullify enemy forces and combatants.

8.3.1 Application to CNS Acting Chemical Agent Weapons

Proponents of CNS-acting chemical agent weapons have promoted their research and development and subsequent employment in specific law enforcement operations where there is deemed to be an urgent requirement to speedily and totally incapacitate specific individuals or group of individuals without causing them permanent disability or fatality. Whilst the legitimacy of CNS-acting chemical agent weapons is controversial and highly contested, a number of legal scholars have contended that their use may potentially be permitted under IHRL in certain, very limited and extreme law enforcement situations where the State is compelled to employ potentially lethal force in order to end urgent, life-threatening situations because less violent and dangerous methods have failed, are impractical or have a low chance of success.

However, the International Committee of the Red Cross (ICRC), in its 2012 *Synthesis* paper, underlined the dangers of the use of CNS-acting chemical agent weapons and the very limited circumstances under which the use of such "potentially lethal force" may potentially be permissible under IHRL:

> "In light of the certainty that bystanders will also come to harm, the question to be asked is whether such a means is absolutely necessary to save the lives of those who are threatened, that is whether there are any other means available that would achieve the same aim while posing less of a danger to life; and whether this is an unavoidable measure of last resort, the State having exhausted all feasible less harmful means before it resorts to this means."[34]

In addition, even in these extreme situations, the obligations for States under IHRL to protect the right to life still apply, with contingent constraints upon employment of the agent as well as the requirement to take appropriate remedial measures. Aceves, for example, has stated that:

> "The right to life norm places strict limits on the use of force, which includes the use of incapacitating biochemical weapons ... States must, therefore, act with due diligence in all cases involving these

weapons. The use of these weapons must be carefully regulated and cannot cause indiscriminate harm. Their use must be proportionate to the perceived threat and must be justified under the circumstances."[35]

Similarly, Fidler has argued that:

"The inability to control dosage or exposure environment in extreme law enforcement emergencies heightens governmental responsibility to ensure all precautions are taken to minimize harm to innocent people and to provide immediate and adequate medical attention to those exposed and perhaps adversely affected."[36]

Case law in this area is limited. To date, there has been only one well-documented instance of a large-scale use of an aerosolised CNS-acting chemical agent weapon in an extreme law enforcement situation, *i.e.*, by Russia to end the Moscow Theatre siege in October 2002.[37] Two UN human rights authorities subsequently issued statements on this case. In a January 2003 report, the UN Special Rapporteur on Extra-Judicial, Summary or Arbitrary Executions expressed concern "about the actions by Russian police/security forces" and stated that he "has been collecting information from various sources about the incident and plans to take the issue up in 2003 with the Government of the Russian Federation".[38] However, to date, the results of the Rapporteur's actions on this issue have not been made public.

Subsequently, in June 2003, the UN Human Rights Committee declared that: "[W]hile acknowledging the serious nature of the hostage-taking situation, the Committee cannot but be concerned at the outcome of the rescue operation in the Dubrovka theatre in Moscow on 26 October 2002".[39] The Committee further expressed "its concern that there has been no independent and impartial assessment of the circumstances, regarding medical care of the hostages after their liberation and the killing of the hostage-takers".[40] It called upon Russia to "ensure that the circumstances of the rescue operation in the Dubrovka theatre are subject to an independent, in-depth investigation, the results of which are made public, and, if appropriate, prosecutions are initiated and compensation paid to the victims and their families".[41] Again, to date, no further reference appears to this incident in subsequent reports of the UN Human Rights Committee. While the interventions by UN human rights bodies were limited, the relevant regional human rights body – the European Court of Human Rights (ECtHR) – issued a significant, and potentially precedent-setting, ruling on this case.

8.3.1.1 European Court of Human Rights: Finogenov and Others vs. Russia

In August 2003, a group of 64 former hostages and relatives filed a complaint before the ECtHR, claiming that their right to life (protected under Article 2 of the European Convention on Human Rights [ECHR]) had been violated by the actions of the Russian authorities surrounding the Moscow theatre siege. In December 2007, this case was formally accepted by the Court, and in December 2011, the Court announced its ruling in this matter.[42]

The Court rejected the Russian "official explanation of the mass death of the hostages on 26 October 2002... [*i.e*] ... that all the persons who had died were weakened by the siege or seriously ill". According to the Court:

> "It is unthinkable that 125 people of different ages and physical conditions died almost simultaneously and in the same place because of various pre-existing health problems. Equally, the mass death of hostages cannot be attributed to the conditions in which they had been held for three days, during which none of them had died, despite prolonged food and water deprivation, immobility, psychological stress, *etc.*"[43]

Furthermore, the Court also did not accept Russian statements that the CNS-acting chemical agent weapon had been "harmless", and that, "there was no 'direct causal link' between the death of those 125 people and the use of the [sic] gas, and that the [sic] gas was just one of many factors which led to such a tragic outcome".[44]

The Court found that even if the CNS-acting chemical agent weapon was not a "lethal force" but rather a "non-lethal incapacitating weapon", the weapon was, "at best, potentially dangerous for an ordinary person, and potentially fatal for a weakened person. [Furthermore] .... .it is safe to conclude that the gas [sic] remained a primary cause of the death of a large number of the victims".[45] Consequently, the Court confirmed that "the situation is thus covered by Article 2 of the Convention".[46]

However, with respect to the decision to employ a CNS-acting chemical agent weapon and invade the building, the Court stated:

> "...that the authorities were not in control of the situation inside the building. In such a situation the Court accepts that difficult and agonising decisions had to be made by the domestic authorities. It is prepared to grant them a margin of appreciation[‡], at least in so far as the military and technical aspects of the situation are concerned, even if now, with hindsight, some of the decisions taken by the authorities may appear open to doubt."[47]

In this particular case, the Court believed there had been a real, serious and immediate risk of mass human losses and the authorities had every reason to believe that a forced intervention had been "the lesser evil".

[‡]The term "margin of appreciation" refers to the space for manoeuvre that the Court and other Strasbourg organs are willing to grant national authorities, in fulfilling their obligations under the European Convention on Human Rights.

The Court considered that the solution, using a "dangerous and even potentially lethal" chemical agent, had put at risk the lives of hostages and hostage-takers alike; but "it was not used 'indiscriminately' as it had left the hostages a high chance of survival, which depended on the efficiency of the authorities' rescue effort".[48] Indeed, the Court believed that the use of the CNS-acting chemical agent weapon facilitated the liberation of the hostages and reduced the likelihood of an explosion. The Court therefore concluded "that the use of [a CNS-acting chemical agent weapon] during the storming was not in the circumstances a disproportionate measure, and, as such, did not breach Article 2 of the Convention".[49]

However, the Court found that, as a whole, the Russian authorities had not taken all feasible precautions to minimize the loss of civilian life. "The rescue operation. . .was not sufficiently prepared, in particular because of the inadequate information exchange between various services, the belated start of the evacuation, limited on-the-field coordination of various services, lack of appropriate medical treatment and equipment on the spot, and inadequate logistics". Consequently, the Court concluded "that the State breached its positive obligations under Article 2 of the Convention".[50] Furthermore, the Court stated that it "had sufficient evidence to conclude that the investigation into the authorities' alleged negligence in this case was neither thorough nor independent, and, therefore, not 'effective'". The Court consequently concluded "that there was a breach of the State's positive obligation under Article 2 of the Convention on this account".[51]

The ICRC,[52] and a number of arms control and legal experts[53] subsequently raised their concerns about the Court's judgement. And the current authors believe that there are a number of elements of the Court's ruling that are questionable or would benefit from additional legal analysis and discourse. These include the inexplicable failure of the Court to assess the legality of the use of the CNS-acting chemical agent weapon under the CWC;[54] the Court's decision to make a ruling without being given adequate information by the Russian authorities concerning the chemical composition, nature and consequently the likely effects of the CNS-acting chemical agent weapon used; and the Court's contention that it was "impossible for [it] to establish whether or not the gas [sic] was a 'conventional weapon' or to identify the rules for its use".[55] The lack of technical knowledge by the Court is exemplified by the referral in the Court's ruling to the CNS-acting agent as a "gas" when it was in fact an aerosol.

A further potential worrying aspect of the ruling, discussed by Hoffberger-Pippan,[56] concerned the principle of legality, which requires States to adopt national legislation regulating the use of force and firearms. In this case, Russia had not enacted any domestic laws regulating the use of the CNS-acting weapon within anti-terror operations. The only law applicable to such operations was the so-called Russian Anti-Terrorism Act, which did not clarify exactly whether the use of this CNS-acting weapon was permitted in such scenarios. Moreover, the Russian Anti-Terrorism Act required that the relevant type of weapon used in an anti-terror scenario shall not be made public. The Court stated that the Chechen attack was "totally unpredictable, exceptional and required a tailor-made response". Therefore, the absence of "sufficiently detailed internal regulations" did not lead to a violation of the principle of legality.[57] Consequently, it appears that the requirements established by the Court in this case regarding the principle of legality are less restrictive when it comes to anti-terror operations or other "totally unpredictable and exceptional events" than in other situations. In contrast, the Court has applied stricter standards when it comes to the deployment of other 'less lethal' weapons that are (potentially) lethal in custodial centres, including tear gas. The *Finogenov* judgment raises concerns that Russia or other States could use this Court ruling to argue for the excessive and disproportionate use of a CNS-acting chemical agent weapon (or other lethal force) in future anti-terror operations.

However, despite all the preceding concerns, the judgement of the Court, though subject to an appeal, was subsequently made final in June 2012.[58] It was certainly an important legal determination which is likely to inform all future judgements on the use of CNS-acting chemical agent weapons for law enforcement purposes; it may also influence whether and how States seek to develop, acquire and employ such weapons in the future. It is notable that in its ruling the Court considered the use of the CNS-acting chemical agent weapon – although labelled as a 'non-lethal agent' by Russia – as definitely falling under the ambit of the ECHR and that such use could potentially be a violation of Article 2 of the Convention. It appears that the use of such weapons would only be considered consistent with the Convention in very limited, extreme situations (*i.e.*, where there was an immediate and direct threat to life and where there was no recourse to other measures to resolve the situation with less risk of injury or death), and only when such use occurred in conjunction with proper planning and the provision of adequate medical care and remedial support. Although the Court gave Russia some "margin of appreciation" regarding its decision to use a CNS-acting chemical agent weapon with then-unknown effects in this "anti-terror operation", the consequences of such action have now been shown to include the deaths of a large proportion of the hostages and long-term injury for many of the survivors. Given such empirical data, it is at least an open question as to whether another court would give such a "margin of appreciation" or range of discretion to Russia or any other State for the use of a CNS-acting chemical agent weapon in similar situations in the future.

Furthermore, and with more general application, this case has highlighted the potential utility of regional human rights mechanisms as a means that victims of human rights violations (including those inappropriately targeted with CNS-acting chemical agent weapons) can employ to hold the relevant authorities to account (and also, in the case of the ECHR, to obtain financial compensation for the wrongs done to them).[59] Although the relevant regional courts[60] have a number of limitations and only have jurisdiction over those States that have ratified the relevant Conventions and Protocols, they deliver legally binding judgements. In contrast, while the UN Human Rights

Committee can also consider a case raised through individual petition, it can only reach non-binding conclusions in such instances.[61]

In addition, Hampson has highlighted that: "[A]n important feature of these enforcement mechanisms is that, almost uniquely in international law, they can be directly or indirectly triggered by individuals ... who are far more likely to bring such complaints than are foreign States, particularly where the claim concerns something that the State has done on its own territory".[62]

8.3.1.2 Employment of CNS-acting Chemicals in Judicial Executions

Although the inherent right to life is enshrined and protected under IHRL and the arbitrary ending of life by the State is prohibited, capital punishment, *i.e.* State-sanctioned judicial execution, is not currently proscribed.[63] However such practices are subject to important restrictions. Notably, the ICCPR requires that:

> "...Sentence of death may be imposed only for the most serious crimes in accordance with the law in force at the time of the commission of the crime and not contrary to the provisions of the present Covenant and to the Convention on the Prevention and Punishment of the Crime of Genocide. This penalty can only be carried out pursuant to a final judgement rendered by a competent court."[64]

The ICCPR prohibits the imposition of capital punishment "for crimes committed by persons below eighteen years of age" and specifies that it "shall not be carried out on pregnant women".[65]

'Lethal injection' – the intravenous administration of a lethal dose of certain pharmaceutical chemicals (normally including certain CNS-acting chemicals) has been provided for as a method of execution in China,[66] Guatemala,[67] the Maldives,[68] Papua New Guinea,[69] Taiwan,[70] Thailand,[71] Vietnam,[72] and the US.[73] However during 2023 it appears that only three States China, Vietnam and the US carried out executions using this method.[74]

In China, the number of death sentences issued and executions conducted is a State secret. However, Amnesty International, as of May 2024, claimed that China "remained the world's lead executioner" and had continued to sentence to death and execute several thousand people annually; which was more than the rest of the world combined.[75] Since 1997, China's Criminal Procedure Law has specified that the only methods of execution are 'lethal injection' and shooting.[76] Although China's authorities have not confirmed the composition of the pharmaceutical chemicals used in 'lethal injections', the mixture is reportedly similar, if not identical, to the three-drug cocktail that has been used in many States of the US, comprising a barbiturate, muscle relaxant and potassium chloride.[77] In China, 'lethal injection' executions have been conducted in prisons and also in specially designed mobile 'death vans'. Here, according to Amnesty International "The windowless execution chamber at the back contains a metal bed on which the prisoner is strapped down. Once the needle is attached by the doctor, a police officer presses a button and an automatic syringe inserts the lethal drug cocktail into the prisoner's vein".[78] In December 2003, according to Amnesty International "the People's Supreme Court reportedly urged all courts throughout China to purchase mobile execution chambers 'that can put to death convicted criminals immediately after sentencing'".[79] The number of 'death vans' now in use is not known, although a US newspaper reported in 2006 that more than 40 were deployed.[80] A further concern has been the associated State-sanctioned practice of harvesting organs – notably heart and lungs – from those executed by 'lethal injection', without their consent. Human Rights Watch reported on such practices and stated "'[L]ethal injection' is clearly the most convenient method of execution to use for purposes of organ harvesting, since it minimizes trauma to the body".[81] Although non-consensual organ harvesting of executed prisoners was previously denied, the Chinese authorities finally admitted the practice, and subsequently promised to end it by 1 January 2015.[82] Concerns about such practices increased when Australian National University researchers analysing Chinese medical papers found 71 cases between 1980 and 2015 where prisoner organ harvesting had taken place *before* brain death had been declared, and concluded "that the removal of the heart during organ procurement must have been the proximate cause of the donor's death". The researchers were unable to identify how the "donors were incapacitated before procurement, and the data is consistent with multiple plausible scenarios... These range from a bullet to the prisoner's head...or a general anesthetic delivered in the operating room directly before procurement...or a 'lethal injection', with execution completed by organ procurement".[83] A 2019 UK-based tribunal on forced organ harvesting in China found that there was "no evidence that the significant infrastructure associated with China's transplantation industry has been dismantled".[84]

In the US, 'lethal injection' was introduced as a legal method of execution in May 1977, in Oklahoma and Texas, with the first such execution taking place in Texas on 7th December 1982.[85] Its use has subsequently spread widely; today all 27 US States that retain the death penalty sanction 'lethal injection' executions, and this is also the execution method authorized by the US military and the US Government.[86] According to the Death Penalty Information Center (DPIC), as of 17 December 2024, 89% (1422 out of 1602) of the prisoners executed since the death penalty was reintroduced in 1977, were killed by 'lethal injection'.[87]

Until 2009, most US States used a three-drug combination for 'lethal injections' which comprised an anaesthetic CNS-acting chemical agent (usually sodium thiopental), pancuronium bromide (a paralytic agent), and potassium chloride which stops the heart and causes death. However, due to resistance from the pharmaceutical industry at the misuse of their medicinal products (notably sodium thiopental) for execution, and the introduction of human rights-based trade controls by the EU (see Section 8.9), which collectively cut off the supply of sodium thiopental to executioners in US correctional facilities, certain US States adopted new 'lethal injection' methods. Many replaced sodium thiopental with another anaesthetic CNS-acting chemical agent, either pentobarbital or midazolam, in either three-drug or two-drug execution protocols.[88] Alternatively, eight States have used a lethal dose of an anaesthetic CNS-acting chemical agent (pentobarbital or midazolam) as a single-drug execution method[89] and six other States have at some time announced plans to use a similar one-drug protocol, but have not yet carried out such an execution.[90] Other CNS-acting drugs have also been tried by certain States, with Nebraska in 2018 being the first to use fentanyl in a four-drug execution method together with diazepam, cisatracurium and potassium chloride[91] and Nevada announcing that it will use fentanyl in combination with other drugs to carry out executions in the future.

In addition to the moral and human rights arguments against the death penalty *per se*, international and regional human rights organisations and bodies, civil society human rights organisations and health professional associations have raised specific additional concerns about 'lethal injection' executions. These include: the potential for this method to cause intense, sometimes prolonged physical suffering of the victim due to botched procedures, with the consequent associated risk that this suffering may be hidden from observers because of the action of a paralysing agent in the 'lethal injection' mixture;[92] the necessary involvement of doctors, nurses and other health personnel, directly or indirectly, in the execution process with the consequent erosion of medical ethics;[93] and the associated non-consensual harvesting of prisoner organs in certain countries.[94]

8.4 The Rights to Freedom of Association and Assembly

The rights to freedom of association and assembly have been enshrined in a wide range of regional,[95] and international,[96] human rights instruments. Notably, the ICCPR states:

> "The right of peaceful assembly shall be recognised. No restrictions may be placed on the exercise of this right other than those imposed in conformity with the law and which are necessary in a democratic society in the interests of national security or public safety, public order (ordre public), the protection of public health or morals or the protection of the rights and freedoms of others."[97]

In addition, the UNBP states: "[I]n the dispersal of assemblies that are unlawful but non-violent, law enforcement officials shall avoid the use of force or, where that is not practicable, shall restrict such force to the minimum extent necessary".[98]

The UN Human Rights Council subsequently adopted a series of Resolutions, including, in 2010, 2012 and 2013, on "freedom of peaceful assembly and of association";[99] and further Resolutions in 2012, 2013 and 2014, on the "promotion and protection of human rights in the context of peaceful protests".[100] Resolution, 25/38 of April 2014, called upon States to "promote a safe and enabling environment" for people to exercise their rights to freedom of peaceful assembly, expression and association, including by "ensuring that their domestic legislation and procedures ... clearly and explicitly establish a presumption in favour of the exercise of these rights, and that they are effectively implemented".[101] Resolution 25/38 also "urge[d] all States to avoid using force during peaceful protests and to ensure that, where force is absolutely necessary, no one is subject to excessive or indiscriminate use of force".[102]

In a February 2016 joint report, the UN Special Rapporteur on the rights to freedom of peaceful assembly and of association, and the UN Special Rapporteur on extrajudicial, summary or arbitrary executions, presented their recommendations on the proper management of assemblies.[103] Although the report did not specifically address CNS-acting chemical agent weapons or other law enforcement weapons, its findings are relevant to whether and how such weapons can be employed. Under a section entitled: "[F]orce shall not be used unless it is strictly unavoidable, and if applied it must be done in accordance with international human rights law", the Rapporteurs argued that:

> "States are required to procure less lethal weapons for use in appropriate situations, with a view to increasingly restraining the application of means capable of causing death or injury. Less-lethal weapons must be subject to independent scientific testing and approval, and used responsibly by well-trained law enforcement officials, as such weapons may have lethal or injurious effects if not used correctly or in compliance with international law and human rights standards. States should work to establish and implement international protocols for the training on and use of less-lethal weapons."[104]

One area of particular pertinence to future potential CNS-acting weapons highlighted by the Rapporteurs was their concerns regarding: "[A] growing range of weapons that are remote controlled [that] are becoming available, particularly in the context of the policing of assemblies". The Rapporteurs argued that: "[G]reat caution should be exercised in this regard. Where advanced technology is employed, law enforcement officials must, at all times, remain personally in control of the actual delivery or release of force".[105]

8.5 The Rights to Liberty and Security

The rights to liberty and security of the person have been included in a range of international,[106] and regional,[107] human rights instruments. According to the ICCPR: "[E]veryone has the right to liberty and security of person. No one shall be subjected to arbitrary arrest or detention. No one shall be deprived of his liberty except on such grounds and in accordance with such procedure as are established by law".[108]

Casey-Maslen has contended that the concepts of liberty and security are "typically considered broadly, and could potentially cover the use of NKE [non-kinetic energy] weapons that prevent a person from moving .... Certain NKE weapons may cause paralysis in human beings. This paralysis may last only a few seconds or it may be more prolonged".[109] He subsequently has argued "the use of such weapons by security officials might, in certain circumstances, be deemed to violate a person's right to liberty and security of person in addition to other human rights they may infringe".[110]

The potential significance of the application of these rights to the development and employment of CNS-acting chemical agent weapons intended to immobilize or incapacitate their targets for lengthy periods is underlined by NATO's previous definition of an "incapacitating agent" as: "[A] chemical agent which produces temporary disabling conditions which (unlike those caused by riot control agents) can be physical or mental and persist *for hours or days after exposure to the agent has ceased*... (emphasis added)". This definition appears to have been introduced in 1982 and was replaced in 2018 with a similar though more generally applicable definition not limited to chemical agents.[111] Although such considerations are worthy of further analysis they do not appear to have been applied by any international human rights body, to date, with reference to CNS-acting chemical agents.

8.6 The Prohibition Against Torture and Other Ill-treatment

The prohibition against torture and other cruel, inhuman or degrading treatment or punishment (other ill-treatment) has been enshrined in numerous international,[112] and regional,[113] human rights agreements. It is a customary norm, that is applicable in all circumstances and at all times, including during armed conflict. It is a fundamental human right, and one of the limited number for which no derogation has been permitted.[114] According to the UNCoC:

> "No law enforcement official may inflict, instigate or tolerate any act of torture or other cruel, inhuman or degrading treatment or punishment, nor may any law enforcement official invoke superior orders or exceptional circumstances such as a state of war or a threat of war, a threat to national security, internal political instability or any other public emergency as a justification of torture or other cruel, inhuman or degrading treatment or punishment."[115]

The UN Convention against Torture, under Article 1 defines torture as:

> "Any act by which severe pain or suffering, whether physical or mental, is intentionally inflicted on a person for such purposes as obtaining from him or a third person information or a confession, punishing him for an act he or a third person has committed or is suspected of having committed, or intimidating or coercing him or a third person, or for any reason based on discrimination of any kind, when such pain or suffering is inflicted by or at the instigation of or with the consent or acquiescence of a public official or other person acting in an official capacity. It does not include pain or suffering arising only from, inherent in or incidental to lawful sanctions."[116]

Aspects of "other ill-treatment" have been delineated and described in further pertinent legal texts. Within the legal framework of war crimes, "inhuman treatment" has been defined under the Elements of Crimes for the International Criminal Court as the infliction of "severe physical or mental pain or suffering".[117] The European Commission of Human Rights considers that the concept of "degrading treatment" covers treatment or punishment that "grossly humiliates the victim before others or drives the detainee to act against his/her will or conscience".[118] Principle 6 of the UN Body of Principles for the Protection of All Persons under Any Form of Detention or Imprisonment prohibits torture and other forms of ill-treatment. The accompanying legal Commentary explains:

> "The term 'cruel, inhuman or degrading treatment or punishment' should be interpreted so as to extend the widest possible protection against abuses, whether physical or mental, including the holding of a detained or imprisoned person in conditions which deprive him, temporarily or permanently, of the use of any of his natural senses, such as sight or hearing, or of his awareness of place and the passing of time."[119]

Former UN Special Rapporteur on Torture, Nowak, previously indicated that the prohibition on ill-treatment places limits on the lawful use of force. He emphasised that the use of force must be regulated by the principles of proportionality and that the disproportionate exercise of police powers might constitute ill-treatment in certain circumstances.[120]

8.6.1 Non-consensual Application of CNS-acting Chemical Agents to Detainees in Prison/Law Enforcement Context

Important constraints upon the use of CNS-acting chemical agent weapons arise from obligations to prevent torture and other ill-treatment. Such obligations are of particular relevance where such weapons are being considered for use against prisoners or detainees. After a review of the relevant law, Aceves concluded that:

> "The prohibition against cruel, inhuman or degrading treatment places significant restrictions on the use of incapacitating biochemical weapons. These weapons are designed to impair the physical and mental integrity of the individual. Depending on the nature, duration and long-term effects of this impairment, the use of incapacitating biochemical weapons can give rise to a claim of cruel, inhuman, or degrading treatment."[121]

In its report exploring the potential use and misuse of neuroscience, the Royal Society highlighted the European Commission of Human Rights' definition of degrading treatment and considered it to be of "particular importance in considering the potential applications of neuroscience that could, for example, manipulate behaviour or thought processes".[122] A supporting note observed that: "the use of potential militarised agents including noradrenaline antagonists such as propranolol to cause selective memory loss, cholecystokinin B agonists to cause panic attacks, and substance P agonists to induce depression could all be considered violations of the prohibition against degrading treatment".[123]

Fidler has stated that: "non-consensual, non-therapeutic use of any chemical or biochemical against detained individuals would constitute degrading treatment and could, constitute cruel or inhumane treatment and perhaps even torture".[124] He did, however, believe that there may be situations where use of CNS-acting chemical agent weapons may be in conformity with IHRL, for example when a detainee was an immediate, violent threat to themself (for example attempting suicide) or to the safety and order of the detention facility (for example they were attacking guards or were engaging in riots).[125]

Over a 30 year period (1960s–1990s), the totalitarian regimes in the Soviet Union[126] and China[127] imprisoned significant numbers of political prisoners and other sane detainees in mental institutions, despite clear prohibitions upon such practices under IHRL. Reports subsequently emerged that prisoners held in certain of these institutions were forcibly given a range of psychotropic substances as a means of punishment or control rather than for therapeutic purposes.[128] There are indications that such practices subsequently took place in certain States – albeit at a far lower level. A review by the authors of relevant documentation produced by UN and human rights bodies and non-governmental organisations in a 10 year period (2004–2013) highlighted examples of the reported non-consensual, non-therapeutic application of psychoactive chemicals against political prisoners, detainees or patients in the Russian Federation,[129] Vietnam,[130] and Uzbekistan,[131] apparently with the intention to intimidate or punish the victims.

8.6.1.1 Behavioural Modification – 'Chemical Restraint' in Non-correctional/Law Enforcement Contexts

A related area of concern, outside of the setting and context of correctional facilities and law enforcement, has been the inappropriate application, without informed consent, of pyscho-active (CNS-acting) chemicals to patients in hospitals, to children with developmental disabilities in care centres, or to elderly people often suffering from dementia or similar illnesses living in aged care facilities. For example, from 2015–2019 UN and regional human rights monitors, medical oversight bodies and non-governmental human rights organisations documented cases including in Australia,[132] Brazil,[133] Kazakhstan,[134] Nigeria,[135] Russia,[136] Serbia,[137] Somaliland[138] and the US[139] where such non-consensual application has not been for therapeutic purposes but rather to control the behaviour of patients, a practice known as 'chemical restraint', which the relevant organisations consider to be abusive in nature.

There appears to be some divergence in terminology between certain human rights organisations describing 'chemical restraint' as a *de facto* abusive practice, while others define the practice in neutral terms. For example, the European Committee for the Prevention of Torture and Inhuman or Degrading Treatment or Punishment (CPT) defines 'chemical restraint' as "forcible administration of medication for the purpose of controlling a patient's behaviour". Consequently the CPT appears to accept the use of chemical restraint in limited circumstances for *i.e.* violent patients, patients at risk of doing harm to themselves, other patients or staff, agitated patients, provided it is strictly regulated.[140] Consequently, whilst the CPT reports the use of chemical restraint from 2015–2019, in 26 European countries,[141] the organisation considered such practices to be abusive in a more limited number of cases including for example in Azerbaijan,[142] Greece,[143] Italy,[144] Portugal,[145] and Serbia.[146]

In certain countries abusive 'chemical restraint' practices appear to have been systematised in at least parts of the medical profession with significant numbers of highly vulnerable patients suffering ill-treatment and potentially torture as a consequence.

8.6.1.1.1 Kazakhstan. In 2019, Human Rights Watch (HRW) issued a report on conditions in some of Kazakhstan's 19 Special Social Services Centers for Children where more than 2000 children with mental health conditions and developmental disabilities are inmates. In 2018, an HRW research team visited three such centres in Almaty, Shymkent and Karaganda. They found that in all three institutions staff admitted using psychotropic drugs usually medically prescribed to treat schizophrenia, sleep disorders, and strong pain to instead sedate children, in some cases for up to 24 hours. Human Rights Watch documented the experiences of former inmates from several institutions who described the effects of such 'chemical restraint' upon themselves or witnessing it being used against other children. Bakhyt, a 24-year-old woman who grew up in a special children's centre, claimed that institution staff punished children by "giving shots to make [us] sleep". She also stated that the institution staff had punished her for running around in the bedroom by giving her an injection and had sent her to a psychiatric hospital for 2 months. Asel, 22, who grew up in the same institution, also stated that staff punished children by giving them sedatives. "When they give you a shot", she said, "then you sleep for one whole day".[147]

In his May 2013 report to the UN Human Rights Council, the UN Special Rapporteur on Torture,[148] Juan Mendez, explored the application of the concept of torture or ill-treatment to the range of abuses committed in health care settings. Mendez noted that "medical treatments of an intrusive and irreversible nature, when lacking a therapeutic purpose, may constitute torture or ill-treatment when enforced or administered without the free and informed consent of the person concerned". He warned that this was "particularly the case when intrusive and irreversible, nonconsensual treatments are performed on patients from marginalized groups, such as persons with disabilities, notwithstanding claims of good intentions or medical necessity". In this regard he specifically highlighted the administration of "non-consensual medication" which was "often claimed as being a necessary treatment for the so-called best interest of the person concerned".[149] Consequently, he called on all States to "[I]mpose an absolute ban on all forced and non-consensual medical interventions against persons with disabilities, including the non-consensual administration of... mind-altering drugs such as neuroleptics, the use of restraint and solitary confinement, for both long- and short-term application".[150]

In addition to the prohibitions against torture and other ill-treatment under CAT and the ICCPR, the UN Convention on the Rights of People with Disabilities (CRPD) is of potential relevance in such circumstances,[151] as it protects the rights of people who "have long-term physical, mental, intellectual or sensory impairments which in interaction with various barriers may hinder their full and effective participation in society on an equal basis with others".[152] The UN Committee on the Rights of Persons with Disabilities, which monitors State compliance with the CRPD, has previously criticized the use of 'chemical restraints' in relation to States' obligations to prohibit torture and ill treatment. For example, in its 2013 review of Australia it noted that, "persons with disabilities, particularly those with intellectual impairment or psychosocial disability, are subjected to unregulated behaviour modification or restrictive practices such as chemical, mechanical and physical restraints and seclusion, in various environments" and called on Australia to end such restrictive practices.[153]

8.6.1.2 Behavioural Modification – Use of 'Truth Serums' in Interrogation

Certain States have reportedly employed CNS-acting chemical agents (such as sodium thiopental, sodium amytal or scopolamine) as so-called 'truth drugs' or 'truth serums' upon detainees with or without their consent, for the purposes of interrogation by law enforcement, security or military officials – a practice known as narcoanalysis. Dando and Furmanski,[154] and Perry Robinson,[155] have documented the historical attempts by the US and the Soviet Union during the Cold War to develop such chemical aids to interrogation. More recently, such methods have reportedly been employed in India,[156] Turkmenistan,[157] Turkey,[158] and in the US detention centre at Guantanamo Bay.[159]

Chemical-facilitated interrogations of prisoners are contrary to the prohibitions against torture and other ill-treatment contained in numerous IHRL instruments including the CAT, ICCPR[160] and the UN Body of Principles for the Protection of All Persons under Any Form of Detention or Imprisonment (UN Principles for the Protection of Detainees), Principle 6 of which states:

> "No person under any form of detention or imprisonment shall be subjected to torture or to cruel, inhuman or degrading treatment or punishment. No circumstance whatever may be invoked as a justification for torture or other cruel, inhuman or degrading treatment or punishment."[161]

Such practices would also breach provisions against self-incrimination[162] and those concerning the conduct of interrogations under Principle 21 of the UN Principles for the Protection of Detainees which states:

> "1. It shall be prohibited to take undue advantage of the situation of a detained or imprisoned person for the purpose of compelling him to confess, to incriminate himself otherwise or to testify against any other person.
>
> 2. No detained person while being interrogated shall be subject to violence, threats or methods of interrogation which impair his capacity of decision or his judgement."[163]

Furthermore, the involvement of health professionals in any form of torture or other ill-treatment (including the non-consensual application of CNS-acting chemical agents such as 'truth drugs' in interrogations) is prohibited under the UN Principles of Medical Ethics.[164] Under these standards it is deemed a contravention of medical ethics for health personnel, particularly physicians: "to be involved in any professional relationship with prisoners or detainees the purpose of which is not solely to evaluate, protect or improve their physical and mental health",[165] or "to apply their knowledge and skills in order to assist in the interrogation of prisoners and detainees in a manner that may adversely affect the physical or mental health or condition of such prisoners or detainees".[166] The UN Principles of Medical Ethics have also stated that "there may be no derogation from the foregoing principles on any ground whatsoever, including public emergency".[167]

In addition to IHRL, as discussed in Chapter 9, important restrictions upon the non-consensual use of CNS-acting chemicals to facilitate the interrogation of prisoners may also arise from obligations under IHL (notably the Geneva Conventions and Additional Protocols); such provisions potentially becoming applicable in circumstances where military and law enforcement activities overlap. For example, the reported application of CNS-acting chemicals to detainees held in Guantanamo Bay by US military officials, apparently with the intention of facilitating their interrogation appeared to contravene regulations established under the US Army Field Manual.[168]

Further constraints or outright prohibitions upon narco-analysis have been established at the national level in certain States. In May 2010, the Indian Supreme Court ruled that the forced administration of CNS-acting chemicals to detainees to facilitate their interrogation, violated the Indian Constitution and was illegal as it constituted cruel, inhuman or degrading treatment.[169] The Court stated that:

> "Even though 'the right against cruel, inhuman and degrading punishment' cannot be asserted in an absolute sense, there is a sufficient basis to show that Article 21 [of the Indian Constitution which protects 'personal liberty'] can be invoked to protect the 'bodily integrity and dignity' of persons who are in custodial environments. This protection extends not only to prisoners who are convicts and under trials, but also to those persons who may be arrested or detained in the course of investigations in criminal cases."[170]

Long-standing concerns regarding narcoanalysis have been raised by US legal scholars and jurists, consequently the use of 'truth serums' or 'truth drugs' is not recognized as an authorized method of interrogation by US courts. Under US law, confessions made under the influence of 'truth serums' are not considered as 'voluntary' and are consequently inadmissible as evidence.[171] The employment of CNS-acting chemicals to facilitate interrogations also appears to be contrary to the Inter-American Convention to Prevent and Punish Torture. Critically, the Convention under Article 2 defines torture as including: "the use of methods upon a person intended to obliterate the personality of the victim or to diminish his physical or mental capacities, even if they do not cause physical pain or mental anguish".[172]

8.6.1.3 Behavioural Modification – 'Brain-washing'

In addition to the prohibitions against torture and other ill-treatment, further important potential constraints upon the non-consensual application of CNS-acting chemical agents to detainees relate to obligations to ensure respect for the detainee's right to freedom of opinion. For example, Article 19 of the ICCPR has declared that "everyone shall have the right to hold opinions without interference".[173] In his legal commentary to the Convention, Nowak stated that this provision consequently "obligates the States Parties to refrain from any interference with freedom of opinion (by indoctrination, 'brainwashing', influencing the conscious or unconscious mind with psychoactive drugs or other means of manipulation) and to prevent private parties from doing so".[174]

8.6.1.4 Behavioural Modification – 'Conversion Therapy'

'Conversion therapy' is a pseudo-therapeutic practice purportedly intended to change an individual's gender identity or sexual orientation from homosexual or bisexual to heterosexual.[175] Major medical and mental health associations have condemned its use because it is harmful, lacks scientific basis and is unethical. In September 2015 twelve UN agencies, including the World Health Organisation issued a statement calling for an end to "unethical and harmful so-called 'therapies' to change sexual orientation",[176] and the World Psychiatric Association, in 2016, condemned the practice as "wholly unethical".[177] However, according to the International Rehabilitation Council for Torture Victims (IRCT), at least 68 countries continue to allow health professionals to conduct 'conversion therapy', and in at least 14 countries[178] medication including anti-anxiety medication, anti-psychotics, anti-depressants and sedatives (*i.e.*, CNS-acting chemical agents) is used as part of such practices. According to the IRCT in at least three countries – China,[179] Russia and Vietnam[180] – such 'therapy' reportedly incorporates the enforced administration of medication or injections, and there are reports that such medication is "administered by force or threat".[181] According to a 2017 Human Rights Watch report, in China such 'therapy' included forced medication and injection of unknown chemicals which appeared to include CNS-acting chemicals (*e.g.* anti-anxiety medication and anti-depressants), as well as the use of electroshocks. Human Rights Watch found that 'conversion therapy' often took place in government-run public hospitals. In a few cases, 'conversion therapy' was conducted in privately owned psychiatric or psychological clinics, licensed and supervised by the National Health and Family Planning Commission.[182] The IRCT believes that for the administration of medication "in other countries it is likely that an individual's consent is provided in coercive settings or is not duly informed".[183]

8.6.2 Torture in Non-custodial Contexts

While torture and other ill-treatment have traditionally been seen as occurring in places of detention, in recent years there has been a growing recognition that such violations can occur in a range of non-custodial settings. In 2017 then UN Special Rapporteur on Torture, Nils Melzer, examined the phenomenon in extra-custodial settings.[184] These include incidents on the street, where the policing of public protest is carried out in a deliberately punitive fashion. Melzer concluded that:

> "Any extra-custodial use of force that is intended to inflict pain or suffering on a 'powerless' person (that is, a person who is under direct

physical or equivalent control and is unable to escape or resist) as a vehicle for achieving a particular purpose amounts to an aggravated form of cruel, inhuman or degrading treatment or punishment, irrespective of considerations of lawful purpose, necessity and proportionality and irrespective of what else, if anything, might be required for such use of force to constitute torture under the respective treaty instruments."[185]

Melzer consequently examined the use and misuse of 'less lethal' weapons and concluded that:

> "Any extra-custodial use of an otherwise permissible weapon, irrespective of its lethal or less lethal design, in order to intentionally and purposefully inflict pain or suffering on a powerless person, always amounts to an aggravated form of cruel, inhuman or degrading treatment or punishment or even torture."[186]

These considerations are particularly important given the previous largescale use in a non-custodial context by Russia of a CNS-acting chemical agent weapon in its anti-terrorist operation in 2002. Although the legitimacy of the specific Russian employment of this weapon under such extreme law enforcement circumstances is contested, this action highlights the wider risk that Russia or other States may in future employ such CNS-acting weapons to facilitate torture and other ill-treatment against mass protests and other public assemblies. Similar concerns arise from the development by Chinese companies of CNS-acting chemical agent weapons intended for use against individuals, their possession by the Chinese PLA and their reported use by Chinese security forces to facilitate the arrest of villagers protesting Chinese Government policies.

8.7 Health Considerations and the Right to Health

The universal right to health has been clearly recognized under the International Covenant on Economic, Social and Cultural Rights (ICESCR):[187]

> "1. The States Parties to the present Covenant recognize the right of everyone to the enjoyment of the highest attainable standard of physical and mental health.
>
> 2. The steps to be taken by the States Parties to the present Covenant to achieve the full realization of this right shall include those necessary for:
>
> d. The creation of conditions which would assure to all medical service and medical attention in the event of sickness".[188]

The Committee on Economic, Social and Cultural Rights (CESCR), which is the body that monitors this Covenant, has stated, in a formal General Comment, that violations of the obligation to respect the right to health "are State actions, policies or laws that contravene standards set out in Article 12 [of the IESCR] and are likely to result in bodily harm, unnecessary morbidity and preventable mortality".[189] In addition, the CESCR stated that violations of the obligation to protect the right to health "follow from the failure of a State to take all necessary measures to safeguard persons within their jurisdiction from infringements of the right to health by third parties".[190]

Casey-Maslen has explored the potential application of the right to health in addressing the use of certain weapons, but notes that to date this area of IHRL "does not appear to have been tested".[191] He does, however, argue that "a challenge to certain NKE weapons on the basis of their health effects, including by prisoners or patients at a mental health institution, merits consideration".[192]

Indeed, such considerations appear to be particularly apposite for CNS-acting chemicals, given the previous history of medicalised misuse of such chemicals to torture and ill-treat prisoners, detainees and patients, as well as the long-standing concerns voiced by some in the medical community regarding the immediate and long-term health effects of their use as weapons in non-custodial contexts.[193]

In addition, disquiet has also been voiced by medical professionals and human rights organisations regarding the potential involvement of medical personnel in the development or application of CNS-acting chemical agent weapons,[194] as well as the failure of the State to provide adequate medical treatment following the use of such weapons.[195]

In terms of the inappropriate application of CNS-acting chemicals in custodial settings, a significant intervention was made by five UN Special Rapporteurs (including the UN Special Rapporteur on the right of everyone to the enjoyment of the highest attainable standard of physical and mental health) in a 2006 UN Human Rights Commission report into conditions in Guantanamo Bay. The five Rapporteurs raised concerns relating to the "right to health" with regard to the forcible application of drugs to detainees.[196] The UN Special Rapporteur on the right to health stated that he had "received serious and credible reports of violations of the right to health – both health care and the underlying determinants of health – at Guantánamo Bay".[197] The reports he received alleged, *inter alia*, that "detainees have been subjected to non-consensual treatment, including drugging and force-feeding".[198] The UN Special Rapporteur on the right to health also raised his concerns that reports indicated that "some health professionals have been complicit in abusive treatment of detainees detrimental to their health. Such unethical conduct violates the detainees' right to health, as well as the duties of health professionals arising from the right to health".[199] Related concerns have arisen regarding the apparent non-consensual application of CNS-acting chemicals by health professionals, at the request of law enforcement officials, to subdue individuals in order to facilitate their arrest and/or transport to places of detention (as highlighted below).

8.7.1 US: Application of CNS-acting Chemicals for Law Enforcement Chemical Restraint

From the early 2010s media, human rights organisations and health professionals have raised concerns regarding the employment of pharmaceutical CNS-acting chemicals to enforce compliance in law enforcement contexts in certain US States. The circumstances of such application in certain cases have been contested, as has whether such use is appropriate or may constitute ill-treatment. In 2018, the *Star Tribune* newspaper obtained an unreleased draft report from the Minneapolis Office of Police Conduct Review, that detailed a pattern of police urging paramedics to sedate dozens of 'agitated' or 'aggressive' people with the CNS-acting agent, ketamine. Certain individuals exhibiting such behaviour have sometimes been diagnosed as suffering from the contested medical condition of 'excited delirium'. The number of documented ketamine injections during Minneapolis police calls increased from three in 2012 to 62 in 2017. In many cases, the individual being detained or arrested was not only handcuffed, but strapped down on a stretcher in an ambulance before receiving ketamine.[200] The *Star Tribune* later reported that some of the patients sedated were then enrolled in a ketamine drug trial conducted by Hennepin Healthcare "without their prior consent", and that in some cases the existence of the ketamine study may have influenced whether they had been administered with the drug.[201] Similar practices have been uncovered in other US States. In Colorado, from January 2018 till June 2020, at least 902 people suffering 'excited delirium' were injected with ketamine by paramedics and first responders, with serious health consequences arising in 17% of these incidents.[202] In August 2020, the Associated Press highlighted the continuing lack of US regulation in this area: "there are no [US] Federal standards for law enforcement or emergency medical personnel on the drug's use. State policies and reporting requirements vary, so it's not clear how regularly it's used during police encounters and why".[203]

In July 2020 the American Society of Anaesthesiologists stated that it "firmly opposes the use of ketamine or any other sedative/hypnotic agent to chemically incapacitate someone for a law enforcement purpose and not for a legitimate medical reason".[204] In June 2021, the American Medical Association adopted formal policy clarifying that it: "opposes [use of 'excited delirium' as an official diagnosis] until a clear set of diagnostic criteria has been established; denounces 'excited delirium' as a sole justification for law enforcement use of excessive force; [and] opposes the use of sedative/hypnotic and dissociative drugs—including ketamine—as an intervention for an agitated individual in a law enforcement setting, without a legitimate medical reason".[205]

8.8 Considerations Regarding Means of Delivery and Dispersal of CNS-acting Chemical Agents

Although this chapter has concentrated upon the constraints imposed by IHRL and related standards on the use of weaponised CNS-acting chemical agents for law enforcement purposes, a further dimension needs to be considered when examining the potential legitimacy of such weapons, namely the mechanisms employed to deliver or disperse these agents to their targets. The nature of such mechanisms will raise important additional considerations regarding safety, lethality and discrimination, leading potentially to determinations that they are inherently incompatible with IHRL or that they may be compatible with IHRL when used in certain restricted circumstances. To date two distinct types of CNS-acting chemical agent weapons have been developed.

First, *CNS-acting chemical agent weapons that are intended to be employed against specific targeted individuals or small groups of individuals, normally by law enforcement officials in close proximity to the targets.* A range of delivery mechanisms have been explored and tested by State bodies or commercial companies, including injection devices; dart projectiles, agent-soaked felt pad projectiles, frangible projectiles, together with associated launchers; and hand thrown grenades. To date the only CNS-acting chemical agent delivery systems that have been produced and marketed to the international law enforcement community have been so-called narcosis dart guns promoted by certain Chinese companies.

In addition to these specifically developed CNS-acting chemical agent delivery systems, there are a wide range of commercially available RCA projectiles and delivery systems which have a narrow dispersal area and emit a limited quantity of RCA. These narrow area delivery mechanisms include hand-thrown or weapons-launched RCA canisters and grenades, and hand-held RCA spray disseminators. Such devices in future could be adapted so that they were filled with a relevant CNS-acting chemical agent rather than an RCA. As discussed in Chapter 5 of this publication, the relevance of this development route has been underlined by the recent reported attempts of Iranian scientists to adapt RCA grenades and weapons-launched RCA projectiles so that they contain the CNS-acting chemical agent, medetomidine, rather than the RCA, CS, as the dispersed agent.

The limited area RCA delivery systems discussed above are currently widely deployed by law enforcement officials, notably in the policing of protests and other public order situations, as well as in certain custodial settings. If such devices have been properly designed, manufactured, tested and trialled, their use for RCA dispersal should not raise undue concerns, as long as it is in strict accordance with the relevant IHRL and standards, notably the UNBP, the UNCoC, and the UN Standard Minimum Rules for the Treatment of Prisoners.

However, *any* device or delivery system originally developed or subsequently adapted for *CNS-acting chemical agent* delivery against individuals needs to undergo its own specific review against the previous IHRL standards highlighted. Specific medico-legal analysis will need to be undertaken with regard to the immediate and long-term effects upon the target of the specific CNS-acting chemical agent in the amounts delivered to the individual in a range of tactical situations. In such reviews consideration will need to be given to the limited circumstances, if any, under which such a weapon could be lawfully employed and the constraints upon such use.

Second, *CNS-acting chemical agent weapons that are intended to be employed against large groups of individuals, the agents are delivered over wide areas and normally by law enforcement officials at some distance from the targets*. A range of delivery mechanisms have been explored including *ad hoc* aerosol delivery in buildings; aircraft or helicopter delivered cluster munitions; airburst non-lethal munitions; wide area aircraft dispersal mechanisms; and drone delivery systems. To date one State – the US – previously developed, manufactured and stockpiled large quantities of aerial delivered BZ cluster munitions which it did not use and subsequently destroyed; a second State – Russia – developed and employed an *ad hoc* aerosol delivery mechanism to disperse fentanyl analogues through a building's ventilation system.

In addition to these specifically developed CNS-acting chemical agent delivery systems, a range of 'wide area' RCA delivery mechanisms have been developed and/or deployed for crowd control and dispersal that deliver far larger amounts of RCAs over wide areas or extended distances. These include: fixed internal building RCA dispersal systems, water cannon, multiple barrel launchers, automatic grenade launchers, 'less lethal' area denial systems, delivery systems on drones or unmanned ground vehicles, as well as unmanned weapons system. A number of these devices raise safety concerns as well as questions about the feasibility of their discriminate use, with the consequent danger of affecting bystanders, in contravention of IHRL. Furthermore, an additional range of 'wide area' RCA delivery systems have previously been developed and promoted which are inherently unacceptable for use in law enforcement activities and would breach IHRL. These include munitions containing RCAs which have military utility, such as cluster munitions, rocket propelled grenades, aerial bombs, mortar rounds and artillery shells.

Once again, any device or dispersal system originally developed or subsequently adapted for wide area delivery of CNS-acting chemical agent needs to undergo its own specific medico-legal review against the previously highlighted IHRL standards. Many if not all such systems would raise significant, if not insurmountable, IHRL concerns including with regard to: their indiscriminate nature with attendant risks of affecting bystanders; dangers to the health and life of those targeted; potential misuse for torture and other ill-treatment on a large scale, *etc.* It would be hard to justify the employment of such wide area CNS-acting chemical agent weapons in any law enforcement scenario, that would be consistent with IHRL and related standards.

8.9 Obligations Regarding Design, Review, Training and Monitoring Use of 'Less Lethal' Weapons

As discussed in Chapter 9, States are required under Article 36 of Additional Protocol I to the Geneva Conventions to undertake reviews of all "new weapons" intended for use in warfare.[206] Unfortunately, there is no corresponding binding obligation under IHRL explicitly requiring States to carry out comparable reviews of "new weapons" developed or acquired solely for the purpose of law enforcement. However, the norm-setting UNBP has provided that: "[T]he development and deployment of 'non-lethal' incapacitating weapons should be carefully evaluated in order to minimize the risk of endangering uninvolved persons, and the use of such weapons should be carefully controlled".[207]

The issue has also previously been raised by the UN Human Rights Council in its Resolution 25/38 of April 2014, which: "underline[d] the importance of thorough, independent and scientific testing of non-lethal weapons prior to deployment to establish their lethality and the extent of likely injury, and of monitoring appropriate training and use of such weapons".[208] The Resolution further "encourage[d] States to make protective equipment and non-lethal weapons available to their officials exercising law enforcement duties, while pursuing international efforts to regulate and establish protocols for the training and use of non-lethal weapons".[209]

The 2020 *UN Human Rights Guidance on Less-Lethal Weapons in Law Enforcement*, builds upon and elaborates these recommendations. In fact, it goes one step further by first outlining State obligations with regard to design and production of such weapons:

> "Less-lethal weapons and related equipment intended for use in law enforcement shall be designed and produced to meet legitimate law enforcement objectives and to comply with international human rights law. This duty applies to the State and its agents, and also to companies that manufacture weapons for law enforcement.
>
> Public and private manufacturers of less-lethal weapons and related equipment should bring specific risks to the attention of the purchaser and/or user and to that of the public at large. States, law enforcement agencies and manufacturers should be transparent about the technical specifications of weapons in use, while respecting manufacturers' intellectual property rights. At a minimum, all material safety data sheets should be provided by the manufacturers for the purchaser and/or user. Publicly available data should include each weapon's design features and parameters with a view to facilitating medical treatment and public acceptance. Manufacturers should also identify and release all medical studies and the names of the experts who have contributed to safety analyses, indicating those who have received compensation for promoting their products.

> The nature of law enforcement places special constraints on the extent to which force may be delivered remotely. Among other reasons, this is because distance is likely to reduce substantially the potential for assessing a situation that requires a law enforcement intervention (such as the arrest of a criminal suspect), and then for resolving it peacefully. In any event, an appropriate measure of human control over the use of force in law enforcement should be ensured. This has significant implications for the design of less-lethal weapons and related equipment".[210]

With regard to State obligations to review and test "less lethal" weapons, the UN Guidance states:

> "States shall ensure that, prior to the procurement of, and equipping of law enforcement officials with, less-lethal weapons or items of related equipment, a legal review is conducted to determine whether this would, in some or all circumstances, be prohibited by any rule of international or domestic law, in particular human rights law.
>
> As part of the legal review, testing should be conducted independently of the manufacturer and in accordance with recognized standards. Testing should take into full account both the required and the potential capabilities and effects of the weapons, and should be based on impartial legal, technical, medical and scientific expertise and evidence. Testing should evaluate the effects of all reasonably likely or expected uses of the weapons. Particular consideration should be given to assessing the potential effects of using less-lethal weapons and related equipment against individuals who may be especially vulnerable.
>
> Less-lethal weapons and related equipment whose designated, expected or intended use does not comply with the rules governing law enforcement, or which presents undue risk of loss of life or serious injury to anyone, including criminal suspects, bystanders or law enforcement officials themselves, shall not be authorized for procurement, deployment or use."[211]

In addition to requiring that States "carefully evaluate" development and deployment of 'non-lethal' or 'less lethal' weapons, the UNBP also incorporated obligations upon States to ensure effective control of the use of force and firearms through the introduction of rules and regulations, which would certainly be applicable to such 'less lethal' weapons.[212] These provisions have been supplemented by requirements relating to the training of law enforcement officials "with a view to limiting the use of force and firearms", and the review of "training programmes and operational procedures in the light of particular incidents".[213]

Once again these have been elaborated in the UN Guidance with regard to monitoring:

> "States and law enforcement agencies shall monitor the use and effects of all the less-lethal weapons and related equipment they procure, deploy and use for law enforcement purposes.
>
> Monitoring should include contextual information about the circumstances of use. Relevant data on those against whom force is used should, to the extent possible, be disaggregated, for example by age, sex/gender, disability (where that exists) and ethnic group. The results of monitoring should be made public in accordance with the transparency requirements set out below. This should include publicly available national statistics on deaths and serious injuries relating to different categories of less-lethal weapons."[214]

With regard to training the UN Guidance states:

> "Law enforcement officials shall receive appropriate initial and refresher training in the use and effects of any less-lethal weapons with which they have been equipped or deployed. The training should be to a high standard, and trainees should be authorized to carry and use only those weapons in which they have been trained successfully. Command and supervisory officials should also receive ongoing training on their obligations, arising from their command functions, in areas relating to the use of force.
>
> Training should include information on the particular vulnerabilities of certain individuals to the effects of a particular weapon, and on how to identify especially vulnerable individuals. Trainees shall be made aware not only of the primary risk of injury arising from use of any less-lethal weapons with which they may be equipped but also of the secondary injuries that may result (for example, if a person against whom the weapon is used falls from an elevated position or onto a hard surface). These effects and risks should also be reflected in standard operating procedures."[215]

With regard to medical assistance the UN Guidance states:

> "Medical assistance shall be rendered to any injured or affected person at the earliest possible moment. The duty to assist applies without discrimination of any kind. Thus, assistance shall be duly provided, irrespective of whether the injured person is a suspected offender. Distinctions with regard to the medical assistance provided are justifiable only on medical grounds, such as the urgency of medical needs. Access to medical assistance should therefore be provided on the basis of the severity of the injury."[216]

The UNBP required that: "[G]overnments shall ensure that arbitrary or abusive use of force and firearms by law enforcement officials is punished as a criminal offence under their law".[217] The Human Rights Council, in its 2014 Resolution on the "promotion and protection of human rights in the context of peaceful protests", has "call[ed] upon States to investigate any death or significant injury committed during protests, including those resulting from the discharge of firearms or the use of nonlethal weapons by officials exercising law enforcement duties".[218]

The preceding recommendations on design, review, testing, acquisition, use and monitoring of 'less lethal' weapons are clearly relevant to CNS-acting chemical agent weapons. They are drawn primarily from the UNBP and the UN Guidance on Less Lethal Weapons in Law Enforcement which are both soft law instruments that reflect best practice, but which are not legally binding upon States.

8.10 Trade Agreements Incorporating IHRL and IHL Criteria

Over the last 30 years there have been calls from a wide range of UN and regional human rights bodies, civil society arms control, humanitarian and human rights organisations that the trade in conventional arms, law enforcement equipment including 'less lethal' weapons, and relevant dual use goods needs to come under international regulation and be conditioned by State obligations under IHRL and IHL. In response, there has been a growing recognition by States of their responsibilities in this area. Advances have subsequently been made both through regional and international instruments, although this has been somewhat fragmented and uneven with regard to the goods covered.

8.10.1 Conventional Arms and Dual Use Goods

From the 1990s onwards, a number of regional and pluri-lateral bodies began to develop instruments or other measures regulating the trade in conventional arms, certain 'less lethal' weapons and associated dual-use goods including a range of toxic chemicals. Certain of these instruments incorporate human rights considerations to varying degrees and could potentially be employed to prevent or restrict the transfer of CNS-acting chemical agents and related means of delivery to end users likely to misuse them for human rights violations or to breach IHL.

8.10.1.1 Wassenaar Arrangement

The Wassenaar Arrangement (WA) was initially formed by a group of 33 States[219] in 1996 with the expressed intention of contributing to "regional and international security and stability, by promoting transparency and greater responsibility in transfers of conventional arms and dual-use goods and technologies, thus preventing destabilising accumulations".[220] To date the WA has expanded to comprise 42 Participating States who have agreed to: maintain national export controls on a range of commonly agreed listed items; report on transfers and denials of specified controlled items to destinations outside the Arrangement; exchange information on sensitive dual-use goods and technologies; and be guided by agreed Best Practices, Guidelines or Elements.[221] All WA decision making is undertaken by consensus and its deliberations are confidential.

The range of materials regulated by the Wassenaar Arrangement is specified in the WA Dual Use Goods and Technologies List and the WA Munitions List (last updated 1 December 2023).[222] CNS-acting chemical agents are incorporated in the WA Munitions List under ML7.b as: "b.3. CW incapacitating agents, *such as*: 3-Quinuclidinyl benzilate (BZ) (CAS 6581-06-2)"; (emphasis added).[223] Although only BZ has been specifically mentioned in the control list to date, the wording of ML7.b implies that further CNS-acting chemical agents could also be added to this list, but were not explicitly controlled at present. There are ambiguities regarding the scope of this category and the range of agents controlled as the descriptive phrase "CW incapacitating agents" has not been defined in the WA Munitions List (or in other WA documents). The current wording of ML7.a of the WA Munitions List covers: "Biological agents" or radioactive materials selected or modified to increase their effectiveness in producing casualties in humans or animals, degrading equipment or damaging crops or the environment";[224] Once again the descriptive phrase "biological agents" has not been defined and furthermore, unlike the chemical agents section, there has been no listing of specific agents covered. Although this issue has not been clarified to date by WA Participating States, ML7.a does appear to incorporate 'less lethal' biological agents, and thus would potentially include CNS-acting chemical agents, or other incapacitating chemical agents of biological origin such as SEB. The current version is notable in removing a previous restrictive application only to those "biological agents" that are "adapted for use in war",[225] so that it would now cover a broader range of agents potentially including those adapted for use in law enforcement. The current version of ML7.a is also notable in extending its coverage to include both *selected* "biological agents" as well as *modified* "biological agents" intended to "increase their effectiveness in producing human casualties", which is a welcome recognition of the risks posed by the malign application of advances in relevant chemical and life sciences facilitating weaponization of modified biological agents.

All measures with respect to the WA have been taken in accordance with Participating States' national legislation and policies, and have been "implemented on the basis of national discretion".[226] Consequently, although the range of goods covered under Participating States' export controls has been determined by the WA lists, the practical implementation varies amongst States in accordance with national procedures.[227] Furthermore, and crucially, "the decision to transfer or deny transfer of any item will be the sole responsibility of each Participating State".[228] Participating States, have, however, agreed certain guidelines, elements and procedures which will inform their decision-making through the application of their own national legislation and policies. These include the advisory, non-binding "Elements for Objective Analysis and Advice Concerning Potentially Destabilising Accumulations of Conventional Weapons".[229] This has recommended that WA Participating States consider a range of factors when making decisions on relevant transfers, including an "assessment of motivation" of the potential recipient state, and specifically whether there is a "clearly identifiable risk that the weapons would be used to commit or facilitate the violation and suppression of human rights and fundamental freedoms or the laws of armed conflict",[230] or be used "offensively against another country or in a manner inconsistent with the UN Charter...".[231] Consequently, although the WA was not developed as a focused human rights transfer control regime, IHRL and IHL are amongst the factors that States should consider when deciding whether to allow trade in relevant dual use goods including CNS-acting chemical agents.

8.10.1.2 EU Common Position 2008/944/CFSP

In December 2008, the EU Member States adopted an EU Council Common Position 2008/944/CFSP, "defining common rules governing control of exports of military technology and equipment".[232] In the Common Position pre-ambulatory paragraphs, Member States declared themselves "determined to set high common standards which shall be regarded as the minimum for the management of, and restraint in, transfers of military technology and equipment by all Member States",[233] and to "prevent the export of military technology and equipment which might be used for internal repression or international aggression or contribute to regional instability".[234]

The Common Position requires that each EU Member State "assess the export licence applications made to it for items on the EU Common Military List ... on a case-by-case basis against the criteria of Article 2",[235] namely: international obligations and commitments; human rights and international humanitarian law; the internal situation of the country; regional peace, security and stability; national security; terrorism and respect for international law; the risk of diversion or re-export under undesirable conditions; and the compatibility of the exports with the technical and economic capacity of the recipient country.[236] Article 2, Criterion 2, covering international human rights law and international humanitarian law, is of particular relevance. Under this Criterion, Member States shall "deny an export licence if there is a clear risk that the military technology or equipment to be exported might be used for internal repression".[237] Member States are also required to: "exercise special caution and vigilance" in issuing licences "on a case-by-case basis" and taking account of the nature of the military technology or equipment, to countries where serious violations of human rights have been established by competent bodies of the EU, Council of Europe or UN.[238] The Common Position also requires Member States to assess "the recipient country's attitude towards relevant principles established by instruments of international humanitarian law",[239] and to "deny an export licence if there is a clear risk that the military technology or equipment to be exported might be used in the commission of serious violations of international humanitarian law".[240]

Whilst the Common Position instituted relevant operative provisions and the criteria by which export applications will be judged, the EU Common Military List,[241] and EC Regulation 1334/2000,[242] established the recommended range of items covered.[243] Examination of the EU Common Military List has shown that the relevant text effectively replicates the Wassenaar Arrangement Control List wording with regard to CNS-acting chemical agents. The EU Common Military List has been in place since July 2000, and has been reviewed on a regular basis[244] with the most recent version adopted by the Council on 25 February 2025.[245] It should be noted that, over this period, while the category ML.b.3(a) – *i.e.* "CW incapacitating agents, such as: 3-Quinuclidinyl benzilate (BZ) (CAS 6581-06-2)"[246] has remained unchanged, the category ML7.a has been altered to " 'Biological agents' or radioactive materials *selected or modified to increase their effectiveness in* producing casualties in humans or animals, degrading equipment or damaging crops or the environment; (changes emphasised)"[247] reflecting the changes in the relevant WA Control List category.

8.10.1.3 Arms Trade Treaty

Building upon the existing patchwork of national, regional and pluri-lateral measures regulating the transfer of dual use goods and conventional arms, in 1996 a coalition of civil society arms control and human rights organisations and Nobel Peace Laureates began to advocate for the development of a set of explicit standards based on relevant international law which all UN Member States would apply to the international transfer of conventional arms and associated law enforcement equipment including 'less lethal' weapons. Following the backing of an increasingly large number of States from all regions, a formal process to develop such standards through the UN-mandated negotiation of an international Arms Trade Treaty (ATT) commenced in 2012. In April 2013, the UNGA adopted the ATT,[248] and the Treaty came into force on 24 December 2014. The ATT was signed by 130 States and currently 110 have ratified the Treaty.[249] Unfortunately, during the negotiations, the range of goods covered by the instrument was significantly reduced so that it only included: battle tanks; armoured combat vehicles; large-calibre artillery systems; combat aircraft; attack helicopters; warships; missiles and missile launchers; and small arms and light weapons. Consequently, the transfer of CNS-acting chemical agents would not be covered by its provisions, although certain relevant means of delivery may fall within its scope. Although there are mechanisms to expand the scope of goods covered by the ATT, it is highly unlikely that such an expansion will take place in the near future, and if it does, that CNS-acting chemical agents will be added.

8.10.2 Anti-torture Trade Instruments

The UN Human Rights Council, successive UN Special Rapporteurs on Torture, regional organisations such as the EU, CoE and African Union, as well as civil society arms control, torture prevention and broader human rights organisations, have advocated the development of measures to regulate the transfer of a range of policing and security equipment and associated goods including certain CNS-acting chemical agents, so as to prevent them being utilized for torture, other ill-treatment and the death penalty.

8.10.2.1 EU Anti Torture Regulation

European Council (EC) Regulation No. 1236/2005 Concerning trade in goods which could be used for capital punishment, torture or other cruel, inhuman or degrading treatment or punishment (the EU Anti-Torture Regulation) was agreed in July 2005 and came into force on 30 July 2006.[250] It was the world's first multilateral instrument in this area, is legally binding, and is directly applicable in all 27 EU Member States. The Regulation is a 'living document' and has subsequently been revised and strengthened in light of technological and industry developments; a consolidated version, Regulation (EU) 2019/125, came into force on 20 February 2019.[251] The EU Anti-Torture Regulation established a harmonised system across the EU for:

- Prohibiting the trade (import, export, and transit) into, from, or through, all EU Member States of equipment and products with "no other practical use than capital punishment, torture or other cruel, inhuman or degrading treatment". These prohibited goods are listed in Annex II of the Regulation. Further, the Regulation prohibits the provision of components; related technical assistance; brokering of trade deals between third countries; and promotion of such goods at trade fairs or exhibitions, on TV, radio, or the internet.
- Regulating and licensing the trade in law enforcement equipment that could be misused for torture and other ill-treatment, including principal components and assembly kits for certain goods. As part of this, States are required to refuse export authorisation "when there are reasonable grounds to believe that the goods might be used for torture or other [ill-treatment] including judicial corporal punishment". All such controlled goods are listed in Annex III of the Regulation.
- Regulating and licensing the trade in certain pharmaceutical CNS-acting chemicals to ensure they are not transferred for use in 'lethal injection' executions. This process has been designed to not impede trade of such chemicals for legitimate medical, veterinary, or other purposes.

> These goods are listed in Annex IV of the Regulation and currently include amobarbital, pentobarbital, secobarbital, thiopental, and their sodium salts.

Although the EU Anti-Torture Regulation currently controls trade of certain CNS-acting chemicals solely with the intention of halting the transfer of such drugs for 'lethal injection' executions, some of these chemicals have reportedly been employed in torture and other ill-treatment, including for narcoanalysis, or have previously been explored (though discounted) as potential CNS-acting chemical agent weapons by certain States. In the future, if EU Member States were so minded, the EU Anti Torture Regulation could potentially be amended to specifically cover these and other CNS-acting chemical agents when intended for law enforcement purposes beyond the death penalty, with a prohibition on their transfer to end users likely to misuse them for torture and other ill-treatment.

8.10.2.2 Council of Europe Committee of Ministers Recommendation CM/REC (2021)2

In March 2021, the Committee of Ministers of the Council of Europe adopted Recommendation CM/REC(2021)2 to member States on measures against the trade in goods used for the death penalty, torture and other cruel, inhuman or degrading treatment or punishment.[252] The CoE Recommendation adopts a similar categorisation to the EU Anti-Torture Regulation covering: (a) goods that have no legitimate law enforcement purpose; (b) law enforcement goods that are misused to inflict torture or other ill-treatment; and (c) pharmaceutical CNS-acting chemical agents that can be used in the death penalty. Whilst the CoE Recommendation incorporates very similar lists of prohibited and controlled goods to those established under the EU Anti-Torture Regulation, it does expand the latter list to include commonly employed tear gases notably CS, CR, and CN.

Unlike the EU Anti Torture Regulation, the CoE Recommendation is not directly legally binding upon its 46 CoE member States. Instead it provides recommendations that the States are urged to undertake, namely: to ensure their national legal frameworks and administrative measures prohibit the trade (import, export, transit) of those goods that have no practical use other than the death penalty, and torture and other ill-treatment, as well as related technical assistance; to regulate and licence the export and transit of pharmaceutical CNS-acting chemicals to ensure that they are not used for the death penalty; and to establish national licensing measures and export and transit control measures with respect to weapons and equipment that can have a legitimate law enforcement purpose but that can be misused to inflict torture or other ill-treatment.[253] The Recommendation invites CoE member States to review their national legislation and practice relating to the trade to assess whether they comply with these measures. As with the EU

Anti-Torture Regulation, the CoE Recommendation currently controls the trade of certain CNS-acting chemicals solely with the intention of halting their transfer for 'lethal injection' executions. If CoE member States were so minded they could in future amend the Recommendation to cover these and other CNS-acting chemical agents when intended for law enforcement purposes beyond the death penalty, with a prohibition on their transfer to end users likely to misuse them for torture and other ill-treatment.

8.10.2.3 International Measures to Promote Torture-free Trade

8.10.2.3.1 Alliance for Torture-free Trade. On 18 September 2017, the EU, Argentina and Mongolia launched the Alliance for Torture Free Trade in the margins of the United Nations General Assembly in New York. To date, 63 States from all regions of the world[254] have signed the Alliance's Political Declaration, acknowledging that "the availability of goods used for capital punishment, torture and other cruel, inhuman or degrading treatment or punishment enables such practices", and committing themselves to "act together to further prevent, restrict and end trade" of such goods. The States specifically commit to "take effective measures, *inter alia* through legislation and effective enforcement where appropriate, for the restriction of the trade". To date the Alliance has not put forward collective recommendations as to the range of goods it seeks to prohibit and those whose trade it seeks to control. The inclusion of capital punishment makes it likely that CNS-acting chemicals will at least be considered as goods falling within the latter category.

8.10.2.3.2 Advancing a UN Process. In 2019, due in large part to effective collective international advocacy by the Alliance, UNGA adopted Resolution 73/304 Towards torture free trade,[255] asking the Secretary-General to (1) prepare a report on feasibility, scope and parameters for possible common international standards and (2) then establish a group of governmental experts (GGE) to examine the issue. Both the UN Secretary General's Report[256] and the subsequent GGE Report[257] were supportive of moving forward on international Torture-Free Trade initiatives. One option the GGE recommended was the development of a legally binding instrument covering both goods that have no practical use other than for torture or other ill-treatment, and also goods that could be misused for such practices. Unfortunately, there has been no substantive advance in this UN process since the UN GGE Report was published in June 2022.

8.10.2.3.3 UN Special Rapporteur on Torture's Thematic Report on the Torture Trade. In October 2023, the UN Special Rapporteur on Torture, Dr Alice Edwards, presented a thematic report at the UNGA Third Committee analysing the global trade in law enforcement equipment used for torture.[258] The Special Rapporteur's report, included one annex identifying specific types of equipment that are inherently abusive and should be completely prohibited,[259] and a second annex identifying law enforcement equipment that could be readily misused for torture and whose trade should be stringently controlled.[260] Whilst neither annex incorporated CNS-acting chemical agents, the Special Rapporteur noted that the annexes were not intended to be exhaustive and could be revised in the future, so the potential for subsequent inclusion of CNS-acting chemicals remains. The Special Rapporteur's report called for the negotiation of a legally binding international instrument to regulate the trade in all these goods. These Recommendations were subsequently endorsed by the UN Special Rapporteur on Extrajudicial Executions and the UN Special Rapporteur on Freedom of Assembly and Association.

8.11 Conclusions

This chapter explored the constraints upon the use of CNS-acting chemical agent weapons and CNS-acting agents more broadly that are imposed by human rights law derived from regional and international agreements, as well as that arising from customary international law. Human rights law was found to impose significant constraints upon the use of force by the State. Such limitations consequently have important implications as to whether, in what circumstances, by whom and how CNS-acting chemical agents could be lawfully employed.

Human rights law is particularly relevant to the discussion of the regulation of CNS-acting chemical agent weapons as it potentially covers the full 'use of force' spectrum from law enforcement activities through to armed conflict, including 'grey areas' such as counter-terrorist, counter-insurgency and military operations outside armed conflict where use of these chemical agents has been proposed. While several human rights norms may be applicable to the regulation of these weapons, the rights to life; to liberty and security; to freedom from torture and other ill-treatment; to freedom of association and assembly; and to health, together with the associated obligations on States to restrict the use of force, are the most relevant.

A significant potential strength of human rights law is the existence of a number of international and regional mechanisms to monitor adherence to relevant treaties and *jus cogens* norms. Hampson has noted that two of these mechanisms – the UN Special Rapporteur on Torture and the UN Special Rapporteur on Extrajudicial, Summary or Arbitrary Executions – "can scrutinize relevant conduct of any UN member. They are not limited to those that have ratified a particular treaty".[261] Furthermore, Hampson noted that the two Special Rapporteurs and two of the international treaty bodies (the Human Rights Committee and the Committee against Torture): "could monitor the use of unlawful weapons or the unlawful use of potentially lawful weapons".[262] In addition to the monitoring mechanisms, certain treaties have established enforcement mechanisms that can be accessed by individuals, and these can result in legally binding judgements upon those

States that are party to the relevant instrument (as was seen in the ECtHR rulings on the use of a CNS-acting chemical agent weapon by Russia).

However, it must be recognized that these monitoring and enforcement mechanisms have limited preventative value (though they may have some deterrent effect), as they would only be initiated after a potential misuse of a CNS-acting chemical agent weapon had occurred. Furthermore, there are currently no internationally accepted procedures, under IHRL, for evaluating new CNS-acting chemical agent weapons developed or otherwise acquired by States, or for monitoring their subsequent employment at a national level. Despite the limitations highlighted, it is clear that IHRL imposes significant constraints upon the use of CNS-acting chemical agent weapons and CNS-acting chemical agents more broadly for law enforcement purposes.

One emerging and potentially impactful application of IHRL in recent years has been the development of regional and international trade controls on conventional arms, dual use goods and law enforcement equipment which incorporate human rights criteria. Certain of these mechanisms notably the EU Common Position 2008/944/CFSP, Wassenaar Arrangement, EU Anti-Torture Regulation and CoE Committee of Ministers Recommendation currently cover CNS-acting chemical agents within their scope and could potentially (and to varying degrees) be used to prevent transfers of such chemicals to those intending to employ them for certain human rights violations. Unfortunately, at present the only existing international measure, the Arms Trade Treaty, does not cover and is unlikely to be revised to cover CNS-acting chemicals within its scope. Currently, human rights civil society organisations and 'progressive' States are advocating for the development of an international Torture-Free Trade Treaty (TFTT) through a suitable UN process. Depending on its nature and scope, such a TFTT, if eventually negotiated, could prove a powerful mechanism to constrain the misuse of CNS-acting chemical agents notably for torture, other ill-treatment and the death penalty.

References

1. This Chapter draws upon, updates and expands the authors' previous research and publications notably including: M. Crowley, *Chemical Control*, Palgrave MacMillan, pp. 166–189, 2016, which is used with permission from Palgrave MacMillan, Copyright 2016.
2. The Use of Force in Law Enforcement Operations, ICRC Advisory Service, 2015. www.icrc.org/en/download/file/12747/the-use-of-force-in-law-enforcement-icrceng.pdf (accessed 8 November 2024).
3. International Court of Justice, *Legal Consequences of the Construction of a Wall in the Occupied Palestinian Territory, Advisory Opinion*, 9th July 2004, paragraphs 107–112.
4. UN, Human Rights Committee, *General Comment 31 on The Nature of the General Legal Obligation on the States Parties to the International*

Covenant on Civil and Political Rights, CCPR/C/21/Rev.1/Add.13, 26th May 2004, paragraph 11.
5. E. Hoffberger-Pippan, *Less-Lethal Weapons under International Law, a three-dimensional perspective*, Cambridge University Press, Cambridge, 2021, p. 89.
6. F. Hampson, International law and the regulation of weapons, in *Incapacitating Biochemical Weapons*, ed. A. Pearson, M. Chevrier and M. Wheelis, Lexington Books, Lanham, MD., 2007, pp. 244–245, pp. 231–260.
7. United Nations, *Code of Conduct for Law Enforcement Officials [UNCoC]*, adopted by United Nations General Assembly Resolution 34/169 of 17th December 1979.
8. United Nations, *Basic Principles on the Use of Force and Firearms by Law Enforcement Officials [UNBP]]*, 7th September 1990, adopted by the Eighth United Nations Congress on the Prevention of Crime and the Treatment of Offenders, Havana, Cuba, 27th August–7th September 1990.
9. United Nations, Standard Minimum Rules for the Treatment of Prisoners (the Nelson Mandela Rules), General Assembly resolution 70/175, annex, adopted on 17 December 2015, 2015.
10. African Commission on Human and Peoples' Rights, *Guidelines for the Policing of Assemblies by Law Enforcement Officials in Africa*, 4 March 2017.
11. African Commission on Human and Peoples' Rights, *Principles and Guidelines on Human and Peoples' Rights while Countering Terrorism in Africa*, August 2016.
12. Inter-American Commission on Human Rights, *Report on Citizen Security and Human Rights*, OEA/Ser.L/V/II. Doc. 57, 31 December 2009.
13. Council of Europe, *European Code of Police Ethics*, Recommendation Rec (2001) 10 adopted by the Committee of Ministers of the Council of Europe on 19 September 2001 and explanatory memorandum, March 2002.
14. For further discussion see: S. Casey-Maslen and S. Connolly, *Police Use of Force under International Law*, Cambridge University Press, 2017, pp. 79–107.
15. United Nations, *Code of Conduct for Law Enforcement Officials [UNCoC]*, adopted by United Nations General Assembly Resolution 34/169 of 17th December 1979, Article 3.
16. United Nations, *Basic Principles on the Use of Force and Firearms by Law Enforcement Officials [UNBP]]*, 7th September 1990, adopted by the Eighth United Nations Congress on the Prevention of Crime and the Treatment of Offenders, Havana, Cuba, 27th August–7th September 1990, Principle 5 (a).
17. E. Hoffberger-Pippan, *Less-Lethal Weapons under International Law, a three-dimensional perspective*, Cambridge University Press, Cambridge, 2021, p. 94.

18. United Nations, *Basic Principles on the Use of Force and Firearms by Law Enforcement Officials [UNBP]]*, 7th September 1990, adopted by the Eighth United Nations Congress on the Prevention of Crime and the Treatment of Offenders, Havana, Cuba, 27th August–7th September 1990, Principle 7.
19. United Nations, *Basic Principles on the Use of Force and Firearms by Law Enforcement Officials [UNBP]]*, 7th September 1990, adopted by the Eighth United Nations Congress on the Prevention of Crime and the Treatment of Offenders, Havana, Cuba, 27th August–7th September 1990, Principle 3.
20. According to a report commissioned by the UN Office of the United Nations High Commissioner for Human Rights (OHCHR), less lethal weapons are "designed or intended for use on individuals or groups of individuals and which, in the course of expected or reasonably foreseen use, have a lower risk of causing death or serious injury than firearms." See: OHCHR, Guidance on less lethal weapons in law enforcement, 2020, p. 45. Such weapons have sometimes been called non-lethal, although this term is now becoming obsolete in recognition that the use of any weapon can have fatal consequences. Acknowledging the continuing contested discourse over the nature, scope and application of the terms non-lethal and less lethal weapon, the term "less lethal" will be placed in quotation marks when used by the authors during this publication.
21. United Nations, *Basic Principles on the Use of Force and Firearms by Law Enforcement Officials [UNBP]]*, 7th September 1990, adopted by the Eighth United Nations Congress on the Prevention of Crime and the Treatment of Offenders, Havana, Cuba, 27th August–7th September 1990, Principle 4.
22. United Nations, *Basic Principles on the Use of Force and Firearms by Law Enforcement Officials [UNBP]]*, 7th September 1990, adopted by the Eighth United Nations Congress on the Prevention of Crime and the Treatment of Offenders, Havana, Cuba, 27th August–7th September 1990, Principle 5 (a)–(c).
23. This term is not defined in the United Nations, Basic Principles on the Use of Force and Firearms by Law Enforcement Officials. However, it can be considered synonymous with 'less lethal' weapons. Common types of "less lethal" weapon employed by law enforcement officials include: police batons, hand-held chemical irritant dispensers, chemical irritants launched at a distance, conducted electrical weapons ("Tasers"), kinetic impact projectiles and launchers, dazzling weapons, water cannons, and acoustic weapons and devices. For further discussion of these "less lethal" weapon types and their regulation under IHRL see: OHCHR, Guidance on less lethal weapons in law enforcement, 2020, p. 45.
24. United Nations, *Basic Principles on the Use of Force and Firearms by Law Enforcement Officials [UNBP]]*, 7th September 1990, adopted by the Eighth United Nations Congress on the Prevention of Crime and the

Treatment of Offenders, Havana, Cuba, 27th August–7th September 1990, Principle 2.

25. United Nations, *Basic Principles on the Use of Force and Firearms by Law Enforcement Officials [UNBP]]*, 7th September 1990, adopted by the Eighth United Nations Congress on the Prevention of Crime and the Treatment of Offenders, Havana, Cuba, 27th August–7th September 1990, Article 3.
26. United Nations, *Guidance on Less-Lethal Weapons in Law Enforcement*, 2020. www.ohchr.org/Documents/HRBodies/CCPR/LLW_Guidance.pdf (accessed 8 November 2024). The expert group comprised international lawyers, law enforcement officials and human rights experts inter alia from the Geneva Academy, the Institute for International and Comparative Law in Africa, ICRC, UN Office on Drugs and Crime, OHCHR and regional organisations, as well as academics and NGOs.
27. United Nations, *Guidance on Less-Lethal Weapons in Law Enforcement*, 2020. www.ohchr.org/Documents/HRBodies/CCPR/LLW_Guidance.pdf (accessed 8 November 2024), pp. 13–19.
28. (a) See, *e.g.*, United Nations, Universal Declaration of Human Rights, adopted and proclaimed by UN General Assembly Resolution 217 A (III), 10th December 1948, 1948, Article 3; (b) United Nations, International Covenant on Civil and Political Rights [ICCPR], adopted on 16th December 1966, 1966. Article 6.
29. See, *e.g.*, European Convention on the Protection of Human Rights and Fundamental Freedoms, signed on 4th November 1950, 1950, Article 2. The obligations deriving from such regional instruments are binding upon those States party to the relevant agreements, and their application has subsequently been clarified following judgements made by the relevant regional legal institutions.
30. United Nations, International Covenant on Civil and Political Rights [ICCPR], adopted on 16th December 1966, 1966. Article 6(1).
31. Human Rights Committee, *General Comment No. 6: The right to life*, 16th Session, 30th April 1982, 1982, paragraph 1.
32. Human Rights Committee, *General Comment No. 6: The right to life*, 16th Session, 30th April 1982, 1982, paragraph 3.
33. Human Rights Committee, *General Comment No. 6: The right to life*, 16th Session, 30th April 1982, 1982, paragraph 1.
34. ICRC, Toxic chemicals as weapons for law enforcement: A threat to life and international law?, *Synthesis*, 2012, 3.
35. W. Aceves, Human rights law and the use of incapacitating biochemical weapons, in *Incapacitating Biochemical Weapons*, ed. A. Pearson, M. Chevrier and M. Wheelis, Lexington Books, Lanham, MD., 2007, p. 286, pp. 261–284.
36. D. Fidler, Incapacitating chemical and biochemical weapons and law enforcement under the chemical weapons convention, in *Incapacitating Biochemical Weapons*, ed. A. Pearson, M. Chevrier and M. Wheelis, Lexington Books, Lanham, MD, 2007, p. 175, pp. 171–194.
37. For further discussion of this case see Chapter 2 of this publication.

38. UN, Commission on Human Rights, *Report of the Special Rapporteur, Asma Jahangir*, submitted pursuant to Commission on Human Rights Resolution 2002/36, UN doc. E/CN.4/2003/3, 13th January 2003, 2003, p. 15, paragraph 34.
39. UN, Human Rights Commission, *Concluding observations of the Human Rights Committee: Russian Federation*, UN doc. CCPR/CO/79/RUS, 6th November 2003, p. 4, paragraph 14.
40. UN, Human Rights Commission, *Concluding observations of the Human Rights Committee: Russian Federation*, UN doc. CCPR/CO/79/RUS, 6th November 2003, p. 4, paragraph 14.
41. UN, Human Rights Commission, *Concluding observations of the Human Rights Committee: Russian Federation*, UN doc. CCPR/CO/79/RUS, 6th November 2003, p. 4, paragraph 14.
42. See European Court of Human Rights (ECtHR), *Finogenov and others v. Russia*, application numbers 18299/03 and 27311/03, Judgement, 20th December 2011, 2011.
43. See European Court of Human Rights (ECtHR), *Finogenov and others v. Russia*, application numbers 18299/03 and 27311/03, Judgement, 20th December 2011, 2011, paragraph 201.
44. See European Court of Human Rights (ECtHR), *Finogenov and others v. Russia*, application numbers 18299/03 and 27311/03, Judgement, 20th December 2011, 2011, paragraph 201.
45. See European Court of Human Rights (ECtHR), *Finogenov and others v. Russia*, application numbers 18299/03 and 27311/03, Judgement, 20th December 2011, 2011, paragraph 202.
46. See European Court of Human Rights (ECtHR), *Finogenov and others v. Russia*, application numbers 18299/03 and 27311/03, Judgement, 20th December 2011, 2011, paragraph 203.
47. See European Court of Human Rights (ECtHR), *Finogenov and others v. Russia*, application numbers 18299/03 and 27311/03, Judgement, 20th December 2011, 2011, paragraph 213.
48. See European Court of Human Rights (ECtHR), *Finogenov and others v. Russia*, application numbers 18299/03 and 27311/03, Judgement, 20th December 2011, 2011, paragraph 232.
49. See European Court of Human Rights (ECtHR), *Finogenov and others v. Russia*, application numbers 18299/03 and 27311/03, Judgement, 20th December 2011, 2011, paragraph 236.
50. See European Court of Human Rights (ECtHR), *Finogenov and others v. Russia*, application numbers 18299/03 and 27311/03, Judgement, 20th December 2011, 2011, paragraph 266. [For further discussion see paragraphs 243–262.]
51. See European Court of Human Rights (ECtHR), *Finogenov and others v. Russia*, application numbers 18299/03 and 27311/03, Judgement, 20th December 2011, 2011, paragraph 282. [For further discussions see paragraphs 277, 279 and 281.]

52. ICRC, Toxic chemicals as weapons for law enforcement: A threat to life and international law?, *Synthesis*, 2012, 3.
53. (a) A. Kelle, The message from strasbourg, *Bull. At. Sci.*, 23 February 2012. (b) A. Kelle, Legally Incapacitated, Politically Outmaneuvered, *Bull. At. Sci.*, 7 June 2012. (c) E. Hoffberger-Pippan, *Less-Lethal Weapons under International Law, a three-dimensional perspective*, Cambridge University Press, Cambridge, 2021, pp. 102–103.
54. See in particular, See European Court of Human Rights (ECtHR), *Finogenov and others v. Russia*, application numbers 18299/03 and 27311/03, Judgement, 20th December 2011, 2011, paragraphs 162–164 and 228–229.
55. See European Court of Human Rights (ECtHR), *Finogenov and others v. Russia*, application numbers 18299/03 and 27311/03, Judgement, 20th December 2011, 2011, paragraph 229.
56. E. Hoffberger-Pippan, *Less-Lethal Weapons under International Law, a three-dimensional perspective*, Cambridge University Press, Cambridge, 2021, pp. 108–109.
57. See European Court of Human Rights (ECtHR), *Finogenov and others v. Russia*, application numbers 18299/03 and 27311/03, Judgement, 20th December 2011, 2011, paragraph. 230.
58. See: EctHR, Press release issued by the Registrar of the Court, ECHR 270 (2012), 27th June 2012.
59. In the case of Finogenov and others v. Russia, the Court ruled that Russia was to pay all 64 applicants a total award – as regards non-pecuniary damage – of €1,254,000, and €30,000, jointly, for costs and expenses. See European Court of Human Rights (ECtHR), Finogenov and others v. Russia, application numbers 18299/03 and 27311/03, Judgement, 20th December 2011, 2011, paragraphs 285–296.
60. The African Court on Human and Peoples' Rights, the European Court of Human Rights, and the Inter-American Courts of Human Rights.
61. F. Hampson, International law and the regulation of weapons, in *Incapacitating Biochemical Weapons*, ed. A. Pearson, M. Chevrier and M. Wheelis, Lexington Books, Lanham, MD., 2007, p. 243.
62. F. Hampson, International law and the regulation of weapons, in *Incapacitating Biochemical Weapons*, ed. A. Pearson, M. Chevrier and M. Wheelis, Lexington Books, Lanham, MD., 2007, p. 243.
63. There is, however, an ongoing attempt, initiated by Italy in 2007, to establish an international moratorium on the use of the death penalty with a view to progressing towards abolition. Repeated UNGA Resolutions, supported by an increasing number of Member States, call upon all States that still maintain the death penalty to progressively restrict the use of the death penalty, reduce the number of offences for which it may be imposed, and establish a moratorium on executions with a view to abolishing the death penalty altogether. The most recent being *UNGA Resolution A/RES/77/222*, adopted by the UN General Assembly on 15th December 2022.

64. United Nations, International Covenant on Civil and Political Rights [ICCPR], adopted on 16th December 1966, 1966, Article 6(2).
65. United Nations, International Covenant on Civil and Political Rights [ICCPR], adopted on 16th December 1966, 1966, Article 6(5).
66. Amnesty International, *Death Sentences and Executions in 2023*, ACT 50/7952/2024, May 2024.
67. Amnesty International, *Execution by Lethal Injection*, ACT 50/007/2007, October 2007, p. 38.
68. Amnesty International, *Death Sentences and Executions in 2013*, Act 50/001/2014, 2013.
69. Amnesty International, *Death Sentences and Executions in 2013*, Act 50/001/2014, 2013.
70. Amnesty International, *Death Sentences and Executions in 2009, ACT 50/001/2010*, March 2010, p. 6.
71. Amnesty International, *Death Sentences and Executions in 2009, ACT 50/001/2010*, March 2010, p. 6.
72. Amnesty International, *Death Sentences and Executions in 2023*, ACT 50/7952/2024, May 2024.
73. Amnesty International, *Death Sentences and Executions in 2023*, ACT 50/7952/2024, May 2024.
74. Amnesty International, *Death Sentences and Executions in 2023*, ACT 50/7952/2024, May 2024.
75. Amnesty International, *Death Sentences and Executions in 2023*, ACT 50/7952/2024, May 2024.
76. Criminal Procedure Law of the People's Republic of China, 1st July 1979, Article 212, as cited in: Cornell University Law School, *Death Penalty Worldwide* website/Way Back Machine, https://web.archive.org/web/20150503175139/http://www.deathpenaltyworldwide.org/country-search-post.cfm?country=China (accessed 17 December 2024).
77. (a) C. MacLeod, China Makes Ultimate Punishment Mobile, USA Today, 14th June 2006, 2006; (b) C. Segura, China Injects "Humanity" into Death Sentence, Asia Times Online, 16th December 2009, 2009, Way Back Machine https://web.archive.org/web/20091218062406/http://www.atimes.com/atimes/China/KL16Ad01.html (accessed 17 December 2024).
78. Amnesty International, *Execution by Lethal Injection*, ACT 50/007/2007, October 2007, p. 17.
79. AFP, Chinese Courts Purchasing Mobile Execution Units, 18th December 2003, 2003.
80. C. MacLeod, China Makes Ultimate Punishment Mobile, *USA Today*, 14th June 2006, 2006.
81. Human Rights Watch, *China: Organ procurement and judicial execution in China*, August 1994.
82. BBC, *China to stop harvesting executed prisoners' organs*, World News, 4 December 2014. https://www.bbc.co.uk/news/world-asia-china-30324440 (accessed 17 December 2024).

83. M. Robertson and J. Lavee, Execution by organ procurement: Breaching the dead donor rule in China, *Am. J. Transplant.*, 2022, **22**(7), 1804–1812.
84. R. Hurley, China's forced organ harvesting constitutes crimes against humanity, informal London tribunal finds, *Br. Med. J.*, 2019, **18**, 14287.
85. Amnesty International, *Execution by Lethal Injection*, ACT 50/007/2007, October 2007, p. 38.
86. Death Penalty Information Center, Death Penalty Information Center, *Methods of Execution, Authorized Methods*, http://www.deathpenaltyinfo.org/methods-execution (accessed 17 December 2024) The 27 States are: Alabama, Arizona, Arkansas, California, Florida, Georgia, Idaho, Indiana, Kansas, Kentucky, Louisiana, Mississippi, Missouri, Montana, Nebraska, Nevada, North Carolina, Ohio, Oklahoma, Oregon, Pennsylvania, South Carolina, South Dakota, Tennessee, Texas, Utah, Wyoming. In addition although New Hampshire has abolished the death penalty, it is unclear whether this repeal applies retroactively, potentially leaving a prisoner on death row facing execution by lethal injection.
87. Death Penalty Information Center, *Methods of Execution, Authorized Methods*, http://www.deathpenaltyinfo.org/methods-execution (accessed 17 December 2024).
88. Death Penalty Information Center, *Overview of Lethal Injection Protocols*, https://deathpenaltyinfo.org/executions/lethal-injection/overview-of-lethal-injection-protocols (accessed 17 December 2024).
89. Arizona, Georgia, Idaho, Missouri, Ohio, South Dakota, Texas, and Washington. Death Penalty Information Center, *Overview of Lethal Injection Protocols*, https://deathpenaltyinfo.org/executions/lethal-injection/overview-of-lethal-injection-protocols (accessed 17 December 2024).
90. Arkansas, California, Kentucky, Louisiana, North Carolina, and Tennessee. Death Penalty Information Center, *Overview of Lethal Injection Protocols*, https://deathpenaltyinfo.org/executions/lethal-injection/overview-of-lethal-injection-protocols (accessed 17 December 2024).
91. BBC, *Nebraska first in US to use opioid fentanyl in execution*, World News, 14 August 2018.
92. For further discussion see, for example, Amnesty International, *Execution by Lethal Injection*, ACT 50/007/2007, October 2007, p. 38.
93. (a) World Medical Association, Resolution on Physician Participation in Capital Punishment Adopted by the 34th World Medical Assembly Lisbon, Portugal, 28th September–2nd October 1981; (b) International Council of Nurses, Torture, Death Penalty and Participation by Nurses in Executions, 2006, adopted 1998, revised 2006.
94. World Medical Association, *WMA Resolution on Organ Donation in Prisoners*, adopted by the 173rd WMA Council Session, Divonne-les-Bains, France, May and reaffirmed by the 203rd WMA Council Session, Buenos Aires, Argentina, April 2016 and revised by the 226th WMA

Council Session, Seoul, Korea, April 2024. and adopted by the 75th WMA General Assembly, Helsinki, Finland, October 2024.

95. *Council of Europe, European Convention on the Protection of Human Rights and Fundamental Freedoms (1950)*, signed on 4th November 1950, Articles 10 and 11.
96. (a) See, *e.g.*, United Nations, Universal Declaration of Human Rights, adopted and proclaimed by UN General Assembly Resolution 217 A (III), 10th December 1948, 1948, Articles 19 and 20; (b) See, *e.g.*, United Nations, International Covenant on Civil and Political Rights [ICCPR], adopted on 16th December 1966, 1966, Articles 19, 21 and 22.
97. United Nations, International Covenant on Civil and Political Rights [ICCPR], adopted on 16th December 1966, 1966, Article 21.
98. United Nations, *Basic Principles on the Use of Force and Firearms by Law Enforcement Officials [UNBP]]*, 7th September 1990, adopted by the Eighth United Nations Congress on the Prevention of Crime and the Treatment of Offenders, Havana, Cuba, 27th August–7th September 1990, Principle 13.
99. See for example: United Nations Human Rights Council Resolution, 15/21: *The rights to freedom of peaceful assembly and of association*, adopted on 30th September 2010.
100. See for example: United Nations Human Rights Council, Resolution, *25/38: Promotion and protection of human rights in the context of peaceful protests*, adopted on 11th April 2014.
101. See for example: United Nations Human Rights Council, Resolution, *25/38: Promotion and protection of human rights in the context of peaceful protests*, adopted on 11th April 2014, paragraph 3.
102. United Nations Human Rights Council, Resolution, *25/38: Promotion and protection of human rights in the context of peaceful protests*, adopted on 11th April 2014, paragraph 9.
103. United Nations Human Rights Council, *Joint report of the Special Rapporteur on the rights to freedom of peaceful assembly and of association and the Special Rapporteur on extrajudicial, summary or arbitrary executions on the proper management of assemblies*, A/HRC/31/66, 4 February 2016.
104. United Nations Human Rights Council, *Joint report of the Special Rapporteur on the rights to freedom of peaceful assembly and of association and the Special Rapporteur on extrajudicial, summary or arbitrary executions on the proper management of assemblies*, A/HRC/31/66, 4 February 2016, paragraph 55.
105. United Nations Human Rights Council, *Joint report of the Special Rapporteur on the rights to freedom of peaceful assembly and of association and the Special Rapporteur on extrajudicial, summary or arbitrary executions on the proper management of assemblies*, A/HRC/31/66, 4 February 2016, paragraph 56.
106. (a) See, *e.g.*, United Nations, Universal Declaration of Human Rights, adopted and proclaimed by UN General Assembly Resolution 217 A (III),

10th December 1948, 1948, Article 3; (b) See, *e.g.*, United Nations, International Covenant on Civil and Political Rights [ICCPR], adopted on 16th December 1966, 1966, Article 9.

107. Council of Europe, *European Convention on the Protection of Human Rights and Fundamental Freedoms (1950)*, signed on 4th November 1950, Article 5.
108. United Nations, International Covenant on Civil and Political Rights [ICCPR], adopted on 16th December 1966, Article 9(1).
109. S. Casey-Maslen, Non-kinetic-energy weapons termed "non-lethal", *A Preliminary Assessment under International Humanitarian Law and International Human Rights Law*, Geneva Academy, 2010, p. 34.
110. S. Casey-Maslen, Non-kinetic-energy weapons termed "non-lethal", *A Preliminary Assessment under International Humanitarian Law and International Human Rights Law*, Geneva Academy, 2010, p. 34.
111. See for example NATO, *Standardisation Agency, Glossary of Terms and Definitions*, AAP-6(2010), p. 2-1-1, 22nd March 2010. Although the same term is retained, the definition removes reference to chemical agents, ie. "incapacitating agent" defined as "An agent that produces temporary physical and/or mental disabling conditions which may persist for hours or days after exposure has ceased." See: NATO *Glossary of Terms and Definitions*, AAP-06 Edition 2020, undated.
112. (a) See, *e.g.*, United Nations, Universal Declaration of Human Rights, adopted and proclaimed by UN General Assembly Resolution 217 A (III), 10th December 1948, 1948, Article 5; (b) United Nations, International Covenant on Civil and Political Rights [ICCPR], adopted on 16th December 1966, 1966, Article 7; (c) United Nations, Convention against Torture and Other Cruel, Inhuman or Degrading Treatment or Punishment, adopted by UNGA Resolution 39/46, 10th December 1984.
113. See for example: *European Convention for the Prevention of Torture and Inhuman or Degrading Treatment or Punishment*, Strasbourg, 26th November 1987.
114. (a) See, *e.g.*, United Nations, International Covenant on Civil and Political Rights [ICCPR], adopted on 16th December 1966, 1966, Articles 4 and 7; (b) United Nations, Human Rights Committee, General Comment Number 29: States of Emergency (Article 4), UN doc. CCPR/C/21/Rev.1/Add.11, 21 August 2001, paragraph 7.
115. United Nations, *Code of Conduct for Law Enforcement Officials [UNCoC]*, adopted by United Nations General Assembly Resolution 34/169 of 17th December 1979, Article 5.
116. United Nations, *Convention against Torture and Other Cruel, Inhuman or Degrading Treatment or Punishment*, adopted by UNGA Resolution 39/46, 10th December 1984, Article 1(1).
117. Rome Statute of the International Criminal Court, (1998) UN doc. A/CONF.183/9 of 17th July 1998. Article 8(2) (a) (ii), as cited, in *Customary International Humanitarian Law Study*, ed. J. M. Henckaerts and

L. Doswald-Beck, Cambridge University Press., 2005, vol. 1, Rules, Rule 90.

118. European Commission of Human Rights, Greek case, *(cited in Vol. II, Ch. 32, §1339), as cited in*, J. M. Henckaerts and L. Doswald-Beck, Cambridge, Cambridge University Press., 2005, vol. 1, Rules, Rule 90.
119. United Nations, *Body of Principles for the Protection of All Persons under Any Form of Detention or Imprisonment*, adopted by UNGA Resolution 43/173 of 9th December 1988, Principle 6, Commentary.
120. United Nations, *Torture and other cruel, inhuman or degrading treatment*, Report of the Special Rapporteur on the question of torture, Manfred Nowak, UN doc. E/CN.4/2006/6, 23rd December 2005, paragraph 38.
121. W. Aceves, Human rights law and the use of incapacitating biochemical weapons, in *Incapacitating Biochemical Weapons*, ed. A. Pearson, M. Chevrier and M. Wheelis, Lexington Books, Lanham, MD., 2007, p. 271
122. Royal Society, *Brain Waves Module 3, Neuroscience, Conflict and Security*, Royal Society, London, 2012, p. 24.
123. Royal Society, *Brain Waves Module 3, Neuroscience, Conflict and Security*, Royal Society, London, 2012, p. 24, note 78.
124. D. Fidler, Incapacitating chemical and biochemical weapons and law enforcement under the chemical weapons convention, in *Incapacitating Biochemical Weapons*, ed. A. Pearson, M. Chevrier and M. Wheelis, Lexington Books, Lanham, MD, 2007, p. 176.
125. D. Fidler, Incapacitating chemical and biochemical weapons and law enforcement under the chemical weapons convention, in *Incapacitating Biochemical Weapons*, ed. A. Pearson, M. Chevrier and M. Wheelis, Lexington Books, Lanham, MD, 2007, p. 176.
126. (a) See, *e.g.*, British Medical Association, *Medicine Betrayed: The Participation of Doctors in Human Rights Abuses*, Zed Books, London, 1992, pp. 64–72; (b) Amnesty International, *Prisoners of Conscience in the USSR: Their Treatment and Conditions*, AI Publications, London, 1980; (c) S. Bloch and P. Reddaway, *Soviet Psychiatric Abuse: The Shadow over World Psychiatry*, Gollancz, London, 1984.
127. Human Rights Watch and Geneva Initiative on Psychiatry, *Dangerous Minds: Political Psychiatry in China Today and its Origins in the Mao Era*, 2002.
128. Human Rights Watch and Geneva Initiative on Psychiatry, *Dangerous Minds: Political Psychiatry in China Today and its Origins in the Mao Era*, 2002.
129. United Nations Human Rights Council, *Seventh Session, Report of the Special Rapporteur on torture and other cruel, inhuman or degrading treatment or punishment, Manfred Nowak, Addendum, Summary of information, including individual cases, transmitted to Governments and replies received*, A/HRC/7/3/Add.1, 19 February 2008, paragraph 123, p. 149.
130. Amnesty International, Fear for safety/torture/ill-treatment/arbitrary detention: Bui Thi Kim Thanh, Urgent Action 316/06, ASA 41/011/2006, 24 November 2006.

131. (a) Human Rights Watch, List of Imprisoned Uzbek Defenders and Activists, 11th December 2008; (b) Human Rights Watch, Uzbekistan: Imprisoned Activists' Health in Danger UN Body Should Urge their Immediate Release, End to Repression, 11th December 2008.
132. Human Rights Watch, *"Fading Away", How Aged Care Facilities in Australia Chemically Restrain Older People with Dementia*, 15 October 2019.
133. Human Rights Watch, *"They Stay until They Die" A Lifetime of Isolation and Neglect in Institutions for People with Disabilities in Brazil*, 23 May 2018.
134. Human Rights Watch, *Kazakhstan: Children in Institutions Isolated, Abused*, 17 July 2019.
135. Human Rights Watch, *Nigeria: People with Mental Health Conditions Chained, Abused, 11 November* 2019.
136. Y. Gorbunova and Human Rights Watch, *Held in Chains in Russia, Investigate Abuse Allegations* at the 'Mercy' Wing, 8 August 2017.
137. Human Rights Watch, "It is My Dream to Leave This Place", Children with Disabilities in Serbian institutions, 8 June 2016.
138. Human Rights Watch, *"Chained Like Prisoners", Abuses Against People with Psychosocial Disabilities in Somaliland*, 25 October 2015.
139. (a) Human Rights Watch, Why Are Nursing Homes Drugging Dementia Patients Without Their Consent? (also published in Washington Post), 13 August 2018; (b) L. Labaki and Human Rights Watch, Migrant kids are being traumatized, not treated for mental health needs, (also published in The Sacramento Bee), 25 July 2018.
140. European Committee for the Prevention of Torture and Inhuman or Degrading Treatment or Punishment (CPT), Means of restraint in psychiatric establishments for adults (Revised CPT standards), CPT/Inf (2017)6, 21 March 2017, paragraph 37.
141. Albania, Armenia, Azerbaijan, Bosnia and Herzegovina, Croatia, Cyprus, Czech Republic, Denmark, Germany, Greece, Hungary, Iceland, Italy, Latvia, Lithuania, Moldova, Montenegro, Netherlands, Norway, Poland, Portugal, Russian Federation, Serbia, Slovak Republic, Ukraine, and the UK. For details of relevant CPT country reports see CPT *HUDOC Database*, available at https://www.coe.int/en/web/cpt/database.
142. CPT, *Report to the Azerbaijani Government on the visit to Azerbaijan carried out by the CPT from 29 March to 8 April 2016*, CPT/Inf (2018) 35, 18 July 2018, paragraph 196.
143. CPT, *Report to the Greek Government on the visit to Greece carried out by the CPT from 10 to 19 April 2018*, CPT/Inf (2019) 4, 19 February 2019, paragraph 36.
144. CPT, *Report to the Italian Government on the visit to Italy carried out by the CPT from 8 to 21 April 2016*, CPT/Inf (2017) 23, 8 September 2017, paragraph 118.

145. CPT, *Report to the Portuguese Government on the visit to Portugal carried out by the CPT from 27 September to 7 October 2016*, CPT/Inf (2018) 6, 27 January 2018, paragraph 66.
146. CPT, *Report to the Government of Serbia on the visit to Serbia carried out by the CPT from 26 May to 5 June 2015*, CPT/Inf (2016) 21, 24 June 2016, paragraphs 170 & 190.
147. Human Rights Watch, *Kazakhstan: Children in Institutions Isolated, Abused*, 17 July 2019.
148. UN Human Rights Council, *Report of the special rapporteur on torture and other cruel, inhuman or degrading treatment or punishment*, Juan E. Méndez, A/HRC/22/53, 1 February 2013.
149. UN Human Rights Council, *Report of the special rapporteur on torture and other cruel, inhuman or degrading treatment or punishment*, Juan E. Méndez, A/HRC/22/53, 1 February 2013, paragraph 32.
150. UN Human Rights Council, *Report of the special rapporteur on torture and other cruel, inhuman or degrading treatment or punishment*, Juan E. Méndez, A/HRC/22/53, 1 February 2013, paragraph 89 (b).
151. UN, *Convention on the Rights of Persons with Disabilities (CRPD)*, adopted 13 December 2006, UN General Assembly Resolution 61/106, A/61/49, 2006.
152. UN, *Convention on the Rights of Persons with Disabilities (CRPD)*, adopted 13 December 2006, UN General Assembly Resolution 61/106, A/61/49, 2006, Article 1.
153. UN Committee on the Rights of Persons with Disabilities, Concluding observations on the initial report of Australia, adopted by the Committee at its tenth session, 2-13 September, CRPD/C/AUS/CO/1, 2013, paragraphs 35 & 36.
154. M. Dando and M. Furmanski, Midspectrum incapacitant programs, in *Deadly Cultures: Biological Weapons since 1945*, ed. M. Wheelis, L. Rózsa and M. Dando, Harvard University Press, Cambridge, MA, 2006, pp. 243–244.
155. J. Perry Robinson, *Disabling Chemical Weapons: A Documented Chronology of Events, 1945–2003*, Unpublished monograph, 1st November 2003, copy given to author. For US research see also US Senate, Committee on Intelligence and Human Resources Subcommittee on Health and Scientific Research, Joint Hearing, *Project MKULTRA, the CIA's Program of Research in Behavioral Modification*, 3rd August 1977.
156. (a) S. Acharya, Is Narco Analysis a Reliable Science? Present Legal Scenario in India, Legal Service India, 19th February 2008, http://www.legalserviceindia.com/article/l176-Narco-Analysis.html (accessed 14th January 2011); (b) Mumbai Attacks: Militant Kept in Underwear to Prevent Suicide, Daily Telegraph, 8th December 2008.
157. (a) Amnesty International, Turkmenistan: Individuals Continue to be at Risk of Violations in Turkmenistan, AI Index: EUR 61/001/2009, 12th February, 2009; (b) Human Rights Watch, Turkmenistan: Open Letter from a Coalition of Human Rights Organizations, 17th July 2006.

158. Human Rights Watch, "*We Need a Law for Liberation" Gender, Sexuality, and Human Rights in a Changing Turkey*, May 2008.
159. (a) J. Warrick, Detainees allege being drugged, Questioned. U.S. Denies Using Injections for Coercion, Washington Post, 22nd April 2008; (b) UN Commission on Human Rights, Situation of Detainees at Guantánamo Bay, 27th February 2006, UN doc. E/CN.4/2006/120. The report was prepared jointly by: the Chairperson-Rapporteur of the Working Group on Arbitrary Detention, the Special Rapporteur on the independence of judges and lawyers, the Special Rapporteur on torture and other cruel, inhuman or degrading treatment or punishment, the Special Rapporteur on freedom of religion or belief and the Special Rapporteur on the right of everyone to the enjoyment of the highest attainable standard of physical and mental health. For details of the US Government position on this issue, see: United States Department of Defense, Deputy Inspector General for Intelligence, Investigation of Allegations of the Use of Mind-Altering Drugs to Facilitate Interrogations of Detainees, 23rd September 2009. https://media.defense.gov/2017/Oct/31/2001836211/-1/-1/1/09-INTEL-13.PDF (accessed on 15 February 2025); (c) J. Kaye and J. Leopold, DoD Report Reveals Some Detainees Interrogated While Drugged, Others "Chemically Restrained", Truthout, 11 July 2012; (d) see also: R. Nevin, Mass administration of the antimalarial drug mefloquine to Guantánamo detainees: A critical analysis, *Trop. Med. Int. Health*, 2012, **17**(10), 1281–1288. (e) CBS News, CIA considered potential truth serum to force terror suspects to talk, 13 November 2018, https://www.cbsnews.com/news/project-medication-cia-considered-using-versed-to-force-terror-suspects-to-talk/ (accessed 15 February 2025).
160. United Nations, *International Covenant on Civil and Political Rights [ICCPR]*, adopted on 16th December 1966, 1966, Article 7.
161. United Nations, Body of Principles for the Protection of All Persons under Any Form of Detention or Imprisonment (UN Principles for the Protection of Detainees), adopted by UN General Assembly, UNGA Resolution 43/173 on 9 December 1988, 2004, Principle 6.
162. Such practices would also breach United Nations, *International Covenant on Civil and Political Rights [ICCPR]*, adopted on 16th December 1966, 1966, Article 14 (g) protecting against self-incrimination.
163. United Nations, Body of Principles for the Protection of All Persons under Any Form of Detention or Imprisonment (UN Principles for the Protection of Detainees), adopted by UN General Assembly, UNGA Resolution 43/173 on 9 December 1988, 2004, Principle 21.
164. UN, UNGA Resolution 37/194, Principles of Medical Ethics, adopted on 18th December 1982.
165. UN, UNGA Resolution 37/194, Principles of Medical Ethics, adopted on 18th December 1982, Principle 3.
166. UN, UNGA Resolution 37/194, Principles of Medical Ethics, adopted on 18th December 1982, Principle 4.

167. UN, UNGA Resolution 37/194, Principles of Medical Ethics, adopted on 18th December 1982, Principle 6.
168. US Army, *Field Manual FM 2–22.3 (FM 34–52): Human Intelligence Collector Operations*, Headquarters, Department of the Army, 2006, p. 102.
169. Supreme Court of India, Criminal Appellate Jurisdiction, Selvi v. *State of Karnataka & Anr.*, Criminal Appeal Number 1267 of 2004, 5th May 2010, paragraph 223. In its ruling the Court did, however "leave room for the voluntary administration of the impugned techniques in the context of criminal justice, provided that certain safeguards are in place" and in very limited circumstances. The Court ruled that even when the subject had given consent to undergo narcoanalysis, "the test results by themselves cannot be admitted as evidence because the subject does not exercise conscious control over the responses during the administration of the test". However, the Court did rule that "any information or material that is subsequently discovered with the help of voluntary administered test results can be admitted". Although this is a narrow loophole, it does appear that the practice of narcoanalysis where voluntary consent has been given continues. See for example: Lie Test Hints at Internal Plot Behind, *The Indian Express*, 18th July 2013; Murder Case: Accused to Seek Narco-Analysis Test, *The Hindu*, 22nd March 2015.
170. Supreme Court of India, Criminal Appellate Jurisdiction, Selvi v. *State of Karnataka & Anr.*, Criminal Appeal Number 1267 of 2004, 5th May 2010, paragraph 195.
171. See: US Supreme Court, *Townsend v. Sain*, 372 U.S. 293 (1963), pp. 307–309.
172. Organisation of American States, Inter-American Convention to Prevent and Punish Torture, adopted at Cartagena de Indias, Colombia, on 9 December 1985, at the fifteenth regular session of the OAS General Assembly.
173. United Nations, *International Covenant on Civil and Political Rights [ICCPR]*, adopted on 16th December 1966, 1966, Article 19.
174. M. Nowak, *U.N. Covenant on Civil and Political Rights: CCPR Commentary*, 2nd edn, N. P. Engel, Kehl, 2005, p. 340.
175. It has also been termed ex-gay therapy, gay cure, reparative therapy, reintegrative therapy, reorientation therapy, and sexual orientation change effort (SOCE).
176. United Nations, Ending violence and discrimination against lesbian, gay, bisexual, transgender and intersex people, 15 September 2015.
177. D. Bhugra, K. Eckstrand, P. Levounis, A. Kar and K. R. Javate, WPA position statement on gender identity and same-sex orientation, attraction and behaviours, *World Psychiatry*, 2016, **15**(3), 299–300.
178. China, Ecuador, El Salvador, France, India, Iran, Panama, Russia, Sri Lanka, Turkey, Uganda, United Arab Emirates, United States of America, Vietnam, see: International Rehabilitation Council for Torture

Victims (IRCT), *Its torture not therapy, A global overview of conversion therapy: practices, perpetrators and the role of States*, 2020, p. 8.

179. Human Rights Watch, *"Have You Considered Your Parents' Happiness?" Conversion Therapy Against LGBT People in China*, 15 November, 2017.
180. United Nations Development Programme and United States Agency for International Development, *Being LGBT in Asia: The Vietnam Country Report*, 2014.
181. International Rehabilitation Council for Torture Victims (IRCT), *Its torture not therapy, A Global overview of conversion therapy: practices, perpetrators and the role of States*, 2020, p. 8.
182. Human Rights Watch, *"Have You Considered Your Parents' Happiness?" Conversion Therapy Against LGBT People in China*, 15 November, 2017.
183. International Rehabilitation Council for Torture Victims (IRCT), *Its torture not therapy, A Global overview of conversion therapy: practices, perpetrators and the role of States*, 2020, p. 8.
184. UN General Assembly, Extra-custodial use of force and the prohibition of torture and other cruel, inhuman or degrading treatment or punishment,2017A/72/178, 20 July 2017. https://digitallibrary.un.org/record/1302624/files/A_72_178-EN.pdf.
185. UN General Assembly, Extra-custodial use of force and the prohibition of torture and other cruel, inhuman or degrading treatment or punishment, 2017A/72/178, 20 July 2017. https://digitallibrary.un.org/record/1302624/files/A_72_178-EN.pdf, Paragraph 62(d).
186. UN General Assembly, Extra-custodial use of force and the prohibition of torture and other cruel, inhuman or degrading treatment or punishment, 2017A/72/178, 20 July 2017. https://digitallibrary.un.org/record/1302624/files/A_72_178-EN.pdf, paragraph 57.
187. See, *e.g.*, United Nations, *Universal Declaration of Human Rights*, adopted and proclaimed by UN General Assembly Resolution 217 A (III), 10th December 1948, 1948, Article 25.
188. UN, International Covenant on Economic, Social and Cultural Rights, adopted by UNGA Resolution 2200 (XXI), 1966, 16th December 1966, Article 12.
189. UN, *Committee on Economic, Social and Cultural Rights, General Comment No. 14: The Right to the Highest Attainable Standard of Health*, E/C.12/2000/4, 11th August 2000, paragraph 50.
190. UN, *Committee on Economic, Social and Cultural Rights, General Comment No. 14: The Right to the Highest Attainable Standard of Health*, E/C.12/2000/4, 11th August 2000, paragraph 51.
191. S. Casey-Maslen, Non-kinetic-energy weapons termed "non-lethal", *A Preliminary Assessment under International Humanitarian Law and International Human Rights Law*, Geneva Academy, 2010, p. 35.
192. S. Casey-Maslen, Non-kinetic-energy weapons termed "non-lethal", *A Preliminary Assessment under International Humanitarian Law and International Human Rights Law*, Geneva Academy, 2010, p. 36.

193. See, for example, British Medical Association, *The Use of Drugs as Weapons*, May 2007.
194. See, for example, British Medical Association, *The Use of Drugs as Weapons*, May 2007.
195. Human Rights Watch, press release: *Independent Commission of Inquiry Must Investigate Raid on Moscow Theater,* 2002.
196. UN Commission on Human Rights, Situation of Detainees at Guantánamo Bay, 27th February 2006, UN doc. E/CN.4/2006/120.
197. UN Commission on Human Rights, Situation of Detainees at Guantánamo Bay, 27th February 2006, UN doc. E/CN.4/2006/120. paragraph 70.
198. (a) UN Commission on Human Rights, Situation of Detainees at Guantánamo Bay, 27th February 2006, UN doc. E/CN.4/2006/120. paragraph 70. Although the Commission report did not elaborate on the nature of the drugs administered nor the purpose of such actions, a number of former detainees have alleged that injections of unknown drugs occurred before interrogation and were intended to coerce confessions; (b) see: J. Warrick, Detainees allege being drugged, Questioned. U.S. Denies Using Injections for Coercion, Washington Post, 22nd April 2008.
199. UN Commission on Human Rights, Situation of Detainees at Guantánamo Bay, 27th February 2006, UN doc. E/CN.4/2006/120. paragraph 75.
200. A. Mannix, At urging of Minneapolis police, Hennepin EMS workers subdued dozens with a powerful sedative, *Star Tribune*, 2018.
201. A. Mannix, Patients sedated by ketamine were enrolled in Hennepin Healthcare study, *Star Tribune*, 2018.
202. R. Solomon and M. de Yoanna, Medics in Colorado dosed 902 people with ketamine for "excited delirium" in 2.5 years, including Elijah McClain, *The Colorado Sun*, 23 June 2020. Available at https://coloradosun.com/2020/07/23/ketamine-use-paramedics-elijah-mclain/ (accessed 30 January 2023).
203. P. Neiberg and Associated Press, Ketamine that's injected during arrests draws new scrutiny, 22 August 2020.
204. American Society of Anesthesiologists, *Statement on the Use of Ketamine for a Non-medical Purpose*, 15 July 2020. Available at https://www.asahq.org/about-asa/newsroom/news-releases/2020/07/asa-statement-on-the-use-of-ketamine-for-a-non-medical-purpose (accessed 15 February 2025).
205. American Medical Association (AMA), *New AMA policy opposes "excited delirium" diagnosis, Press Release*, 14 June 2021. Available at https://www.ama-assn.org/press-center/press-releases/new-ama-policy-opposes-excited-delirium-diagnosis (accessed 15 February 2025).
206. Protocol Additional to the Geneva Conventions of 12th August 1949, and relating to the *Protection of Victims of International Armed Conflicts [Protocol I]*, 8th June 1977, Article 36.

207. United Nations (1990) *Basic Principles on the Use of Force and Firearms by Law Enforcement Officials [UNBP]]*, 7th September 1990, adopted by the Eighth United Nations Congress on the Prevention of Crime and the Treatment of Offenders, Havana, Cuba, 27th August–7th September, Principle 3.
208. See for example: United Nations Human Rights Council, Resolution, *25/38: Promotion and protection of human rights in the context of peaceful protests*, adopted on 11th April 2014, paragraph 15.
209. See for example: United Nations Human Rights Council, Resolution, *25/38: Promotion and protection of human rights in the context of peaceful protests*, adopted on 11th April 2014, paragraph 14.
210. United Nations, *Guidance on Less-Lethal Weapons in Law Enforcement*, 2020. www.ohchr.org/Documents/HRBodies/CCPR/LLW_Guidance.pdf (accessed 8 November 2024). p. 13.
211. United Nations, *Guidance on Less-Lethal Weapons in Law Enforcement*, 2020. www.ohchr.org/Documents/HRBodies/CCPR/LLW_Guidance.pdf (accessed 8 November 2024). p. 14.
212. United Nations (1990) *Basic Principles on the Use of Force and Firearms by Law Enforcement Officials [UNBP]]*, 7th September 1990, adopted by the Eighth United Nations Congress on the Prevention of Crime and the Treatment of Offenders, Havana, Cuba, 27th August–7th September. *Principle 1*. Which reads: "Governments and law enforcement agencies shall adopt and implement rules and regulations on the use of force and firearms against persons by law enforcement officials. In developing such rules and regulations, Governments and law enforcement agencies shall keep the ethical issues associated with the use of force and firearms constantly under review."
213. United Nations, *Basic Principles on the Use of Force and Firearms by Law Enforcement Officials [UNBP]]*, 7th September 1990, adopted by the Eighth United Nations Congress on the Prevention of Crime and the Treatment of Offenders, Havana, Cuba, 27th August–7th September 1990, Principle 20.
214. United Nations, *Guidance on Less-Lethal Weapons in Law Enforcement*, 2020. www.ohchr.org/Documents/HRBodies/CCPR/LLW_Guidance.pdf (accessed 8 November 2024). p. 15
215. United Nations, *Guidance on Less-Lethal Weapons in Law Enforcement*, 2020. www.ohchr.org/Documents/HRBodies/CCPR/LLW_Guidance.pdf (accessed 8 November 2024). p. 17.
216. United Nations, *Guidance on Less-Lethal Weapons in Law Enforcement*, 2020. www.ohchr.org/Documents/HRBodies/CCPR/LLW_Guidance.pdf (accessed 8 November 2024). p. 18.
217. United Nations, *Basic Principles on the Use of Force and Firearms by Law Enforcement Officials [UNBP]]*, 7th September 1990, adopted by the Eighth United Nations Congress on the Prevention of Crime and the Treatment of Offenders, Havana, Cuba, 27th August–7th September 1990, Article 7.

218. See for example: United Nations Human Rights Council, Resolution, *25/38: Promotion and protection of human rights in the context of peaceful protests*, adopted on 11th April 2014, paragraph 12.
219. Wassenaar Arrangement (undated) Frequently Asked Questions, https://www.wassenaar.org/about-us/#faq
220. Wassenaar Arrangement, *Basic Document, Compiled by WA Secretariat*, January 2015. All WA documents cited are available from the WA website, https://www.wassenaar.org/
221. Wassenaar Arrangement, website, *Introduction*, undated.
222. Wassenaar Arrangement, *List of Dual Use Goods and Technologies and Munitions List*, WA-LIST WA-LIST (23) 1, 1 December 2023.
223. Wassenaar Arrangement, *List of Dual Use Goods and Technologies and Munitions List*, WA-LIST WA-LIST (23) 1, 1 December 2023, ML 7.b.3.a
224. Wassenaar Arrangement, *List of Dual Use Goods and Technologies and Munitions List*, WA-LIST WA-LIST (23) 1, 1 December 2023, ML 7.a.
225. Wassenaar Arrangement, *List of Dual Use Goods and Technologies and Munitions List*, WA-LIST WA-LIST (23) 1, 1 December 2023, ML 7.a.
226. Wassenaar Arrangement, website, *Introduction*, undated.
227. Wassenaar Arrangement, (undated) Frequently Asked Questions, https://www.wassenaar.org/about-us/#faq
228. Wassenaar Arrangement, *Guidelines & and Procedures, including the Initial Elements*, Section II (scope), 2014, paragraph 3.
229. Wassenaar Arrangement, *Elements for Objective Analysis and Advice Concerning Potentially Destabilising Accumulations of Conventional Weapons*, WA Plenary approved 3rd December 1998 and amended by the WA Plenary in December 2004 and December 2011.
230. Wassenaar Arrangement, *Elements for Objective Analysis and Advice Concerning Potentially Destabilising Accumulations of Conventional Weapons*, WA Plenary approved 3rd December 1998 and amended by the WA Plenary in December 2004 and December 2011, 1.e.
231. Wassenaar Arrangement, *Elements for Objective Analysis and Advice Concerning Potentially Destabilising Accumulations of Conventional Weapons*, WA Plenary approved 3rd December 1998 and amended by the WA Plenary in December 2004 and December 2011, 1.c.
232. Council of the European Union, Common Position 2008/944/CFSP adopted on 8th December 2008, *Official Journal of the European Union L 335/99*, 13th December 2008.
233. Council of the European Union, Common Position 2008/944/CFSP adopted on 8th December 2008, *Official Journal of the European Union L 335/99*, 13th December 2008, paragraph 3.
234. Council of the European Union, Common Position 2008/944/CFSP adopted on 8th December 2008, *Official Journal of the European Union L 335/99*, 13th December 2008, paragraph 4.
235. Council of the European Union, Common Position 2008/944/CFSP adopted on 8th December 2008, *Official Journal of the European Union L 335/99*, 13th December 2008, Article 1.

236. Council of the European Union, Common Position 2008/944/CFSP adopted on 8th December 2008, *Official Journal of the European Union L 335/99*, 13th December 2008, Article 2.
237. Council of the European Union, Common Position 2008/944/CFSP adopted on 8th December 2008, *Official Journal of the European Union L 335/99*, 13th December 2008, Article 2, Criterion 2.
238. Council of the European Union, Common Position 2008/944/CFSP adopted on 8th December 2008, *Official Journal of the European Union L 335/99*, 13th December 2008, Article 2, Criterion 2.
239. Council of the European Union, Common Position 2008/944/CFSP adopted on 8th December 2008, *Official Journal of the European Union L 335/99*, 13th December 2008, Article 2.2.
240. Council of the European Union, Common Position 2008/944/CFSP adopted on 8th December 2008, *Official Journal of the European Union L 335/99*, 13th December 2008, Article 2.2.c.
241. Council of the European Union, *Common Military List of the European Union* (equipment covered by Council Common Position 2008/944/ CFSP defining common rules governing the control of exports of military technology and equipment), (*CFSP*) (*2020/C 85/01*), adopted by the Council on 17 February 2020.
242. European Communities, EC Regulation (EC) No 1334/2000 of 22nd June 2000 setting up a Community regime for the control of exports of dual-use items and technology, *Official Journal of the European Union*, EN L 159/1, 30th June 2000.
243. It should be noted that Article 12 of the EU Common Position stipulates that the EU Common Military List shall act as a reference point for Member States' national military technology and equipment lists, but shall not directly replace them.
244. Council of the European Union, *Council Common Position 2008/944/ CFSP*, 2008, paragraph 16.
245. Council of the European Union, *Common Military List (2025)*, adopted by the Council on 25 February 2025, COARM 17 CFSP/PESC 92
246. Council of the European Union, *Common Military List (2025)*, adopted by the Council on 25 February 2025, COARM 17 CFSP/PESC 92, ML.b.3(a).
247. Council of the European Union, *Common Military List (2025)*, adopted by the Council on 25 February 2025, COARM 17 CFSP/PESC 92, ML7. (a).
248. United Nations, *UNGA, Resolution* 67/234, 2nd April 2013.
249. United Nations, Office for Disarmament Affairs, *Arms Trade Treaty*, https://disarmament.unoda.org/convarms/arms-trade-treaty-2/ (accessed 19 December 2024).
250. EC Regulation, 1236/2005 of 27th June 2005 concerning trade in goods which could be used for capital punishment, torture or other cruel, inhuman or degrading treatment or punishment, published in *Official Journal of the European Union*, L200/1, 30th July 2005.
251. European Union, *Regulation (EU)* 2019/125 of 16 January 2019 concerning trade in certain goods which could be used for capital

punishment, torture or other cruel, inhuman or degrading treatment or punishment ('Anti-Torture Regulation'), 2019.

252. Council of Europe, *Recommendation CM/Rec (2021)2* of the Committee of Ministers to member States on measures against the trade in goods used for the death penalty, torture and other cruel, inhuman or degrading treatment or punishment (Adopted by the Committee of Ministers on 31 March 2021 at the 1400th meeting of the Ministers' Deputies), 2021.
253. Council of Europe, *Recommendation CM/Rec (2021)2* of the Committee of Ministers to member States on measures against the trade in goods used for the death penalty, torture and other cruel, inhuman or degrading treatment or punishment (Adopted by the Committee of Ministers on 31 March 2021 at the 1400th meeting of the Ministers' Deputies), 2021.
254. *See Statement on behalf of the Global Alliance for Torture-Free Trade by H.E. Enkhbold Vorshilov, Ambassador, Permanent Representative of Mongolia*, 22nd June 2022. The Statement lists the following Alliance members: Albania, Argentina, Armenia, Australia, Austria, Belgium, Bosnia and Herzegovina, Brazil, Bulgaria, Canada, Cape Verde, Chile, Colombia, Costa Rica, Croatia, Cyprus, Czech Republic, Denmark, Ecuador, El Salvador, Estonia, Finland, North Macedonia, France, Georgia, Germany, Greece, Hungary, Iceland, Ireland, Italy, Latvia, Liechtenstein, Lithuania, Luxembourg, Madagascar, Malta, Mexico, Moldova, Mongolia, Montenegro, The Netherlands, New Zealand, Nicaragua, Norway, Palau, Panama, Paraguay, Poland, Portugal, Romania, Serbia, Seychelles, Slovenia, Slovakia, Spain, Sweden, Switzerland, Ukraine, United Kingdom, Uruguay, Vanuatu and European Union.
255. UNGA, *Resolution* 73/304. Towards torture-free trade: examining the feasibility, scope and parameters for possible common international standards, adopted by the General Assembly on 28 June 2019.
256. UN, *Towards torture-free trade: examining the feasibility, scope and parameters for possible common international standards: Report of the Secretary-General*, A/74/969, 28 July 2020.
257. UN, A/76/850: *Towards torture-free trade: Examining the feasibility, scope and parameters for possible common international standards - Report of the Group of Governmental experts*, A/76/850, 30 May 2022.
258. United Nations, *Torture and other cruel, inhuman or degrading treatment or punishment.* Note by the Secretary-General, thematic study on the global trade in weapons, equipment and devices used by law enforcement and other public authorities that are capable of inflicting torture and other cruel, inhuman or degrading treatment or punishment, Special Rapporteur on torture and other cruel, inhuman or degrading treatment or punishment, Alice Jill Edwards, Advanced Unedited Version, A/78/324, 23 August 2023.
259. United Nations, *Torture and other cruel, inhuman or degrading treatment or punishment.* Note by the Secretary-General, thematic study on the global trade in weapons, equipment and devices used by law enforcement and other public authorities that are capable of inflicting torture and other cruel, inhuman or degrading treatment or punishment,

Special Rapporteur on torture and other cruel, inhuman or degrading treatment or punishment, Alice Jill Edwards, Advanced Unedited Version, A/78/324, 23 August 2023, Appendix 1, Category A Goods.

260. United Nations, *Torture and other cruel, inhuman or degrading treatment or punishment*. Note by the Secretary-General, thematic study on the global trade in weapons, equipment and devices used by law enforcement and other public authorities that are capable of inflicting torture and other cruel, inhuman or degrading treatment or punishment, Special Rapporteur on torture and other cruel, inhuman or degrading treatment or punishment, Alice Jill Edwards, Advanced Unedited Version, A/78/324, 23 August 2023, Appendix 2, Category B Goods.

261. F. Hampson, International law and the regulation of weapons, in *Incapacitating Biochemical Weapons*, ed. A. Pearson, M. Chevrier and M. Wheelis, Lexington Books, Lanham, MD., 2007, p. 243.

262. F. Hampson, International law and the regulation of weapons, in *Incapacitating Biochemical Weapons*, ed. A. Pearson, M. Chevrier and M. Wheelis, Lexington Books, Lanham, MD., 2007, p. 243.

CHAPTER 9

International Humanitarian Law Applicable to CNS-acting Chemical Agent Weapons

9.1 Introduction:[1] Nature of Scope of IHL

International humanitarian law (IHL) is the body of law that applies during situations of armed conflict with the aim of protecting civilians and others who are no longer participating in hostilities; and regulating the conduct of such hostilities. Amongst its provisions are those regulating the means of conflict (including the weapons employed) and also the methods of warfare (how such weapons are employed). These IHL provisions have relevance to the permissibility of development, manufacture, acquisition and use of CNS-acting chemical agent weapons. Indeed, whilst the Geneva Protocol, BTWC and CWC constitute the *lex specialis* regulating chemical and biological weapons (including CNS-acting chemical agent weapons), they do not erase the obligations deriving from IHL.[2] Such obligations will underpin and can inform the implementation of the specific arms control and disarmament instrument, particularly where such implementation is contested.

As with IHRL (discussed in Chapter 8), IHL encompasses two distinct legal components – treaty law and customary law. Treaty law is binding *only* on those States that are party to the specific treaty or binding legal international instrument concerned, and these States Parties would only be bound by such instruments if they have undertaken two separate processes of first signing and then ratifying that treaty. Although a signature alone merely indicates a future intent by a State to be bound to a treaty, any State that has signed a specific treaty is not free to undermine the objects and purposes of that treaty.[3] Treaty ratification – which can vary considerably between States – is the process whereby the Signature State incorporates approval for the treaty

Preventing Weaponization of CNS-acting Chemicals: A Holistic Arms Control Analysis
By Michael Crowley and Malcolm Dando

Published by the Royal Society of Chemistry, www.rsc.org

at the domestic level and enacts all of the necessary legislation to give domestic effect to that treaty, The second body of IHL is derived from customary international law. The Statute of the International Court of Justice has described customary international law as "a general practice accepted as law".[4] According to the international legal scholars, Henckaerts and Doswald-Beck:

> "...It is generally agreed that the existence of a rule of customary international law requires the presence of two elements, namely State practice *(usus)* and a belief that such practice is required, prohibited or allowed, depending on the nature of the rule, as a matter of law *(opinio juris sive necessitatis)...*"[5]

Where customary international law has been deemed to be established it is binding on *all* States whether or not they are parties to relevant treaties.[6] In 2005 a comprehensive and authoritative study and codification of customary IHL, undertaken by Henckaerts and Doswald-Beck, was published by the International Committee of the Red Cross (ICRC); it was subsequently transformed into a publicly accessible database which is regularly updated to reflect evolving State practice in this area.[7]

IHL is considered to apply to all international armed conflicts, that is to say to all armed conflicts that occur between two or more States, even in cases where a formal state of war has not been declared or where it has not been recognized by one of the parties. IHL is also considered to apply to every partial or total occupation of the territory of a State, even when this occupation faces no armed resistance. IHL is also applicable to armed conflicts that are of a non-international character. These are defined as conflicts between the armed forces of a State and organized armed groups, provided that such groups are under responsible command, that they control territory and that they carry out sustained and concerted military operations. Alternatively, these latter conflicts can arise between two or more such organized armed groups, providing they fulfil the conditions listed. An armed conflict is not deemed to include other circumstances involving tension, instability and turmoil within a country such as isolated and sporadic acts of violence, looting, rioting and other similar disturbances. The determination of whether a specific violent situation taking place within a country should be considered as an internal disturbance requiring law enforcement or as an internal armed conflict can be highly contentious and on occasion has been contested by the parties involved.[8] In addition to controlling the activities of States in armed conflicts, IHL is also applicable to armed non-State actors that meet the requisite criteria.[9] This extensive coverage of IHL enhances its potential applicability to CNS-acting chemical agent weapons which may be used in a variety of relevant contexts ranging from large scale military confrontations; counter-insurgency operations; halting violent disturbances in Prisoner of War (POW) camps; addressing violent unrest by the population in occupied territories; or small-scale employment against individuals. In many of these scenarios, particularly those away from direct combat on the battlefield, and where civilians will be present, international human rights law (IHRL) may also be applicable or indeed be the predominant legal framework. The inter-relation between IHL and IHRL in these areas is complicated, often contested and will be dependent on the specific (often rapidly changing) circumstances under which a particular CNS-acting chemical agent weapon is used.[10]

According to the international legal scholar, Francoise Hampson, although international armed conflicts are subject to both "more rules and more precise rules" than are non-international armed conflicts, a significant body of customary international law has developed to regulate non-international conflicts.[11] Particularly important in such developments have been the Statute of the International Criminal Court and the judgements of the *ad hoc* criminal tribunals for the former Yugoslavia and Rwanda. In this regard, the *Tadic* decision of the Appeals Chamber of the International Criminal Tribunal for the former Yugoslavia has particular significance. This stated that:

> "Elementary considerations of humanity and common sense make it preposterous that the use by States of weapons prohibited in armed conflicts between themselves be allowed when States try to put down rebellion by their own nationals on their own territory. What is inhumane, and consequently proscribed, in international wars, cannot but be inhumane and inadmissible in civil strife."[12]

9.2 Over-arching IHL Obligations Constraining Development and Use of Weapons

In addition to the arms control and disarmament agreements prohibiting development, possession and use of chemical and biological weapons (*i.e.*, Geneva Protocol, BTWC and CWC), there are a number of generally applicable IHL instruments – dealing more broadly with the conduct of armed conflict – that are of potential relevance to CNS-acting chemical agent weapons. Of particular importance are the four Geneva Conventions, which principally safeguard a range of 'protected persons' in the power of the opposing party, including wounded combatants, prisoners of war, medical personnel and civilians;[13] and two Additional Protocols[14] which address certain aspects of the conduct of armed conflict.[15] Key relevant obligations deriving from these instruments, or from related customary international law, are summarised below.

9.2.1 Discrimination and Distinction

The inter-related prohibitions against: indiscriminate weapons; attacks that do not discriminate between civilians and military objectives; and deliberate attacks on civilians, are contained in both treaty law (*e.g.*, Additional

Protocol I[16]) and customary law,[17] and are considered fundamental to IHL. These prohibitions are applicable to both international and non-international armed conflicts.

9.2.1.1 Indiscriminate Weapons

The use of all weapons which are determined by their nature to be inherently indiscriminate is banned. Whilst there is no established list of such weapons, the ICRC highlights examples of weapons cited by States as in practice being indiscriminate, in certain or all contexts, including chemical, biological and nuclear weapons; anti-personnel landmines; cluster bombs; booby traps; and certain rockets and missiles.[18] Certain aerosolised CNS-acting chemical agent weapons would arguably be considered as indiscriminate depending on the nature and precision (or lack thereof) of the relevant dispersal mechanism, quantities of agent dispersed, and area coverage. Previous reported CNS-acting chemical agent weapons and activities of concern in this regard include US BZ cluster munitions production (1960s), unconfirmed South African aerial dispersal of CNS-agents in Mozambique (1992) and the Russian development and use of an aerosolised CNS-acting weapon in the Moscow theatre siege (2002).[19] The contemporary development and proliferation of certain wide area RCA delivery mechanisms including large projectiles; aircraft, helicopter and drone dispersal systems; multi-barrel launchers capable of firing large projectile salvoes, all raise associated concerns given their potential utility for dispersing CNS-acting chemical agents at distance and/or over large areas, to affect large crowds without distinction.

9.2.1.2 Indiscriminate Attacks

For any weapons – including those promoted as being ‘less lethal’ – not considered inherently indiscriminate, their use in an indiscriminate attack is prohibited. Such attacks are defined under Article 51 of Additional Protocol I as those which: “(a) ...are not directed at a specific military objective; (b) ...employ a method or means of combat which cannot be directed at a specific military objective; or (c)... employ a method or means of combat the effects of which cannot be limited”.[20] They include attacks *inter alia* where such use “may be expected to cause incidental loss of civilian life, injury to civilians, damage to civilian objects, or a combination thereof, which would be excessive in relation to the concrete and direct military advantage anticipated”.[21] However, as Casey-Maslen notes, “different States have understood and applied this principle in markedly different ways”.[22] The use of aerosolised or gaseous CNS-acting weapons with a delivery system that disperses the agent over wide areas, particularly in situations where combatants and civilians are inter-dispersed or in close proximity, could potentially be considered indiscriminate.

9.2.1.3 Distinction

Under Article 48 of Protocol I, all those engaged in armed conflict must "at all times distinguish between the civilian population and combatants, and between civilian objects and military objectives and accordingly shall direct their operations only against military objectives".[23] The International Court of Justice has stated that the principle of distinction is one of the "cardinal principles" of international humanitarian law and one of the "intransgressible principles of international customary law".[24] Arms control and legal scholars have emphasised that in the context of armed conflict, direct attacks against civilians would be prohibited regardless of whether the weapons used are labelled lethal, 'non-lethal' or 'less lethal'.[25] Despite the comprehensive nature of this principle, Massingham has warned of the "very real possibility" that the availability of 'less lethal' weapons as an option for commanders may "contribute to a weakening of [the] prohibition on targeting civilians". This danger has been exacerbated still further by the "blurring [of] the lines between combatant and civilian… [due to the involvement in armed conflict] …of non-state actors, militia, 'terrorists', and private military and security companies".[26]

Such concerns are further heightened by attempts from proponents of CNS-acting chemical agent weapons to argue that the obligations and restrictions arising from IHL, specifically with regard to distinction, should be re-examined and potentially applied in a different manner for CNS-acting chemical agent weapons than for conventional arms.[27] These arguments could potentially undermine a comprehensive interpretation of IHL, leading to its partial or contingent implementation, thereby weakening the protection afforded to civilians in armed conflict. Highlighting the dangers of such an approach with regard to CNS-acting chemical agent weapons, Herby, then-Head of the Arms Unit in the Legal Division of the ICRC, in 2007 observed that in deliberations concerning possible scenarios for employment of these weapons, "situations in which civilians are interspersed with combatants are consistently mentioned". He consequently warned that "It is likely that the use of [CNS-acting chemical agent] incapacitants will lower the threshold for attacks that affect civilians and combatants without distinction, with an inherent risk that this rule will be undermined".[28]

9.2.2 Superfluous Injury or Unnecessary Suffering

The prohibition against superfluous injury or unnecessary suffering (SIRUS) has been enunciated in a number of IHL treaties[29] and is considered to be part of customary IHL.[30] According to this rule, "it is prohibited to cause unnecessary suffering to combatants: it is accordingly prohibited to use weapons causing them such harm or uselessly aggravating their suffering".[31] Unnecessary suffering has in turn been defined as "a harm greater than that unavoidable to achieve legitimate military objectives".[32] The 2005 ICRC study noted that, although the existence of the SIRUS prohibition is not contested, “views differ as to whether the rule itself renders a weapon illegal or whether a weapon is illegal only if a specific treaty or customary rule prohibits its use”.[33] Furthermore, there is no international consensus regarding an objective means of determining what constitutes SIRUS, nor on the criteria which can be used to judge whether specific weapons potentially breach the SIRUS prohibition.[34]

Whether a weapon causes “superfluous injury” or “unnecessary suffering” can in theory be established in two different ways. The ‘design-based approach’ refers to the design and intended purpose of a weapon as such without taking into consideration the harm actually inflicted on the military target. In contrast the ‘effects-based approach’ refers exclusively to the effects caused by a weapon used in a particular scenario and the concrete injury inflicted, independent of the actual or purported intent for which the weapon was developed. According to Hoffberger-Pippan “Over the course of time…the effects-based approach prevailed and is nowadays considered to reflect customary law”.[35]

In 1996, the ICRC established the SIRUS Project, which attempted to develop objective criteria in this area. Where previous application of the SIRUS prohibition had examined the weapon’s intended design or nature, or the purported intent of States employing that weapon, the SIRUS Project proposed examining a weapons’ actual effects, drawn from analysis of data on available medical treatment and survival rates. In 1997, the SIRUS Project proposed that:

> “[W]hat constituted ‘superfluous injury and unnecessary suffering’ should be determined by design-dependent, foreseeable effects of weapons when they are used against human beings and cause[d]:
>
> - specific disease, specific abnormal physiological state, specific abnormal psychological state, specific and permanent disability or specific disfigurement; or
> - field mortality of more than 25% or hospital mortality of more than 5%; or
> - Grade 3 wounds as measured by the Red Cross wound classification; or,
> - effects for which there is no well-recognized and proven treatment.”[36]

A number of States were highly critical of the SIRUS Project’s methodology[37] and in particular criticized the study for not adequately taking into account other relevant aspects, such as military necessity. Consequently, the SIRUS Project was terminated in 2001.[38] It did however gain important backing from elements of the medical community, notably the World Medication Association,[39] and the present authors believe that its methodology and conclusions merit re-consideration. Consequently, States should formally assess whether development (or acquisition) and use of CNS-acting chemical agent weapons would breach the SIRUS prohibition, and we believe should give particular attention to the applicability of Criteria 1 and 4. In doing so, States would need to comprehensively examine the serious and potentially life-threatening health risks to those targeted or otherwise affected. The ICRC has previously underlined the SIRUS prohibition's significance when considering the lawfulness of CNS-acting chemical agent weapons, and the importance of examining the likely effects of these weapons in practice. In 2007, Herby, argued that "'non-lethal' biochemical weapons" [*i.e.*, CNS-acting chemical agent weapons] should not be simply assumed to "merely incapacitate by making a person sleep".[40] Instead, he highlighted the risks that such weapons, if they resulted in effects such as "lifelong epileptic convulsions, permanent damage to internal organs, long-term and severe vomiting, or an extended coma" would violate the SIRUS prohibition.[41]

Though State application of SIRUS to CNS-acting chemical agent weapons is unclear and definitely not uniform, it appears that certain States may consider the SIRIUS prohibition covering these weapons. According to an analysis by Hoffberger-Pippan:

> "The military manuals of Australia of 1994 and Germany of 1992 cite *inter alia* chemical weapons as examples of weapons causing superfluous injury or unnecessary suffering. Their use is considered as prohibited and under no circumstances justifiable by invoking any considerations of military necessity... According to the military manuals of Australia and Germany, chemical weapons do not require a weighing between the injuries inflicted and the relevant military advantage anticipated. *This principle at least applies to [CNS-acting chemical agent] incapacitants...* (emphasis added)."[42]

9.2.3 Identification and Protection of Persons Considered *Hors de Combat*

According to the *ICRC Case Book*, under IHL a combatant is *hors de combat* if: he is in the power of an adverse party; he clearly expresses an intention to surrender; or he has been rendered unconscious or is otherwise incapacitated by wounds or sickness, and is therefore incapable of defending himself. Furthermore, "provided that in any of these cases he abstains from any hostile act and does not attempt to escape, he may not be made the object of attack. A fundamental rule of international humanitarian law is that persons who are *hors de combat* must not be attacked and must be treated humanely".[43]

The ICRC has raised concerns that the use in armed conflict of CNS-acting chemical agent weapons which have less visible signs of injury than conventional weapons, would make it very difficult and in some cases impossible to determine whether a combatant is *hors de combat* and should consequently be given "protection and assistance".[44] The effects of certain 'less lethal' weapons – including CNS-acting chemical agent weapons – may also prevent the person who is incapacitated from indicating their intention to surrender. This situation may be further complicated by uncertainties in determining the extent and duration that these individuals will remain incapacitated, and whether on recovery they will re-engage in hostilities. In practice, given such uncertainties over their status and intentions, these incapacitated individuals risk continued attack.[45] Consequently, the ICRC has argued in the case of CNS-acting chemical agent weapons that "[I]t would be difficult to train soldiers to distinguish whether an enemy were incapacitated or remained a threat. The resulting combination of incapacitants and lethal force *could significantly increase the lethality of armed conflicts.*" (Emphasis added)[46]

A further concern has been the use of CNS-acting chemicals against former combatants and civilians under the power of a military force. Under Common Article 3 of the Geneva Conventions "[P]ersons taking no active part in the hostilities, including members of armed forces who have laid down their arms and those placed *hors de combat* by sickness, wounds, detention, or any other cause, shall in all circumstances be treated humanely".[47] Furthermore, Common Article 3 specifically prohibits certain actions against such persons, including "(a) violence to life and person, in particular murder of all kinds, mutilation, cruel treatment and torture; (b) taking of hostages; (c) outrages upon personal dignity, in particular humiliating and degrading treatment".[48] There is also a positive obligation to ensure that "the wounded and sick shall be collected and cared for".[49] The provisions of Common Article 3 are judged by the International Court of Justice to reflect "elementary considerations of humanity";[50] they are deemed to be part of international customary humanitarian law.[51]

The UK Royal Society has voiced its disquiet concerning the possible application of CNS-acting chemicals in the coercive treatment of enemy combatants that have been rendered *hors de combat*. The Royal Society underlined IHL's significance in preventing such activities, noting that the prohibition against "outrages upon personal dignity, in particular humiliating and degrading treatment [as enunciated under both Common Article 3 of the Geneva Conventions and also under Additional Protocol I] could extend ... to the use of neuropharmacological agents to control or alter behaviour".[52] In addition, the Society contended that "the use of neuropharmacological agents or stimulation technologies to coerce or interrogate prisoners of war" would also be prohibited under the Third Geneva Convention.[53]

Such concerns have also been addressed by certain States. For example, the *US Army Field Manual,* while describing what it considers legitimate techniques for interrogation of prisoners of war, stressed that:

> "The psychological techniques and principles in this manual should neither be confused with, nor construed to be synonymous with,

unauthorized techniques *such as brainwashing, physical or mental torture, including drugs that may induce lasting or permanent mental alteration or damage.* Physical or mental torture and coercion revolve around eliminating the source's free will, and are expressly prohibited by [the Geneva Conventions]."[54] (Emphasis added).

9.2.4 Preventing Malign Application of Medicine

Article 11 of Additional Protocol I protects "the physical or mental health and integrity of persons who are in the power of the adverse Party or who are interned, detained or otherwise deprived of liberty...". It specifically prohibits such individuals from being subjected to "any medical procedure which is not indicated by the state of health of the person concerned and which is not consistent with generally accepted medical standards".[55] This prohibition notably includes "medical or scientific experiments", even if such procedures were carried out with the person's consent.[56] The strength of these prohibitions is underlined under Article 11.4. which states that "Any wilful act or omission which seriously endangers the physical or mental health or integrity of any person who is in the power of a Party...shall be a grave breach this Protocol".[57] Further protection to the provision of appropriate medical care is provided by Article 10 of Additional Protocol II, which states that "Persons engaged in medical activities shall neither be compelled to perform acts or to carry out work contrary to ... the rules of medical ethics or other rules designed for the benefit of the wounded and sick, or this Protocol".[58] Drawing upon these prohibitions and protections, the Royal Society has reasoned that "[T]he involvement of medical professionals in the administration of neuropharmacological agents for purposes other than those consistent with generally accepted medical practice is therefore prohibited by international law".[59] Consequently, medical involvement in administering CNS-acting chemicals in forced interrogations or for other ill-treatment against individual or groups of prisoners or detainees would be prohibited. Furthermore, it can be argued that medical involvement in the development and employment of CNS-acting chemical agent weapons against such protected individuals would also be prohibited under IHL.

9.2.5 Respect of International Humanitarian Law

Common Article 1 to the four Geneva Conventions, and Article 1 of Additional Protocol I, states that: "The High Contracting Parties undertake to respect and to ensure respect" for the relevant treaties "in all circumstances".[60] However, concerns have been raised by medical and scientific organisations that the employment of certain 'less lethal' weapons and technologies (notably those incorporating a range of CNS-acting or neuropharmacological agents) could potentially impair or degrade the cognitive ability of enemy combatants, with the consequent dangers of their committing serious breaches of IHL. In this regard, the Royal Society has maintained that "degrading the cognitive abilities of an adversary such that they are unable to distinguish between military targets and civilians, which often require a high degree of concentration, will undermine this requirement".[61] This is because this cognitive impairment could easily result in an unintended attack on one's own civilians or other persons or places that are specifically protected by IHL. In addition, the Royal Society has argued that "[S]uch attacks could not be prosecuted because the perpetrators will have been rendered mentally incapable of being responsible for the offences".[62]

9.2.6 The Martens Clause

The "Martens clause", named after Fyodor Fyodorovich Martens, a diplomat and jurist of the Russian Empire, who introduced it for the first time into the Preamble of the 1899 Hague Convention,[63] is today incorporated, in whole or in part, into a number of IHL treaties.[64] It has now acquired the status of customary international law, as recognized notably by the International Court of Justice.[65] As formulated in Additional Protocol I, the clause declares that: "[I]n cases not covered by this Protocol or by other international agreements, civilians and combatants remain under the protection and authority of the principles of international law derived from established custom, from the principles of humanity and from dictates of public conscience".[66]

According to the 2006 *ICRC Guide to the legal review of new weapons, means and methods of warfare*, "A weapon which is not covered by existing rules of international humanitarian law would be considered contrary to the Martens clause if it is determined *per se* to contravene the principles of humanity or the dictates of public conscience".[67] The International Court of Justice, in its 1996 Advisory Opinion on the *Legality of the Threat or Use of Nuclear Weapons,* confirmed the importance of the Martens clause, "whose continuing existence and applicability is not to be doubted",[68] and further noted that the clause "had proved to be an effective means of addressing rapid evolution of military technology".[69]

Although the Martens clause is considered one of the cornerstones of IHL, the interpretation, applicability (including with regard to review of new weapons) and substantive force of the clause are contested.[70] Consequently, the current authors would argue that the international community should first seek to employ existing weapons-specific arms control and disarmament mechanisms and regimes, to effectively prohibit or severely constrain the development, acquisition and employment of CNS-acting chemical agent weapons. However if such routes have been tried and found to be unsuccessful, then application of the Martens clause together with relevant IHL principles examined in this chapter could, *in extremis*, provide a possible mechanism to prevent, or at least limit, malign application in armed conflict of the relevant rapidly evolving chemical, life and associated sciences and technologies, until more comprehensive measures can be established and implemented.

9.3 Reviewing Legality of 'New' Weapons

Article 36 of Additional Protocol I to the Geneva Conventions, states that:

> "In the study, development, acquisition or adoption of a new weapon, means or method of warfare a High Contracting Party is under an obligation to determine whether its employment would, in some or all circumstances, be prohibited by this Protocol or by any other rule of international law applicable to the High Contracting Party."[71]

The corresponding ICRC Commentary explains the nature of this undertaking. Namely that States Party are obliged:

> "To determine the possibly unlawful nature of a new weapon both with regard to the provisions of the Protocol, and with regard to any other applicable rule of international law. The determination is to be made on the basis of normal use of the weapon as anticipated at the time of evaluation. If these measures are not taken, the State will be responsible in any case for any wrongful damage ensuing."[72]

To date, there are 174 States party to Additional Protocol I, which are all explicitly bound by this requirement.[73] In addition, the ICRC,[74] Boothby[75] and Copeland,[76] have argued that this obligation is applicable to every State, including those that are not party to Additional Protocol I. However, certain legal observers, including Jevglevskaja, have disputed whether the requirement to conduct legal reviews of new weapons has become established as part of customary international law, particularly given the relatively small number of States that have established formal review processes.[77]

9.3.1 Scope of New Weapons to be Reviewed

The *ICRC Guide* has asserted that the material scope of States' Article 36 legal reviews should be very broad, and should "cover weapons of *all types* – be they *anti-personnel* or anti-materiel, 'lethal', *'non-lethal' or 'less lethal'* – and weapons systems" (Emphases added).[78] The obligation to include weapons purported to be 'non lethal' or 'less lethal' within the scope of such reviews has been endorsed by leading legal scholars such as Lawland[79] and Boothby.[80] Lawland, ICRC legal advisor and principal author of the *ICRC Guide*, has specifically highlighted the importance of including CNS-acting chemical agent weapons in such reviews, arguing that: "[E]stablishing national mechanisms to review the legality of new weapons is *especially relevant and urgent in view of emerging new weapons technologies such as directed energy, [chemical] incapacitants, behaviour change agents, acoustics* and nanotechnology, to name but a few" (Emphasis added).[81]

Although publicly available information regarding current national weapons review processes is scarce, there appears to be some variation in the description of weapons covered. Certain States including Belgium, Sweden, and the US explicitly include 'non-lethal' or 'less lethal' weapons within their scope,[82] others such as Norway, Switzerland and the UK review all weapons,[83] and some employ vague though potentially applicable categories such as Germany's inclusion of weapons intended to "eliminate or reduce their [opponents] attack or defence capabilities".[84] The Netherlands does not define coverage in terms of "weapons", but broadens review scope to any "instrument" intended to cause harm or damage, regardless of the anticipated degree of harm, potentially capturing certain instruments that may not be formally described by their proponents as 'less lethal' weapons. However, the Netherlands requirement that such instruments are intended to "cause harm or damage", and the similar requirement in New Zealand's review scope to weapons or munitions designed or adapted to "cause harm", may lead to interpretations by certain States that potentially exclude certain purported 'non-lethal' or 'less lethal' weapons,[85] including certain CNS-acting chemical agent weapons.

In addition to ensuring a comprehensive scope of weapon types studied, the *ICRC Guide* recommended that legal reviews should assess:

- the ways in which these weapons were to be used pursuant to military doctrine, tactics, rules of engagement, operating procedures and countermeasures;
- all weapons to be acquired, be they procured further to research and development on the basis of military specifications, or purchased 'off-the shelf';
- a weapon which the State was intending to acquire for the first time, without necessarily being 'new' in a technical sense;
- an existing weapon that was modified in a way that altered its function, or a weapon that had already passed a legal review but that was subsequently modified;
- an existing weapon where a State had joined a new international treaty which may affect the legality of the weapon.[86]

Although not specifically called for in Article 36, Daoust, Coupland and Ishoey have argued that it "would be desirable" for States also to examine the legality of weapons they intend to export. This would be in line with their obligation under Article 1, common to the four Geneva Conventions of 1949 and Additional Protocol I, "to respect and ensure respect" for these treaties.[87]

According to the *ICRC Guide* the temporal application of Article 36 is very broad. Consequently it "requires" that States undertake:

> [A]n assessment of the legality of new weapons at the stages of their *"study, development, acquisition or adoption."* This covers all the stages of the weapons procurement process, in particular the initial stages of the research phase (*i.e.* conception, study), the development phase (*i.e.* development and testing of prototypes) and the acquisition phase (including 'off-the-shelf' procurement)." [italics as in original][88]

The current authors therefore argue that all States that are presently conducting research into, and/or undertaking development of new CNS-acting chemical agent weapons intended for use in armed conflict, should have already undertaken a formal Article 36 legal review of these weapons before such activities were initiated. If this legal review has not taken place, all research and development should be halted until this review has been completed, and these weapons found to be in accordance with IHL.

Likewise, any State seeking to purchase or otherwise obtain CNS-acting chemical agent weapons, from either a second State or from the commercial marketplace, should undertake a formal Article 36 legal review "at the stage of the study of the weapon proposed for purchase, and certainly before entering into the purchasing agreement".[89] Furthermore, a range of legal experts, notably the authors of the *ICRC Guide*, have emphasised that "the purchasing State is under an obligation to conduct its own review of the weapon it is considering to acquire, and cannot simply rely on the analysis of the vendor or manufacturer as to the legality of the weapon, nor on another State's evaluation".[90]

The scope of Article 36 is not, however, open-ended. Boothby notes that Article 36 "does not extend to scientific research that has yet to amount to the study of a weapon".[91] Where this dividing is to be drawn with regard to 'dual use' chemical and life science research applicable to CNS-acting weapons development will no doubt be contested on a case-by-case basis. Boothby does highlight the responsibilities of "the relevant elements of the scientific community [to] consider both the potential dangers and benefits associated with lines of research when determining where scientific endeavour ought properly to be focused".[92]

9.3.2 Criteria for Review

While Article 36 did not establish specific criteria for States to employ in their review of new weapons, legal scholar William Boothby has argued that States should incorporate the following criteria:[93]

- whether, in its normal or intended circumstances of use, the weapon was of a nature to cause SIRUS, and as part of this test, consideration should be given to the alternative weapons or methods of choice for accomplishing the military purpose intended for the weapon under review;
- whether the weapon is by its nature indiscriminate (*i.e.*, whether it is capable of being used in compliance with the discrimination rule);
- whether the weapon was intended, or might be expected, to cause widespread, long-term and severe damage to the natural environment;
- whether there are any specific rules of treaty or customary law that would prohibit or restrict the use of the weapon;

- whether there are any likely future developments in the law of armed conflict that might be expected to affect the weapon [Although this final test criterion is not required by customary or treaty law, Boothby argues that it would be "sensible" to employ it, as "weapons are usually acquired to meet a capability requirement over a period...often a substantial period...of time"].

9.3.2.1 Applicability of Human Rights Considerations

A number of international legal scholars have contended that States should consider relevant international human rights law (IHRL) obligations when they undertake formal Article 36 weapons reviews. Casey-Maslen has argued that the reference in Article 36 to "any other rule of international law applicable" to a State Party to the Protocol, would also encompass IHRL.[94] Similarly, Lawland has reasoned that:

> "In reviewing the legality of new weapons, states may also need to consider the rules of international human rights law applicable to the use of force in situations not amounting to armed conflict. This is especially important in view of the increased involvement of armed forces in peace support operations, where troops are more likely to be involved in law enforcement than warfare."[95]

The incorporation of such IHRL considerations into Article 36 weapons reviews would be particularly relevant and important for any new CNS-acting chemical agent weapons, given previous argumentation by some of their proponents that military personnel should use these weapons in certain counter-insurgency, counter-terrorism, peacekeeping and other military operations other than armed conflict. The limited publicly available information again indicates a variation in how States approach this issue, with certain States including Belgium, the Netherlands, Sweden, Switzerland and the UK incorporating IHRL into their review process, while others, apparently, do not.[96]

9.3.2.2 Health Considerations

The *ICRC Guide* has recommended that States, when undertaking Article 36 legal reviews of any new weapon, should take into account relevant health-related considerations. Specifically, when reviewing any new weapon that injures "by means other than explosive or projectile force, or otherwise causes health effects that are qualitatively or quantitatively different from those of existing lawful weapons and means of warfare" the *ICRC Guide* advises that States could consider the following factors[97]

> "[W]hether all relevant scientific evidence pertaining to the foreseeable effects on humans has been gathered; [H]ow the mechanism of injury is expected to impact on the health of victims; [W]hen used in the

context of armed conflict, what is the expected field mortality and whether the later mortality (in hospital) is expected to be high; [W]hether there is any predictable or expected long term or permanent alteration to the victims' psychology or physiology; [W]hether the effects would be recognised by health professionals, be manageable under field conditions and be treatable in a reasonably equipped medical facility."

The *ICRC Guide* has argued that "[T]hese and other health-related considerations are important to assist the reviewing authority in determining whether the weapon in question can be expected to cause superfluous injury or unnecessary suffering".[98]

This approach, which is particularly relevant to CNS-acting chemical agent weapons, has been broadly supported by the 28th Conference of the Red Cross and Red Crescent, which encouraged States "to review with particular scrutiny all new weapons, means and methods of warfare that cause health effects with which medical personnel are unfamiliar".[99]

9.3.3 Restrictions on the Use of New Weapons

Article 36 requires that a State must determine whether the use of a new weapon "in some or all circumstances" would be prohibited by the protocol or international law. According to the *ICRC Guide:* "A weapon or means of warfare cannot be assessed in isolation from the method of warfare by which it is to be used. It follows that the legality of a weapon does not depend solely on its design or intended purpose, but also on the manner in which it is expected to be used on the battlefield".[100] Furthermore, the *ICRC Guide* has warned that a weapon "used in one manner may 'pass' the Article 36 'test', but may fail it when used in another manner".[101] However, the ICRC's Commentary on Additional Protocol I, Article 36 has observed that the scope of these considerations was not unlimited. A State would only need to determine "whether the employment of a weapon for its *normal or expected use* would be prohibited under some or all circumstances. A State is *not required to foresee or analyse all possible misuses of a weapon*, for almost any weapon can be misused in a way that would be prohibited". (Emphasis added).[102] Other international legal experts have placed slightly greater emphasis on States obligations in this area. Fry notably has argued that States conducting Article 36 legal reviews should consider anticipated uses of new weapons beyond those that are considered "normal".[103]

The *ICRC Guide* has sought to provide practical recommendations for States in cases when the reviewing authority has found that some, but not all, of the expected methods of use of the new weapon were determined to be unlawful. In such circumstances, the *Guide* has recommended that "the authority could either approve the weapon on condition that restrictions be placed on its operational use, in which case such restrictions should be incorporated into the rules of engagement or standard operating procedures, or it could request modifications to the weapon which must be met before approval can be granted".[104]

9.3.4 Mechanism of Review

Unfortunately, Additional Protocol 1 did not specify in what manner and under what authority Article 36 legal reviews of new weapons should be constituted. According to the *ICRC Guide* "it is the responsibility of each State to adopt legislative, administrative, regulatory and/or other appropriate measures to effectively implement this obligation".[105] The *Guide* has argued that "At a minimum, Article 36 requires that each State Party set up a formal procedure and, in accordance with Article 84 of Additional Protocol I, other States parties to the Protocol may ask to be informed about this procedure".[106] Furthermore, "[T]he establishment of a formal procedure implies that there be a standing mechanism ready to carry out reviews of new weapons whenever these are being studied, developed, acquired or adopted".[107] Reviews of State practice, previously undertaken by the Danish Red Cross,[108] and Daoust, Coupland and Ishoey,[109] documented a variety of approaches employed in terms of size, nature and location of the reviewing body, and critically whether it possessed an effective veto power or was instead merely an advisory body. A range of legal experts, notably those connected to the ICRC, have recommended that such review mechanisms should be multidisciplinary in nature and should be able to draw upon relevant legal, medical, military and environmental expertise.[110]

9.3.4.1 Information Exchange and Transparency

As noted above, States are required under Additional Protocol 1 to provide information, on request, to other High Contracting Parties, regarding the *process* of their Article 36 weapons review. However, there is no explicit obligation placed on States to disclose information regarding the *conclusions* of their reviews as to the legality of a particular new weapon.[111] To the current authors this appears to be a very unsatisfactory state of affairs which limits the utility of these legal reviews to prevent the development and proliferation of potentially unsuitable, if not illegal, new weapons. The rationale behind this extremely limited transparency apparently is that the substantive technical information examined by a State in its weapons review is perceived as potentially important for that State in retaining its military relevant technological advantage, preventing the development of counter-measures, as well as in preserving commercial confidentiality. Consequently, limited information of the results of such Article 36 Reviews have become public. Dienelt in 2018 reported that "depending on the domestic regulation, some States release the outcome of a weapons review upon public request, whereas others do not make the results public".[112]

The current authors believe that a strong case can be made that States should provide information to all High Contracting Parties of any Article 36 reviews they have conducted that have resulted in a determination that a specific new weapon, means or method of warfare would be in breach of the Additional Protocol or "any other rule of international law applicable". The provision of this information would appear to be in conformity with, and in fact be required by, the reviewing State's obligation under Article 1, common to the four Geneva Conventions of 1949 and Additional Protocol I, "to respect and ensure respect" for these treaties. The current authors recognise that the determination by one State that the employment of a particular new weapon is prohibited would not be binding upon any other State.[113] However, we argue that the exchange of this information between High Contracting Parties could potentially be an important factor in alerting the international community to, and thereby helping to close down or at least severely limit research and development programmes that might – if allowed to proceed unchecked – lead to the production, proliferation and use of weapons that would be contrary to international law.

Certain arms control and humanitarian bodies – notably the ICRC – have called for greater transparency with regard to Article 36 reviews. For example, the 27th Conference of the Red Cross and Red Crescent encouraged States "to promote, wherever possible, exchange of information and transparency in relation to these mechanisms, procedures and evaluations".[114] Furthermore, the 28th Conference of the Red Cross and Red Crescent "invited" States that have review procedures in place to cooperate with the ICRC, with a view to facilitating the voluntary exchange of experience on review procedures.[115]

In recent years, scholars, civil society organisations and States have explored activities to promote increased information exchange regarding Article 36 and the development of best practices for carrying out such reviews.[116] In 2015, the UK Ministry of Defence, Development, Concepts and Doctrine Centre (DCDC) "held the first ever international Weapons Review Forum (WRF)" which comprised States, academics, the ICRC, the British Red Cross and key NGOs. According to the UK, "the Forum enabled fourteen States, who were the primary participants, to discuss the weapons review process between themselves, as well as in open sessions with selected academics and other parties involved in the process, including procurement teams and defence industry representatives. The aim was to share and learn good practice between States and see how other States fulfil their Article 36 obligations".[117] A second WRF meeting was held in October 2016, co-hosted by the UK DCDC and Stockton Center for the Study of International Law at the U.S. Naval War College, looking at Article 36 reviews of highly automated or autonomous technologies.[118]

Greater transparency with regard to the determinations of Article 36 reviews of new CNS-acting chemical agent weapons – particularly when the evaluation concluded the weapon to be prohibited in some or all circumstances – could play an important role in increasing the effectiveness of the regulation of relevant chemical, life and associated sciences and technologies, particularly given the rapidity and scale of advances in many diverse fields and their potential military applicability.

9.3.5 Article 36 Implementation

The importance of all States undertaking reviews of new weapons has been voiced by a number of pluri-lateral and multilateral bodies, notably by the ICRC, including at International Conferences of the Red Cross and Red Crescent.[119] Furthermore, the ICRC continues to play a leading role both in promoting the importance of legal reviews and through the provision of practical guidance and support to facilitate State action in this area. In addition, on repeated occasions, the Conference of the Convention on Certain Conventional Weapons (CCCW) has encouraged State action in this area.[120] At the Fifth Review Conference of the CCCW in 2016, States declared "Their determination to *urge and support* States which do not already do so to conduct reviews to determine whether any new weapon, means or methods of warfare would be prohibited under international humanitarian law or other rules of international law applicable to them". (Emphasis added).[121]

However, despite the regular endorsement of the obligations established under Article 36, and some attempts to increase information exchange, the international community has collectively failed to establish any mechanisms to systematically monitor or facilitate the implementation of these obligations. Consequently, and unsurprisingly, despite the repeated espousal of States' obligations to conduct legal reviews of new weapons, it appears that implementation has continued to be very low. In 2016 Giacca, Legal Advisor at the ICRC, declared that of the 174 States that ratified Additional Protocol I, only "between 15 and 20 States are known to conduct their own legal reviews".[122] Such States include Australia, Belgium, Canada, France, Germany, the Netherlands, New Zealand, Norway, Sweden, Switzerland and the UK.[123] In addition, two militarily important States not party to the Additional Protocol, Israel and the United States, systematically carry out weapons reviews.[124] In contrast, large numbers of States including some that have previously conducted dual use chemical and life science research of concern or have developed CNS-acting chemical agent weapons – notably China, India, Iran, Russia – have not confirmed the establishment of national mechanisms for systematic review of new weapons.

Although it is likely that further States, including at least some military significant ones, do assess the legality at some stage in the process of determining whether to acquire or introduce new weapons, without public disclosure it will remain uncertain whether such assessments do take place and if so whether they are adequate to fulfil States' obligations under Additional Protocol I or customary international law. Giacca believes that many States without dedicated formal national mechanisms will instead "rely on manufacturer information and on the reviews conducted by others".[125] There are obvious dangers in reliance on such 'second hand' analysis, as State interpretation of relevant international law and its application to specific weapons will vary, whilst assessments conducted by commercial manufacturers into their own products may well not be sufficiently objective or rigorous.

The apparent very limited number of States that conduct formal review processes, together with the lack of transparency regarding the nature and stringency of both formal and informal review processes, raise serious concerns regarding the current effectiveness of Article 36, as a tool to prevent development, proliferation and use of indiscriminate or inherently abusive CNS-acting chemical agent weapons or constrain the misuse of limited area CNS-acting chemical agent weapons during armed conflict.

9.4 Implementation of IHL: Investigation and Enforcement Measures

The Geneva Conventions and Additional Protocols have established certain investigation and enforcement measures that could potentially be utilised in cases involving the alleged use of prohibited CNS-acting chemical agent weapons or the use of any permitted CNS-acting chemical agent weapons in armed conflict in a manner that breaches IHL. For example, under the Four Geneva Conventions, each High Contracting Party must "enact any legislation necessary to provide effective penal sanctions for persons committing, or ordering to be committed, any of the grave breaches of the Convention...search for persons alleged to have committed, or to have ordered to be committed, grave breaches; bring such persons, regardless of their nationality before its own courts or hand them over for trial to another High Contracting Party concerned, provided it has made a *prima facie* case".[126] If there is dispute about the circumstances of an alleged violation, a Party to the conflict can request an enquiry, in a manner to be decided between the interested Parties. And subsequently "[O]nce the violation has been established, the Parties to the conflict shall put an end to it and shall repress it with the least possible delay".[127] Enhanced investigation and enforcement mechanisms were developed under Additional Protocol I, with the establishment of an independent fact-finding Commission. Following an allegation by a Party to a conflict, this Commission can "enquire into any facts alleged to be a grave breach ... or other serious violation of the Conventions or of this Protocol",[128] and can "facilitate ... the restoration of an attitude of respect for the Conventions and this Protocol".[129] Unfortunately, all such investigation and enforcement mechanisms can only be initiated by High Contracting Parties, and normally in the first instance by the Parties to the specific conflict. If such Parties are unwilling to trigger these mechanisms there are no routes by which individuals – such as the civilian targets of inherently abusive CNS-acting chemical agent weapons or of abusive attacks involving potentially permissible CNS-acting chemical agent weapons – can initiate such procedures directly.

9.5 Conclusions

International Humanitarian Law (IHL), notably the Four Geneva Conventions and associated Additional Protocols, as well as corresponding international customary law, constitutes a significant body of rules that places very important restrictions on development, acquisition, stockpiling or employment of any weapon in armed conflict. Its constraints upon State action should reinforce, inform and in certain cases go beyond those established by relevant arms control and disarmament instruments, notably the BTWC, CWC and Geneva Protocol in case of CNS-acting chemical agent weapons.

It is clear from the preceding analysis that the development and/or employment of certain CNS-acting chemical agent weapons in armed conflict, notably those dispersing aerosolised or gaseous agents over wide-areas or that caused severe "abnormal physiological" or "abnormal psychological" states, would, because of their inherent design, inevitably breach relevant IHL prohibitions (*e.g.*, the prohibitions on indiscriminate weapons or SIRUS). Furthermore, even a limited area CNS-acting chemical agent weapon (for example one targeting specific individuals rather than groups) which did not inherently breach these prohibitions could be employed in a manner which breached other IHL prohibitions (*e.g.*, prohibition on deliberate attack on civilians) or could undermine relevant IHL obligations (*e.g.*, the duty to project persons considered *hors de combat* or the duty to respect and ensure respect of IHL). Consequently, in theory, IHL should be an important mechanism to prevent the development and use of certain (*i.e.*, gaseous or aerosolised) CNS-acting chemical agent weapons in armed conflict and, at the least, constrain the development and use of other CNS-acting chemical agent weapons.

However, despite the potential utility of the Geneva Conventions and Additional Protocols in this area, these IHL instruments do have important limitations. Firstly, although the Conventions and Protocols do incorporate investigation and enforcement procedures in response to allegations of grave breaches; as with the majority of international legal instruments, these can only be initiated by High Contracting Parties, and normally in the first instance by the Parties to the specific conflict. If such Parties are unwilling to trigger these mechanisms there are no routes by which individuals – such as the civilian targets of indiscriminate or inherently abusive CNS-acting chemical agent weapons or of indiscriminate or abusive attacks involving such weapons – can initiate such procedures directly. This is in contrast to certain relevant IHRL instruments which do allow a mechanism for such individual petition (as explored in Chapter 8). Secondly, although States are required – under Article 36 of Additional Protocol I – to undertake reviews of the legality under IHL of all new weapons, including CNS-acting chemical agent weapons, developed or acquired for use in armed conflict, the number of States carrying out Article 36 reviews, the nature of the review process and crucially the findings of such reviews are largely unknown to the public.

References

1. This Chapter draws upon, updates and expands the authors' previous research and publications, notably: M. Crowley, Chemical Control,

Palgrave MacMillan, 2016, pp. 152–165, which is used with permission from Palgrave MacMillan, Copyright 2016.

2. For a wider discussion of IHL in context of 'less lethal' weapons see: E. Hoffberger-Pippan, *Less-Lethal Weapons under International Law: A Three-Dimensional Perspective*, Cambridge University Press, 2021, p. 51.
3. (a) UN, Vienna Convention on the Law of Treaties, Article 18. For further discussion see: F. Hampson, International law and the Regulation of Weapons, ed. A. Pearson, M. Chevrier and M. Wheelis, in *Incapacitating Biochemical Weapons*, Lexington Books, Lanham, Maryland, USA, 2007, pp. 231–260; (b) J. M. Henckaerts and L. Doswald-Beck, *Customary International Humanitarian Law: Rules*, Cambridge University Press, Cambridge, 2005, vol. I.
4. International Court of Justice, *Statute of the International Court of Justice*, Article 38(1)(b), 1945.
5. J. M. Henckaerts and L. Doswald-Beck, *Customary International Humanitarian Law: Rules*, Cambridge University Press, Cambridge, 2005, vol. I, pp. xxxi–xxxii.
6. J. M. Henckaerts and L. Doswald-Beck, *Customary International Humanitarian Law: Rules*, Cambridge University Press, Cambridge, 2005, vol. I, pp. xxxi–xxxii.
7. ICRC, *International Humanitarian Law Databases, Customary IHL*, https://ihl-databases.icrc.org/en/customary-ihl (accessed 13 February 2025).
8. For example, in the mid-1990s, Russia denied that its military operations in Chechnya were part of an internal armed conflict and therefore covered by IHL, even though the Russian Constitutional Court had characterized the situation as coming within Additional Protocol II of the Geneva Conventions. See: *Judgement of the Constitutional Court of the Russian Federation, 31st July 1995, on the constitutionality of the Presidential Decrees and the Resolutions of the Federal Government concerning the situation in Chechnya*. European Commission for Democracy through Law of the Council of Europe, CDL-INL (96), 1, as cited by F. Hampson, International law and the Regulation of Weapons, in *Incapacitating Biochemical Weapons*, ed. A. Pearson, M. Chevrier and M. Wheelis, Lexington Books, Lanham, Maryland, USA, 2007, p. 251.
9. A. Clapham, *Human Rights Obligations of Non-State Actors*, Oxford University Press, Oxford, 2006, p. 271.
10. (a) For further discussion see: E. Hoffberger-Pippan, *Less-Lethal Weapons under International Law: A Three-Dimensional Perspective*, Cambridge University Press, 2021; (b) ICRC/G. Gaggioli, Expert meeting report: The use of force in armed conflicts – interplay between the conduct of hostilities and law enforcement paradigms, November 2013; (c) N. Davison, New Weapons: Legal and Policy Issues Associated with Weapons Described as 'Non-lethal', in *International Humanitarian Law and the Changing Technology of War*, ed. D. Saxon, Martinus Nijhoff Publishers., 2013, pp. 281–314.

11. F. Hampson, International law and the Regulation of Weapons, in *Incapacitating Biochemical Weapons*, ed. A. Pearson, M. Chevrier and M. Wheelis, Lexington Books, Lanham, Maryland, USA, 2007, p. 238.
12. International Criminal Tribunal for the former Yugoslavia, *Prosecutor v. Tadic*, Decision on the Defence Motion for Interlocutory Appeal on Jurisdiction (Appeals Chamber), 2nd October 1995, Case no. IT-94-1, paragraphs 119 and 127.
13. Geneva Conventions (I)-(IV), Adopted by Diplomatic Conference for the Establishment of International Conventions for the Protection of Victims of War, Geneva, 12 August 1949.
14. (a) Protocol Additional to the Geneva Conventions of 12th August 1949, and relating to the Protection of Victims of International Armed Conflicts [Protocol I], 8 June 1977. (b) Protocol Additional to the Geneva Conventions of 12th August 1949, and relating to the Protection of Victims of Non-International Armed Conflicts [Protocol II], 8 June 1977.
15. A number of further IHL instruments dealing with the methods and means of warfare may also have relevance such as Hague Convention (II) on the Laws and Customs of War on Land, 1899. For further details see: ICRC, *International Humanitarian Law Databases*, https://ihl-databases.icrc.org/en/ihl-treaties/treaties-and-states-parties (accessed 16 December 2024).
16. Additional Protocol I, Protocol Additional to the Geneva Conventions of 12th August 1949, and relating to the Protection of Victims of International Armed Conflicts [Protocol I], 8 June 1977, Articles 48 and 51.
17. J. M. Henckaerts and L. Doswald-Beck, *Customary International Humanitarian Law: Rules – Volume 1*, Cambridge University Press, Cambridge, 2005, Rule 1 [pp. 3–8], Rule 7 [pp. 25–29], Rules 9 & 10 [pp. 25–34], Rules 11 & 12 [pp. 37–43].
18. J.-M. Henckaerts and L. Doswald-Beck, *Customary International Humanitarian Law: Rules – Volume 1*, Cambridge University Press, Cambridge, 2005, Rule 71 [pp. 249–250].
19. See Chapter 2 of this publication for further discussion.
20. Additional Protocol I, Protocol Additional to the Geneva Conventions of 12th August 1949, and relating to the Protection of Victims of International Armed Conflicts [Protocol I], 8 June 1977, Article 51, 4(a–c).
21. (a) Additional Protocol I, Protocol Additional to the Geneva Conventions of 12th August 1949, and relating to the Protection of Victims of International Armed Conflicts [Protocol I], 8 June 1977, Article 51(5) (b); (b) See also in *Customary International Humanitarian Law – Volume 1*, ed. J.-M. Henckaerts and L. Doswald-Beck, 2005, Rules, Rule 14, p. 46.
22. S. Casey-Maslen, *Non-kinetic-energy weapons termed 'non-lethal', A Preliminary Assessment under International Humanitarian Law and International Human Rights Law*, Geneva Academy of International Humanitarian Law and Human Rights, October 2010, p. 19.
23. (a) Additional Protocol I, Protocol Additional to the Geneva Conventions of 12th August 1949, and relating to the Protection of Victims of

International Armed Conflicts [Protocol I], 8 June 1977, Article 48; (b) J. M. Henckaerts and L. Doswald-Beck, *Customary International Humanitarian Law – Volume 1*, Cambridge University Press, Cambridge, 2005, Rule 1, pp. 3–8.
24. International Court of Justice, Nuclear Weapons case, Advisory Opinion, 1996, as cited by J. M. Henckaerts and L. Doswald-Beck, *Customary International Humanitarian Law – Volume 1*, Cambridge University Press, Cambridge, 2005, Rule 1, p. 5.
25. (a) See for example, E. Massingham, Conflict without casualties: a note of caution: non-lethal weapons and international humanitarian law, *Int. Rev. Red Cross*, 2012, **94**(886), 679. (b) N. Davison, New Weapons: Legal and Policy Issues Associated with Weapons Described as 'Non-lethal', in *International Humanitarian Law and the Changing Technology of War*, ed. D. Saxon, Martinus Nijhoff Publishers., 2013, pp. 293–294.
26. E. Massingham, Conflict without casualties: a note of caution: non-lethal weapons and international humanitarian law, *Int. Rev. Red Cross*, 2012, **94**(886), 679.
27. See for example: G. Fenton, Current and Prospective Military and Law Enforcement Use of Chemical Agents for Incapacitation, in *Incapacitating Biochemical Weapons*, ed. A. Pearson, M. Chevrier and M. Wheelis, Lexington Books, Lanham, Maryland, USA, 2007, pp. 103–120.
28. P. Herby, Protecting and Reinforcing Humanitarian Norms: The Way Forward, in *Incapacitating Biochemical Weapons*, ed. A. Pearson, M. Chevrier and M. Wheelis, Lexington Books, Lanham, Maryland, USA, 2007, p. 286.
29. See for example: Additional Protocol I, Protocol Additional to the Geneva Conventions of 12th August 1949, and relating to the Protection of Victims of International Armed Conflicts [Protocol I], 8 June 1977, Article 35.
30. J. M. Henckaerts and L. Doswald-Beck, *Customary International Humanitarian Law – Volume 1*, Cambridge University Press, Cambridge, 2005, Rule 70, pp. 237–244.
31. International Court of Justice, *Legality of The Threat or Use of Nuclear Weapons*, Advisory Opinion of 8 July 1996, paragraph 78.
32. International Court of Justice, *Legality of The Threat or Use of Nuclear Weapons*, Advisory Opinion of 8 July 1996, paragraph 238.
33. J. M. Henckaerts and L. Doswald-Beck, *Customary International Humanitarian Law – Volume 1*, Cambridge University Press, Cambridge, 2005, Rule 70, p. 242.
34. (a) For contrasting discussions see: S. Casey-Maslen, Non-kinetic-energy weapons termed 'non-lethal', A Preliminary Assessment under International Humanitarian Law and International Human Rights Law, Geneva Academy of International Humanitarian Law and Human Rights, October 2010, 2010, p.19; (b) S. Watts, Regulation-Tolerant Weapons, Regulation-Resistant Weapons and the Law of

War, International Law Studies, 91, US Naval War College, 2015, pp. 540–621.
35. E. Hoffberger-Pippan, *Less-Lethal Weapons under International Law: A Three-Dimensional Perspective*, Cambridge University Press, 2021, p. 66.
36. *The SIRUS Project, Towards a Determination of Which Weapons Cause "Superfluous Injury or Unnecessary Suffering"*, ed. R. Coupland, ICRC, Geneva, 1997, p. 23.
37. D. Verchio, Just Say No! The SIrUS Project: Well-Intentioned, But Unnecessary and Superfluous, *AFL Rev.*, 2001, **51**, 183–228.
38. J. McClelland, The Review of Weapons in Accordance with Article 36 of Additional Protocol 1, *Int. Rev. Red Cross*, 2003, **85**, 400.
39. WMA, *Resolution on the SIrUS Project*, adopted by the 50th World Medical Assembly, Ottawa, Canada, October 1998.
40. P. Herby Protecting and Reinforcing Humanitarian Norms: The Way Forward, in *Incapacitating Biochemical Weapons*, ed. A. Pearson, M. Chevrier and M. Wheelis, Lexington Books, Lanham, Maryland, USA, 2007, p. 286.
41. P. Herby Protecting and Reinforcing Humanitarian Norms: The Way Forward, in *Incapacitating Biochemical Weapons*, ed. A. Pearson, M. Chevrier and M. Wheelis, Lexington Books, Lanham, Maryland, USA, 2007 p. 286.
42. E. Hoffberger-Pippan, *Less-Lethal Weapons under International Law: A Three-Dimensional Perspective*, Cambridge University Press, 2021, pp. 69–70.
43. ICRC Casebook, *hors de combat*, https://casebook.icrc.org/a_to_z/glossary/hors-de-combat.
44. (a) ICRC, Statement of the International Committee of the Red Cross, First CWC Review Conference, 28th April–9th May 2003. (b) See alsoN. Davison, New Weapons: Legal and Policy Issues Associated with Weapons Described as 'Non-lethal', in *International Humanitarian Law and the Changing Technology of War*, ed. D. Saxon, Martinus Nijhoff Publishers., 2013, pp. 295–296.
45. (a) ICRC, Statement of the International Committee of the Red Cross, First CWC Review Conference, 28th April–9th May 2003; (b) E. Massingham, Conflict without casualties: a note of caution: non-lethal weapons and international humanitarian law, *Int. Rev. Red Cross*, 2012, **94**(886), 677. (c) N. Davison, New Weapons: Legal and Policy Issues Associated with Weapons Described as 'Non-lethal', in *International Humanitarian Law and the Changing Technology of War*, ed. D. Saxon, Martinus Nijhoff Publishers., 2013, pp. 295–296.
46. ICRC, *Statement of the International Committee of the Red Cross, First CWC Review Conference*, 28th April–9th May 2003.
47. Geneva Conventions, *Adopted by Diplomatic Conference for the Establishment of International Conventions for the Protection of Victims of War*, Geneva, 12 August 1949, Common Article 3.1.

48. Geneva Conventions, *Adopted by Diplomatic Conference for the Establishment of International Conventions for the Protection of Victims of War*, Geneva, 12 August 1949, Common Article 3.1.a–c.
49. Geneva Conventions, *Adopted by Diplomatic Conference for the Establishment of International Conventions for the Protection of Victims of War*, Geneva, 12 August 1949, Common Article 3.2.
50. International Court of Justice, Military and Paramilitary Activities in and against Nicaragua (Nicaragua v. United States of America), *ICJ Rep.*, 1986, 14–112.
51. For discussion see J. M. Henckaerts and L. Doswald-Beck, *Customary International Humanitarian Law – Volume 1*, Cambridge University Press, Cambridge, 2005, Rule 87, pp. 306–308; Rule 110, pp. 400–403; Rule 111, pp. 403–405.
52. Royal Society, Science Policy Centre, in *Brain Waves Module 3, Neuroscience, Conflict and Security*, Royal Society publications, London, 2012, p. 19.
53. (a) Royal Society, Science Policy Centre, in *Brain Waves Module 3, Neuroscience, Conflict and Security*, Royal Society publications, London, 2012, p. 19. (b) Specifically, Under: Third Geneva Convention Convention (III) relative to the Treatment of Prisoners of War, Geneva, 12 August 1949, Article 17.
54. US Army, *Field Manual FM 2–22.3 (FM 34–52): Human Intelligence Collector Operations*, Headquarters, Department of the Army, September 2006, p. 102.
55. Additional Protocol I, *Protocol Additional to the Geneva Conventions of 12th August 1949, and relating to the Protection of Victims of International Armed Conflicts [Protocol I]*, 8 June 1977, Article 11.1.
56. Additional Protocol I, *Protocol Additional to the Geneva Conventions of 12th August 1949, and relating to the Protection of Victims of International Armed Conflicts [Protocol I]*, 8 June 1977, Article 11.2. (b).
57. Additional Protocol I, *Protocol Additional to the Geneva Conventions of 12th August 1949, and relating to the Protection of Victims of International Armed Conflicts [Protocol I]*, 8 June 1977, Article 11.4.
58. Additional Protocol II, *Protocol Additional to the Geneva Conventions of 12th August 1949, and relating to the Protection of Victims of International Armed Conflicts [Protocol I]*, 8 June 1977, Article 10.2.
59. Royal Society, Science Policy Centre, in *Brain Waves Module 3, Neuroscience, Conflict and Security*, Royal Society publications, London, 2012, p. 19.
60. Geneva Conventions (I)-(IV), Adopted by Diplomatic Conference for the Establishment of International Conventions for the Protection of Victims of War, Geneva, 12 August 1949. Common Article 1; Additional Protocol I, Protocol Additional to the Geneva Conventions of 12th August 1949, and relating to the Protection of Victims of International Armed Conflicts [Protocol I], 8 June 1977, Article 1.

61. Royal Society, Science Policy Centre, in *Brain Waves Module 3, Neuroscience, Conflict and Security*, Royal Society publications, London, 2012, p. 20.
62. For an example of this principle, see the Statute of the International Criminal Court, 1998, Article 31, paragraphs 1(a) and (b). [Cited in Royal Society, Science Policy Centre, in *Brain Waves Module 3, Neuroscience, Conflict and Security*, Royal Society publications, London, 2012, p. 20.]
63. Convention (II) with Respect to the Laws and Customs of War on Land and its annex: Regulations concerning the Laws and Customs of War on Land, The Hague, 29 July 1899.
64. See for example, Additional Protocol 1, Protocol Additional to the Geneva Conventions of 12th August 1949, and relating to the Protection of Victims of International Armed Conflicts [Protocol I], 8 June 1977, Article 1(2).
65. International Court of Justice, Legality of the Threat or Use of Nuclear Weapons, Advisory Opinion, 8th July 1996, paragraph 84.
66. Additional Protocol I, Protocol Additional to the Geneva Conventions of 12th August 1949, and relating to the Protection of Victims of International Armed Conflicts [Protocol I], 8 June 1977, Article 1(2).
67. ICRC, *A Guide to the Legal Review of New Weapons, Means and Methods of Warfare, Measures to Implement Article 36 of Additional Protocol 1 of 1977*, January 2006, p. 17.
68. International Court of Justice, Legality of the Threat or Use of Nuclear Weapons, Advisory Opinion, 8th July 1996, paragraph 87.
69. International Court of Justice, Legality of the Threat or Use of Nuclear Weapons, Advisory Opinion, 8th July 1996, paragraph 78.
70. (a) For a range of positions on this issue see: W. Boothby, *Weapons and the law of armed conflict*, Oxford University Press, Oxford, 2nd edn, 2016, pp. 14, 57–58, 94, 351, 369; (b) A. Cassese, The Martens Clause: Half a Loaf or Simply Pie in the Sky?, *Eur. J. Int. Law*, 2000, **11**, 187–216; (c) E. Crawford, *The Modern Relevance of the Martens Clause, Legal Studies Research Paper Number11/27, Sydney Law School*, University of Sydney, 2011; (d) T. Meron, The Martens Clause, Principles of Humanity, and Dictates of Public Conscience, *Am. J. Int. Law*, 2000, **91**(1), 78–89; (e) R. Ticehurst, The Martens Clause and the Laws of Armed Conflict, *Int. Rev. Red Cross*, 1997, (317), 125–134; (f) S. Watts, Regulation-Tolerant Weapons, Regulation-Resistant Weapons and the Law of War, International Law Studies, 91, US Naval War College, 2015, pp. 256–259.
71. Additional Protocol I, Protocol Additional to the Geneva Conventions of 12th August 1949, and relating to the Protection of Victims of International Armed Conflicts [Protocol I], 8 June 1977, Article 36.
72. ICRC, *Humanitarian Law Database, Additional Protocol I, Commentary of 1987*, p.423, paragraph 1466. https://ihl-databases.icrc.org/en/

ihl-treaties/api-1977/article-36/commentary/1987?activeTab=default (accessed 12 December 2024).
73. ICRC Humanitarian Law Database, *Additional Protocol I*, https://ihl-databases.icrc.org/en/ihl-treaties/api-1977/state-parties (accessed 12 February 2025).
74. K. Lawland, *A Guide to the Legal Review of New Weapons, Means and Methods of Warfare, Measures to Implement Article 36 of Additional Protocol I of 1977*, ICRC, January 2006, p. 4.
75. W. Boothby, *Weapons and the law of armed conflict*, Oxford University Press, Oxford, 2nd edn, 2016, pp. 342–343.
76. D. Copeland, Legal Review of New Technology Weapons, in *New Technologies and the Law of Armed Conflict*, ed. H. Nasu and R. McLaughlin, T.M.C. Asser Press, 2014.
77. N. Jevglevskaja, Weapons Review Obligation under Customary International Law, *Int. Law Studies*, 2018, **94**, 187–221.
78. ICRC, *A Guide to the Legal Review of New Weapons, Means and Methods of Warfare, Measures to Implement Article 36 of Additional Protocol 1 of 1977*, January 2006, p. 9.
79. K. Lawland, Reviewing the legality of new weapons, means and methods of warfare, *Int. Rev. Red Cross*, 2006, **88**(864), 926.
80. W. Boothby, *Weapons and the law of armed conflict*, Oxford University Press, Oxford, 2nd edn, 2016, pp. 233–346.
81. K. Lawland, Reviewing the legality of new weapons, means and methods of warfare, *Int. Rev. Red Cross*, 2006, **88**(864), 926.
82. V. Boulanin and M. Verbruggen, *Compendium on Article 36 Reviews*, Background paper, SIPRI, 2017, pp. 3, 15, 20–25.
83. V. Boulanin and M. Verbruggen, *Compendium on Article 36 Reviews*, Background paper, SIPRI, 2017, pp. 13, 16 & 17.
84. V. Boulanin and M. Verbruggen, *Compendium on Article 36 Reviews*, Background paper, SIPRI, 2017, p. 6.
85. V. Boulanin and M. Verbruggen, *Compendium on Article 36 Reviews*, Background paper, SIPRI, 2017, pp. 8 & 11.
86. ICRC, *A Guide to the Legal Review of New Weapons, Means and Methods of Warfare, Measures to Implement Article 36 of Additional Protocol 1 of 1977*, January 2006, p. 9.
87. I. Daoust, R. Coupland and R. Ishoey, New wars, new weapons? The obligation of States to assess the legality of means and methods of warfare, *Int. Rev. Red Cross*, 2002, **84**(846), 345–363.
88. ICRC, *A Guide to the Legal Review of New Weapons, Means and Methods of Warfare, Measures to Implement Article 36 of Additional Protocol 1 of 1977*, January 2006, p. 23.
89. ICRC, *A Guide to the Legal Review of New Weapons, Means and Methods of Warfare, Measures to Implement Article 36 of Additional Protocol 1 of 1977*, January 2006, p. 24.
90. ICRC, *A Guide to the Legal Review of New Weapons, Means and Methods of Warfare, Measures to Implement Article 36 of Additional Protocol 1 of*

1977, January 2006, p. 24; see also J. McClelland, The Review of Weapons in Accordance with Article 36 of Additional Protocol 1, *Int. Rev. Red Cross*, 2003, 85, 411.

91. W. Boothby, *Weapons and the law of armed conflict*, Oxford University Press, Oxford, 2nd edn, 2016, p. 369.
92. W. Boothby, *Weapons and the law of armed conflict*, Oxford University Press, Oxford, 2nd edn, 2016, p. 369.
93. W. Boothby, *Weapons and the law of armed conflict*, Oxford University Press, Oxford, 2016, pp. 347–348.
94. S. Casey-Maslen, Non-kinetic-energy weapons termed 'non-lethal', A Preliminary Assessment under International Humanitarian Law and International Human Rights Law, Geneva Academy of International Humanitarian Law and Human Rights, October 2010, p. 22.
95. K. Lawland, Reviewing the legality of new weapons, means and methods of warfare, *Int. Rev. Red Cross*, 2006, **88**(864), 926.
96. V. Boulanin and M. Verbruggen, *Compendium on Article 36 Reviews*, Background paper, SIPRI, 2017, pp. 3, 9, 16, and 18.
97. ICRC, *A Guide to the Legal Review of New Weapons, Means and Methods of Warfare, Measures to Implement Article 36 of Additional Protocol 1 of 1977*, January 2006, pp. 18–19.
98. ICRC, *A Guide to the Legal Review of New Weapons, Means and Methods of Warfare, Measures to Implement Article 36 of Additional Protocol 1 of 1977*, January 2006, pp. 18–19.
99. ICRC, *Final Goal 2.5 of the Agenda for Humanitarian Action* adopted by the 28th International Conference of the Red Cross and Red Crescent, Geneva, 2nd–6th December 2003, paragraph 2.5.2.
100. ICRC, *A Guide to the Legal Review of New Weapons, Means and Methods of Warfare, Measures to Implement Article 36 of Additional Protocol 1 of 1977*, January 2006, p. 10.
101. ICRC, *A Guide to the Legal Review of New Weapons, Means and Methods of Warfare, Measures to Implement Article 36 of Additional Protocol 1 of 1977*, January 2006, p. 10.
102. ICRC, Commentary on the Additional Protocols, paragraph 1469, as cited in ICRC, A Guide to the Legal Review of New Weapons, Means and Methods of Warfare, Measures to Implement Article 36 of Additional Protocol 1 of 1977, January 2006, p. 10.
103. J. Fry, Contextualized Legal Reviews for the Methods and Means of Warfare: Cave Combat and International Humanitarian Law, *Colomb. J. Transnational Law*, 2006, **44**(2), 470–471.
104. ICRC, *A Guide to the Legal Review of New Weapons, Means and Methods of Warfare, Measures to Implement Article 36 of Additional Protocol 1 of 1977*, January 2006, p. 26.
105. ICRC, *A Guide to the Legal Review of New Weapons, Means and Methods of Warfare, Measures to Implement Article 36 of Additional Protocol 1 of 1977*, January 2006, p. 20.

106. ICRC, *A Guide to the Legal Review of New Weapons, Means and Methods of Warfare, Measures to Implement Article 36 of Additional Protocol 1 of 1977*, January 2006, p. 20.
107. ICRC, *A Guide to the Legal Review of New Weapons, Means and Methods of Warfare, Measures to Implement Article 36 of Additional Protocol 1 of 1977*, January 2006, p. 20.
108. Danish Red Cross, *Reviewing the Legality of New Weapons*, December 2000, 2006.
109. I. Daoust, R. Coupland and R. Ishoey, New wars, new weapons? The obligation of States to assess the legality of means and methods of warfare, *Int. Rev. Red Cross*, 2002, **84**(846), 354–360.
110. (a) See for example: I. Daoust, R. Coupland and R. Ishoey, New wars, new weapons? The obligation of States to assess the legality of means and methods of warfare, *Int. Rev. Red Cross*, 2002, **84**(846), 352–353; (b) K. Lawland, Reviewing the legality of new weapons, means and methods of warfare, *Int. Rev. Red Cross*, 2006, **88**(864), 926; (c) ICRC, Final Goal 2.5 of the Agenda for Humanitarian Action adopted by the 28th International Conference of the Red Cross and Red Crescent, Geneva, 2nd–6th December 2003, paragraph 2.5.1.
111. (a) ICRC Commentary on the Additional Protocols, paragraph 1469, as cited in: ICRC, A Guide to the Legal Review of New Weapons, Means and Methods of Warfare, Measures to Implement Article 36 of Additional Protocol 1 of 1977, January 2006, p. 10; (b) W. Boothby, *Weapons and the law of armed conflict*, Oxford University Press, Oxford, 2nd edn, 2016, p. 345.
112. A. Dienelt, The shadowy existence of the weapons review and its impact on disarmament, *Secur. Peace*, 2018, **36**(3), 129.
113. ICRC, *Commentary on the Additional Protocols*, paragraphs 1469 and 1481.
114. ICRC, *Final Goal 1.5 of the Plan of Action for the years 2000–2003* adopted by the 27th International Conference of the Red Cross and Red Crescent, Geneva, 31st October to 6th November 1999.
115. ICRC, *Final Goal 2.5 of the Agenda for Humanitarian Action* adopted by the 28th International Conference of the Red Cross and Red Crescent, Geneva, 2nd–6th December 2003, paragraph 2.5.3.
116. (a) V. Boulanin, Implementing Article 36 weapons reviews in the light of increasing autonomy in weapons systems, SIPRI insights on peace and security, No. 2015/1, November 2015; (b) V. Boulanin and M. Verbruggen, Compendium on Article 36 Reviews, Background paper, SIPRI, December 2017.
117. UK Ministry of Defence, UK Weapons Reviews, 2015, p. 1.
118. J. Farrant and C. Ford, Autonomous Weapons and Weapon Reviews: The UK Second International Weapon Review Forum, *Int. Law Stud.*, 2017, **93**, 389–422.

119. See for example: 30th International Conference of the Red Cross and Red Crescent, Geneva, 26th–30th November 2007, Paragraph 19, Resolution 3.
120. (a) See for example: Final Declaration of the Second Review Conference of the States Parties to the Convention on Certain Conventional Weapons, Geneva, 21 December 2001, CCW/CONF.II/2, p. 11; (b) Final Declaration of the Fourth Review Conference of the States Parties to the Convention on Certain Conventional Weapons, Geneva, 15 December 2011, CCW/CONF.IV/4/Add.1, p. 4.
121. United Nations, Final Document, Fifth Review Conference of the High Contracting Parties to the Convention on Certain Conventional, Weapons, CCW/CONF.V/10, 23 December 2016. P.7.
122. ICRC, Legal review of new weapons: scope of the obligation and best practices, *Humanitarian Law and Policy*, 6 October 2016, http://blogs.icrc.org/law-and-policy/2016/10/06/legal-review-new-weapons/ (accessed 16 December 2024).
123. (a) ICRC, A Guide to the Legal Review of New Weapons, Means and Methods of Warfare, Measures to Implement Article 36 of Additional Protocol 1 of 1977, January 2006; (b) V. Boulanin and M. Verbruggen, Compendium on Article 36 Reviews, Background paper, SIPRI, 2017.
124. N. Jevglevskaja, Weapons Review Obligation under Customary International Law, *Int. Law Studies*, 2018, **94**, 209.
125. ICRC, Legal review of new weapons: scope of the obligation and best practices, *Humanitarian Law and Policy*, 6 October 2016. http://blogs.icrc.org/law-and-policy/2016/10/06/legal-review-new-weapons/ (accessed 16 December 2024).
126. Geneva Convention (I), Adopted by Diplomatic Conference for the Establishment of International Conventions for the Protection of Victims of War, Geneva, 12 August 1949, Article 49; Geneva Convention (II), Adopted by Diplomatic Conference for the Establishment of International Conventions for the Protection of Victims of War, Geneva, 12 August 1949, Article 50; Geneva Convention (III), Adopted by Diplomatic Conference for the Establishment of International Conventions for the Protection of Victims of War, Geneva, 12 August 1949, Article 129; Geneva Convention (IV), Adopted by Diplomatic Conference for the Establishment of International Conventions for the Protection of Victims of War, Geneva, 12 August 1949, Article 146.
127. Geneva Convention (I) Adopted by Diplomatic Conference for the Establishment of International Conventions for the Protection of Victims of War, Geneva, 12 August 1949, Article 52; Geneva Convention (II), Adopted by Diplomatic Conference for the Establishment of International Conventions for the Protection of Victims of War, Geneva, 12 August 1949, Article 53; Geneva Convention (III), Adopted by Diplomatic Conference for the Establishment of International Conventions for the Protection of Victims of War, Geneva, 12 August 1949,

Article 132; Geneva Convention (IV), Adopted by Diplomatic Conference for the Establishment of International Conventions for the Protection of Victims of War, Geneva, 12 August 1949, Article 149.

128. Additional Protocol I, Protocol Additional to the Geneva Conventions of 12th August 1949, and relating to the Protection of Victims of International Armed Conflicts [Protocol I], 8 June 1977, Article 90.2.(c).(i).

129. Additional Protocol I, Protocol Additional to the Geneva Conventions of 12th August 1949, and relating to the Protection of Victims of International Armed Conflicts [Protocol I], 8 June 1977, Article 90.2.(c).(ii).

CHAPTER 10

Application of the United Nations Drug Control Conventions to CNS-acting Chemical Agent Weapons

10.1 Introduction

As earlier chapters of this publication have detailed, a range of naturally occurring and synthetic substances have previously been investigated by States as potential CNS-acting chemical agent weapons. This has included a number of narcotic drugs that cause drowsiness or unconsciousness and relieve pain, and also certain psychotropic substances that affect a person's mental state. Many of these chemicals can have legitimate medical, veterinary, or other scientific applications, some substances however have been widely employed by the general population as so-called 'recreational drugs', often such use has resulted in detrimental consequences for the individuals concerned and wider society, notably drug dependency and addiction.

Following the development and signing of the International Opium Convention in 1912,[1] the international governmental community adopted a series of additional instruments[2] intended to combat the illicit trade and abuse of such 'recreational drugs', while at the same time protecting the supply and use of these substances for legitimate purposes. However, according to Rabbat (a former official at the International Narcotics Control Board (INCB) Secretariat),[3] this resulted in a "patchwork drug-control legal framework" that was built upon sectoral legal instruments having limited material, geographical and jurisdictional scope. Consequently, the growing number of legal norms contained within these instruments also contributed to making the drug-control framework "unduly cumbersome and unclear".[4]

Preventing Weaponization of CNS-acting Chemicals: A Holistic Arms Control Analysis
By Michael Crowley and Malcolm Dando

Published by the Royal Society of Chemistry, www.rsc.org

Recognizing the limitations of this patchwork framework, the international community negotiated and adopted the Single Convention on Narcotic Drugs (SCND) intended to streamline, clarify and update existing legal norms related to drug control and develop an effective regulatory framework. This was subsequently widened and strengthened by the adoption of the UN Convention on Psychotropic Substances (CPS) and the UN Convention against Illicit Traffic in Narcotic Drugs and Psychotropic Substances (CITNDPS).

This chapter will assess the potential applicability of the two most relevant instruments – namely the SCND and the CPS – to the regulation of CNS-acting chemical agents that are intended to be employed: as weapons for law enforcement or military operations; as aids in interrogation; for torture and other ill-treatment of detainees, or for the death penalty.[5] It will end by examining civil society and State discourse regarding the relevancy of such instruments to the regulation of CNS-acting chemical agent weapons.

10.2 Single Convention on Narcotic Drugs

The initial version of the SCND was signed in 1961; a protocol amending this was subsequently agreed in 1972, and this revised text came into force in 1975.[6] To date, the Single Convention has 186 States Parties.[7] The SCND codified all previous multilateral treaties on narcotic drug control and also extended the existing control systems to include the cultivation of plants that were grown as raw materials for narcotic drugs.[8] The principal objectives of this Convention are to restrict possession, use, trade in, distribution, import, export, manufacture and production of narcotic drugs exclusively to medical and scientific purposes,[9] and also to address narcotic drug trafficking through international cooperation.

In accordance with the SCND, the International Narcotics Control Board (INCB) was established in 1968 to monitor State implementation of this Convention (and subsequently that of the CPS and the CITNDPS).[10] The INCB is an independent quasi-judicial control organ composed of 13 members elected by the UN Economic and Social Council (ECOSOC), 10 members being nominated by Governments and 3 by the World Health Organization (WHO).

The INCB undertakes the following activities:

a. As regards the licit manufacture of, trade in and use of drugs, the INCB endeavours, in cooperation with Governments, to ensure that adequate supplies of drugs are available for medical and scientific uses and that the diversion of drugs from licit sources to illicit channels does not occur. The INCB also monitors Governments' control over chemicals used in the illicit manufacture of drugs and assists them in preventing the diversion of those chemicals into the illicit traffic;

b. As regards the illicit manufacture of, trafficking in and use of drugs, the INCB identifies weaknesses in national and international control systems and contributes to correcting such situations. The INCB is also responsible for assessing chemicals used in the illicit manufacture of drugs, in order to determine whether they should be placed under international control.[11]

The specific narcotic drugs that come under the scope of the SCND have been detailed in four Scheduled lists. These lists are regularly reviewed and can be updated following recommendations to the UN Secretary General by the Convention's States Parties or the WHO.[12] The current Schedules, which were last updated in June 2024,[13] include a number of narcotic drugs that have reportedly[14] been explored as potential CNS-acting chemical agent weapons in previous State weaponization programmes, including cannabis[15] carfentanil,[16] cocaine,[17] etorphine,[18] fentanyl,[19] morphine,[20] remifentanil[21] and sufentanil.[22] Some of the SCND Scheduled narcotic drugs (for example, cannabis[23] and a combination of morphine[24] and scopolamine) were previously reportedly investigated by certain countries, notably the US, for use in drug-assisted interrogations.[25] In addition, certain US States have authorised the SCND Schedule 1 drugs alfentanil and fentanyl for 'lethal injection' executions.[26]

Article 4 of the SCND, which sets out the instrument's "General Obligations", states that:

> "The parties shall take such legislative and administrative measures as may be necessary.... (c) Subject to the provisions of this Convention, to *limit exclusively to medical and scientific purposes* the production, manufacture, export, import, distribution of, trade in, use and possession of drugs."[27] (Emphasis added)

The clear obligation upon States Parties established by Article 4 to "limit" narcotic drugs "exclusively to medical and scientific purposes",[28] appears to put into doubt the legitimacy of any research, development, manufacture, transfer, acquisition, stockpiling, and use of narcotic drugs – or at the very least, the specific narcotic drugs that are listed in the current SCND's Schedules – by States Parties for other purposes, notably employment as CNS-acting chemical agent weapons in armed conflict, military operations other than war, or in law enforcement operations. Similarly, the development, acquisition, *etc.,* of these substances for use as chemical aids to interrogation, for torture and other ill-treatment or for the death penalty, would also appear to be in contravention of Article 4 of this Convention. However, the applicability of the SCND to such activities is uncertain as the Convention was developed as a crime control instrument and not as a means of arms control and disarmament or protection of human rights. To date, there is no public record of the SNCD being applied by any of its States Parties to such activities. In addition, even if the Convention could be applied by its States Parties for such purposes, it is uncertain which activities would be considered permissible and which prohibited, and how such determinations would be made.

Furthermore, the SCND does allow States Parties to retain so-called "special stocks" of narcotic drugs, which are defined under Article 1 as: "drugs held in a country or territory by the Government of such country or territory for special government purposes and to meet exceptional circumstances".[29] Although neither "special government purposes" nor "exceptional circumstances" are explicitly defined under the SNCD, their meaning is discussed in the UN Secretary General's Commentary to the Convention.[30] This Commentary explains that "'[E]xceptional circumstances' is meant to cover such catastrophic events as large-scale epidemics and major earthquakes".[31] The Commentary also notes that: "'[S]pecial government purposes' can cover all purposes other than the usual needs of the civilian population or 'exceptional circumstances'. *In practice the phrase is generally synonymous with 'military purposes'*". (Emphasis added)[32] The Commentary further elaborates that the distinction between 'stocks' and 'special stocks' was "motivated by the desire to protect confidential information which might have military value".[33]

A very important caveat and clarification to this interpretation has subsequently been provided by Rabbat, who states that although "'[S]pecial government purposes' includes use by the military... this does not imply that the use is to be military in nature (such non-military uses may for example include medical treatment of military personnel)".[34] And he also argues that: "'[S]pecial stocks' remain subject to the obligation to limit use of controlled substances to medical and scientific purposes".[35]

While the application of the Convention as a potential means to constrain or prevent development and use of CNS-acting chemical agent weapons has still to be tested by its States Parties, the SCND has established certain reporting and investigatory mechanisms that could be employed by a State Party wishing to obtain information of potential benefit in the regulation of certain CNS-acting chemicals that have been or could be weaponised. Article 19 of the Convention requires that States Parties provide the INCB with annual estimates of drug requirements,[36] and drug production,[37] for scheduled chemicals. The relevant statistical information provided by States Parties is then included in INCB Annual Reports which are subsequently made public.[38] However, this transparency is limited as the SCND does not require that States Parties furnish statistical returns respecting "special stocks".[39] A review by the authors of all published *INCB Annual Reports* could find no specific details of research, development, manufacture, transfer, acquisition, stockpiling or use of weapons employing CNS-acting chemical agents with narcotic properties intended for law enforcement or other purposes, and it is unknown whether States Parties have provided details of such activities to the INCB.[40]

Concerns about the privileged nature of "special stocks" are reinforced by the extremely curtailed nature of the INCB's ability to monitor or otherwise assess such stocks. The UN Secretary General's Commentary stated that: "[T]he Single Convention allows Governments to maintain 'special stocks' whose management is not subject to the control of the Board, nor its examination, its enquires or its rights of criticism".[41]

The SCND does include certain procedures, under Article 14, that allow the INCB to engage in consultations with relevant States Parties and investigate potential circumstances of concern in order to resolve them.[42] If the INCB considers that the aims of the Convention were being "seriously endangered" and it had not been possible to "resolve the matter satisfactorily in any other way", particularly when "the situation is serious" and required "cooperative action at the international level", it could alert the States Parties, the Council and the Commission of the matter. Subsequently, if it were deemed necessary, the Council could "draw the attention of the UN General Assembly to the matter".[43] However, once again, given the extreme limitations in State reporting, and in the oversight measures afforded to the INCB with regard to "special stocks", it is uncertain whether, in what circumstances, and how effectively the INCB could in practice investigate the use of "special stocks" for weaponization of CNS-acting chemical agents, and whether and how the States Parties would address these issues.

Consequently, it is not known whether the INCB has received information or investigated any reports of research, development, manufacture, acquisition, stockpiling, transfer, or use of CNS-acting chemical agent weapons employing Scheduled chemicals by States Parties for military or law enforcement purposes. For example, there is no publicly available information documenting any INCB deliberations or actions taken specifically in response to the Russian Federation's use of a CNS-acting chemical agent weapon employing derivatives of fentanyl to end the Moscow theatre siege in October 2002, nor of Russia's associated presumptive production and/or acquisition and stockpiling of these Scheduled chemicals intended for this purpose.[44]

To date, although these issues have yet to be publicly addressed by the SCND States Parties; the INCB, in its *2003 Annual Report*, did contain a subsection entitled "Use of internationally controlled drugs by military and police forces".[45] This appears to be the only time that this subject has been addressed, albeit in a very circumspect manner, in any *INCB Annual Report*. It is noteworthy that the *2003 Annual Report*, published in February 2004, was written following the Russian use of a CNS-acting chemical agent weapon – although this incident was not explicitly addressed in the text.

The INCB *2003 Annual Report* firstly noted previous differing uses of certain drugs in armed conflict and law enforcement:

> "The use of narcotic drugs and psychotropic substances in military warfare and for law enforcement purposes has a long history. During the Second World War...Pervitin (methamphetamine), together with alcohol, was distributed to soldiers in the armed forces. During the same period, amphetamines were widely used in the Japanese armed forces to increase soldiers' performance...[46]

The INCB then went on to raise the following concerns:

> "The Board is aware that drugs scheduled under the 1961 Convention or the 1971 Convention, mainly drugs of the amphetamine-type group, continue to be used by some military forces, for example during armed conflict, *and that research into further possible uses is taking place. The Board is of the opinion that this type of drug use may not be in line with the international drug control conventions, which require Governments to limit the use of narcotic drugs to medical and scientific purposes only.* The Board appeals to Governments to ensure that the military and law enforcement sectors follow the principles of sound medical practice in their use of internationally controlled substances and that the international drug control conventions are respected in those sectors."[47] (Emphasis added).

Whilst it is somewhat unclear to what extent the INCB was referring to the potential use of Scheduled drugs (under either the SCND or the CPS) as CNS-acting chemical agent weapons, there are intimations in the highlighted text that this was part of its concerns. Furthermore, the Board statement is very important in confirming that law enforcement and military forces are not exempt from the obligation on States to restrict the use of drugs covered by the relevant drug control conventions solely to "scientific and medical purposes".

As Rabbat notes, "[D]espite the reduced scrutiny to which 'special stocks' are subject to in the Conventions, the Board does not consider that the provisions in question relieve States Parties of the obligation to limit their use of controlled substances to medical and scientific purposes – a position clearly enunciated in the 2003 INCB Annual Report".[48] Subsequently, the legal scholar Pippen Hofberger has argued that "[E]ven though the [INCB 2003 Annual Report] merely constitutes soft law, it reflects a clear tendency toward a prohibition of incapacitants for law enforcement purposes".[49]

10.3 The UN Convention on Psychotropic Substances

The UN CPS was signed in 1971 and came into force in 1976. To date, it has 184 States Parties and 34 Signatory States.[50] This Convention established an international control system intended to limit the diversion and abuse of certain psychotropic substances, such as central nervous stimulants, sedative-hypnotics and hallucinogens, which had resulted in public health and social problems in some countries. It introduced graduated controls on certain psychotropic substances according to their potential for abuse on the one hand and their therapeutic value on the other.[51] The drugs covered by the CPS have been elaborated in four Scheduled lists which can be updated following recommendations by States Parties or the WHO to the UN Secretary General.[52] The current Schedules, which were last updated in December 2024,[53] include a number of psychotropic drugs that have reportedly[54] previously been explored as CNS-acting chemical agent weapons for potential use in certain law enforcement or military operations including: amphetamine,[55] dexamphetamine,[56] LSD,[57] mescaline,[58] methamphetamine,[59] midazolam,[60] phencyclidine,[61] and psilocybine.[62] The Schedules also include drugs reportedly[63] explored or utilized in drug-assisted interrogations such as amobarbital,[64] LSD, mescaline, secobarbital,[65] and temazepam.[66] In addition, certain US States have authorised and employed the CPS Scheduled drugs pentobarbital,[67] midazolam and diazepam[68] for 'lethal injection' executions,[69] whilst amobarbital[70] and secobarbital[71] have reportedly been proposed by certain US States but never authorised to date for such purposes.[72]

Under Article 5 of the Convention, each State Party is required to "limit by such measures as it considers appropriate the manufacture, export, import, distribution and stocks of, trade in, and use and possession of, substances in Schedules II, III and IV to medical and scientific purposes".[73] The restrictions for Schedule I substances are even stricter, with States Parties required, under Article 7 to: "prohibit all use except for scientific and very limited medical purposes by duly authorized persons, in medical or scientific establishments which are directly under the control of their Governments or specifically approved by them",[74] and to "require that manufacture, trade, distribution and possession be under a special licence or prior authorization".[75]

In contrast to the SCND, the CPS contains no ambiguously worded provisions allowing the State Parties to hold "special stocks" of drugs to be used for "special government purposes" or in "exceptional circumstances". Instead, Article 4 of the CPS permits the use of Schedule II, III and IV psychotropic substances for three specific activities – "the carrying by international travellers of small quantities of preparations for personal use"; during the industrial "manufacture of non-psychotropic substances"; and for "the capture of animals".[76] The CPS contains no specific exemptions allowing for the use of any substances contained in its Schedules for the purpose of law enforcement or for military activities *i.e.* armed conflict.

Although the applicability of this Convention to the employment of psychotropic substances as CNS-acting chemical agent weapons in military or law enforcement operations is uncertain, the restrictions elaborated under the Convention do appear to put into question the legitimacy of research, development, manufacture, acquisition, stockpiling, transfer and use of CPS scheduled psychotropic drugs by States Parties for such activities. Similarly, the research, development, manufacture, acquisition, stockpiling, transfer or use of psychotropic substances in drug-assisted interrogations and for 'lethal injection' executions are not specifically provided for in any of the Convention's exemptions and therefore appear to be prohibited under the CPS. To date, there are no documents publicly available from the CPS States Parties, the INCB or other relevant organs of the CPS clarifying these issues.[77]

As with the SCND, the CPS incorporates a range of reporting and investigation provisions that could potentially be utilized to provide information of benefit in the regulation of CNS-acting chemical agent weapons employing CPS scheduled chemicals. Under the Convention, States Parties are required to provide the INCB with an annual report of quantities of each Schedule I, II, III and IV psychotropic substance manufactured, imported and exported by the country, as well as details of stocks held of Schedule I and II chemicals – however State Parties are not required to provide information on the intended use of such chemicals.[78] The statistical information provided by States Parties has subsequently been incorporated into *INCB Annual Reports*, which have been made publicly available.[79] To date, a review of all published *INCB Annual Reports* could find no details of contemporary research, development, manufacture, transfer, acquisition, stockpiling or use of CNS-acting chemical agent weapons employing specific named scheduled psychotropic substances for law enforcement or military activities, and it is unknown whether States Parties have provided details of such activities to the INCB.[80] However once again, it is important to note the Board's apparent raising of its concerns regarding CNS-acting chemical agent weapons, in its *2003 Annual Report*. Although this would appear to relate to the 2002 Russian weaponization of fentanyl derivatives, the INCB's concerns appeared to extend also to the use of unnamed CPS scheduled drugs by the military for unspecified purposes, possibly including performance enhancement: "The Board is aware that drugs scheduled under the 1961 Convention *or the 1971 Convention, mainly drugs of the amphetamine-type group, continue to be used by some military forces, for example during armed conflict, and that research into further possible uses is taking place."* [Emphasis added].[81]

Under Article 19 of the CPS, if the INCB, following examination of information supplied to it by Governments or relevant UN organs,[82] had reason to believe that the aims of the Convention were "being seriously endangered" by the actions of a country or region, it could seek "explanations from the Government of the country or region in question".[83] If the matter was satisfactorily resolved at this stage then these processes and the information provided would be treated as confidential.[84] However, if the Board were unsatisfied by the State's response, it could bring the matter to the States Parties, the Council and the Commission, and could recommend that the States Parties "stop the export, import, or both, of particular psychotropic substances, from or to the country or region concerned" for a "designated period" or until the INCB was "satisfied as to the situation in that country or region".[85] It is not known whether the INCB has formally received information or investigated any reports of research, development, manufacture, transfer, acquisition, stockpiling or use of CNS-acting weapons employing CPS Scheduled chemicals by States Parties for law enforcement or military purposes.[86]

10.4 Civil Society and State Discourse Concerning Potential Relevancy of SCND and CPS to Regulation of CNS-acting Chemical Agent Weapons

Certain civil society observers and international organisations have previously examined the potential applicability of the UN drug control conventions to CNS-acting chemical agent weapons. In 2010 and 2012 the

International Committee of the Red Cross (ICRC) held two Expert Meetings on "incapacitating chemical agents".[87] In the first meeting which explored the implications for international law, Crowley examined a range of potentially relevant international legal regimes including and beyond those concerned with arms control and disarmament, and highlighted the applicability of the SCND and the CPS. Whilst examining the explicit restriction by both Conventions of the relevant drugs "exclusively to medical and scientific purpose", he recognised the limited publicly available information as to their application by State Parties to CNS-acting chemical agent weapons.[88] The SCND, CPS and broader drug control framework were subsequently examined in more technical detail by Rabbat in the ICRC's second expert meeting[89] and his analysis has informed this chapter. Drawing on the findings of the two expert meetings and further analysis, the ICRC in September 2012 concluded:

> "The 1961 Single Convention on Narcotic Drugs and the 1971 Convention on Psychotropic Substances place strict controls on some of the toxic chemicals that have been considered as weapons for law enforcement. They require that the production, manufacture, export, import, distribution of, trade in, use and possession of these controlled substances must be limited exclusively to 'medical and scientific purposes."[90]

In 2018, Mathews, former Australian Government Scientific Advisor and OPCW SAB member, highlighted the potential synergies between the CWC and the SCND and CPS, and called for greater cooperation between relevant international organisations with regard to addressing CNS-acting chemical agents:

> "There are already a number of international agreements and other regulatory frameworks which have responsibilities relating to CNS-acting chemicals, including the UN Single Convention on Narcotic Drugs, the UN Convention on Psychotropic Substances.... It has become clear that there will be considerable benefits if the OPCW is able to work in a collaborative and cooperative manner with other international agencies, for example, the UNODC [United Nations Office on Drugs and Crime] which already have responsibilities related to opiates such as fentanyl and its analogs. However, the discussions that have taken place so far have been hampered because of the lack of decision by the CWC Member States to discuss these issues within the OPCW policy-making organs. This is most unfortunate as the OPCW has several issues in common with other international agencies, including at least partial overlap in the scope of the various prohibitions and regulatory frameworks, and similar requirements related to safety issues, the detection of relevant materials in operational environments, laboratory analysis of relevant substances, medical treatment and decontamination."[91]

Subsequently certain States, whilst not directly addressing CNS-acting chemical agent weapons, have taken a technocratic approach to the broader issue of addressing CNS-acting chemicals, highlighting the potential benefits for the OPCW in greater interaction with the UNODC and the INCB with regard to the regulation of narcotic and psychoactive drugs. For example, in its 2023 paper on "Convergence of science between international treaties and agreements", circulated at the CWC 5th Review Conference the UK stated that "An enhanced understanding of the decisions made in other fora on toxic chemicals through consultations, exchanges of information and documents, technical cooperation, as well as cross-representation at relevant meetings would improve the [OPCW's] decision-making". Consequently, the UK saw "benefit from the OPCW Technical Secretariat... enhancing external engagement to increase understanding of issues relevant to the Convention".[92] One of the regimes examined was that established under the SCND and CPS. The UK noted "[O]f particular relevance to the [CWC] is the ongoing increase in international controls on chemicals that fall under the banner of CNSAC [CNS-acting chemicals], and precursors to these chemicals". And notably that "[I]n response to a rapidly evolving proliferation of synthetic opioids, the regular additions to the List of Narcotic Drugs under International Control of the Single Convention have enabled it to adapt to changing trends". The UK also highlighted the INCB, and noted that "[I]n many respects, there are similarities with how the OPCW policy-making organs and advisory boards are constituted and act".[93]

Similarly in its Report on Developments in Science and Technology for the 5th CWC Review Conference, the SAB recommended that "the Director-General consider establishing a [Temporary Working Group] to study current developments concerning CNS-acting chemicals" and it further recommended "that this TWG include the participation and expertise of other relevant international institutions (*e.g.*, the United Nations Commission on Narcotic Drugs)".[94]

Another, perhaps surprising, voice arguing strongly for greater collaboration and consultation between the OPCW and the UN drug control regimes, specifically with regard to CNS-acting chemical agent weapons has been Russia. As discussed in previous chapters, Russia has consistently opposed attempts by other CWC States Parties to determine whether the use of aerosolised CNS-acting chemicals for law enforcement purposes was permissible under the CWC. In a November 2018 paper issued during the 4th CWC Review Conference, Russia stated that use of aerosolised CNS-acting chemicals for law enforcement purposes "is not regulated under the [Chemical Weapons] Convention".[95] When proposals were moving ahead during 2021 for the tabling of a resolution at CSP-26 for an OPCW "[U]nderstanding" effectively prohibiting use of aerosolised CNS-acting chemicals for law enforcement purposes, Russia issued a "questions and answers" paper for the March 2021 Executive Council meeting "regarding the proposal to prohibit the aerosolised use of central nervous system acting chemicals for law enforcement purposes".[96] According to the Russians, their paper was "aimed at analysing some key aspects of the draft decision... [so as to] ... understand the aim of the draft decision, the possible consequences of its adoption by the OPCW as submitted (including for States Parties to the Convention), and what could be done so that the idea conveyed in the draft decision could be acceptable to the ...OPCW".[97]

One issue highlighted in the Russian paper was the implications of existing coverage of narcotic and psychotropic chemicals under the UN drug conventions:

> "During several presentations arranged in 2019 on the sidelines of the OPCW events by the co-sponsors of [the "Understanding"], examples of so-called CNS-acting chemicals mostly included anaesthetic agents, sedative agents, and analgesics, including fentanyl and its analogues. However, there was no reference to the fact that such chemicals fall within the scope of other international treaties, such as [the SCND, CPS and CITNDPS]. The aforementioned conventions contain Schedules of narcotic drugs and psychotropic substances. These chemicals are subject to control and verification mechanisms both at the national level by States Parties, and at the international level directly by the International Narcotics Control Board (INCB) and the World Health Organization (WHO). These schedules are updated and supplemented on an ongoing basis."[98]

The Russians consequently highlighted the lack of definition of CNS-acting chemicals in the draft resolution, warning:

> "It should be stressed from the outset that the absence of the necessary definitions and a schedule of chemicals that fall under the draft decision creates a conflict of interest of the relevant international organisations implementing the objects and purposes of the aforementioned international treaties (conventions), including provisions for international verification of compliance with those treaties and conventions."[99]

And Russia consequently recommended that work should be undertaken [presumably by the OPCW Executive Council], *inter alia* to:

> "2. Prepare recommendations on the key terms and definitions to be used in the context of the Convention, in particular:
>
> - 'central nervous system-acting chemicals' (including aerosolised);
>
> - 'law enforcement purposes';
>
> - 'temporary incapacitating effect'; and
>
> - others at the discretion of the States Parties.

3. Compile a list of chemicals to be covered by the draft decision.

4. Work should be initiated among the OPCW, the INCB, the United Nations Office of Drugs and Crime, the United Nations Commission on Narcotic Drugs and the WHO, harmonising the areas of responsibility of these organisations in terms of the chemicals that can fall under the proposed draft decision."[100]

Subsequently, on 11 March 2021, when the "Understanding" was discussed by the Executive Council, Russian Ambassador Shulgin stated that: "[W]e must not forget that the control and verification mechanisms for most CNS-acting chemicals are provided for by other international treaties". And argued against "putting an entire class of chemical compounds under the control of the [CWC] Convention, without addressing the matter with the United Nations Commission on Narcotic Drugs, the United Nations Office on Drugs and Crime, the International Narcotics Control Board, or the World Health Organization".[101]

Following the adoption of the "Understanding" by CSP-26 in December 2021,[102] it may be worthwhile revisiting the March 2021 Russian paper. Although the intervention of Russia may have been seen by certain States and observers as part of a strategy to confuse, delay or halt OPCW action to prohibit use of aerosolised CNS-acting chemicals in law enforcement; certain issues raised, and suggested areas for clarification and action, made by Russia are worthy of subsequent examination. This includes the recommended engagement with the INCB, the United Nations Office of Drugs and Crime, the United Nations Commission on Narcotic Drugs and the WHO.

10.5 Conclusion

It is clear that, although both the SCND and the UN CPS were developed as crime prevention and control instruments and not as mechanisms for disarmament or arms control, these treaties do restrict the use of a range of narcotic and psychotropic drugs previously investigated as potential CNS-acting chemical agent weapons to "medical and scientific purposes". Further technical analysis and engagement with the relevant treaty bodies are required to establish the implications of these restrictions upon current or future attempts by States to research, develop, manufacture, transfer, acquire, stockpile, or use such drugs as CNS-acting chemical agent weapons. Similarly, to date there does not seem to have been any concerted attempt, at least publicly, by any States that have previously sought to address the issue of CNS-acting chemical agent weapons in the OPCW, to highlight this issue in relevant meetings of the SCND and CPS, or to bring this issue to the attention of the INCB. Such State actions would help to clarify the nature and scope of SCND and CPS applicability and also the current willingness of SCND and CPS State Parties to engage with these previously unexplored issues.

Given the SCND and CPS focus upon preventing the criminal misuse of narcotic and psychotropic drugs by non-State actors, there may be greater potential capability in both Conventions' mechanisms and implementing bodies, and more immediate willingness amongst States Parties to explore the two treaties' applicability to preventing and addressing CNS-acting chemical agent weapons research, development, manufacture, transfer, acquisition, stockpiling, transfer and use when undertaken by non-State actors including criminal organisations, terrorists, paramilitaries, private military and security companies, and armed opposition forces.

Furthermore, regardless of whether the limitations on use established by one or both treaties were deemed, by their States Parties, to be applicable to Scheduled drugs employed as CNS-acting chemical agent weapons, the reporting (and potentially the investigatory) mechanisms of both Conventions should be considered as potential routes for obtaining information relevant to the regulation of such substances. Similarly, there is great potential for profitable consultation, cooperation and collaboration between the relevant treaty bodies and international organisations concerned with preventing the criminal misuse of narcotic and psychotropic drugs (*i.e.*, the INCB and UNODC) and those concerned with preventing the development, stockpiling, transfer and use of chemical weapons utilizing such substances (*i.e.* the OPCW).

To date, civil society and State discourse in this area has been limited and superficial; further technical and legal research into the applicability of both Conventions to CNS-acting chemical agent weapons, and engagement with their States Parties, should be prioritized as part of a broad HAC regulatory strategy.

References

1. League of Nations, International Opium Convention, signed at the Hague on 23rd January 1912, Treaty Series, vol. VIII, number 222.
2. Relevant instruments included: Agreement concerning the Manufacture of, Internal Trade in, and use of Prepared Opium (1925); International Opium Convention (1925); Convention for Limiting the Manufacture and Regulating the Distribution of Narcotic Drugs (1931); Agreement for the Control of Opium Smoking in the Far East (1931); Convention for the Suppression of the Illicit Traffic in Dangerous Drugs (1936); 1946 Lake Success Protocol; 1948 Paris Protocol; and the 1953 New York Opium Protocol.
3. Rabbat was previously Drug Control Officer with the Narcotics Control and Estimates Section at the International Narcotics Control Board Secretariat.
4. P. Rabbat, Control of substances under the international drug control Conventions, in International Committee of the Red Cross, Expert Meeting, *"Incapacitating Chemical Agents", Law Enforcement, Human*

Rights Law and Policy Perspectives, Montreux, Switzerland, 24th–26th April 2012, ICRC, Geneva, January 2013, p. 80.

5. The UN Convention against Illicit Traffic in Narcotic Drugs and Psychotropic Substances, 1988, establishes the framework for the international control of precursors. It provides comprehensive measures against drug trafficking, including provisions against money laundering and the diversion of precursors chemicals. It provides for international cooperation through, for example, extradition of drug traffickers, controlled deliveries and transfer of proceedings. Text of the CITNDPS and a Commentary is available on International Narcotics Control Board website: at https://www.incb.org/incb/en/precursors/1988-convention.html (accessed 21 February 2025).
6. UN, Single Convention on Narcotic Drugs, 1961, as amended by the Protocol amending the Single Convention on Narcotic Drugs, New York, 8th August 1975.
7. For details of States Parties to the SCND, see: UN Treaty Collection Database, https://treaties.un.org/pages/ViewDetails.aspx?src=TREATY&mtdsg_no=VI-18&chapter=6&clang=_en (accessed 21 February 2025).
8. Text of SCND available on International Narcotics Control Board, available at: https://www.incb.org/documents/Narcotic-Drugs/1961-Convention/convention_1961_en.pdf (accessed 21 February 2025).
9. UN, Single Convention on Narcotic Drugs, Article 4, 1961.
10. INCB website, mandate and functions, https://www.incb.org/incb/en/about/mandate-functions.html (accessed 21 February 2025).
11. INCB website, mandate and functions, https://www.incb.org/incb/en/about/mandate-functions.html (accessed 21 February 2025).
12. UN, Single Convention on Narcotic Drugs, Article 3, 1961.
13. International Narcotics Control Board, Yellow List, *List of narcotic drugs under international control*, Annex to Forms A, B and C, 63rd edition, July 2024 https://www.incb.org/incb/en/narcotic-drugs/Yellowlist/yellow-list.html (accessed 21 February 2025).
14. (a) In addition to Chapters 1 and 2 of this publication see: F. Sidell, E. T. Takafuji and D. R. Franz, Military Aspects of Chemical and Biological Warfare, Office of the Surgeon General, U.S. Army, Washingon D.C., 1997; (b) J. Perry Robinson, Disabling Chemical Weapons: A Documented Chronology of Events, 1945–2003, Unpublished monograph, 2003; (c) M. Wheelis, L. Rózsa and M. Dando, *Deadly Cultures: Biological Weapons Since 1945*, 2006; (d) A. Pearson, M. Chevrier and M. Wheelis, *Incapacitating Biochemical Weapons*, Lexington Books, Lanham, MD, 2007.
15. INCB, Yellow List, List of narcotic drugs under international control, 2024, Schedule 1.
16. INCB, Yellow List, List of narcotic drugs under international control, 2024, Schedule 4.
17. INCB, Yellow List, List of narcotic drugs under international control, 2024, Schedule 1.

18. INCB, Yellow List, List of narcotic drugs under international control, 2024, Schedule 4.
19. INCB, Yellow List, List of narcotic drugs under international control, 2024, Schedule 1.
20. INCB, Yellow List, List of narcotic drugs under international control, 2024, Schedule 1.
21. INCB, Yellow List, List of narcotic drugs under international control, 2024, Schedule 1.
22. INCB, Yellow List, List of narcotic drugs under international control, 2024, Schedule 1.
23. INCB, Yellow List, List of narcotic drugs under international control, 2024, Schedule 1.
24. INCB, Yellow List, List of narcotic drugs under international control, 2024, Schedule 1.
25. (a) C. Muehlberger, Interrogation under Drug Influence, *J. Crim. Law Criminol.*, 1951, **42**(4), 513–528. (b) Files Show Tests For Truth Drug Began in O.S.S., New York Times, 5th September 1977, 1977.
26. Death Penalty Information Centre, State-by-State Execution Protocols, 2025. https://deathpenaltyinfo.org/executions/methods-of-execution/state-by-state-execution-protocols (accessed 25 April 2025).
27. UN, Single Convention on Narcotic Drugs, Article 4, 1961.
28. UN, Single Convention on Narcotic Drugs, Article 4, 1961.
29. UN, Single Convention on Narcotic Drugs, Article 1 (1) (w), 1961.
30. UN, Commentary on the Single Convention on Narcotic Drugs, 1961, prepared by the Secretary-General in accordance with Paragraph 1 of ECOSOC Resolution 914.D (XXXIV) of 3rd August 1962, published 1973. https://www.unodc.org/documents/commissions/CND/Int_Drug_Control_Conventions/Commentaries-OfficialRecords/1961Convention/1961_COMMENTARY_en.pdf (accessed 21 February 2025).
31. UN, Commentary on the Single Convention on Narcotic Drugs, 1961, prepared by the Secretary-General in accordance with Paragraph 1 of ECOSOC Resolution 914.D (XXXIV) of 3rd August 1962, published 1973, p. 33, paragraph 6.
32. UN, Commentary on the Single Convention on Narcotic Drugs, 1961, prepared by the Secretary-General in accordance with Paragraph 1 of ECOSOC Resolution 914.D (XXXIV) of 3rd August 1962, published 1973, p. 33, paragraph 5.
33. UN, Commentary on the Single Convention on Narcotic Drugs, 1961, prepared by the Secretary-General in accordance with Paragraph 1 of ECOSOC Resolution 914.D (XXXIV) of 3rd August 1962, published 1973, p. 34, paragraph 8.
34. P. Rabbat, Control of substances under the international drug control Conventions, in International Committee of the Red Cross, Expert Meeting, *"Incapacitating Chemical Agents", Law Enforcement, Human Rights Law and Policy Perspectives*, Montreux, Switzerland, 24th–26th April 2012, ICRC, Geneva, January 2013, p. 83.

35. P. Rabbat, Control of substances under the international drug control Conventions, in International Committee of the Red Cross, Expert Meeting, *"Incapacitating Chemical Agents", Law Enforcement, Human Rights Law and Policy Perspectives*, Montreux, Switzerland, 24th–26th April 2012, ICRC, Geneva, January 2013, p. 83.
36. UN, Single Convention on Narcotic Drugs, Article 19.1 and 19.2, 1961.
37. UN, Single Convention on Narcotic Drugs, Article 20, 1961.
38. UN, Single Convention on Narcotic Drugs, Article 15, 1961.
39. UN, Single Convention on Narcotic Drugs, Article 20.3, 1961.
40. An archive containing each Report of the International Narcotics Board from 1968 to 2023 is available at http://www.incb.org/incb/en/publications/annual-reports/annual-report.html (accessed 25 April 2025).
41. UN, Commentary on the Single Convention on Narcotic Drugs, 1961, prepared by the Secretary-General in accordance with Paragraph 1 of ECOSOC Resolution 914.D (XXXIV) of 3rd August 1962, published 1973, p. 233, paragraph 2.
42. UN, Single Convention on Narcotic Drugs, Article 14.1, 1961.
43. UN, Single Convention on Narcotic Drugs, Article 14.1, 1961.
44. A review was undertaken of relevant documents publicly available on the INCB website on 25 April 2025.
45. International Narcotics Control Board, *Report of the International Narcotics Control Board for 2003*, United Nations, New York, 2004.
46. INCB, *Report of the International Narcotics Control Board for 2003.* New York: United Nations. paragraph 215, 2004.
47. INCB, *Report of the International Narcotics Control Board for 2003.* New York: United Nations. paragraph 216, 2004.
48. P. Rabbat, Control of substances under the international drug control Conventions, in International Committee of the Red Cross, Expert Meeting, *"Incapacitating Chemical Agents", Law Enforcement, Human Rights Law and Policy Perspectives*, Montreux, Switzerland, 24th–26th April 2012, ICRC, Geneva, January 2013, p. 84.
49. E. Hoffberger-Pippan, *Less-Lethal Weapons under International Law, a three-dimensional perspective*, Cambridge University Press, 2021, p. 44.
50. For details of States Parties to the Convention on Psychotropic Substances, see: United Nations Treaty Collection Database, available at https://treaties.un.org/pages/viewdetails.aspx?src=treaty&mtdsg_no=vi-16&chapter=6&clang=_en (accessed 21 February 2025).
51. See: United Nations, Convention on Psychotropic Substances, 1971, available at https://www.incb.org/documents/Psychotropics/conventions/convention_1971_en.pdf (accessed 21 February 2025); Introduction to Convention on INCB website.
52. United Nations, Convention on Psychotropic Substances, Article 2, 1971.
53. International Narcotics Control Board (2024) Green List, List of Psychotropic Substances under International Control, 35th edition, Annex to the annual statistical report on psychotropic substances (form P),

December. https://www.incb.org/incb/en/psychotropics/green-list.html (accessed 21 April 2025).

54. (a) In addition to Chapters 1 and 2 of this publication see: F. Sidell, E. T. Takafuji and D. R. Franz, Military Aspects of Chemical and Biological Warfare, Office of the Surgeon General, U.S. Army, Washingon D.C., 1997; (b) J. Perry Robinson, Disabling Chemical Weapons: A Documented Chronology of Events, 1945–2003, Unpublished monograph, 2003; (c) M. Wheelis, L. Rózsa and M. Dando, *Deadly Cultures: Biological Weapons Since 1945*, 2006; (d) A. Pearson, M. Chevrier and M. Wheelis, *Incapacitating Biochemical Weapons*, Lexington Books, Lanham, MD, 2007.
55. INCB, Green List, List of Psychotropic Substances under International Control, 35th edition, 2024, Schedule 2.
56. INCB, Green List, List of Psychotropic Substances under International Control, 35th edition, 2024, Schedule 2.
57. INCB, Green List, List of Psychotropic Substances under International Control, 35th edition, 2024, Schedule 1.
58. INCB, Green List, List of Psychotropic Substances under International Control, 35th edition, 2024, Schedule 1.
59. INCB, Green List, List of Psychotropic Substances under International Control, 35th edition, 2024, Schedule 2.
60. INCB, Green List, List of Psychotropic Substances under International Control, 35th edition, 2024, Schedule 4.
61. INCB, Green List, List of Psychotropic Substances under International Control, 35th edition, 2024, Schedule 2.
62. INCB, Green List, List of Psychotropic Substances under International Control, 35th edition, 2024, Schedule 1.
63. (a) In addition to Chapters 1 and 2 of this publication see: F. Sidell, E. T. Takafuji and D. R. Franz, Military Aspects of Chemical and Biological Warfare, Office of the Surgeon General, U.S. Army, Washingon D.C., 1997; (b) J. Perry Robinson, Disabling Chemical Weapons: A Documented Chronology of Events, 1945–2003, Unpublished monograph, 2003; (c) M. Wheelis, L. Rózsa and M. Dando, *Deadly Cultures: Biological Weapons Since 1945*, 2006; (d) A. Pearson, M. Chevrier and M. Wheelis, *Incapacitating Biochemical Weapons*, Lexington Books, Lanham, MD, 2007.
64. INCB, Green List, List of Psychotropic Substances under International Control, 35th edition, 2024, Schedule 3.
65. INCB, Green List, List of Psychotropic Substances under International Control, 35th edition, 2024, Schedule 2.
66. INCB, Green List, List of Psychotropic Substances under International Control, 35th edition, 2024, Schedule 4.
67. INCB, Green List, List of Psychotropic Substances under International Control, 35th edition, 2024, Schedule 3.
68. INCB, Green List, List of Psychotropic Substances under International Control, 35th edition, 2024, Schedule 4.

69. Death Penalty Information Centre, State-by-State Execution Protocols, 2025. https://deathpenaltyinfo.org/executions/methods-of-execution/state-by-state-execution-protocols (accessed 25 April 2025).
70. INCB, Green List, List of Psychotropic Substances under International Control, 35th edition, 2024, Schedule 3.
71. INCB, Green List, List of Psychotropic Substances under International Control, 35th edition, 2024, Schedule 2.
72. S. Riley, Navigating the new era of assisted suicide and execution drugs, *J. Law Biosci.*, 20172017, **4**(2), 424–434.
73. United Nations, Convention on Psychotropic Substances, Article 5, 1971.
74. United Nations, Convention on Psychotropic Substances, Article 7.a, 1971.
75. United Nations, Convention on Psychotropic Substances, Article 7.b, 1971.
76. United Nations, Convention on Psychotropic Substances, Article 4.a–c, 1971.
77. A review was undertaken of relevant documents publicly available on the INCB website on 25 April 2025.
78. United Nations, Convention on Psychotropic Substances, Article 16, 1971.
79. United Nations, Convention on Psychotropic Substances, Article 18, 1971.
80. An archive containing each Report of the International Narcotics Board from 1968 to 2023 is available at http://www.incb.org/incb/en/publications/annual-reports/annual-report.html (accessed 25 April 2025).
81. INCB, *Report of the International Narcotics Control Board for 2003*. New York: United Nations, paragraph 216, 2004.
82. Unlike the SCND, there is no provision in the CPS for the INCB to formally receive information from inter-governmental organizations or international NGOs.
83. United Nations, Convention on Psychotropic Substances, Article 19.1.a, 1971.
84. United Nations, Convention on Psychotropic Substances, Article 19.1.a, 1971.
85. United Nations, Convention on Psychotropic Substances, Article 19.1.c and Article 19.2, 1971.
86. A review was undertaken of relevant documents publicly available on the INCB website on 21 February 2025.
87. (a) International Committee of the Red Cross, Incapacitating chemical agents: implications for international law, expert meeting, Montreux, Switzerland, 24–26 March 2010; (b) International Committee of the Red Cross, Expert Meeting: "Incapacitating Chemical Agents", Law Enforcement, Human Rights Law and Policy Perspectives, Montreux, Switzerland, 24–26 April 2012.

88. M. Crowley, Potential implications for disarmament and other areas of international law: in: International Committee of the Red Cross, *Incapacitating chemical agents: implications for international law, expert meeting*, Montreux, Switzerland, 24–26 March 2010, pp. 51–52.
89. P. Rabbat, Control of substances under the international drug control Conventions, in International Committee of the Red Cross, Expert Meeting, *"Incapacitating Chemical Agents", Law Enforcement, Human Rights Law and Policy Perspectives*, Montreux, Switzerland, 24th–26th April 2012, ICRC, Geneva, January 2013, p. 80.
90. International Committee of the Red Cross, *Toxic Chemicals as Weapons for Law Enforcement, A Threat to Life and International Law?*, ICRC, Geneva, September 2012.
91. R. Mathews, Central Nervous System-acting chemicals and the Chemical Weapons Convention: A former Scientific Adviser's perspective, *Pure Appl. Chem.*, 2018, **90**(10), 1559–1575.
92. United Kingdom, *Convergence of science between international treaties and agreements*, RC-5/WP.2, OPCW, The Hague, April 2023, p. 1.
93. United Kingdom, *Convergence of science between international treaties and agreements*, RC-5/WP.2, OPCW, The Hague, April 2023, p. 4.
94. SAB, *Report by the Director-General, Report of the Scientific Advisory Board on developments in science and technology to the 5th Review Conference*, RC-5/DG.1. OPCW, The Hague, February 2023, p.8, paragraph 31.
95. Russian Federation, *Aerosolisation of Central Nervous System-Acting Chemicals for Law Enforcement Purposes*, RC-4/NAT.9. OPCW, The Hague, November 2018, p. 3.
96. Russian Federation, *Questions and answers, regarding the proposal to prohibit the aerosolised use of central nervous system-acting chemicals for law enforcement purposes*, EC-96/NAT.6, OPCW, The Hague, March 2021.
97. Russian Federation, *Questions and answers, regarding the proposal to prohibit the aerosolised use of central nervous system-acting chemicals for law enforcement purposes*, EC-96/NAT.6, OPCW, The Hague, March 2021, p. 1.
98. Russian Federation, *Questions and answers, regarding the proposal to prohibit the aerosolised use of central nervous system-acting chemicals for law enforcement purposes*, EC-96/NAT.6, OPCW, The Hague, March 2021, pp. 1–2.
99. Russian Federation, *Questions and answers, regarding the proposal to prohibit the aerosolised use of central nervous system-acting chemicals for law enforcement purposes*, EC-96/NAT.6, OPCW, The Hague, March 2021, p. 2.
100. Russian Federation, *Questions and answers, regarding the proposal to prohibit the aerosolised use of central nervous system-acting chemicals for law enforcement purposes*, EC-96/NAT.6, OPCW, The Hague, March 2021, p. 9.

101. Russian Federation, *Statement by H.E. Ambassador A.V. Shulgin, Permanent Representative of the Russian Federation to the OPCW*, EC-96/NAT.63. OPCW, The Hague, March 2021.
102. Conference of States Parties, *Decision: Understanding regarding the aerosolised use of central nervous system-acting-chemicals for law enforcement purposes*, C-26/DEC.10. OPCW, The Hague, 1 December 2021.

CHAPTER 11

The Scientific and Medical Communities' Role in Preventing Weaponization of CNS-acting Chemical Agents and Broader Misuse of Neuroscience

11.1 Introduction

In its 2004 public statement "*Preventing hostile use of the life sciences*", the International Committee of the Red Cross (ICRC) declared:

> "If measures to prevent the hostile use of advances in the life sciences are to work, a culture of responsibility is necessary among individual life scientists. This applies whether these scientists are working in industry, academia, health, defense or in related fields such as engineering and information technology."[1]

According to the ICRC, such a culture of responsibility was also needed "within the institutions that employ scientists and fund research in the life sciences."[2] Similar calls to the scientific (and medical) communities have also been made by the States Parties to the BTWC and the CWC at Review Conferences.[3]

The following sections of the chapter will explore the current range of initiatives being undertaken by those in the scientific and medical

Preventing Weaponization of CNS-acting Chemicals: A Holistic Arms Control Analysis
By Michael Crowley and Malcolm Dando

Published by the Royal Society of Chemistry, www.rsc.org

communities to nurture a 'culture of responsibility'. First, we discuss the growing recognition of the 'dual-use dilemma' and the consequent requirement for effective collective oversight of research, as illustrated through certain State Human Brain Projects. This is followed by a discussion of the potential utility of oaths, codes and pledges and the parallel processes of education and awareness-raising in building the appropriate norms of behaviour for the scientific and biomedical communities. The difficult practical application of such principles by professional bodies, as well as by individual scientists including through non-participation and 'whistle-blowing' of unethical practices, will be explored.

11.2 Regulating 'Dual-use' Research

Whilst scientific discoveries and their application are intrinsic to the advance of civilisation, most science to varied degrees is 'dual use' or 'multi-use'[4] in nature – it has the potential for benign and malign pathways and applications. A balance must be achieved between allowing scientific progress and open-source disclosure, whilst ensuring against potential misuse of that knowledge. To date, much of the discourse amongst the life science community concerning how best to combat the proliferation and misuse of chemical and biological weapons has concentrated on regulating the actions of individual life scientists conducting 'dual-use' research of potential concern namely – "research that, based on current understanding, can be reasonably anticipated to provide knowledge, products, or technologies that could be directly misapplied by others to pose a threat to public health and safety, agriculture, plants, animals, the environment, or material".[5] Highly influential in this discourse have been the 2004 *Fink Report*[6] and the 2006 *Lemon Relman Report*[7] both produced under the auspices of the National Research Council of the US National Academies. The two reports highlighted the importance of taking a comprehensive approach to analysing 'dual-use' research of potential concern. Among its recommendations the Lemon Report highlighted the need to:

> "2a. Recognize the limitations inherent in any agent-specific threat list and consider instead the intrinsic properties of pathogens and toxins that render them a threat and how such properties have been or could be manipulated by evolving technologies.
>
> 2b. Adopt a broadened awareness of threats beyond the classical 'select agents' and other pathogenic organisms and toxins, so as to include, for example, *approaches for disrupting host homeostatic and defense systems and for creating synthetic organisms.*"[8] (Emphasis added)

The *Lemon Relman Report* in its Executive Summary also stated that:

> "...The accelerating pace of discovery in the life sciences has fundamentally altered the threat spectrum. *The immune, neurological, and*

> *endocrine systems are particularly vulnerable to disruption by manipulation of bioregulators*. . .[These]. . . small, biologically active compounds, pose an increasingly apparent dual-use risk. This risk is magnified by improvements in targeted delivery technologies that have made the potential dissemination of these compounds much more feasible than in the past." (Emphasis added)[9]

The broad threat spectrum enunciated by the Lemon Relman Committee and its specific reference to bioregulators appears to capture CNS-acting agents and other types of incapacitating chemical agents within their scope.

As part of its comprehensive proposals for addressing dual use research, the Lemon Committee recommended "the adoption and promotion of a common culture of awareness and a shared sense of responsibility within the global community of life scientists" which would:

> "4a. Recognize the value of formal international treaties and conventions, including the. . . BWC and the. . .CWC.
>
> 4b. Develop explicit national and international codes of ethics and conduct for life scientists.
>
> 4c. Support programs promoting beneficial uses of technology in developing countries.
>
> 4d. Establish globally distributed, decentralized, and adaptive mechanisms with the capacity for surveillance and intervention in the event of malevolent applications of tools and technologies derived from the life sciences."[10]

In order to effectively regulate chemical and life science 'dual-use' research it has been recognised that a wider circle of additional relevant actors and communities, beyond individual life scientist researchers, need to be sensitised, informed of their responsibilities and encouraged to actively prevent 'dual use' of concern with consequent voluntary and mandatory regulatory measures introduced. For example, Garfinkel, Endy, Epstein and Friedman defined three major points for potential policy intervention namely: "commercial firms that sell synthetic DNA (oligionucleotides, genes or genomes) to users; owners of laboratory 'bench-top' DNA synthesizers, with which users can produce their own DNA; the users (consumers) of synthetic DNA themselves and the institutions that support and oversee their work".[11]

As a result of the Fink and Lemon Relman Reports, and the work of others, a range of oversight structures and processes were established notably by US and European governments, scientific bodies, academic institutions, funders and publishers to review 'dual-use' research of potential concern, to assess the risks and benefits of such research and determine whether the proposals needed to be modified or withdrawn.[12]

However, following analysis of the application of such oversight measures in practice, in 2010 Rappert concluded that: "such procedures rarely conclude that manuscripts, grant applications or experiment proposals should not be undertaken or restricted".[13] Similarly, Van Aken and Hunger analysed the application of biosecurity policies agreed by a group of 32 influential science journals under which manuscripts could be modified or rejected where "the potential harm of publication outweighs the potential societal benefits".[14] Despite such policies having been established in 2003, Van Aken and Hunger found in 2009 that no manuscript had ever been rejected on security grounds.[15] Rappert contended the "same could be said" of those funders that have established submission-oversight systems.[16] Furthermore, Rappert has stated that "even more notable with these review processes is the infrequency with which they have identified items 'of concern' in the first place".[17] Whilst information relating to the research controls of government departments (especially defence-related ones) is not readily available, Rappert argued that in relation to universities and other publicly funded agencies "it seems justifiable to conclude that – barring dramatic changes – oversight processes will identify little research as posing security concerns and will stop next to nothing".[18]

Others have criticised the voluntary nature of the existing controls on life science 'dual-use' research. For example, commenting upon the release of a draft of the National Science Advisory Board for Biosecurity (NSABB) Proposed Framework for the Oversight of Dual Use Life Sciences Research: Strategies for Minimizing the Potential Misuse of Research Information,[19] the Sunshine Project Director, Hammond stated that the "NSABB is divorced from reality if its members believe that another set of voluntary NIH [National Institutes of Health] guidelines is sufficient, and would be remotely effective, at preventing dual-use disasters".[20] Research conducted by the Sunshine Project from 2004 to 2007 indicated that many US organisations obliged to follow NIH guidelines did not do so.[21] A 2007 Sunshine Project survey discovered that 18 of the top 20 US biotechnology companies did not comply with existing voluntary NIH biotechnology guidelines.[22] Instead of a voluntary approach, Hammond stated: "Effective federal management of dual-use risks requires making safety and security oversight truly mandatory and subject to the sobering light of public scrutiny".[23] The debate on what kinds of measures of scientific self-regulation were required continued through the following years with several adjustments and attempts to strengthen these systems being agreed within the United States.

However, in June 2024, the *Bulletin of the Atomic Scientists* carried an article titled "MIT researchers ordered and combined parts of the 1918 pandemic influenza virus. Did they expose a security flaw?" In a detailed account linked to the article the authors of the research stated that:[24]

> "We performed an FBI-overseen experiment to determine whether 38 synthesis providers would ship gene fragments that could be misused. Appending a harmless sequence as camouflage yielded DNA capable of

> generating infectious pandemic 1918 influenza virus or *ricin toxin* from 24 providers; adding correctable non-synonymous mutations led to shipment by another 12. Constructs were readily assembled following receipt, underscoring the importance of adopting and auditing improved screening systems to safeguard biotechnology." (Emphasis added)

The article in the *Bulletin of the Atomic Scientists* noted that people involved in the system for checking such orders questioned the conclusion that the system was inadequate as the laboratory making the order was checked to ensure its legitimacy as well as the material requested. The senior author of the original report disagreed suggesting to the *Bulletin* that "[I]f the system was designed to ship DNA sufficient to generate a pandemic virus that once killed over 50 million people to anyone loosely affiliated with a laboratory. . .then we may want to design a better system".

In 2004, the influential US National Academies Fink Committee Report on "*Biotechnology Research in an Age of Terrorism*" had highlighted seven classes of dual use life science research that "illustrate the types of endeavors or discoveries that will require review and discussion by informed members of the scientific and medical community before they are undertaken or, if carried out, before they are published in full detail".[25] One of these seven classes was experiments that "Would enable the weaponization of a biological agent or toxin". Twenty years later, such 'dual-use' research still did not appear to be under adequate control in the State that had probably endeavoured most to develop and implement such measures. However, the situation, in the US at least may be about to change. Following the COVID-19 pandemic it initially seemed that the most likely cause of the outbreak was from a natural spillover of the virus from an animal species. However, more recently numerous scientists have argued that a laboratory accidental release was also possible and maybe the more likely cause of the pandemic. At present it appears probable that the US will introduce mandatory legal control of some kinds of dangerous experimentation.[26]

As well as concerns about the implementation of both voluntary and mandatory oversight systems in practice, a further concern relates to the limited range of issues being considered by such bodies. The discourse and much of the current activity appear to be concentrated upon preventing the diffusion of "dual use" knowledge, skills and materials to various non-State actors with malign intent, principally terrorist organisations. Insufficient attention has been given to utilising existing "dual use" monitoring mechanisms or adopting additional process to specifically combat the misuse of "dual use" expertise in State programmes, even though national chemical and biological weapon research and development programmes arguably pose a greater danger to the CWC and BTWC than the limited activities of non-State actors in this area. This can be illustrated in the field of neuroscience and the Brain projects.

11.2.1 'Dual-use' Neuroscience

As explored in Chapters 3 and 4, during this century, the technologies available to neuroscientists have developed at an extremely rapid pace and made it ever more possible to understand the neuronal circuits in the central nervous system (CNS) that produce our behaviour, and the roles played by bioregulators in such processes. The benefits of such research in treating people with brain dysfunctions have led certain States and regions in recent years to initiate large-scale brain research projects notably in China, the EU and the US. In some States, notably China and the US, scientists from military medical institutions and other defence-related facilities have been involved in much of the research, raising potential dual-use concerns. Whilst effective State transparency, monitoring, oversight and accountability measures to prevent such research being employed in CNS-acting agent weaponization are essential, the development of a strong 'culture of responsibility' within the neuroscience community will provide another barrier against individual neuroscientist involvement in such unethical activities. The differing approaches taken by the Chinese, EU and US Brain projects to such dual use research are explored below.

11.2.1.1 China Brain Project and Potential Dual-use Concern

A number of areas of the China Brain Project (CBP) and work of Chinese scientists on neurological systems and associated bioregulators (including noradrenaline, 5-HT, and orexin) in simple animal models—such as fruit flies, zebrafish and mice—have potential 'dual-use' applicability. However, in order to better understand more complex human behaviours and the mechanisms that produce them, large-scale work is also being carried out in China on the CNS and responses of non-human primates (NHPs), like macaque monkeys.[27] This work has potential dual-use implications as previous State-level biological weapons programmes did not limit their activities to manipulation of basic functions of the brain, but also sought to influence human emotions, cognition and behaviour. Consequently, as Palchik *et al.* have highlighted, this:

> "[R]aises concerns about tacit capabilities... and the yoking of NHP studies and findings to military agendas under programs of dual- or direct-use.... Of particular note in such efforts is *Junweikejiwei*, the newly developed Chinese research agency that conjoins efforts of the CAS [Chinese Academy of Sciences] and China's Ministry of Defense, and which is modelled after the United States' Defense Advanced Research Projects Agency (DARPA), to engage rapid, high-risk/high-return approaches to bioscience and technology".[28]

The CBP includes studies of higher functions of the brain using advanced biotechnology with NHP as test subjects. A presentation from one of China's leading researchers in 2016 highlighted the "Goals of non-human primate research" which included "1. Studies of the cognitive functions using non-human primate as the animal model" and "2. Generation of genetically modified [including transgenic] monkeys as animal models of human brain disorders and for basic neurobiology research". This research explored the neural basis of cognition including how cells establish synaptic contacts and generate neural circuit activities.[29] Whilst the purposes of such research may be benign, there are dangers that it could be abused to aid the development of bioregulator weapons attacking human cognition.

It is not easy to discover how 'dual-use' issues are dealt with in China, but it is not difficult to find reasons for dual-use concerns. For example, a recent article by Chinese scientists was titled *Military Brain Science – How to influence future wars*. This had a section on "Interfering with the brain causing brain dysfunction and a loss of control with the 'smokeless'" method which stated that:[30]

> "According to the physiological basis of the transfer of brain nerve potentials and the transmission of chemical transmitters, nervous system incapacitating agents and body incapacitating agents will be developed, including brainwave interference weapons.... The strategy of interfering with the brain can affect mentality, influence thinking, and affect decision making to create a whole new 'brain war' combat style and redefine the battlefield."

It is hardly surprising in that kind of context that concerns about the China Brain Project have been aroused.

11.2.1.2 US BRAIN Project and Potential Dual-use Concern

The US Brain Research Through Advancing Innovative Neurotechnologies (BRAIN) Initiative was launched in 2013, as a public-private partnership with about $100 million in the President's Fiscal Year 2014 Budget.[31] It is notable that approximately half of the US funding came from the Department of Defense (DoD) and Defense Advanced Research Projects Agency (DARPA).[32] The potential for investigating the neural circuits producing human behaviour was highlighted in the scoping report for the US BRAIN project which stated:[33] "In considering these goals and the current state of neuroscience, the working group identified the analysis of circuits of interacting neurons as being particularly rich in opportunity with potential for revolutionary advances".

The report further stated that:

> "Truly understanding a circuit requires identifying and characterizing the component cells, defining their specific connections with one another, observing their dynamic patterns of activity as the circuit functions *in vivo* during behaviour, *and perturbing these patterns to test their*

significance. It also requires an understanding of the algorithms that govern information processing within a circuit and between interacting circuits in the brain as a whole." (Emphasis added).

A range of research subsequently funded by the BRAIN initiative has potential 'dual-use' applicability, such as that relating to the functioning of opioid receptors,[34] and that examining the modulation of thalamo-cortico-striatal circuits of the brain in pain processing and reward learning.[35] Although these studies were undertaken with apparently benign objectives, these same receptors are those that can be attacked by fentanyl and related agents that are certainly of current concern as potential CNS-acting chemical agent weapons. Likewise, work on circuits underlying the sleep/wake cycle and arousal mechanisms that could be of dual-use interest includes a paper[36] on the neuropeptide hypocretin (also known as orexin), the loss of which is a cause of narcolepsy. The paper examined the role of this neuropeptide as a hub for arousal and motivation. Another paper examined how rhythms of the sleeping brain can affect the formation of memory.[37] Similarly, it is not difficult to find projects funded by the BRAIN initiative that are concerned with the circuits underlying more complex behaviour such as anxiety.[38] Likewise, detailed investigations of key structures within such circuits as the amygdala[39] and the LC (Locus Coeruleus)[40] could aid the development of the means to alleviate suffering from overactive fear processing in post-traumatic stress, but such knowledge could also be misused to induce fear.

In the middle of the 10 years of the US BRAIN Initiative a review was made of both the scientific and the neuroethics aspects of the project.[41] As is clear from the parts of the report of the review shown in Box 11.1, dual-use issues were expected to be encountered in the scientific research being funded.

And the report sets out initial steps that should be taken to deal with the problem of 'dual use' through better integration of neuroethics with neuroscience.[42] However, the question that has to be asked is whether neuroethics is at present up to the task of dealing with the problem of dual use.

While neuroethicists internationally have recently stated that they were ready to engage with such societal implications of their work,[43] there is reason to doubt that they would easily be able to do so. For example, in May 2024 a paper[44] was published that attempted to analyse the "abstracts presented at the International Neuroethics Society (INS) annual meetings and published as top abstracts in *AJOB Neuroscience* in the last decade (2011–2020)". The aim of the authors was to use this data to "evaluate *who neuroethics scholarship addresses and what it recommends*" (emphasis added).

The authors concluded that "[R]oughly 77% of all abstracts did not explicitly address a target audience, yet nearly all of these were implicitly directed at other scholars" and that "[O]nly a subset of abstracts provided practical or policy recommendations (19.7%)". They argued that there was a good reason for these findings as: "[T]his likely reflects the fact that neuroethics emerged from scholarly traditions in the social sciences and

Box 11.1 Extracts from the neuroethics report of the BRAIN project.[42]

Dual use of BRAIN Initiative neuroscience What do we know?
'Dual-use' technologies are those that can be used for both peaceful and military purposes. Neurotechnology is a dual-use technology.... Neuroethics analysis of military use of neuroscience suggests that the question is not if the military will use neuroscience, but how – and thus ethical accountability is critical...

What could we learn? Neuroethics research opportunities
Most importantly, there is a need for more precise evidence on what types of neurotechnology are actually being developed for military national security purposes. Additionally, greater transparency and collaboration can be developed between the neuroscience and neuroethics communities...

Potential next steps: Implementable goals
First, neuroscientists and neuroethicists should convene to consider the potential for dual use of fundamental BRAIN Initiative-supported research...

Second, neuroscientists and neuroethicists should expand dialogue between NIH staff, the NIH-associated neuroethics community and other entities exploring the challenge of dual-use neurotechnology both domestically and internationally...

Finally, more specific guidance should be provided to BRAIN Initiative-funded researchers on potential ways their research could be used for military or national-security purposes. At present, The BRAIN Initiative® does not have a formal dual-use education and awareness training program. Development of such materials could be integrated into BRAIN 2.0.

humanities, which have a more robust tradition of critique and argumentation rather than practical application". And furthermore, they suggest that their findings give support "to critiques expressed by us and others, that neuroethics and bioethics ...do not often attend to considerations of audience and downstream applications". In that context it is hardly surprising that no specific mention of biosecurity and dual use issues can be found in the summaries of the biannual meetings of the Neuroethics Working Group of the US Brain Initiative from January 2020 to January 2024.[45] It is therefore unlikely that the issues identified in Box 11.1 will have been seriously addressed within the US BRAIN Initiative.

11.2.1.3 European Brain Project

In October 2013 the European Union launched the Human Brain Project (HBP) – a 1 billion Euro European Commission Future and Emerging Technologies (FET) Flagship project. The HBP aims to put in place a cutting-edge,

ICT-based scientific Research Infrastructure for brain research, cognitive neuroscience and brain-inspired computing. This 10-year project involves scientists and developers at 116 universities and research centres across 19 European countries.[46] The objectives of this Flagship project were stated to be to:

> "Create and operate a European scientific Research Infrastructure for brain research, cognitive neuroscience, and other brain-inspired sciences;
>
> Gather, organise and disseminate data describing the brain and its diseases;
>
> Simulate the brain;
>
> Build multi-scale scaffold theory and models of the brain;
>
> Develop brain-inspired computing, data analytics and robotics;
>
> Ensure that the HBP's work is undertaken responsibly and that it benefits society."

Several HBP projects raised potential dual use concerns including Sub-project 3 on Systems and Cognitive Neuroscience the goal of which is to:

> "...[U]ncover neural mechanisms underlying cognitive processes .... *How is consciousness generated by the brain?* .... Similarly, how can disparate phenomena such as *sleep and wakefulness* emerge from the same cortico-thalamic systems in the brain? .... SP3 also investigates *brain mechanisms of memory* .... In summary, by its cross-disciplinary approach to multiple levels of neural and brain organization, we will work to elucidate mind–brain relationships that have eluded explanation for centuries. (Emphases added)"[47]

Co-Design Project 6 on "Modelling Drug Discovery" has the objective to:

> "... Develop new strategies for more effective drug treatments of major brain diseases such as Alzheimer's, Schizophrenia, Epilepsy, Parkinson's, glioblastoma and rare diseases using computational models... Innovative neuromedicine approaches require a detailed understanding of the molecular and systems-level organisation of the human brain, the causes and mechanisms of diseases, their progression, and the response to treatments."[48]

Dando has highlighted further studies in some of the EU HBP Partnering Projects on brain networks and rhythms are clearly related to issues of interest to those who might be seeking to develop novel CNS-acting chemical agent weapons and other types of ICA weapons. These include projects:

> "Shedding light on the multiscale organisation of cortical computation by integrating neuronal and population activities with inter-areal interactions
>
> Motor Control and Timing in the Cerebellum: Spatio-Temporal Integration in Complex Neuronal Networks;
>
> Slow Wave Dynamics: from experiments, analysis and models to rhythm restoration;"[49]

So, like both China and the US, the EU Human Brain Project incorporates a range of dual use research projects of potential concern. It does appear, however, that the EU HBP has sought to identify and address dual-use concerns as part of its study of the ethical and social issues involved in the project's activities. This is carried out in Subproject 12 "Ethics and Society" of the core project which focuses on the implications of HBP research:

> "...Our methods include conceptual and philosophical analyses, organizing workshops and webinars, and performance of interviews with experts. We aim to develop better foresight and ethical analysis, to promote engagement with broader audiences, and to raise awareness among researchers on how to recognise ethical issues and the possibilities for action when they arise."[50]

Another welcome aspect of the European HBP intervention on dual use has been its position that the concept of dual use should not be restricted to just considering misuse of civil research in the military domain but can also take place in other domains. Therefore, they state that:

> "To help in clarifying matters, we argue that policy makers and regulators need to identify and focus on specific harmful or undesirable uses in the following four domains: political, security, intelligence and military (PSIM)..."[51]

The European HBP appears to have more explicitly sought to engage with these 'dual use' considerations and in 2018 developed a concrete series of recommendations as set out in Box 11.2.[52]

While there were some imaginative attempts to implement these recommendations and efforts to find ways to support them through biosecurity education,[53] it proved impossible to achieve such integrated and comprehensive objectives. However, there may still be ways in which cooperation between civil society and government can improve biosecurity.[54] One recent development that may well be very useful in the future is the elaboration and

Box 11.2 Recommendations for dealing with dual-use neuroscience.[49]

1. Establishment of a permanent working group dedicated to addressing and making assessments of HBP [Human Brain Project] activities with potential for DURC [Dual use research of concern]/military-warfare uses
2. Engagement of this working group in proactive discourse with relevant interest groups
3. Development of an educational program that provides material addressing DURC/military-warfare use of HBP efforts
4. Formulation of an updated mission statement that affirms the intent of all HBP research for civilian biomedical purposes and proscribes intent for specific dual use(s) of concern, inclusive of military-warfare applications
5. Establishing requirements that HBP research personnel:
   (a) participate in these educational activities on a defined, regular basis;
   (b) explicitly affirm intent of purpose;
   (c) explicate such intent in a formal statement to be included in any/all publications and proposals;
   (d) be prohibited from receiving military funding both during their tenure in an HBP-supported activity, and for a defined period of time subsequent to termination of such involvement
6. Development of white papers and position statements to more effectively guide and govern dual use(s) of both HBP research and brain science, writ large.

implementation of self-assessment systems to check potentially 'dual-use' projects[55] and wider issues of institutional care of responsible conduct.[56] These systems obviously called upon much previously published work by a wide range of authors within civil society.[57] However, to date such self-assessment systems have been largely applied to questions involved in the manipulation of microorganisms and have not been systematically applied to other potential issues of 'dual use', such as within neuroscience.

11.3 Oaths, Codes and Pledges for the Chemical and Life Science Communities

Another approach to building a culture of responsibility has been through the development of a range of non-binding ethical codes, codes of conduct and oaths or pledges. An early advocate of such activities was Nobel Peace

Prize laureate Professor Joseph Rotblat who declared in his 1995 Nobel acceptance speech that:

> "The time has come to formulate guidelines for the ethical conduct of scientists, perhaps in the form of a voluntary Hippocratic Oath.... At a time when science plays such a powerful role in the life of society, when the destiny of the whole of mankind may hinge on the results of scientific research, it is incumbent on all scientists to be fully conscious of that role, and conduct themselves accordingly. I appeal to my fellow scientists to remember their responsibility to humanity."[58]

In his 2004 review paper, UK sociologist Brian Rappert developed a basic three tier nomenclature to differentiate between the alternative types of codes, which he ascribed as:

- Codes of ethics: Aspirational codes that aim to set realistic or idealistic standards as well as alert individuals to certain issues.
- Codes of conduct: Educational or advisory codes that aim to provide guidelines for action, raise awareness of issues, and foster moral agency.
- Codes of practice: Enforceable codes that prescribe or proscribe certain behaviour.[59]

To date, the majority of attention in both the chemical and life science communities has been focussed on the development and promulgation of voluntary non-binding codes of ethics and codes of conduct – with limited success.

In general terms such codes of conduct and codes of ethics are essential elements of professional culture, as they outline a set of shared principles and norms that practitioners agree to abide by. These principles and norms reflect both the way a given profession has evolved and the way it relates to its broader social milieu. By design, codes of conduct and codes of ethics are self-governance instruments that enable practitioners in different domains to maintain a certain degree of autonomy in their affairs and preserve the integrity of their professional culture.[60] For such codes to be widely accepted and enthusiastically and effectively implemented by the relevant professional community, they need to have been generated by that community. Consequently, the processes of developing and amending existing codes of conduct and codes of ethics must be deliberative processes that primarily involve representatives of the respective professional domain. It therefore follows that the process of developing such codes for the life and chemical science communities (as for health professionals) needs to be undertaken with the active engagement of all the relevant chemical and life science stakeholders in academia, industry, and government.

11.3.1 Ethical Codes for the Life Sciences

Since the 5th BTWC Review Conference in 2002, the BTWC States Parties have collectively encouraged and supported the development of voluntary ethical codes of conduct for life scientists, with each Review Conference integrating this work into the 2003–2005,[61] 2007–2010,[62] 2012–2015[63] and 2018–2020[64] inter-sessional processes. Consequently, initiatives supporting development of such codes have been undertaken by a wide range of scientific associations and organisations including the American Society of Microbiology,[65] National Academy of Sciences, Royal Society,[66] International Centre for Genetic Engineering and Biotechnology, International Union of Biochemistry and Molecular Biology, and the International Council for the Life Sciences. The Inter-Academy Partnership of National Academies has been particularly important in this regard.[67]

A major advance, moving from this patchwork of societal action towards concerted attempts to develop a single set of ethical standards agreed and implemented by the life science community at the international level, occurred at the 8th BTWC Review Conference in 2016 with the submission of a working paper by China and Pakistan.[68] Authored by the Centre for Biosafety Research and Strategy at Tianjin University, the paper contained draft ethical guidelines for life scientists, which were intended to promote responsible science and strengthen biosecurity governance. The *Tianjin Biosecurity Guidelines for Codes of Conduct for Scientists* (or *Tianjin Guidelines* as they became known) were subsequently further cooperatively developed in 2021 by Tianjin University with experts from Johns Hopkins Center for Health Security and the Inter-Academy Partnership.[69] According to those involved, during the drafting process for the *Tianjin Guidelines*, the organizing institutions elicited feedback from scientists and governance experts from more than 20 countries across six continents, revising the guidelines accordingly. The final 10 guidelines represented a consensus set of baseline ethical principles that can inform new and supplement existing codes of conduct for scientists. Since the *Guidelines* were finalized, the organizing team has been working to raise awareness of their existence and provisions.[70]

According to their authors "The guidelines aim to prevent misuse of bioscience research without hindering beneficial outcomes, in accordance with the articles and norms of the [BTWC]".[71] They cover the areas of: ethical standards; laws and norms; responsible conduct of research; respect for research participants; research process management; education and training; research findings dissemination; public engagement on science and technology; role of institutions and international cooperation. Of particular note are the first three principles:

> "*Ethical Standards*. . .Scientists should respect human life and relevant social ethics. They have a special responsibility to use biosciences for peaceful purposes that benefit humankind, to promote a culture of responsible conduct in biosciences and to guard against the misuse of science for malicious purposes, including harm to the environment.
>
> *Laws and Norms*. . .Scientists should be aware of and observe applicable domestic laws and regulations, international legal instruments, and

> norms relating to biological research, including those on the prohibition of biological weapons. Scientists and their professional bodies are encouraged to contribute to the establishment and further development and strengthening of relevant legislation.
>
> *Responsible Conduct of Research*...Scientists should promote scientific integrity and strive to prevent misconduct in research. They should be aware of the multiple applications of biological sciences, including their potential use for developing biological weapons. Measures should be taken to prevent the misuse and negative impacts of biological products, data, expertise, or equipment." (Emphases added)[72]

Given their explicit repeated reference to responsible research and the "prohibition on biological weapons", these guidelines are a potentially important tool that may help to engender an ethical framework that discourages involvement of life scientists in the malign use of their knowledge and expertise, including for the development of CNS-acting or other types of incapacitating chemical agent weapons.

Following their development, the organizing team and supporting organisations and States worked to raise awareness of the *Guidelines*. Subsequently major scientific associations, including the InterAcademy Partnership[73] and the International Union of Biochemistry and Molecular Biology,[74] disseminated the Guidelines to their members. And in July 2021 the InterAcademy Partnership – an organisation comprising over 140 national, regional, or global academies of science, medicine, and engineering as members – formally endorsed the Guidelines.[75]

On 13 December 2023, China, Pakistan, now joined by Brazil and North Macedonia submitted a working paper to the 9th BTWC Review Conference, calling on the Review Conference to:

> "(a) Endorse the Tianjin Guidelines and encourage all stakeholders to voluntarily incorporate elements from the Guidelines in their practices, protocols, and regulations, and to disseminate the Guidelines, as appropriate; and
>
> (b) Task the inter-sessional process to exchange information, experiences and good practices about the dissemination of the Tianjin Guidelines and report the outcomes of these exchanges and dissemination to the Tenth Review Conference."[76]

Unfortunately, because of the negative international political environment, the 9th BTWC Review Conference was a failure and was unable to adopt any substantive decisions. Consequently, no formal measures were agreed collectively by the States Parties on advancing the *Tianjin Guidelines* through the BTWC system.

11.3.2 Ethical Codes for Chemists

In comparison to the time and energy expended by diverse life science bodies, the chemical science community's efforts to develop codes of conduct were more limited and centralised through the International Union of Pure and Applied Chemistry (IUPAC). In 2004, IUPAC's President and the Director-General of the OPCW agreed a joint project on chemistry education, outreach, and the professional conduct of chemists. This led to a joint IUPAC/OPCW international workshop in 2005 which concluded that codes of conduct were needed for all those engaged in science and technology using chemicals so as to "protect public health and the environment and to ensure that [such] activities... are, and are perceived to be, in compliance, with international treaties, national laws and regulations such as those relating to illicit drugs, chemical and biological weapons, banned and severely restricted chemicals".[77] The Conference also concluded that such codes were "complementary to national implementing legislation for the CWC" and would "help to achieve in-depth compliance throughout academia, industry, and government of those engaged in science and technology using chemicals". They would also "extend awareness of the general-purpose criteria of both the CWC and the BTWC and thus help ensure its effective implementation".[78] The Conference also recommended that IUPAC should develop a model code of principles as well as draft elements for codes which might be promulgated to IUPAC national adhering organisations (NAOs) and associate national adhering organisations (ANAOs) urging them to review any existing codes to ensure these elements are included.[79] IUPAC and its Committee on Chemical Research Applied to World Needs (CHEMRAWN) subsequently initiated a project to develop such a code.[80] The project was completed in November 2011, and although a formal IUPAC code of conduct had not been established to date, guiding principles for a code were developed and promoted.[81]

However, as Bowman *et al.* noted in 2018, there were still relatively few codes within the non-governmental chemistry community and the chemical industry that explicitly include provisions related to the CWC or chemical weapons.[82] A web search and analysis in 2015 of chemistry-relevant codes in the English language from the chemistry community or industry yielded only 7 out of over 140 codes from 46 countries that mention chemical weapons, and 3 that mention the CWC.[83] A partial explanation as regards the dearth of codes in the chemical industry may reflect a perception amongst certain companies that whilst their in-house research and development, and the resulting new products, can be dual-use in character they are proprietary and therefore under their control with regard to their practical in-house application. Consequently, the industry's focus has usually been on preventing misuse of their products once they get onto the market – *i.e.*, through implementing export controls and more broadly controls of clients and orders.[84]

To address this lack of concerted action, in December 2014, the German OPCW Ambassador introduced a proposal for a "Hippocratic Oath" for chemists at the 19th Conference of CWC States Parties (CSP-19):

> "...We also have to appeal to the responsibility of individuals – individuals who have the capability to develop and produce chemical weapons. This is the reason why Germany has submitted the proposal of a code of conduct for chemical professionals.... Similar to the Hippocratic oath for medical personnel, this concise text could lay the moral basis for the work of chemical professionals."[85]

Germany's proposal found favour with the States Parties and the CSP ultimately:

> "Welcomed the initiative for a text of ethical guidelines for chemical professionals related to the Convention and invited the Secretariat to inform the Council of its efforts for the advancement of the initiative and its objectives in close collaboration with relevant professional and chemical industry organizations. The Conference encouraged States Parties to discuss the matter further."[86]

Germany subsequently provided dedicated funding for two workshops in 2015 to develop the text for ethical guidelines.[87] According to Bowman *et al.*, the project was notable because, although funded and organized *via* the OPCW, the work was done by an international group of 35 scientists from academia and the chemical industry in 24 countries from all regions of the world. Several members of the TWG with close ties to IUPAC, as well as its current and immediate past presidents, and its president elect, took part in the workshops.[88]

The results of the workshops, namely the *Hague Ethical Guidelines*, were developed by consensus and then subsequently endorsed by all workshop participants.[89] They comprise a set of nine principles that can be used to support the development of new codes, but also as the basis for reviewing existing codes to explore whether they are in line with the goals of the CWC. The *Hague Guidelines* include an initial "core element" principle which states that "Achievements in the field of chemistry should be used to benefit humankind and protect the environment". This is followed by eight principles on sustainability, education; awareness and engagement; ethics; safety and security; accountability; oversight; and exchange of information.[90]

Particularly relevant is the principle on "Awareness and engagement" which states that:

> "Teachers, chemistry practitioners, and policy-makers should be aware of the multiple uses of chemicals, specifically their use as chemical weapons or their precursors. They should promote the peaceful applications of chemicals and work to prevent any misuse of chemicals,

scientific knowledge, tools and technologies, and any harmful or unethical developments in research and innovation. They should disseminate relevant information about national and international laws, regulations, policies, and practices."[91]

However, although the ethical framework presented by the *Guidelines* promotes behaviour in line with, and supportive of, the broad principles of the CWC, the Convention itself is not mentioned in the text. And there is no specific injunction upon individual chemists to abide by the precepts of the Convention and specifically not to participate in the misuse of chemical knowledge, expertise, chemical agents and technology for research, development and production of chemical weapons or for other malign applications. This indirect approach appears to be a conscious strategy as the preamble to the *Hague Guidelines* brochure states:

> "The key elements presented in this text should be incorporated into new and existing codes in order to align with the provisions of the CWC. *A code need not mention chemical weapons or the CWC to support its basic goals, and provisions may need to be tailored for particular sectors or circumstances, while still reflecting the fundamental values.* Taken together, "The Hague Ethical Guidelines" provide the key elements that should be applied universally."[92] (Emphasis added).

This is in contrast with the more explicit approach taken in the *Tianjin Biosecurity Guidelines for Codes of Conduct for Scientists* (see below).

The *Hague Guidelines* were formally announced to the CWC States Parties by the OPCW Director General in his opening remarks to the 80th session of the Executive Council:

> "The Hague Ethical Guidelines have been made available on the OPCW public website. I encourage all States Parties to share the guidelines with their National Authorities, Ministries of Education, educational institutions, scientific communities and other stakeholders to advance understanding of the importance of nurturing responsible and ethical scientific development among chemistry professionals"[93]

The *Hague Guidelines* have been translated into the six official languages of the OPCW, as well as German. They have been actively promoted by the OPCW including through the production and dissemination of a brochure, poster and educational exercise.[94]

The *Guidelines* have subsequently been endorsed by leading chemical bodies. According to the OPCW, IUPAC, the International Council of Chemical Associations (ICCA) and the International Chemical Trade Association (ICTA) have endorsed The *Hague Ethical Guidelines* "to guide the responsible practice of chemistry under the norms of the Chemical Weapons Convention".[95] The *Hague Guidelines* have also inspired the development of national and organisation codes. In 2016, for example, the American Chemical Society held a workshop where a group of 30 chemists representing 18 countries developed a *Global Chemists Code of Ethics*, which explicitly drew upon the *Hague Guidelines*.[96]

11.3.3 Constraints on Effectiveness of Codes

The proponents of codes of ethics and conduct assert their potential utility in helping to sensitise chemical and life scientists to the dangers of "dual use" research, and to reinforce the importance of, and promulgate, ethical "red lines" where the legal prohibitions or normative taboos are already clearly defined and widely accepted. However, the effectiveness of such an approach, to date, has been questioned by a range of scholars.[97] One important limitation of the majority of code-based initiatives guiding the activities of individual researchers is that the resulting instruments are aspirational and non-binding in nature with no clearly identified penalties elaborated for those individuals who breach the prohibitions, or mechanisms established to monitor and enforce such prohibitions.[98]

Recognising the weaknesses of self-governance initiatives to effect change in this area, some have called for codes of conduct to become binding with those breaching such codes facing sanction from their professional community peers. For example, Rotblat in a letter to a Pugwash Workshop on Science, Ethics and Society in 2004 wrote:

> "...Some believe that the search for knowledge overrides all other considerations and that scientists should be entitled to ignore the ethical elements of their work .... The harm to society that has resulted from such attitudes has brought science into disrepute, and action is needed to restore the proper image of science. The introduction of a "hippocratic" oath is our example of such action, but it should perhaps be given more than a symbolic value. *Perhaps the time has come for a binding code of conduct, where only those who abide by the code should be entitled to be practicing scientists, something which applies now to medical practice."*[99] (Emphasis added).

However, examination of the literature reveals that no such binding codes of conduct or ethics prohibiting or severely constraining malign research, development or utilisation of biological or chemical agents (and that would be applicable to CNS-acting chemicals or other incapacitating chemical agents) have been established by any national or international scientific organisation to date.

Rappert has noted that "if codes are to go beyond reiterating platitudes about the abhorrence of using modern biology toward malign ends, then they are likely to confront major issues of controversy. For instance, codes could comment on the acceptability of disputed attempts to develop "non-lethal" incapacitating agents".[100] However, it is just those areas of dispute or controversy, such as research and development of CNS-acting chemical agents and ICA weapons, where codes remain silent or at best provide ambiguous guidance. For whilst numerous codes condemn and seek to prevent the involvement of scientists in the development of biological and chemical weapons by non-State actors, it is questionable whether enough energy has been devoted to targeting the more contentious issues of the involvement of life scientists in State-run weapons programmes including those exploring CNS-acting chemical agent weapons.

11.4 Education and Outreach for Chemical and Life Science Communities

In his 2010, analysis of the life science community's efforts to develop and promote codes of conduct, Rappert concluded that despite the energy and resources expended, such "efforts to devise meaningful codes have largely floundered.... In no small part, this has been due to the lack of prior awareness and attention by researchers as well as science organisations to the destructive applications of the life sciences. Before codes can help teach, education is needed".[101]

11.4.1 Education and Outreach to Chemists

In comparison to the activities in the life sciences (as discussed in the following section), the nurturing of a 'culture of responsibility' for chemists has been far more centralised, with many activities undertaken with the support of the OPCW, often in association with IUPAC. Since at least the early 2000s, the OPCW has repeatedly formally recognised the important roles that education, promulgation and awareness raising initiatives can play in facilitating national implementation of the CWC and in combating the development and use of chemical weapons.[102] In order to facilitate work in this area, the OPCW in 2001 established its *Ethics Project* with the aims of promoting the "development of awareness" among chemistry/engineering professionals of the "object and purpose of the CWC" and a "culture of compliance" with the requirements of the Convention; integrating ethical and scientific aspects of chemical weapons disarmament into chemistry and chemical engineering education;[103] and attempting to establish a "non-proliferation code of conduct" for professionals working with chemicals.[104]

Despite the recognition of the importance of education and awareness-raising by successive CWC Review Conferences, and the existence of OPCW institutional machinery potentially available to assist such initiatives, for many years activity in this area was relatively weak and unsustained,[105] with limited engagement by both governmental and non-governmental scientific communities. As a 2008 IUPAC report acknowledged:

> "Awareness in the scientific and technological communities in all countries about the CWC and its norms, prohibitions, and

implementation requirements remains poor. This calls for increased outreach by States Parties and the OPCW; at the same time, it also proves the need for further efforts to incorporate ethical norms and knowledge about the CWC and the dual-use nature of advances in science and technology into chemistry education.... The requirements and norms of the CWC should become a regular part, at an early stage, of the education of every student of chemistry and chemical engineering."[106]

However, the situation changed positively following the appointment in 2009 as OPCW Director General of Ambassador Üzümcü, who has been a strong advocate for education and outreach. Ambassador Üzümcü underlined the importance of such work, including through social media, to facilitate and promote the OPCW's core objectives in his acceptance speech when receiving the Nobel Peace Prize on behalf of the Organisation, calling on the international community to think about "how [organizations] can employ new communication tools to raise awareness of the need to practice responsible science, to instil the highest ethical standards in our future scientists and researchers".[107]

In 2012 Üzümcü created a Temporary Working Group of the SAB to address education and outreach. It was chaired by Algerian SAB member Djafer Benachour, and included several SAB members previously involved in awareness raising, education and codes of conduct initiatives notably with the IUPAC. In 2014 the TWG released its report on *Education and Engagement: Promoting a Culture of Responsible Chemistry* which described a holistic conception of education and its more centralised role in the work of the OPCW:[108]

"...The past two years have seen a significant paradigm shift in the way in which education and outreach is viewed and how it is approached [by the OPCW]. Traditionally, education and outreach had been seen primarily in terms of sensitising relevant audiences to the prohibitions and obligations of the CWC and raising awareness about the role and activities of the OPCW. While such an approach is clearly appropriate.... the OPCW and its Member States are now adopting a more expansive understanding of education and outreach. In this new conceptualisation, education and outreach is seen as an integral component of efforts to prevent the re-emergence of chemical weapons, which will be one of the key tasks of the OPCW in future years. In this situation, education and outreach is a core component of national implementation of the Convention. Engagement with stakeholders in the chemical industry, the scientific community, academia and the public at large contributes to promoting and expanding a culture of responsible science."

In December 2015, the 20th Conference of States Parties adopted a resolution calling for the establishment of a permanent Advisory Board on

Education and Outreach (ABEO) tasked with providing "specialised advice in areas of education and outreach relevant to the Organisation's mandate".[109] The ABEO was established in 2016 by the Director General and has facilitated the Organisation in its efforts to develop both short and long-term education and outreach strategies to promulgate the Convention and nurture support for its full implementation. The Board's remit is wide, allowing it to consider and facilitate E&O of the CWC in the widest possible sense and not just within the chemistry and disarmament communities. The ABEO is made up of 15 independent experts acting in their personal capacities, with the current post holders expert in a broad range of relevant subjects including education in science, technology, engineering, mathematics and other relevant disciplines, science communication, the chemical industry, dual/multiple-use issues related to chemistry and the life sciences, ethics and the Convention, and related international measures.[110] The ABEO, the SAB and wider OPCW have sought to facilitate and deepen engagement on E&O with relevant national and international chemical societies and organisations notably IUPAC. In late 2014, an agreement was reached to have a staff member from the OPCW serve as an advisory member for IUPAC's standing Committee on Chemistry Education and that IUPAC has permanent observer status on the ABEO.[111]

The ABEO and OPCW leadership have also sought to encourage and facilitate more concerted engagement by the relevant Organisation institutions and by its State Parties. A particularly important means for fruitful awareness raising and deeper cooperation on educational projects at the national level has been through the actions of certain OPCW National Authorities, notably Argentina.[112] The OPCW website contains and promotes resources developed by the OPCW, IUPAC, and also other national and international chemical organisations, for those interested in conducting more effective education and outreach to promote the CWC and a culture of chemical responsibility.[113] Illustrative education initiatives undertaken by or in cooperation with the OPCW are summarised below.

11.4.1.1 *High School Chemistry Student Education Project: Chemistry in Conflict*

The OPCW has collaborated with a group of educationalists including Leiden University, the International School and the International Baccalaureate Organisation to produce the *Chemistry in Conflict Workbook* which introduces the CWC to high school chemistry students.[114] This education module, which is free for public use, is intended as a thematic project for the International Baccalaureate standard level Chemistry curriculum or any comparable national chemistry program. It familiarises students with the history of chemical weapons, the obligations of the Convention, and the role of ethics in science. The workbook facilitates these learning objectives through practical application, which includes lab activities, group projects and independent learning exercises. It uses a multidisciplinary approach and can be applied to a wider curriculum or adapted to incorporate current events to promote the work of the OPCW and its continued relevance. The text is available in English and has been thoroughly tested with both students and educators. The revised third edition, published in 2018, has been distributed through the International Baccalaureate Organisation which works with nearly 4000 schools in 147 countries.[115]

11.4.1.2 *Educating Chemistry Students, General Public and Policy Makers: Multiple Uses of Chemicals*[116]

In 2005 IUPAC established a project to produce educational materials under the leadership of Alastair Hay and Peter Mahaffy to develop a website containing material which introduced students, educators, and policymakers to the topic of the multiple uses of chemicals. Avoiding the term 'dual-use', the project became known as *Multiple Uses of Chemicals* to reflect the fact that chemicals have many uses and that it is the choice of the individual practitioner as to whether the use is for beneficial or harmful purposes.[117] The education material was intended to enable discussion about the control of chemicals, their safe use, and how to prevent misuse. Visitors to the site were introduced to various case studies including ones on illegal drugs and chemical weapons. In 2013–14 with funds from the EU through the OPCW, undergraduate students and faculty at the King's Centre for Visualization in Science, in the UK, directed by Mahaffy and Brian Martin revised the website, in consultation with scientists and educational specialists from IUPAC and the OPCW. The *Multiple Uses of Chemicals* website[118] is highly interactive and uses numerous case studies and role-playing scenarios to communicate information to a range of audiences, having three different portals one designed for secondary and post-secondary chemistry students. Here the aim is to enable discussion of scientific responsibility and integrity around various concepts in general and organic chemistry; the material is also useful for courses on ethics. Ideas on implementing the resource in presentations or classroom discussions, learning outcomes for topics, worksheets for students, and links to other useful websites are contained in a portal for educators and policymakers. This covers seven major topics: multi-use chemicals, responsible choices in chemistry, convergence of chemistry and biology, codes of conduct, CWC, BTWC, and the Australia Group. Like its predecessor, the updated material was piloted at several IUPAC meetings in 2013 and 2014 before a final launch.

11.4.1.3 *OPCW E-learning for OPCW National Authorities*[119]

The OPCW Technical Secretariat has developed a series of online education modules to provide OPCW National Authorities and other stakeholders notably customs officials and those involved in verification, with the requisite tools and knowledge to perform their duties in relation to the implementation of the Convention. There are three clusters of modules, the first dealing with *International Cooperation and Assistance* covering a history of chemical weapons, the CWC and OPCW; Article by Article summary of the Convention; how to establish and run a National Authority; national implementing legislation; implementation of Article X: Assistance and Protection; National Implementing Legislation: initial measures; a guide for National Authorities on submissions under Article X and Article XI: an introduction to chemical safety and security management. The *Customs Modules* cover: customs role in implementation of the CWC; Scheduled Chemicals and their principal uses; transfer provisions of the CWC; identification of Scheduled Chemicals; problems in reporting imports and exports; and practical issues in controlling international trade of Scheduled Chemicals. The final cluster deals with the *Verification Regime* with E-learning modules covering CWC declarations; identification of declarable activities under Article VI; introduction to Article VI declarations; introduction to the electronic declaration information system (EDIS); and preparing for and receiving an Article VI inspection of a plant site.

In recent years, the OPCW has made advances in facilitating education projects to promote ethical chemical research, reinforced by its extensive national and international organisational resources, and its ability to acquire and distribute financial support from State and regional organisation funders. However, it appears that the educational initiatives that have been undertaken to date through the OPCW including those in collaboration with institutional partners notably IUPAC, as well as leading educational establishments (as illustrated above) – have tended to focus upon raising awareness and exploring non-contentious areas of the CWC. Although there has been some (albeit limited) material produced regarding the regulation of RCAs, there do not appear to be any significant attempts to address the contested obligations regarding CNS-acting chemical agents or other incapacitating chemical agents.

11.4.2 Education and Outreach to Life Science Communities

The importance of education and training in the building of a culture of responsibility amongst the life science community was recognised in the *Tianjin Guidelines*, principle 6 of which states:

> "Scientists, along with their professional associations in industry and academia, should work to maintain a well-educated, fully trained scientific community that is well versed in relevant laws, regulations, international obligations and norms. Education and training of staff at all levels should consider the input of experts from multiple fields, including social and human sciences, to provide a more robust understanding of the implications of biological research. Scientists should receive ethical training on a regular basis." (Emphasis added).

Achievement of that level of security education for a large majority of life and associated scientists would indeed be an important step forward, but in present circumstances is likely to require civil society action until State Parties to the BTWC can activate an international response.

Given that the OPCW's Advisory Board on Education and Outreach was in operation, it was perhaps natural that even though many members of civil society thought that strengthening the CWC and BTWC were better considered together,[120] their effort has been concentrated on the security education of life scientists. By far the most important initiative in this regard has been the development by the World Health Organisation (WHO) of a *Global Guidance Framework for the Responsible Use of the Life Sciences* which was organised with assistance from numerous scientists from around the world and published in 2022.[121] According to the WHO, the Global Guidance is a "global, technical and normative framework for informing the development of national frameworks and approaches for mitigating biorisks and governing dual-use research." The Global Guidance is "intended as the starting point for the development and strengthening of biorisk management, which relies on three core pillars: laboratory biosafety, laboratory biosecurity and the oversight of dual-use research."[122] The WHO subsequently appointed a Technical Advisory Group on the Responsible Use of the Life Sciences and Dual-Use Research to support the WHO in advising Member States and other stakeholders on developing and implementing biorisk management in their different contexts.[123] Unfortunately, despite the obvious dangers of dual-use of neuroscience research,[124] there is very limited attention being given to the biosecurity education of neuroscientists; the overwhelming focus of attention is still the dangers of the misuse of microbiology.

A survey of numerous biosecurity education projects that had been carried out over recent years concluded that:[125]

> "...our survey shows that while all of the ideal elements required to effectively implement the Tianjin Guidelines are obviously not yet in place, the efforts of multiple groups over the last two decades has put in place resources and experience that can be fruitfully used in that endeavour over coming years, and that the deficiencies identified here can also be remedied relatively quickly if efficiently addressed."

The survey pointed out some of the obvious problems that need to be overcome, such as the lack of material in languages other than English. That particular problem can be successfully addressed, for example, the use of cartoons that require little translation allowed material to be made available in 12 different languages including all 6 official UN languages.[126] Nevertheless, it is clear that some of the problems such as the lack of funding for sustained development and implementation of courses[127] and their effective evaluation[128] will require considerably greater efforts in the future. The recent production of a one-stop-shop biosecurity textbook[129] and the continued development of active learning systems such as Team-Base

Learning[130] will certainly help, but there is much more that will need to be done to achieve the aims set out in the Tianjin Guidelines. One way forward that is starting to be implemented is the idea of an International Biological Security Education Network (IBSEN) modelled on the International Nuclear Security Education Network (INSEN) and covering to some extent the same kind of integrative educational work of the ABEO.[131] The first aim of the IBSEN is to survey what has been done in key security education initiatives and to spread this knowledge through the production and wide distribution of a *Quarterly Newsletter*. To date the newsletter has had main articles on the INSEN organised by the IAEA, the work of the ABEO organised through the OPCW and the Global Guidance Framework work organised by the WHO. The intention is to continue to research and focus on such initiatives in order to review best practices in security education and spread this information widely. In a second stage the aim will be to learn from this wide-ranging experience and to develop and evaluate a series of model biosecurity education programmes in different settings and for different levels of life science expertise from higher school to continuing professional education.

11.5 Whistle Blowing

Any serious attempt by State or non-State actors to develop chemical or biological weapons including those employing CNS-acting chemical agents or other incapacitating chemical agents would require the involvement of an array of chemical and life scientists, engineers, technicians, ancillary workers and administrators. Whilst such personnel are essential to the development and subsequent production of these weapons, they are also potentially capable of "blowing the whistle" *i.e.* informing on these illicit weapons programmes through public denunciations, leaking information to journalists, or by reporting concerns about potential or realised breaches of national regulations or violations of international treaties directly to the relevant national or international regulatory bodies, notably the OPCW and the ISU/States Parties of the BTWC.

There has been a tradition of independent-minded chemical and life scientists, engineers and technicians alerting their fellow citizens and the international community to government policies and practices that breached health, safety or arms control agreements. This has included those engaged in former Soviet clandestine weapons development programmes – notably Vil Mirzayanov, who reported on the secret chemical weapon activities including its "novichok" binary chemical weapons programme;[132] and the Russian defectors Kanatzhan Alibekov (alias Ken Alibek) and Sergei Popov who revealed the existence of Biopreparat, the extensive network of clandestine Soviet biological weapon research centres whose work included attempts to weaponize a range of toxins and bioregulators including those intended to be CNS-acting agents or other incapacitating agents (see Chapter 2).

Joseph Rotblat the nuclear physicist who withdrew from the Manhattan Project on grounds of conscience and later helped found the Pugwash Movement was awarded the 1995 Nobel Peace Prize together with Pugwash "for their efforts to diminish the part played by nuclear arms in international politics and, in the longer run, to eliminate such arms".[133] In his acceptance speech, Rotblat highlighted the importance of whistle-blowing and the potential consequences for the whistle-blowers:

> "The purpose of some government or industrial research is sometimes concealed, and misleading information is presented to the public. It should be the duty of scientists to expose such malfeasance. 'Whistle-blowing' should become part of the scientist's ethos. This may bring reprisals; a price to be paid for one's convictions. The price may be very heavy. . ."[134]

While whistleblowers living under totalitarian regimes who dare to speak out can face long terms of imprisonment or worse for their actions, those who take similar actions in more open democratic countries still must face a range of threatened consequences. Depending on a range of factors such as their standing, the nature of their posts, the information disclosed and its veracity; whistleblowers can face imprisonment on grounds of national security; disciplinary action from their employers including termination of contracts; tarnishing or destruction of their careers and reputations; being sued for breach of contracts, libel, *etc.*; peer group antagonism and isolation from their communities. This in turn can have severe effects on the whistleblower's physical, mental and emotional well-being, as explored in a 2023 study by van Eijbergen and Siebers, of the experiences of whistleblowers reporting wrongdoing in the Netherlands.[135]

Deiseroth has highlighted the particular vulnerability of whistle-blowers: "[C]ompared with normal citizens, employees are in a special situation because they owe their employer a certain loyalty and, by law, are normally not allowed to disclose internal or confidential information. Whistle-blowers, therefore, need protection if they make a disclosure in good faith and on the basis of reliable evidence".[136] Similarly, Falter has noted: "it is neither realistic nor legitimate to put the full burden of whistle-blowing and potential retaliation on individual scientists and their moral sensibilities".[137]

Whilst it is the duty of individual chemical and life scientists to make known their concerns about the misuse of scientific research for activities that breach ethical standards or international law, it is the responsibility of the entire scientific community to ensure that such whistle-blowers are fully protected. This was recognised by the ICRC in its 2004 statement, which declared that: "[T]hose working in life sciences who voice concern and take responsible action require and deserve political and professional support and protection"[138] and the corresponding action point which was to "ensure that adequate mechanisms exist for voicing such concerns without fear of retribution".[139]

In the recent years there has been a growing movement towards establishing whistleblower protection laws at the national level. According to the 2019 Environmental Law Institute, some whistleblower protections had been enacted in at least 59 countries with 29 having dedicated national laws protecting whistleblowers.[140] In 2023 the Whistleblower International Network website held details of relevant national whistleblower legislation from 60 States.[141] These laws vary considerably in terms of coverage and strength of protections afforded.[142] Indeed, without adequate protections and rewards, many laws still fall short of supporting effective whistleblowing. Moreover, despite the improvements in legislative protection, researchers and activities have shown that the effectiveness of such legislation and its enforcement has been variable.[143]

Furthermore, Martin, an academic and activist who has had long experience of working with, and seeking to protect, whistle-blowers in many different spheres, has previously argued that: "the track record of whistle-blower protection measures – whistle-blower laws, hot-lines, ombudsmen and the like – is abysmal. In many cases, these formal processes give only an illusion of protection. Codes of ethics seem similarly impotent in the face of the problems".[144]

Consequently, it is important that independent scientists, health professionals and professional bodies – in cooperation with human rights, civil liberties and whistle-blowing organisations – promote the establishment of truly effective mechanisms under international and domestic law that provide legal protection against discrimination and criminal prosecution for whistle-blowers. Furthermore, given the failings of the current systems of whistleblower protection, the chemical, life science and biomedical communities have a duty to support those individuals who refuse to participate in what they consider immoral research and development projects, and those who blow the whistle on such activities. A number of scientific associations and activist NGOs such as the Federation of German Scientists[145] and the US Union of Concerned Scientists[146] as well as whistleblower organisations such as the Whistleblowing International Network[147] and the US Government Accountability Project[148] have materials, advice and support measures for scientists who are considering whistleblowing action.

In addition to promoting and protecting the rights of whistleblowers, appropriate mechanisms need to be established in relevant international organisations, notably the OPCW and potentially the BTWC ISU, to enable them to receive and appropriately respond to the information provided by such whistleblowers. In general practice, international organisations do not formally receive compliance-related information from individuals. However, the OPCW has increasingly undertaken its own research and analysed information from open sources related to compliance concerns. In the light of such significant changes in the Organisation's culture and working practices there may be more openness to the OPCW formally receiving information from civil society including whistleblowers. To facilitate this, the Organisation would need to establish a suitable function or body within either the

OPCW itself or an independent international body within the UN framework that has authority to talk to the OPCW. This body could be authorised to act as a kind of "international ombudsman" and a recipient where individuals could file complaints and report issues. This body would then evaluate the information and if found credible launch a formal procedure within the OPCW framework to clarify what is going on. Such a body could also be responsible for protecting the rights and safety of whistle-blowers who had submitted information to it.

11.6 Medical Ethics and Ethical Practice

Doctors share a collective ethical and moral framework that has been steadily developed over millennia from the original Hippocratic Oath of Ancient Greece[149] and the 12th century Oath of Maimonides,[150] which established the relationship and duties of the physician firstly to their patients and more broadly to humanity. In the modern era, of particular importance are ethical codifications developed by the World Medical Association (WMA), the international organisation representing physicians which was established after the Second World War to "ensure the independence of physicians, and to work for the highest possible standards of ethical behaviour and care by physicians, at all times".[151] The WMA has sought to establish and promulgate general standards in medical ethics that are applicable worldwide, and its first task was to update the Hippocratic Oath, which resulted in the 1948 World Medical Association Declaration of Geneva and the 1949 WMA International Code of Medical Ethics. Under the Declaration of Geneva, physicians pledge to consecrate their lives to the service of humanity; ensure that the health of the patient will be their first consideration; maintain the utmost respect for human life; and not to use their medical knowledge to violate human rights and civil liberties, even under threat.[152] In addition to these two foundational ethical statements, the WMA has developed guidance upholding and promoting IHL and IHRL, prohibiting chemical and biological weapons development, and regulating research on human subjects. These and other WMA Declarations and Codes gain their unique strength and credibility from being ethical standards written by doctors for doctors.

11.6.1 Regulating Scientific Research on Human Subjects

During 1946–7, the Nuremberg War Crimes Tribunal tried physicians and scientists accused of participation in the Nazi regime's large-scale programme of human experimentation conducted upon its prisoners. The Tribunal heard evidence of 'experiments' involving intentional infection with diseases and exposure to chemical warfare agents including lewisite, mustard gas, and phosgene; the application of toxic chemicals, irradiation with X-rays, and of surgery designed to forcibly sterilise patients.[153] In response to these horrific practices, the first substantial attempt to formulate universal ethical standards regulating human testing – the *Nuremberg Code* – was incorporated into the Tribunal's judgement of the Doctors Trial. A central tenet of the *Code* was that "the voluntary consent of the human subject is absolutely essential". In addition, the *Code* requires that "the experiment should be so conducted as to avoid all unnecessary physical and mental suffering and injury...[and] no experiment should be conducted, where there is an *a priori* reason to believe that death or disabling injury will occur".[154]

The international medical community subsequently reaffirmed, revised and expanded the *Code's* injunctions in the light of ongoing advances in medicine and science creating a comprehensive ethical framework regulating medical involvement in all forms of scientific research on human subjects, notably with the adoption in 1964 *WMA Declaration of Helsinki*[155] Although not specifically intended to address chemical and biological weapons development, these standards severely constrain medical participation in such research involving human subjects. The importance of such considerations, particularly for any attempted weaponization of CNS-acting chemical agents or other ICAs can be seen in the case study below.

11.6.1.1 UK Chemical Weapons Development and Human Experimentation [1959–early 1970s]

As discussed in Chapter Two, the UK's post-World War II chemical weapons development programme conducted at the Chemical Defence Experimental Establishment at Porton Down, included research into potential ICA weapons and involved research upon volunteer human subjects. The main programme of human studies into such agents, which ran from 1959 to the early 1970s, included trials with physical ICAs such as oripavine derivatives and psychological ICAs including LSD, LAE, BZ, tryptamines and MPIPG.[156] During the mid-1960s, members of Porton Down's Medical Division reportedly developed significant concerns about this chemical weapons research programme following serious ill-effects suffered by test subjects including a case of cardiovascular depression induced by T2636, a catatonic episode produced by T2833 and instances of toxic delirium in the BZ tests.[157] Two members of the Medical Division resigned in November 1964 and December 1964. Another, who played an important part in the programme, consequently fell sick in January 1965 and was admitted to hospital suffering from depression; although he subsequently recovered, he never rejoined Porton Down. His assistant was "very near breaking point" by mid-February 1965 and resigned a few days later.[158]

On February 2nd 1965, the Committee on the Safety of Human Experimentation (COHSE) – Porton Down's internal body for monitoring experiments – questioned the advisability of giving volunteers the ICA doses necessary to induce full incapacitation. The Head of the Medical Division at Porton Down thought that while ICAs were reputed to be safe, there appeared to be "strong reasons" for doubting this. If experiments were dangerous, even if the effects appeared to be innocuous, it was considered unethical and unjustified to conduct human tests with these agents in peacetime. COSHE agreed that very serious consideration should be given as to whether any more work with these agents should be conducted on humans.[159] On 12th February 1965 the Head of the Medical Division banned all further tests with hallucinogenic drugs, the ban being extended to oripavine derivatives in March 1965. The Director of Porton Down was concerned with this "question of medical conscience" and believed that the ban had caused "considerable embarrassment to senior officials in London".[160]

A meeting of senior external medical experts, with the Head of Porton Down's Medical Division, was held in August 1965 to address whether Porton Down medical officers were justified in "deliberately dosing healthy men with drugs specifically designed to induce some malfunction, either physiological or psychological?" The external medical experts were content that this could be done ethically if Porton Down were seeking to develop therapies *against* the chemical agents which enemies might use against the UK military. At this point the Head of the Medical Division intervened "[I] had therefore to disclose ... that we were in fact looking for an agent which we could *use* [as a weapon] under certain circumstances". (Emphasis added). His admission "changed the complexion [of the discussions] very considerably" and it was thought that scientists at Porton Down were being asked to do things that "went far beyond the Medical Research Council rules for human experiments".[161] One concrete result of these discussions was the establishment of the Applied Biology Committee (ABC) – an external group of respected scientists to monitor Porton Down's human experimentation – which should serve as "Father Confessor" for Porton Down staff and "damn the proposals [for experiments] of those who were trying to go too far and too fast".[162] It is notable that these ethical issues were being grappled with by UK medical researchers as the WMA was in the final stages of agreeing and then promoting the Declaration of Helsinki. Although the UK ICA weapons programme subsequently resumed, medical ethical considerations had clearly influenced both the actions of individual medical researchers and also delayed, increased regulation in, and constrained a State CW programme. This is even more notable given that these activities took place well before the coming into force of the BTWC or the CWC, when there was consequently no international legal prohibition on research, development and stockpiling of chemical weapons.[163]

11.6.2 Upholding IHRL and IHL

There is a range of declarations and regulations that guide health professionals in situations of conflict and unrest, and specifically prohibit their involvement in human rights abuses or breaches of international humanitarian law. Of particular importance are the Regulations in Times of Armed Conflict and Other Situations of Violence. Initially adopted by the WMA in

1956 they have been regularly reviewed and strengthened. These state that "Medical ethics in times of armed conflict is identical to medical ethics in times of peace" and that "[I]f, in performing their professional duty, physicians have conflicting loyalties, their primary obligation is to their patients... [and they]... should adhere to international conventions on human rights, international humanitarian law and WMA declarations on medical ethics". Consequently, the Regulations declare that it is "deemed unethical" for physicians to: "[G]ive advice or perform prophylactic, diagnostic or therapeutic procedures that are not justifiable for the patient's health care .... Weaken the physical or mental strength of a human being without therapeutic justification .... Employ scientific knowledge to imperil health or destroy life .... Condone, facilitate or participate in the practice of torture".[164] Furthermore, these Regulations state that "[D]uring times of armed conflict and other situations of violence, standard ethical norms apply, not only in regard to treatment but also to all other interventions, such as research. Research involving experimentation on human subjects is strictly forbidden on all persons deprived of their liberty, especially civilian and military prisoners and the population of occupied countries".[165]

The WMA has additionally adopted more detailed guidance on specific human rights issues of grave and immediate concern to the medical community, enunciated in subsequent Regulations and Declarations, for example with regard to the prevention of torture and other ill-treatment. The WMA Declaration of Tokyo originally adopted in 1975 and most recently revised in 2016, states that "the physician shall not countenance, condone or participate ...[nor] ... provide any premises, instruments, substances or knowledge to facilitate the practice of torture or other forms of cruel, inhuman or degrading treatment or to diminish the ability of the victim to resist such treatment". Furthermore, "the physician shall not use nor allow to be used, as far as he or she can, medical knowledge or skills, or health information specific to individuals, to facilitate or otherwise aid any interrogation, legal or illegal, of those individuals".[166] Such provisions are of particular relevance to the considerations of this publication given historical and contemporary reports of medical involvement in forced application of CNS-acting chemical agents to prisoners and other detainees.

11.6.3 Prohibiting Medical Involvement in Chemical and Biological Weapons Development

The WMA has adopted ethical guidelines specifically banning biomedical participation in the development of chemical and biological weapons. The 1990 Rancho Mirage Declaration on Chemical and Biological Weapons, stated: "the World Medical Association considers that it would be unethical for the physician, whose mission is to provide health care, to participate in the research and development of chemical and biological weapons, and to use his or her personal and scientific knowledge in the conception and manufacture of such weapons".[167] Furthermore, under this Declaration, the WMA: "condemn[ed] the development and use of chemical and biological weapons ... ask[ed] all governments to refrain from the development and use of chemical and biological weapons ... [and] ask[ed] all National Medical Associations to join WMA in actively supporting this Declaration".[168]

In 2002, the WMA General Assembly adopted the Washington Declaration on Biological Weapons, which recommended that:

> "The World Medical Association and National Medical Associations worldwide take an active role in promoting an international ethos condemning the development, production, or use of *toxins* and biological agents that have no justification for prophylactic, protective, or other peaceful purposes ... [and that] ... the World Medical Association urge all who participate in biomedical research to consider the implications and possible applications of their work and to weigh carefully in the balance the pursuit of scientific knowledge with their ethical responsibilities to society." (Emphasis added)[169]

In its 2013 Resolution on the Prohibition of Chemical Weapons, the WMA noted that "development, production, stockpiling and use of chemical weapons is banned under the CWC, and that use of such weapons is regarded by some to be a crime against humanity, regardless of whether the target populations are civilian or military". It consequently urged all relevant parties to "make active efforts to abide by the CWC ban on the development, production, stockpiling and use of chemical weapons"; and supported "[U]N initiatives to identify anyone who is responsible for the use of chemical weapons and to bring them to justice".[170]

National medical bodies have also established structures and procedures to implement ethical standards. These include ethics boards, which have the authority to suspend or disbar physicians from practising medicine in cases of extreme misconduct. In certain countries, such bodies have tried to investigate the alleged participation of health professionals in chemical and/or biological weapons programmes, including their involvement in the development of CNS-acting chemical agent weapons. In those limited instances where such investigations have been attempted, they have proved challenging.

11.6.3.1 Addressing Medical Involvement in Apartheid South Africa's Chemical and Biological Weapons Programme

In 2007, the Health Professions Council of South Africa (HPCSA) issued *Guidelines for Good Practice in the Health Professions: Research, Development and Use of Chemical and Biological Weapons*[171] the preamble of which stated:

> "The South African military authorities under the apartheid government sustained a covert programme for the development of chemical

and biological weapons, and recruited health care practitioners and scientists to staff the programme. The evidence that emerged in the TRC's [Truth and Reconciliation Commission's] investigations into this clandestine project has pointed to the importance of developing clear guidelines for the health professions in regard to participation in such programmes. It was particularly evident that the secrecy surrounding the apartheid government's CBW programme enabled health professional scientists to conduct research outside of any ethical oversight.... It is completely contrary to the fundamental principles of the ethics of the health professions for a health care practitioner to participate in research activities directed at generating materials intended to cause harm to human health and well-being ..."[172]

The *HPCSA Guidelines* established the following ethical obligations upon healthcare professions:

> "All research to develop CBWs designed to inflict harm on humans is unethical and health care practitioners should not participate in such activities. Should health care practitioners find themselves in dual loyalty situations where they are coerced or experience pressure or threats to comply from the military or other authorities, they should appeal to the HPCSA or any other appropriate professional body for support in resisting such pressures."[173]

The *HPCSA Guidelines* recognised the potential for physicians to become involved in legitimate research to "protect military or civilian personnel against CBWs". However, the HPCSA Guidelines stated that such research "should be subject to open peer review and ethical oversight by a suitably appointed independent body...".[174] The HPCSA considered that "the guidelines form an integral part of the standards of professional conduct against which a complaint of professional misconduct will be evaluated".[175] And, furthermore, that: "Health care practitioners who decide not to follow the guidance... must be prepared to explain and justify their actions and decisions to patients and their families, their colleagues and, if necessary, to the courts and the HPCSA".[176]

The HPCSA applied these *Guidelines* during its investigation of the alleged activities of Dr Wouter Basson, the former head of 'Project Coast' which was Apartheid South Africa's CBW research and development programme. The HPCSA initiated its investigation in 2000 following a complaint submitted by over 40 doctors of Dr Basson's unethical behaviour. However, the investigation was delayed pending Basson's criminal trial on related charges.[177] The HPCSA disciplinary hearing, which followed Basson's acquittal in 2002 on all criminal charges, was also blocked following claims by Dr Basson that the HPCSA was biased against him. However, on 10 May 2010, the North

Gauteng High Court of Pretoria found the claim to be unjustified, prompting the continuation of the HPCSA proceedings.[178]

Dr Basson initially faced six charges of alleged professional misconduct as a result of his activities at 'Project Coast' from 1981 to 1993.[179] Although two charges were dropped, the Professional Conduct Committee (PCC) of the HPCSA initiated proceedings against Dr Basson on the remaining four charges – two of which (as highlighted) related directly to CNS-acting chemical agents:[180]

> "*[Coordinating] the production of the following drugs and tear gases on a major scale ... methaqualone ...MDMA... BZ ...* CS ...CR...";
>
> "*weaponiz[ing] thousands of 120mm mortars with tear gas", and of overseeing the filling of "some 120 mm mortars with CR... which ... were supplied by the South African Defence Force to one Savimbi in Angola for use";
>
> **providing "disorientation substances for over the border kidnapping ('grab') exercises, where the substances were used to tranquilise the person to be kidnapped";*
>
> *providing "cyanide capsules to operational officers commanding for distribution to members of its specialised units for suicidal usage". (Emphasis added)[181]

Although the essence of these charges was not in dispute, Dr Basson presented nine arguments as to why his conduct was not "unprofessional"; these were all rejected by the PCC.[182] On 18 December 2013, the PCC ruled that the "conduct under discussion was in breach of established ethical rules" and that "the breaches of medical ethics amount to unprofessional conduct".[183] The Chairperson of the PCC, Professor Jannie Hugo, stated that Dr Basson had contravened international protocols and conventions: "[H]e confused ethics of a doctor with that of a soldier while discharging his duties. A doctor cannot rely on military orders to escape the consequences of his duties".[184] In 2014 Dr Basson applied to the North Gauteng High Court for the recusal of two PCC members. On 27 March 2019, the North Gauteng High Court ruled that there was bias on the part of the Committee members; consequently, the PCC hearing and its finding of "unprofessional conduct" were accordingly set aside.[185]

11.6.4 Developing Medical Ethical Standards Constraining Involvement in Development and Use of CNS-acting Chemical Agent Weapons

The British Medical Association (BMA) has previously taken the lead in attempts to develop ethical guidance for the health community on the issue of

CNS-acting chemical agents and other ICA weapons. In 2006 the BMA explored the issue during a high-level round-table meeting involving leading pharmacologists, toxicologists, anaesthesiologists as well as experts in international law relating to weapons, and representatives from the ICRC and relevant BMA committees. The BMA subsequently conducted research and analysis of the implications for public health of continuing State CNS-acting chemical agent/ICA weapons research, development and use, publishing its findings in the 2007 report: *The Use of Drugs as Weapons*. The "primary conclusion" of this publication was that the use of CNS-acting chemical agent/ICA weapons was "simply not feasible without generating a significant mortality among the target population". The report stated:

> "The agent whereby people could be incapacitated without risk of death in a tactical situation does not exist and is unlikely to in the foreseeable future. In such a situation, it is and will continue to be almost impossible to deliver the right agent to the right people in the right dose without exposing the wrong people, or delivering the wrong dose. Countermeasures may be easy to apply if such an attack is expected. This brings into question whether drugs can be used for law enforcement or any other tactical situation in which deaths and injuries should be minimised..."[186]

The *BMA Report* also examined the potential utility of medical participation in CNS-acting chemical agent/ICA weapons development programmes, recognising that "[M]edical knowledge is required to select an agent and to calculate how the correct dose might be given. Medical knowledge is therefore required in planning an attack". However, the BMA categorically denied any ethical legitimacy for such involvement:

> "Doctors should not knowingly use their skills and knowledge for weapons development for the same reasons that these ethical considerations oppose doctors' involvement in torture and the development of more effective methods of execution. In other words, the duty to avoid doing harm rises above, for instance, a duty to contribute to national security."[187]

The *BMA Report* warned that, if such medical participation were to occur, it could seriously damage the ethos of medicine broadly and endanger the safety of those physicians involved. "[D]octors risk losing their status as protected care-givers and will no longer be perceived as neutral but as active participants in combat".

Consequently, the *BMA Report* called on national organisations that represent healthcare professionals to: "promote the norms prohibiting the use of poisons, and therefore the BTWC and the CWC...[and]...promote understanding that the use of drugs as weapons would violate such norms .... Advocate against the use of drugs as weapons and not be involved in the training of military or law enforcement personnel in the administration of drugs as weapons".[188]

In 2013, as part of its Statement on Chemical Weapons, the WMA addressed the issue of "chemical agents used in policing actions", noting that such agents including when used by the military acting in a policing role, are allowed under the CWC. The WMA also noted that there was a "significant international dialogue underway on the definition of such agents and the situations in which they can be used". And that "the CWC appears to assume such agents will not be lethal, but the use of any agent might have fatal consequences". Consequently, the WMA Statement declared that "[T]hose using them, or authorising their use, must seek to ensure that they are not used in a manner which risks death or serious injury to targeted persons". The WMA consequently urged "[S]tates using chemical agents in riot control and related situations" to "carefully consider and minimise the risks and to, wherever possible, refrain from such use". Furthermore, "any use must follow the establishment of the necessary procedures to reduce the risk of death or serious injury...[and]... they should not be used in a manner, which deliberately increases the risk of injury, harm or death to their targets".[189] However, the 2013 WMA Statement did not specifically identify the chemical agents of concern (*i.e.* whether it referred specifically to CNS-acting chemical agent/ICA weapons or to RCAs, and/or other toxic chemicals) nor did it explicitly establish whether it was permissible or not for health professionals to become involved in the development, testing or employment of weapons employing CNS-acting chemical agents/ICAs, and if so under what circumstances.

11.6.5 Developing and Applying Medical Ethical Standards to Regulation of the Life Sciences

As highlighted in the previous sections of this chapter, there is a long tradition of medical practitioners and their professional associations contributing to efforts to prevent the misuse of the advances in the life and medical sciences and technologies. This tradition continues,[190] but what is significant now is the effort that the World Health Organisation is currently making to implement some advances in this field. Notably, as discussed above, it produced its *Global Framework for the Responsible Use of the Life Sciences*,[191] and is seriously following this publication up in a variety of ways.[192] The *Framework* was preceded by the publication of a forward-looking analysis titled *Emerging technologies and dual-use concerns: a horizon scan for global public health* in 2021.[193] This analysis is of particular interest here because the scan's authors stated that:

> "Research in neurochemistry, neurobiology and neuroscience is vital to finding treatments for various neurological, neurodegenerative and mental and psychiatric disorders. This area has attracted considerable

funding, particularly during the past decade.... New insights into human neurology could, however, also be the basis for problematic applications."

And they noted that:

"Of particular concern is neuroscientific research into assessing or modifying human thought, emotions and actions and means to affect the nervous system and alter cognitive states, behaviour and functions for performance enhancement and degradation..."

Furthermore, they suggested that:

"...This broad topic overlaps substantially with other issues raised in this horizon scan. For instance, nano-technological delivery could erode one of the main barriers to the use of agents that act on the central nervous system... Similarly, understanding of the neurological function of bioregulators could provide another means of exploiting neurological advances for hostile purposes."

While they thought that such problems could take a decade to emerge, and thus give time for better consideration of the problem of 'dual use' amongst the neuroscience community, some significant relevant implications were considered likely in a much shorter timeframe. For example, within 5 years the horizon scan considered that:

"Bioregulators are biochemical compounds, such as peptides, that affect cellular processes. Research has identified a number of bioregulators and synthetic analogues that can modify life processes, including cognition, reproduction and development. Bioregulators can be used therapeutically and for diagnostic purposes. They can also be misused..."

And the authors continued:

"... such compounds can have profound effects within minutes of exposure. While the potential misuses of bioregulators have been known for decades, discoveries in neuroscience and neurochemistry mean that bioregulators could be used for more targeted, hostile purposes, including damage to the central nervous system, soldier enhancement, crowd suppression and behaviour manipulation. Misuse of bioregulators, whether naturally occurring or synthetic, could include their dispersal or manipulation of functions in a target."

It can be expected that the World Health Organisation will continue to assess these potential issues of concern, but there is a clear need for the neuroscience research community and civil society more generally to be alert and engaged in dealing vigorously with the issue of 'dual-use' of CNS-acting chemical agents as soon as possible.

11.7 Conclusion

This chapter illustrates the many ways in which civil society – including professional organisations – has attempted to play an effective role in the prevention of the misuse of CNS-acting chemical agents and neuroscience more broadly in the recent past. This suggests that similar efforts by civil society individuals and organisations will continue in the future. These efforts are important as they will influence what State governments do and what international organisations are able to do. The WHO Framework is one example of a growing national[194] and international[195] attention being given to the prevention of the misuse of the scientific and technological revolution in the life and associated sciences. This development appears unlikely to produce major changes in the governance of the life sciences overnight, but official governmental regulation of the life sciences will surely increase over time. In the meantime, we expect that there will be an important role for the civil sector to assist in such processes,[196] and that there will be an increased input from the social sciences and the humanities in this essential input.[197]

References

1. ICRC, *Preventing hostile use of the life sciences, from ethics and law to best practice*, 11th November 2004. https://web.archive.org/web/20120510002542/http://www.icrc.org/eng/resources/documents/misc/biotechnology-principles-of-practice-111104.htm (accessed 19 February 2025).
2. ICRC, *Preventing hostile use of the life sciences, from ethics and law to best practice*, 11th November 2004. https://web.archive.org/web/20120510002542/http://www.icrc.org/eng/resources/documents/misc/biotechnology-principles-of-practice-111104.htm (accessed 19 February 2025).
3. (a) For example see: Third Review Conference of the Parties to the Convention on the Prohibition of the Development, Production and Stockpiling of Bacteriological (Biological) and Toxin Weapons and on their Destruction, Final Document, Part II, BWC/CONF.III/23, 9th–27th September 1991, p. 3; (b) Report of the Second Special Session of the Conference of the States Parties to the Review the Operation of the Chemical Weapons Convention (Second Review Conference), 7th–18th April 2008, RC-2/4, 18th April 2008.
4. An education project conducted under the auspices of IUPAC has employed the term "multi-use" rather than "dual use" when discussing the use and misuse of chemical agents. See P. Mahaffy, Multiple uses of chemicals, IUPAC, 2007, available on IUPAC website of: Raising Awareness - Multiple Uses of Chemicals and the Chemical Weapons

Convention, IUPAC Project 2005-029-1-050, http://multiple.kcvs.ca/ (accessed 19 February 2025).
5. US National Science Advisory Board for Biosecurity (NSABB), Proposed Framework for the Oversight of Dual Use Life Sciences Research: *Strategies for Minimizing the Potential Misuse of Research Information*, June 2007, p. 17. http://oba.od.nih.gov/biosecurity/pdf/Framework%20for%20transmittal%200807_sept07.pdf (accessed 19 February 2025).
6. National Research Council, *Biotechnology Research in an Age of Terrorism*, (Fink Report), National Academies Press, Washington, DC, 2004.
7. National Research Council, *Globalization, Biosecurity and the Future of the Life Sciences*, (Lemon Relman Report), National Academies Press, Washington, DC, 2006, DOI: 10.17226/11567, (accessed 19 February 2025).
8. National Research Council, *Globalization, Biosecurity and the Future of the Life Sciences (Lemon Relman Report)*, National Academies Press, Washington, DC, 2006, p. 216, DOI: 10.17226/11567, (accessed 19 February 2025)
9. National Research Council, *Globalization, Biosecurity and the Future of the Life Sciences*, (Lemon Relman Report), National Academies Press, Washington, DC, Executive Summary, 2006, p. 5, DOI: 10.17226/11567, (accessed 19 February 2025).
10. National Research Council, *Globalization, Biosecurity and the Future of the Life Sciences*, (Lemon Relman Report), National Academies Press, Washington, DC, 2006, p. 217, DOI: 10.17226/11567, (accessed 19 February 2025).
11. M. S. Garfinkel *et al.*, *Synthetic Genomics: Options for Governance*, J. Craig Venter Institute, CSIS, MIT, 2007.
12. For a discussion of such initiatives see: B. Rappert, The benefits, risks, and threats of biotechnology, *Sci. Public Policy*, 2008, **35**(1), 37–44.
13. B. Rappert, Introduction, in *Education and Ethics in the Life Sciences: Strengthening the Prohibition of Biological Weapons*, Australian National University E-Press, 2010, p. 8. Available at: http://epress.anu.edu.au/apps/bookworm/view/Education+and+Ethics+in+the+Life+Sciences%3A+Strengthening+the+Prohibition+of+Biological+Weapons/202/upfront.xhtml (accessed 1st May 2012).
14. Journal Editors and Authors Group, *Proc. Natl. Acad. Sci. U. S. A.*, 2003, **100**(4), 1464.
15. J. Van Aken and I. Hunger, Biosecurity policies at international life-science journals, *Biosecur. Bioterrorism*, 2009, 7(1), 61–72.
16. B. Rappert, *Education and Ethics in the Life Sciences: Strengthening the Prohibition of Biological Weapons*, Australian National University E-Press, 2010. Rappert cited the UK Biotechnology and Biological Sciences Research Council, the UK Medical Research Council, the Wellcome Trust, the Center for Disease Control, and the Southeast Center of Regional Excellence for Emerging Infectious Diseases and Biodefence.

17. B. Rappert, *Education and Ethics in the Life Sciences: Strengthening the Prohibition of Biological Weapons*, Australian National University E-Press, 2010, p. 9.
18. B. Rappert, *Education and Ethics in the Life Sciences: Strengthening the Prohibition of Biological Weapons*, Australian National University E-Press, 2010, p. 9.
19. US National Science Advisory Board for Biosecurity, *Proposed Framework for the Oversight of Dual Use Life Sciences Research: Strategies for Minimizing the Potential Misuse of Research Information*, June 2007.
20. Sunshine Project, *Earth Calling NSABB: Voluntary Compliance Won't Work, press release*, 17th April 2007.
21. See for example: Sunshine Project, *Mandate for Failure: The State of IBCs in an Age of Bioweapons Research*, 2004, http://www.sunshine-project.org/biodefense/ibcreport.html, (accessed 1st May 2012).
22. Supporting correspondence available at http://www.sunshine-project.org/publications/pr/support/deregistries.pdf (accessed 1st May 2012).
23. Sunshine Project, *Earth Calling NSABB: Voluntary Compliance Won't Work, press release*, 17th April 2007.
24. See R. Edison *et al.*, *Evaluating the robustness of current nucleic acid synthesis screening*, in M. Field, MIT researchers ordered and combined parts of the 1918 pandemic influenza virus. Did they expose a security flaw?, *Bulletin of the Atomic Scientists*, 3 June 2024.
25. National Research Council, *Biotechnology Research in an Age of Terrorism*, (Fink Report), National Academies Press, Washington, DC, 2004, pp. 5–6.
26. See, for example, R. H. Ebright, *Written Testimony*, Submitted for the Record to the US Senate Committee on Homeland Security, Subcommittee on Emerging Threats and Spending Oversight for the Hearing "Origin of COVID-19; An Examination of Available Evidence", 18 June 2024.
27. (a) M.-M. Poo, China Brain Project and non-human primate research in China. Presentation Number 25 at The Brain Forum, Lausanne, 27 May 2016; (b) Z. Quin and X. Li, Non-Human primate models for brain diseases-towards genetic manipulations via innovative technology, *Neurosci. Bull*, 2017, **33**(2), 238–250.
28. G. Palchik, *et al.*, Monkey business? Development, infuence, and ethics of potentially dual-use brain science on the world stage, *Neuroethics*, 2018, **11**, 111–114.
29. M.-M. Poo, *China Brain Project and non-human primate research in China*. Presentation Number 25 at *The Brain Forum*, Lausanne, 27 May 2016, Slide 11.
30. H. Jin, *et al.*, Military Brain Science – How to influence future wars, *Chin. J. Traumatol.*, 2018, **21**, 277–280.
31. The White House, *Fact Sheet: BRAIN Initiative*, 2 April 2013, Available at www.whitehouse.gov/the-press-office/2013/04/02/fact-sheet-brain-initiative.

32. The White House, *Fact Sheet: BRAIN Initiative*, 2 April 2013, p. 2. Available at www.white house.gov/the-press-office/2013/04/02/fact-sheet-brain-initiative.
33. Brain Research Through Advancing Innovative Neurotechnologies (BRAIN) Working Group, Report to the Advisory Committee to the Director, NIH: Brain 2025: A Scientific Vision. National Institutes of Health, 5 June 2014, p. 5; as cited in: M. R. Dando, Advances in Understanding Targets in the Central Nervous System, Chapter 8, in *Preventing Chemical Weapons: Arms Control and Disarmament as the Sciences Converge*, ed. M. M. Crowley *et al.*, Royal Society of Chemistry, London, 2018, p. 230.
34. M. R. Bruchas and B. L. Roth, New Technologies for Elucidating Opioid Receptor Function, *Trends Pharmacol. Sci.*, 2016, **37**(4), 279–289.
35. W. T. Birdsong, *et al.*, Synapse-specific opioid modulation of thalamo-cortico- striatal circuits, *eLife*, 2019, **8**, e45146.
36. S. M. Tyree, Hypocretin as a Hub for Arousal and Motivation, *Front. Neurol.*, 2018, **9**, 413.
37. C. Puentes-Mestril, *et al.*, How rhythms of the sleeping brain tune memory and synaptic plasticity, *Sleep*, 2019, 1–14.
38. BRAIN Initiative Project Number 1U01MH109104-01, *Genetic Tools and Imaging Technology for Mapping Cholinergic Engrams of Anxiety*.
39. S. Ahrens, *et al.*, A Central Extended Amygdala Circuit that Modulates Anxiety, *J. Neurosci.*, 2018, **38**(24), 5567–5583.
40. V. Breton-Provencher, *et al.*, Active control of arousal by a locus coeruleus GABAergic circuit, *Nat. Neurosci.*, 2019, **22**(2), 218–228.
41. Neuroethics Subgroup, *The BRAIN Initiative and Neuroethics: Enabling and Enhancing Neuroethics for Society*, The Brain Initiative, Washington D.C., 2019, ch. 5, pp. 52–54.
42. Neuroethics Subgroup, *The BRAIN Initiative and Neuroethics: Enabling and Enhancing Neuroethics for Society*, The Brain Initiative, Washington D.C., 2019, ch. 6, pp. 59–69.
43. J. Das, *et al.*, Neuroscience is ready for neuroethics engagement, *Front. Commun.*, 2022, **21**, DOI: 10.3389/fcomm.2022.909964.
44. Y. W. Wei, *et al.*, Who does Neuroethics Scholarship address, and What Does it Recommend? A Content Analysis of Selected Abstracts from the International Neuroethics Society Annual Meetings, *Neuroethics*, 2024, **17**, 26.
45. These summaries of the annual January and August meetings of the *Neuroethics Working Group* can be found on the US BRAIN Initiative website.
46. Human Brain Project, HBP facts and figures, HBP in Europe, SGA1 April 2016–March 2018, 2017. https://www.humanbrainproject.eu/en/follow-hbp/sfn2017/ (pp. 3 and 5).
47. Available at https://www.humanbrainproject.eu/.
48. Available at https://www.humanbrainproject.eu/.

49. M. Dando, *Neuroscience and the Problem of Dual Use Neuroethics in the New Brain Research Projects*, Springer Nature, Cham Switzerland, 2020, p. 82.
50. Available at https://www.humanbrainproject.eu/ (accessed 14th September 2018).
51. T. Mahfoud, *et al.*, The limits of dual use. Issues in science and technology, *Summer*, 2018, **73–78**, 73.
52. Adapted from: K. Evers *et al.*, *Dual use in neuroscientific and neurotechnological research: a report on background, developments, and recommendations for ethical address, assessment and guidance of human brain project activities*, Human Brain Project Neuroethics and Philosophy Group, Uppsala University, 31 March 2017, pp. 14, 15 and 40, http://www.crb.uu.se/research/neuroethics/.
53. S. Whitby, Team-Based Learning and Threshold Concepts in Biological Security and Dual-Use, *Creat. Edu.*, 2024, **15**, 1–22.
54. L. Shang, *et al.*, Dual-Use oversight; Is the scientific community fit for purpose? What is to be done if it is not?, *J. Biosaf. Biosecur.*, 2023, **5**, 153–154.
55. L. M. Vennis, *et al.*, Dual-Use Quickscan: A Web-Based Tool to Assess the Dual-Use Potential of Life Science Research, *Front. Bioeng. Biotechnol.*, 2021, **9**, 797076.
56. International Working Group on Strengthening the Culture of Biosafety and Biosecurity, Responsible Conduct in the Life Sciences, *Culture of Biosafety, Biosecurity and Responsible Conduct in the Life Sciences: Self-Assessment Framework*, Washington, D.C, 2020.
57. D. Perkins, *et al.*, The Culture of Biosafety, Biosecurity, and Responsible Conduct in the Life Sciences: A Comprehensive Literature Review, *Appl. Biosaf.*, 2019, **24**(1), 34–45.
58. J. Rotblat, *Remember your humanity*, Nobel Lecture, 10th December 1995, Oslo, Nobelprize.org, http://nobelprize.org/nobel_prizes/peace/laureates/1995/rotblat-lecture.html (accessed 1st May 2012).
59. B. Rappert, *Towards a life sciences code: countering the threats from* biological weapons. Strengthening the Biological Weapons Convention, Briefing Paper 13, Second Series, Department of Peace Studies, University of Bradford, 2004, https://bradscholars.brad.ac.uk/handle/10454/795.
60. T. Novossiolova, *et al.*, Altering an Appreciative System: Lessons from Incorporating Dual-Use Concerns into the Responsible Science Education of Biotechnologists, *Futures*, 2019, **108**, 53–60.
61. Fifth Review Conference of the States Parties to the Convention on the Prohibition of the Development, Production and Stockpiling of Bacteriological (Biological) and Toxin Weapons and on Their Destruction, 2002. *Final Document*, BWC/CONF.V/17, United Nations, Geneva, https://undocs.org/BWC/CONF.V/17.
62. Sixth Review Conference of the States Parties to the Convention on the Prohibition of the Development, Production and Stockpiling of

Bacteriological (Biological) and Toxin Weapons and on Their Destruction, 2006. *Final Document*, BWC/CONF.VI/6, United Nations, Geneva, https://documents-dds-ny.un.org/doc/UNDOC/GEN/G07/600/30/PDF/G0760030.pdf?OpenElement.

63. Seventh Review Conference of the States Parties to the Convention on the Prohibition of the Development, Production and Stockpiling of Bacteriological (Biological) and Toxin Weapons and on Their Destruction, 2011. *Final Document*, BWC/CONF.VII/7, United Nations, Geneva, https://documents-dds-ny.un.org/doc/UNDOC/GEN/G12/600/60/pdf/G1260060.pdf.
64. Biological and Toxin Weapons Convention, Science and Technology under the Biological Weapons Convention, https://www.un.org/disarmament/biological-weapons/science-and-technology, 2021.
65. American Society of Microbiology, *Code of Ethics* (Revised and approved by ASM Council in 2005). Washington, DC, USA: American Society for Microbiology.
66. Royal Society, The Roles of Codes of Conduct in Preventing the Misuse of Scientific Research, RS Policy Document 3/05, Royal Society, UK, London, 2005.
67. InterAcademy Partnership, *Doing Global Science: A Guide to Responsible Conduct in the Global Research Enterprise*, Princeton University Press, Princeton, New Jersey, 2016.
68. China and Pakistan, *Proposal for the development of a model code of conduct for biological scientists under the Biological Weapons Convention*, 8th BTWC Review Conference, BWC/CONF.VIII/WP.30, 15 November 2016.
69. G. Gronvall, *et al.*, The Biological Weapons Convention should endorse the Tianjin Biosecurity Guidelines for Codes of Conduct, *Trends Microbiol.*, 2022, **30**(12), 119–1120.
70. (a) G. Gronvall, *et al.*, The Biological Weapons Convention should endorse the Tianjin Biosecurity Guidelines for Codes of Conduct, *Trends Microbiol.*, 2022, **30**(12), 119–1120. (b) L. Wang *et al.*, The Tianjin Biosecurity Guidelines for Codes of Conduct for Scientists: Enhancing Awareness and Tust for Compliance with the Biological Weapons Convention, in *The Biological Weapons Convention at Fifty: Codifying 100 years of efforts to combat biological weapons*, ed. UNODA, United Nations, New York, 2025.
71. G. Gronvall, *et al.*, The Biological Weapons Convention should endorse the Tianjin Biosecurity Guidelines for Codes of Conduct, *Trends Microbiol.*, 2022, **30**(12), 119–1120.
72. G. Gronvall, *et al.*, The Biological Weapons Convention should endorse the Tianjin Biosecurity Guidelines for Codes of Conduct, *Trends Microbiol.*, 2022, **30**(12), 119–1120, Annex, The Tianjin Biosecurity Guidelines for Codes of Conduct for Scientists, paragraphs 2–4.
73. InterAcademy Partnership, IAP endorses Tianjin Biosecurity Guidelines, *IAP News*, 8 July 2021. https://www.interacademies.org/news/iap-endorses-tianjin-biosecurity-guidelines (accessed 9 Feburary 2025).

74. International Union of Biochemistry and Molecular Biology (undated) *Biosecurity Guideline The Tianjin Biosecurity Guidelines for Codes of Conduct for Scientists*, https://iubmb.org/resources/biosecurity-guideline/ (accessed 9 February 2025).
75. InterAcademy Partnership, Certificate of Endorsement: Tianjin Biosecurity Guidelines for Codes of Conduct for Scientists. Steering Committee of the InterAcademy Partnership, 7 July 2021.
76. China, Pakistan, Brazil and North Macedonia, Advocating Responsible Biological Research: the *Tianjin Biosecurity Guidelines for Codes of Conduct for Scientists. 9th BTWC Review Conference*, BWC/CONF.IX/WP.61, 13 December 2022, paragraph 13 (a) and (b).
77. G. Pearson and P. Mahaffy, Education, outreach, and codes of conduct to further the norms and obligations of the Chemical Weapons Convention (IUPAC Technical Report), *Pure Appl. Chem.*, 2006, **78**, 2186.
78. G. Pearson and P. Mahaffy, Education, outreach, and codes of conduct to further the norms and obligations of the Chemical Weapons Convention (IUPAC Technical Report), *Pure Appl. Chem.*, 2006, **78**, 2186.
79. G. Pearson and P. Mahaffy, Education, outreach, and codes of conduct to further the norms and obligations of the Chemical Weapons Convention (IUPAC Technical Report), *Pure Appl. Chem.*, 2006, **78**, 2186.
80. IUPAC Project: CHEMRAWN XVIII – *Ethics, Science and Development*, http://www.iupac.org/web/ins/2009-013-1-021/ (accessed 19 February 2025).
81. G. Pearson, *et al.*, Why codes of conduct matter, *Chem. Int.*, 2011, **33**(6), 7–11.
82. K. Bowman *et al.*, Role of the Non-governmental Chemical Science Community in Combatting the Development, Proliferation, and Use of Chemical Weapons, in *Preventing Chemical Weapons*, ed. M. Crowley *et al.*, Royal Society of Chemistry, 2018, pp. 517–535.
83. OPCW, Report of the Second Workshop on Ethical Guidelines for the Practice of Chemistry under the Norms of the Chemical Weapons Convention, 17–18 September 2015, 2015, pp. 40–44; as cited in K. Bowman, Role of the Non-governmental Chemical Science Community in Combatting the Development, Proliferation, and Use of Chemical Weapons, in *Preventing Chemical Weapons*, ed. M. Crowley *et al.*, Royal Society of Chemistry, 2018, pp. 517–535.
84. R. Trapp, *Industry and Norms Against Chemical and Biological Weapons*, Working Paper 17, CBWNet, Strengthening the norms against chemical and biological weapons, March 2025.
85. Germany, *Statement by Ambassador Christoph Israng, Permanent Representative of Germany to the OPCW*, 19th Session of the Conference of State Parties, 21 January 2015.
86. OPCW, *Report of the Nineteenth Session of the Conference of the States Parties*, 1–5 December 2014, C-19/5, paragraph 23.3.

87. OPCW, *Report of the Workshop on Guidelines for the Practice of Chemistry under the Norms of the Chemical Weapons Convention*, March 2015.
88. K. Bowman *et al.*, Role of the Non-governmental Chemical Science Community in Combatting the Development, Proliferation, and Use of Chemical Weapons, in *Preventing Chemical Weapons*, ed. M. Crowley *et al.*, Royal Society of Chemistry, 2018, p. 531.
89. J. Husbands and A. Suarez, The Hague Ethical Guidelines: applying the norms of the practice of chemistry to support the Chemical Weapons Convention, *Toxicol. Environ. Chem.*, 2016, **98**(9), 11101114.
90. OPCW, *The Hague Ethical Guidelines*, 2015.
91. OPCW, *The Hague Ethical Guidelines, Awareness and engagement*, 2015.
92. OPCW, *The Hague Ethical Guidelines, Preamble*, 2015.
93. OPCW, *Opening Remarks by the Director General to the Eightieth Session of the Executive Council*. EC-80/DG.2, OPCW, 2015.
94. For text of the Guidelines, associated educational and promotional material, reports of the drafting workshops, etc, see: OPCW, *The Hague Ethical Guidelines*, https://www.opcw.org/hague-ethical-guidelines (accessed 9 February 2025).
95. OPCW, *The Hague Ethical Guidelines*, https://www.opcw.org/hague-ethical-guidelines (accessed 9 February 2025).
96. E. Stoye, Global code of ethics planned for chemists, *Chemistry World*, 2016, available at https://www.chemistryworld.com/news/global-code-of-ethics-planned-for-chemists/1010209.article (accessed 19 February 2025).
97. For further information see: B. Rappert, *Education and Ethics in the Life Sciences: Strengthening the Prohibition of Biological Weapons*, Australian National University E-Press, 2010.
98. This can be contrasted with codes of conduct formally adopted by chemical industry associations and similar chemical industry bodies intended to influence the activities of chemical manufacturing companies. For example, membership of both the International Council of Chemical Associations (ICCA) and the European Chemical Industry Council (CEFIC) requires relevant companies to sign up to and adhere to the provisions of Responsible Care. Responsible Care is the global chemical industry's voluntary initiative to drive continuous improvement in safe chemicals management and achieve excellence in environmental, health, safety and security performance. Responsible Care® has been implemented by 64 chemical associations in more than 70 economies worldwide. The RC provisions require companies to integrate certain practical measures into their management systems; these then translate into company-internal regulations which are enforced within the company. They may not be explicitly linked to the Chemical Weapons Convention but they do include requirements to comply with all regulations and laws of the jurisdiction in which the companies operate (which would include national CWC

implementation laws and regulations), and explicitly include relevant export and trade controls. For further information see https://icca-chem.org/resources/icca-responsible-care-global-overview/.

99. J. Rotblat, Letter to Workshop, *Report of 2nd Pugwash Workshop Science, Ethics and Society*, Ajaccio, Corsica, 10th–12th September 2004. http://www.pugwash.org/reports/ees/corsica2004/corsica2004.htm#attach2 (accessed 1st May 2012).
100. B. Rappert, Biological weapons and the life sciences: the potential for professional codes, *Disarmament Forum*, Science Technology and the CBW regimes, UNIDIR, 2005, p. 56.
101. B. Rappert, *Education and Ethics in the Life Sciences: Strengthening the Prohibition of Biological Weapons*, Australian National University E-Press, 2010, p. 14.
102. (a) See for example: OPCW, Report of the First CWC Review Conference, 28th April – 9th May 2003, RC-1/5, paragraphs 7.79 and 7.83.d; (b) OPCW, Report of the Second CWC Review Conference, RC-2/4, 7th–18th April 2008, paragraph 9.77.
103. R. Trapp, *OPCW activities and perspectives on the content, promulgations, and adoption of codes of conduct for scientists*, Third Meeting of the States Parties of the Biological Weapons Convention, Meeting of Experts, Geneva, 13th–24th June 2005, p. 5.
104. OPCW, Conference of State Parties, Seventh Session C-7/3, 7th–11th October 2002, 10th October 2002, *Report of the OPCW on the implementation of the CWC in the year 2001*, paragraph 3.3 (e).
105. For example, although activities of the Ethics Project are reported up to 2006 on the OPCW website, no further information appears to be publicly available after this date. See http://www.opcw.org/ (accessed 1st May 2012).
106. M. Balali-Mood, *et al.*, Impact of Scientific Developments on the Chemical Weapons Convention, IUPAC Technical Report, *Pure Appl. Chem.*, 2008, **80**, 181.
107. OPCW, *OPCW Today*, 2013, **2**(5), 6.
108. OPCW, *Education and Engagement: Promoting a Culture of Responsible Chemistry*, SAB/REP/2/14, November 2014, pp. 14–15, available at https://www.opcw.org/fileadmin/OPCW/SAB/en/Education_and_Engagement-v2.pdf (accessed 20 February 2025).
109. OPCW, *Decision: Establishment of an Advisory Board on Education and Outreach*, CSP, Twentieth Session, C-20/DEC.9, 3 December 2015.
110. OPCW, *Advisory Board on Education and Outreach, Supporting the OPCW's engagement with external partners*, https://www.opcw.org/about/subsidiary-bodies/advisory-board-education-and-outreach (accessed 20 February 2025).
111. K. Bowman *et al.*, Role of the Non-governmental Chemical Science Community in Combatting the Development, Proliferation, and Use of Chemical Weapons, in *Preventing Chemical Weapons*, ed. M. Crowley *et al.*, Royal Society of Chemistry, 2018, p. 531.

112. For a discussion of the Argentinian CWC National Authority's education and outreach activities see K. Bowman *et al.*, Role of the Non-governmental Chemical Science Community in Combatting the Development, Proliferation, and Use of Chemical Weapons, in *Preventing Chemical Weapons*, ed. M. Crowley *et al.*, Royal Society of Chemistry, 2018, pp. 528–529.
113. OPCW, *Education and Outreach*, https://www.opcw.org/resources/education-and-outreach; and see also OPCW, *Other Resources*, https://www.opcw.org/resources/education-and-outreach/other-resources (accessed 20 February 2025).
114. F. Clark *et al.*, *Chemistry in Conflict Workbook*, an OPCW educational module, revised 3rd edition, OPCW, Leiden University and City Council of the Hague, available from https://www.opcw.org/chemistry-conflict-workbook (accessed 17 February 2025).
115. OPCW, *Chemistry in Conflict*, https://www.opcw.org/chemistry-conflict-workbook (accessed 20 February 2025).
116. OPCW, *Multiple uses of chemistry*, https://www.opcw.org/multiple-uses-chemistry (accessed 21 February 2025)
117. K. Bowman *et al.*, Role of the Non-governmental Chemical Science Community in Combatting the Development, Proliferation, and Use of Chemical Weapons, in *Preventing Chemical Weapons*, ed. M. Crowley *et al.*, Royal Society of Chemistry, 2018, pp. 526–527.
118. King's Centre for Visualization in Science, *Multiple uses of chemistry*, 2014. https://multiple.kcvs.ca/ (accessed 21 February 2025).
119. OPCW, E-Learning, https://www.opcw.org/opcw-e-learning (accessed 20 February 2025).
120. B. Edwards, *et al.*, Meeting the Challenges of Chemical and Biological Weapons: Strengthening the Chemical and Biological Disarmament and Non- Proliferation Regimes, *Front. Political Sci.*, 2022, **4**, 805426.
121. World Health Organisatio, *Global Guidance Framework for the Responsible Use of the Life Sciences*, WHO, Geneva, 2022. https://www.who.int/publications/b/65594
122. WHO, Technical Advisory Group on the Responsible Use of the Life Sciences and Dual-Use Research (TAG-RULSDUR*)*, https://www.who.int/groups/technical-advisory-group-on-the-responsible-use-of-the-life-sciences-and-dual-use-research-(tag-ruls-dur)
123. WHO, Technical Advisory Group on the Responsible Use of the Life Sciences and Dual-Use Research (TAG-RULSDUR). First Report, January 2024, responsiblesciences@who.int.
124. See, for example, K. L. Warmbrod *et al.*, *Advances in Science and Technology in the Life Sciences: Implications for Biosecurity and Arms Control*, UNIDIR, Geneva, 2020.
125. L. Shang, *et al.*, Key issues in the implementation of the Tianjin Biosecurity Guidelines for Codes of Conduct for Scientists; A survey of biosecurity education projects, *Biosaf. Health*, 2022, **4**, 339–346.
126. T. Novossiolova, *et al.*, Strengthening biological security after Covid-19: Using cartoon for engaging life science stakeholders with the Biological

and Toxin Weapons Convention (BTWC), *J. Biosaf. Biosecur.*, 2022, **4**, 68–74.
127. K. Millett, *et al.*, Biological Security Education Needs Constant and Systematic Promotion, *CBW Magaz.: J. Chem. Biol. Weapons*, 2024, 28–34.
128. S. Whitby, Team-Based Learning and Threshold Concepts in Biological Security and Dual-Use, *Creative Education*, 2024, **15**, 1–22.
129. L. Shang, *et al.*, Addressing the biological security education gap, *J. Strategic Trade Control*, 2024, **2**, DOI: 10.25518/2952-7597.104.
130. T. Novossiolova, *Biosecurity Education Handbook: The Power of Team-Based Learning*, University of Bradford, Bradford, 2016.
131. K. Millett, *et al.*, Towards an International Biosecurity Education Network, *CBW Magaz.: J. Chem. Biol. Weapons*, 2024, 4–14.
132. M. Crowley, *Chemical Control: Exploring mechanisms for the regulation of riot control agents, incapacitants and related means of delivery*, PhD Thesis, University of Bradford, December 2012 , pp. 596–598.
133. Nobel Prize Organisation, Nobel Peace Prize 1995, Joseph Rotblat https://www.nobelprize.org/prizes/peace/1995/rotblat/facts/ (accessed 7 February 2025).
134. J. Rotblat, *Remember Your Humanity, Acceptance and Nobel Lecture*, Oslo, 1995, http://www.nobelprize.org/nobel_prizes/peace/laureates/1995/rotblat-lecture.html (accessed 7 February 2025).
135. R. van Eijbergen and V. Siebers, Understanding Whistleblowing in Practice: Experiences in The Netherlands, *Pub. Integ.*, 2025, **27**(1), 22–35.
136. D. Deiseroth, *Societal Verification: Wave of the Future? 2000 Verification Yearbook*, VERTIC, London, 2000, p. 265.
137. A. Falter, Including Civil Society into Confidence Building: Protecting Whistleblowers and Societal Verification, Scientists, in *Assessing the Threat of Weapons of Mass Destruction, NATO Science for Peace and Security Series— E: Human and Societal Dynamics*, ed. J. Finney and I. Slaus, vol. 61, IOS Press, Amsterdam, The Netherlands, 2010, p. 289.
138. International Committee of the Red Cross, *Preventing hostile use of the life sciences, from ethics and law to best practice*, 11th November 2004. https://web.archive.org/web/20120510002542/http://www.icrc.org/eng/resources/documents/misc/biotechnology-principles-of-practice-111104.htm (accessed 19 February 2025).
139. International Committee of the Red Cross, *Preventing hostile use of the life sciences, from ethics and law to best practice*, 11th November 2004. https://web.archive.org/web/20120510002542/http://www.icrc.org/eng/resources/documents/misc/biotechnology-principles-of-practice-111104.htm (accessed 19 February 2025).
140. Environmental Law Institute, *Environmental Rule of Law: First Global Report*, United Nations Environment Programme, Nairobi, 2019, p. 179.
141. Whistleblower International Network, https://whistleblowingnetwork.org/getmedia/594e5777-69a5-4d18-bb86-489823d5da76/List-of-Laws-

and-Additional-Resources-FINAL-01-11-2023.aspx (accessed 7 February 2025).

142. (a) Organisation for Economic Co-operation and Development (OECD), *Committing to Effective Whistleblower Protection*, 2016; (b) Whistleblower Laws Around the World, *National Whistleblower Center*, (undated), https://www.whistleblowers.org/whistleblower-laws-around-the-world/ (accessed 19 August 2025).
143. (a) See for example, R. Calland and G. Dehn, *Whistleblowing around the World*, Open Democracy Advice Centre, Public Concern at Work, Cape Town, London, 2004. (b) D. Deiseroth, *Societal Verification: Wave of the Future? 2000 Verification Yearbook*, VERTIC, London, 2000, p. 265; (c) T. Devine, *The Whistleblower's Survival Guide*, Fund for Constitutional Government, Washington, DC, 1997; (d) T. Devine, Whistleblowing in the United States: The Gap between Vision and Lessons Learned, in *Whistleblowing around the World*, ed. R. Calland and G. Dehn, Open Democracy Advice Centre, Public Concern at Work, Cape Town, London, 2004; (e) B. Martin, *The Whistleblower's Handbook*, Charlbury, UK, Jon Carpenter, 1999; (f) B. Martin, Suppression of Dissent in Science, *Res. Soc. Prob. Public Policy*, 1999, 7, 105–135.
144. B. Martin, Whistle-blowing: Risks and Skills, in *A Web of Prevention*, ed. B. Rappert and C. McLeish, Earthscan, London, 2007.
145. Federation of German Scientists (Vereinigung Deutscher Wissenschaftler) https://vdw-ev.de/about-us/ (accessed 7 February 2023).
146. Union of Concerned Scientists, https://www.ucsusa.org/ (accessed 7 February 2025).
147. Whistleblowing International Network, https://whistleblowingnetwork.org/Home (accessed 7 February 2025).
148. Government Accountability Project, https://whistleblower.org/supporting-science-whistleblowers/ (accessed 7 February 2025).
149. *The Hippocratic Oath: Text, Translation, and Interpretation*, by Ludwig Edelstein. Baltimore: The Johns Hopkins Press, 1943. A modern version of this oath was developed in the 1960s, and it has since been adopted by a number of medical associations. [See: L. Lasagna, *Modern Hippocratic Oath*, School of Medicine, Tufts University, 1964]. Both texts available from Bioethics, Dalhousie University Libraries, https://dal.ca.libguides.com/c.php?g=256990&p=1717826 (accessed 19 September 2025).
150. *The Oath of Maimonides*, 12th Century, text available from Bioethics, Dalhousie University Libraries, https://dal.ca.libguides.com/c.php?g=256990&p=1717826 (accessed 19 September 2025).
151. World Medical Association website, *What is the WMA*, https://www.wma.net/who-we-are/about-us/ (accessed 19 September 2025).
152. World Medical Association, *Declaration of Geneva*, Adopted by the 2nd General Assembly of the World Medical Association, Geneva, Switzerland, September 1948 and editorially revised by the 173rd WMA Council Session, Divonne-les-Bains, France, May 2006. See also the accompanying

Physician's Oath, which was added to the Declaration of Geneva and adopted by the General Assembly of the World Medical Association in September 1948.

153. (a) History and Theory of Human Experimentation The Declaration of Helsinki and Modern Medical Ethics, Franz Steiner Verlag, ed. U. Schmidt and A. Frewer and Stuttgart, 2007, and in particular Schmidt's chapter, The Nuremberg Doctors' Trial; and the Nuremberg Code, pp. 71–116; (b) G. Edwards and M. Grodin, *The Nazi Doctors and the Nuremberg Code: Human Rights in Human Experimentation: Human Rights in Human Experimentation*, Oxford University Press, USA, 1992; (c) V. Spitz, *Doctors from Hell: The Horrific Account of Nazi Experiments on Humans*, Sentient Publications, 2005, Following the main Nuremberg Doctors Trials, additional trials were held of doctors, who though not involved in human experimentation, were accused of other serious ethical breaches such as complicity in sending individuals to death camps.

154. *Trials of War Criminals before the Nuremberg Military Tribunals under Control Council Law* No. 10, volume 2, pp. 181–182, Washington, D.C.: U.S. Government Printing Office, 1949.

155. World Medical Association, *Declaration of Helsinki, Ethical Principles for Medical Research Involving Human Participants*, Adopted by the 18th WMA General Assembly, Helsinki, Finland, June 1964, regularly amended, most recent amended version 75th WMA General Assembly, Helsinki, Finland, October 2024, text available https://www.wma.net/policies-post/wma-declaration-of-helsinki/ (accessed 19 September 2025).

156. (a) See: A. Maclean, Historical survey of the Porton Down volunteer programme, Ministry of Defence, June 2006, Part 4: Human studies with incapacitating agents, pp. 109–142; (b) J. Walker, Inappropriately hilarious: UK and chemical incapacitating chemical agents, *Britain and Disarmament, The UK and Nuclear, Biological and Chemical Weapons Arms Control and Programmes 1956–1975*, Routledge, London, 2012;
(c) U. Schmidt, *Secret Science*, Oxford University Press, Oxford, 2015.

157. A. Maclean, *Historical survey of the Porton Down volunteer programme*, Ministry of Defence, June 2006, Part 4: Human studies with incapacitating agents, pp. 121–2, 134–5, 138.

158. A. Maclean, *Historical survey of the Porton Down volunteer programme*, Ministry of Defence, June 2006, Part 4: Human studies with incapacitating agents, p. 138.

159. A. Maclean, *Historical survey of the Porton Down volunteer programme*, Ministry of Defence, June 2006, Part 4: Human studies with incapacitating agents, p. 138.

160. A. Maclean, *Historical survey of the Porton Down volunteer programme*, Ministry of Defence, June 2006, Part 4: Human studies with incapacitating agents, p. 138.

161. A. Maclean, *Historical survey of the Porton Down volunteer programme*, Ministry of Defence, June 2006, Part 4: Human studies with incapacitating agents, pp. 138–9.
162. A. Maclean, *Historical survey of the Porton Down volunteer programme*, Ministry of Defence, June 2006, Part 4: Human studies with incapacitating agents, pp. 138–9.
163. At this time, the 1925 Geneva Protocol was in force, and there were concerns that certain uses of ICAs would breach this instrument. However, although it prohibits use of chemical and biological weapons in warfare, the Geneva Protocol does not explicitly address their research and development. For further discussion see J. Walker, Inappropriately hilarious: UK and chemical incapacitating chemical agents, Britain and Disarmament, The UK and Nuclear, Biological and Chemical Weapons Arms Control and Programmes 1956–1975, Routledge, London, pp. 17–33.
164. World Medical Association, *Regulations in Times of Armed Conflict And Other Situations of Violence*, Adopted by the 10th World Medical Assembly, Havana, Cuba, October 1956, revised by the 63rd WMA General Assembly, Bangkok, Thailand, October 2012, paragraph 2. These WMA Regulations were reclassified as a 'Statement' by the 63rd General Assembly, Kigali, Rwanda, October 2023, and text changed accordingly.
165. World Medical Association, Regulations in Times of Armed Conflict And Other Situations of Violence, Adopted by the 10th World Medical Assembly, Havana, Cuba, October 1956, revised by the 63rd WMA General Assembly, Bangkok, Thailand, October 2012, paragraph 3. Similarly, the Declaration of Helsinki requires physicians or other researchers seeking informed consent for participation in a research study to be "particularly cautious" if the potential subject is "in a dependent relationship with them or may consent under duress". World Medical Association, *Declaration of Helsinki*, Ethical Principles for Medical Research Involving Human Subjects, Adopted by the 18th WMA General Assembly, Helsinki, Finland, June 1964, regularly amended, most recent amended version 75th WMA General Assembly, Helsinki, Finland, October 2024, text available https://www.wma.net/policies-post/wma-declaration-of-helsinki/ (accessed 19 September 2025), clause 27.
166. WMA Declaration of Tokyo – *Guidelines for Physicians Concerning Torture and other Cruel, Inhuman or Degrading Treatment or Punishment in Relation to Detention and Imprisonment*, Adopted by the 29th World Medical Assembly, Tokyo, Japan, October 1975, and most recently Revised by the 67th WMA General Assembly, Taipei, Taiwan, October 2016.
167. World Medical Association, *Declaration on Chemical and Biological Weapons*, adopted by the 42nd World Medical Assembly Rancho Mirage, CA, USA, October 1990 and rescinded at the WMA General Assembly, Santiago, 2005.

168. World Medical Association, *Declaration on Chemical and Biological Weapons*, adopted by the 42nd World Medical Assembly Rancho Mirage, CA, USA, October 1990 and rescinded at the WMA General Assembly, Santiago, 2005.
169. World Medical Association, Washington *Declaration on Biological Weapons*, adopted by the 53rd WMA General Assembly, Washington, DC, USA, October 2002 and editorially revised by the 164th WMA Council Session, Divonne-les-Bains, France, May 2003, paragraphs 18 and 19. The text was subsequently revised by the 74th WMA General Assembly, Kigali, Rwanda, October 2023. See paragraphs 11–14.
170. World Medical Association, *Resolution on the Prohibition of Chemical Weapons*, Adopted by the 64th General Assembly, Fortaleza, Brazil, October 2013. The Resolution was reaffirmed with minor revisions by the 215th Council session (online), Cordoba, Spain, October 2020.
171. Health Professions Council of South Africa, *Guidelines for Good Practice in the Health Professions: Research, Development and Use of Chemical and Biological Weapons*, 2nd edn, booklet 9, May 2007, Pretoria.
172. HPCSA, *Guidelines for Good Practice in the Health Professions: Research*, Development and Use of Chemical and Biological Weapons, 2nd edn, booklet 9, May 2007, Pretoria, paragraphs 1.2 and 1.3.
173. HPCSA, *Guidelines for Good Practice in the Health Professions: Research*, Development and Use of Chemical and Biological Weapons, 2nd edn, booklet 9, May 2007, Pretoria, paragraph 3.1.
174. HPCSA, *Guidelines for Good Practice in the Health Professions: Research*, Development and Use of Chemical and Biological Weapons, 2nd edn, booklet 9, May 2007, Pretoria, paragraph 3.2. See paragraphs 3.3 and 3.4 for further elaboration.
175. HPCSA, *Guidelines for Good Practice in the Health Professions: Research*, Development and Use of Chemical and Biological Weapons, 2nd edn, booklet 9, May 2007, Pretoria, Introduction: The Spirit of Professional Guidelines.
176. HPCSA, *Guidelines for Good Practice in the Health Professions: Research*, Development and Use of Chemical and Biological Weapons, 2nd edn, booklet 9, May 2007, Pretoria, paragraph 5.
177. *SABC News*, Basson's Application to Bar HPCSA Inquiry Dismissed, 10th May 2010.
178. *HPCSA Press Release*, HSPCA Applauds High Court Ruling on Basson, 10th May 2010.
179. HPCSA, Before the Professional Conduct Committee of the Medical and Dental Professions Board, In the matter of Dr W. J. Basson, accused, *Charge sheet and annexures*, undated.
180. J. Smith, Wouter Basson to Face South Africa Misconduct Hearing, *The Guardian*, 27 January 2012.

181. HPCSA, Dr W. J. Basson, accused, *Charge sheet and annexures* (undated) Guidelines for Good Practice in the Health Professions: Research, Development and Use of Chemical and Biological Weapons, 2nd edn, booklet 9, May 2007, Pretoria, charges 2.2., 4.1–2., 5. & 6.
182. HPCSA, Professional Conduct Committee: *Concerning Dr W. Basson, Reasons of the Committee*, Issued by the HPCSA, 18th December 2013.
183. HPCSA, *Professional Conduct Committee: Concerning Dr W. Basson, Reasons of the Committee*, Issued by the HPCSA, 18th December 2013.
184. HPCSA, *Dr Wouter Basson Found Guilty of Unprofessional Conduct*, http://www.hpcsa-blogs.co.za/dr-wouter-basson-found-guilty-of-unprofessional-conduct/ (accessed 21 March 2017). Obligations of military medical personnel to act as physicians rather than soldiers have been examined and reaffirmed in other contexts and circumstances, see for example: The Baha Mousa Public Inquiry Report, UK House of Commons, 8 September 2011.
185. Z. Venter, *Gauteng Tribunal Committee members ordered to recuse themselves over Basson hearing*, 2019, www.iol.co.za., https://www.iol.co.za/news/south-africa/gauteng/tribunal-committee-members-ordered-to-recuse-themselves-over-basson-hearing-20114302 (accessed 14 February 2025.
186. British Medical Association, *The Use of Drugs as Weapons: The Concerns and Responsibilities of Healthcare Professionals*, BMA, London, 2007.
187. British Medical Association, *The Use of Drugs as Weapons: The Concerns and Responsibilities of Healthcare Professionals*, BMA, London, 2007, p. 20.
188. British Medical Association, *The Use of Drugs as Weapons: The Concerns and Responsibilities of Healthcare Professionals*, BMA, London, 2007, p. 24.
189. WMA Resolution on the Prohibition of Chemical Weapons Adopted by the 64th General Assembly, Fortaleza, Brazil, October 2013.
190. See for example, E. A. Karlsson, *et al.*, We need a global framework for promoting safe handling of high consequence pathogens, *Br. Med. J.*, 2024, **386**, q1855.
191. World Health Organisation, *Global Guidance Framework for the Responsible Use of the Life Sciences*, WHO, Geneva, 2022.
192. See for example, M. M. Guraiib, *et al.*, Oversight of Dual-Use Research: What Role for Ethics Committees?, *Health Secur.*, 2024, **22**(4), 1–13.
193. World Health Organisation, *Emerging technologies and dual-use concerns: a horizon scan for global public health*, World Health Organisation, Geneva, 2021.
194. See for example, National Security Commission on Emerging Biotechnology, *Interim Report*, Washington, D.C., 2023.
195. (a) World Health Organisation, *Laboratory Biosecurity Guidance*, WHO, Geneva, 2024; (b) J. Revill, *Verifying the BWC: A Primer*, UNIDIR, Geneva, 2023, DOI: 10.37559/WMD/23/bio.verification.primer1.

196. L. Shang and M. R. Dando, Rethinking biosecurity in the 21st century; An enhanced role for civil society, *J. Biosaf. Biosecur.*, 2024, **5**(3), 100–106.
197. (a) O. Lev and A. Keren, Addressing the risks of dual use research: who is responsible?, *Res. Ethics*, 2024, 1–19; (b) A. Klimovich-Mickael, *et al.*, Navigating the Nexus of Bioethics and Geopolitics: Implications for Global Health Security and Scientific Collaboration, *J. Bioethical Inquiry*, 2024, DOI: 10.1007/s11673-024-10379-3.

Section IV: Conclusions and Recommendations

CHAPTER 12

Conclusions and Recommendations

12.1 Introduction

As explained in Chapter 1, the approach taken in this publication has been informed by Holistic Arms Control (HAC) which involves the following three-stage process:

- *Stage one:* examination of the nature of the existing weapons and technology to be controlled; current and potential future scenarios of application, together with attendant human security and national security concerns; and an assessment of advances in the relevant science and technology and their implications; as set out in Chapters 2–5, Section II.
- *Stage two:* analysis of a range of potentially applicable international legal instruments (notably including arms control and disarmament instruments) and attendant control regimes, as well as the role of the scientific and medical communities, as set out in Chapters 6–11, Section III.
- *Stage three:* development of a strategy to strengthen existing mechanisms and/or to suggest new mechanisms to facilitate effective regulation or prohibition of the weapon or weapons-related technology of concern.

This concluding chapter will summarise the empirical results and provide attendant analysis, covering stages one and two of the HAC process. Building upon these findings a strategy to prevent and address the misuse of CNS-acting chemical agents and broader neuropharmacology and neuroscience will be proposed in line with HAC stage three.

Preventing Weaponization of CNS-acting Chemicals: A Holistic Arms Control Analysis
By Michael Crowley and Malcolm Dando

Published by the Royal Society of Chemistry, www.rsc.org

12.2 HAC Stage One: Weapons, Technology and Advances in Science

12.2.1 Weapons Development and Use in the Past

Using evidence drawn from 20th century country case studies, Chapter 2 demonstrates how then-available neuroscience and neuropharmacological capabilities were misused for CNS-acting chemical agent weapons development during the Cold War and beyond. In the UK during the 1950–70s despite the then-limited understanding of the CNS, intensive research was carried out with the aim of developing military CNS-acting chemical agent weapons utilising a range of pharmaceutical chemicals with psychological effects and also those with direct physiological effects. In Apartheid South Africa the clandestine 'Project Coast' included research on agents with physiological and psychological effects, and at least one possible South African use was reported of such CNS-acting chemical agent weapons in armed conflict against UNITA forces in Mozambique. The US has a long history of research and development of CNS-acting chemical agent weapons. During the Cold War, the agent BZ was manufactured and weaponised in cluster munitions, though these were never used and all stockpiles were later destroyed. The US also carried out Cold War research and development of toxin and bioregulator agents that could affect the CNS. Subsequent US military-linked research in particular explored μ-opioid receptors, analgesic opioids and their synthetic analogues, notably fentanyl and mechanisms for producing sedation *via* action on the alpha2 adrenergic receptors of the locus coeruleus (LC) system of the mammalian brain by the synthetic drug medetomidine. However, in the last 10 years the US publicly halted and disavowed all CNS-acting chemical agent weapons development. The former Soviet Union had a longstanding interest in weaponizing pharmaceutical CNS-acting agents during the Cold War. The Russian Federation's use of weaponised fentanyl derivatives to end the 2002 Moscow theatre siege demonstrated that it had stocks, operational capabilities and intent to use such agents, and there are indications that further research subsequently continued. The Soviet Union also carried out extensive work weaponizing toxin and bioregulators, including the use of an engineered bacterium to deliver myelin and thus provoke a deadly immune response in the victims. Rather than simply modifying existing naturally occurring agents, the Soviet work indicated then-advancing capabilities to purposefully design *de novo* agents to attack specific physiological targets including the CNS. From at least the mid-1990s, State-owned Chinese companies developed and marketed law enforcement CNS-acting weapons targeting individuals. These were found to be in the possession of China's military and were reportedly used against protestors. Additional dual-use research, from the 1990s onwards, into CNS-acting chemical agents with potential weapons applicability was uncovered in the Czech Republic, India and Iran.

The internal security forces of certain States during the Cold War also focussed on the misuse of novel CNS-acting chemicals originally developed and used to help people with mental illnesses (neuroleptic drugs), for political repression and torture. In the former Soviet Union large numbers of sane people who voiced 'unacceptable' political views were incarcerated in psychiatric establishments and treated with such drugs as if they were ill. A similar distortion of medical practice, on a large scale, also took place in China during the last century, and there is concern that it continues today.

Investigations were carried out by certain States to discover if the new neuroleptic drugs and other CNS-acting chemicals could be used in intelligence operations for interrogation. The full details of what was done by the US CIA under Project MKUltra are not known as much of the information was destroyed, but it was certainly extensive, unethical and involved a wide diversity of research on the impact of the new drugs on human physiological systems. Unfortunately, the misuse of modern neuroscience for torture and ill-treatment has continued in for example Iran and India.

12.2.2 Advances in Neuroscience

At the beginning of Chapter 3 we noted that a goal of neuroscience is to understand the mind at the level of the individual neurons. The chapter analyses our rapidly developing understanding of the basic regulatory and instinctive circuits of the lower reaches of the brain before turning in Chapter 4 to examine how very recent advances are also opening up routes to understand the more complex behaviours that are underpinned by neuronal circuits within the higher reaches of the brain.

12.2.2.1 Advances in Understanding Basic Circuits of the Brain

While the neuron doctrine – that the individual elements of the nervous system are the separate neuronal cells – was established in the early part of the last century, the first chemical neurotransmitter – acetylcholine – was identified only two decades later. Other major neurotransmitters, that are the predominant means of passing information between neurons, such as glutamic acid (the major excitatory transmitter in the brain) and gamma-aminobutyric acid (the major inhibitory neurotransmitter in the brain) were also soon discovered. However, from the 1960s onwards other chemical transmitters such as the catecholamines, dopamine and noradrenaline, and serotonin were also identified. These four chemical transmitters are of particular interest as they were shown to be examples of types of neurons that are located in small groups in the lower reaches of the brain but which have widespread outputs to different areas of the higher parts of the brain. This suggests that they could have a neuromodulatory function by affecting many neurons at the same time.

It appears that these kinds of neurotransmitters, for example amines and amino acids, are likely relatively limited in number, but the more complex neuropeptide neurotransmitters are thought to be larger in number, but still probably identifiable with presently available technology. As these neuropeptides are not yet well known in large numbers, their identification and elucidation of their functions may be a major area of research that could be subject to misuse. Not surprisingly, research on neuropeptides has been identified in official papers at BTWC and CWC meetings as being of dual use concern. The chapter therefore reviews the advances that have been made in research on neuropeptides during this century.

The general theme of having the capability to access, assess and affect specific circuits in the brain comes through very clearly in the following examination of the state of neuronal circuit research in 2023. A forecast of the likely impact of the new technologies was well-illustrated by a 2015 paper that considered what we might be able to discover in the next 50 years. It is argued that understanding of what causes psychiatric and neurological illnesses, how learning and memory work, why we sleep and how we make decisions could surely be issues that might be subject to malign misuse. Clearly also the State-level investments in largescale neuroscience research projects will accelerate both the possible outcomes of new research, but specifically in the further development of new investigative technology. A point illustrated by the full name of the massive United States BRAIN Initiative – Brain Research Through Advancing Innovative Neuro-technologies. As a recent review by the Director of the project stated:[1]

> "...As technologies in this realm advance—likely at a rapid but unpredictable pace—we as a scientific community must take stock not only of the value of these future applications, *but also of the potential for misuse or harm*, even if unintentional..." (Emphasis added)

Similar considerations certainly also apply to the work in the other largescale brain projects, and, one might add, to neuroscience research in general today and increasingly in the future.

Chapter 3 ends by giving more details on three different kinds of simpler circuits that have been subject to successful research: innate behaviours; the sleep/wake cycle; and fear. Innate or instinctive survival behaviours are relatively simple to investigate as they can be called out without the need for learning and were intensively investigated by the early ethologists. Research using modern technologies has enabled studies of such behaviours in diverse species to begin to elucidate general principles such as the way in which the output of neuropeptide modulators can switch between different behaviours and this opens up the prospect of understanding the mechanisms underlying such decision making.

It is necessary for animals and humans to be able to switch between the two very different behaviours of sleeping and being awake, and the sleep/wake cycle has been the subject of intense study since the middle of the last century. Sleep is now known to be a complex behaviour involving different phases rather than just being a passive state, and the switching mechanism and maintenance of the two states are obviously complicated and involve different groups of neurons and different types of neurotransmitters. Neuropeptides are one of the groups of transmitters involved and there has been intensive study of one such neuropeptide – orexin (also called hypocretin) – since its discovery at the turn of the century. It is now known that the previously mysterious devastating illness of Type 1 narcolepsy (that includes attacks of cataplexy) can be caused in mammals and human beings by the total loss of the small number of orexin-producing neurons in the lower parts of the brain and therefore the loss of their contribution to stabilising the wake state. Furthermore, the mechanism by which they operate is being examined in increasing detail.

Clearly, research into many of these innate circuits is and will progress rapidly and obviously in some cases – such as the sleep/wake cycle and narcolepsy – provide much more detailed information than was available in the past. This is particularly likely as modern largescale neuroscience projects are focused on understanding neural circuit function.[2]

12.2.2.2 Disrupting Circuits of the Higher Central Nervous System

In chapter 4 we noted that during the Cold War of the last century States were interested in the possibilities of developing so-called non-lethal weapons, and that certainly drugs such as LSD were investigated as possible agents. However, the level of understanding of the mammalian CNS was not sufficiently developed for effective disruption of the mechanisms underlying complex behaviours to be reliably produced by such drugs and therefore simple sedation was eventually the main objective of such research. Today there are clearly many areas of research in relation to the higher levels of the brain that could be of concern in regard to dual use. We concentrated on two particular fields of intensive current research: the renewed medical interest in the use of psychedelic drugs; and the ongoing revolution in our understanding of the relationship between the immune system and the nervous system.[3]

In the middle of the 20th century there was extensive use of psychedelic drugs for medical research, but concerns about the societal impact of their use as recreational drugs led to the political decision to ban their use. This basically halted the medical research for half a century. It is only in recent years that the developments in other areas of neuroscience such as brain imaging have allowed scientists to look again at the possible uses of psychedelic drugs to assist in the treatment of mental illnesses. What the brain imaging studies have revealed is the possibility to detect largescale functional networks within the brain that operate in different circumstances for example in relation to vision, attention and control. However, in periods of rest many interconnected brain regions are activated, and, in contrast, there is decreased activity in these regions when there is task-dependent attention.

This default mode network (DMN) is considered to be involved in complex functions such as memory, and alterations of the DMN have been implicated in a wide range of mental illnesses.

What is of interest in regard to potential misuse is that psychedelic drugs have been shown both to decrease the connectivity within the DMN and to increase the connectivity of the DMN with other largescale networks. These modulations are at least correlated with the well-known disruption of thinking caused by the use of psychedelics, and obviously that has increased the interest of the medical profession in these phenomena and thus achieving some success in helping people suffering from depression by the use of these drugs. During the Cold War of the last century little was known of the precise actions of psychedelics on the neurons of the brain, but now we know that classic psychedelics such as LSD are agonists of serotonin receptors and that their actions on these receptors are the cause of the psychedelic effects on people.

The serotonin system has been long conserved in evolutionary terms and it has been described as a prototypical neuromodulator. The serotonin system in mammals was first described in the 1960s and is one of the groups of small numbers of neurons in the lower parts of the brain that send outputs that ramify widely in the higher parts of the brain. There have been found to be large numbers of different sub-types of serotonin receptors and one particular subtype of these receptors is of particular interest because the psychedelics have a strong affinity for this type of subtype and where they are located within the brain. The important issue here is that testable hypotheses are being generated about the mechanisms that produce the profound effects of the psychedelic drugs and that such hypotheses are becoming testable with modern techniques. We conclude that in such circumstances it is probable that the mechanisms by which the classic psychedelics affect the operations of the higher reaches of the human brain will be increasingly elucidated, and that a reasonably comprehensive understanding will not take many years to achieve.

While a COVID infection initially affected the respiratory system it also in numerous cases caused a persistent infection – long COVID – in which the nervous system was often also affected. This was not a new finding as post-viral problems were known for other infections, but it did further increase the interest in the possibilities of malfunctions of the immune system that were already evident within the medical profession. There are about 80 autoimmune diseases in which our defence system turns on itself and these affect different parts of the body including the nervous system. Importantly, from the perspective here, some of the outcomes of these malfunctioning attacks of the immune system on the nervous system result in psychiatric disorders. Recalling the Lemon-Relman report in the early years of this century had noted that the problem of dual use was not confined to microbiology and that the body's defence system should also be considered, the chapter attempted to review this issue in some detail.

This approach necessitated a brief review of the basics of the immune system and its division in vertebrates into the primitive innate immune system which forms the first line of defence and the much more specific and powerful, but slower acting, adaptive immune system. During the last century various experiments led to the conclusion that the brain was immune privileged and isolated from the immune system in the rest of the body. It is this doctrine that has been substantially overthrown and the CNS is now known to be intimately connected to the immune system and further to have its own resident immune cells. This connection between the immune system and our brains and behaviour should not really be surprising as we all have experience of the sickness behaviour that accompanies even an infection with the common cold.

The immune system is designed to attack invading organisms and the adaptive immune system is very precise in picking out invasive non-body material to destroy, but not *exactly* precise and therefore things can go wrong when it attacks material that is similar to the invasive material, but is, in fact, its own. In such cases an autoimmune response can occur despite the various mechanisms in place to prevent such malfunctions. The chapter briefly notes that even during the last century such malfunctions were of interest to weaponeers and it then turns to a discussion of what we know about autoimmune diseases today. Well-known examples such as myasthenia gravis (MG), multiple sclerosis (MS), and narcolepsy are used to illustrate the advances being made.

Then the really new consideration was introduced of autoimmune diseases being linked to psychiatric disorders in the discipline of immunopsychiatry, with the idea that our brain and thus behaviour can be governed through peripheral immune mechanisms. This then led on to a review of autoimmune encephalitis in, for example, anti-*N*-methyl-D-aspartate (NMDA) receptor encephalitis in which the major excitatory neurotransmitter glutamate's function is disrupted by antibodies that attack the NMDA receptor and cause similar effects to the drugs ketamine and phencyclidine but through different mechanisms. Again, it seems that hypotheses about the mechanisms involved are being produced and that these can be tested using modern methods. Increasing understanding of the mechanisms underlying complex behaviours seems very likely. The chapter then reviews similar possibilities in regard to the psychiatric implications of neurodegeneration diseases, and finally notes the potential implications of our growing understanding of neuromodulation generally and indeed that, in addition to the synaptic connectome of the brain, there could also be "wireless" neuronal networks dependent on neuromodulatory systems.

It seems very likely that the advances in regard to both the simple and complex networks of the brain will continue and probably at an accelerating pace as more and more powerful technologies are developed. This process, of course, will also be affected by the advances taking place in converging technological areas.

12.2.3 Advances in Other Relevant Science and Technology

Chapter 5 explores how developments and convergences in chemical and life sciences, beyond neuroscience, as well as their interactions with nanoscience, artificial intelligence and machine learning can significantly impact upon research of CNS-acting chemical agents and ICAs with consequent dual-use implications.

Important advances have been made in the discovery of novel naturally occurring chemicals including toxins and bioregulators and in the synthesis of their chemical analogues. While the agents identified during the last century, using then-limited resources, resulted in a restricted range of CNS-acting chemical agents; a very wide range of toxins including from indigenous poisonous plants, amphibians, reptiles, scorpions and marine animals have been examined in recent years, with some shown to affect the CNS and other core physiological processes and having potential utility as weapons. The large increase in capabilities of screening – notably automated *in vitro* high throughput screening systems which can now screen millions of compounds – has facilitated such research.

These developments have been paralleled by advances in analysis of potential chemical agents notably with regard to their modes of activity, interactions with corresponding cell receptor and sub-receptor sites, and their roles in and effect upon the wider operation of the brain, CNS and other core regulatory and physiological systems in humans. These developments occurred at the same time as the capacity to manufacture an ever-wider range of biological substances, notably proteins and peptides including bioregulators, has progressed. Notable advances have been made in the capabilities for biological production of chemicals at scale and the chemical synthesis of biologically-active molecules.

This has been matched by increasing capability to 'rationally design' and synthesise analogues of such biologically active chemicals incorporating modifications resulting in the chemically synthesised substances possessing different physiological properties from their natural forebears. Advances in systems biology as described in Chapters 3 and 4 on the central nervous system could provide targets for which 'rationally-designed' drugs can be developed.

The important manifold applications to such processes of artificial intelligence and machine learning, which can be 'trained' to grasp the 'language' of chemistry and develop new predicted molecules optimised for particular objectives, have become increasingly apparent. A study by civil scientists demonstrated how their machine learning models normally employed for predicting bioactivity of new molecules and enabling design of therapeutic drugs targeting human disease, could be altered to instead generate predictive molecules like the nerve agent VX. Within 6 hours over 40 000 virtual molecules were generated, and many of which were predicted to be more toxic than publicly known chemical warfare agents. The study authors clarified that machine learning models could also be adapted to produce large molecule structures such as peptides. Chapter 5 subsequently highlighted how computer modelling systems, such as AlphaFold, which can predict 3D protein structures from a chain of amino acids, could advance our understanding of the architecture and properties of large protein molecules, and facilitate the development of new protein structures with novel properties. Machine learning, as highlighted by the 2023 SAB report, is also being applied in many ways to facilitate the synthesis routes for the production of organic molecules.

Additionally, the possibilities of biological synthesis of new molecules have been subject to major enhancements, for example by the CRISPRE/Cas9 genome editing techniques which allow almost any genome to be edited and inherited by future organisms. This would enable cells to be turned relatively easily into 'bio-factories' for the production of chemicals for beneficial or malicious use. Biology-orientated synthesis is another enabling tool allowing core structures obtained from natural sources to be used as a scaffold for modification or fragmentation and rearrangement of the original compound. Moreover, the growing capabilities of synthetic biology may well allow quite novel means of manipulating human physiology to be envisioned and developed. It appears that the technical barriers to the artificial design and synthesis of dangerous toxins and bioregulators and of synthetic organisms are falling quickly as these convergent fields of science and technology advance alongside the developments in neuroscience.

Significant technological advances have also been made that can potentially be utilised to overcome the challenge of ensuring controlled delivery of CNS-acting or broader incapacitating chemical agents to targeted individuals, or alternatively facilitating 'wide area' dispersal potentially affecting large numbers of people. Developments in nanotechnology and encapsulation currently being used in a medical context to facilitate delivery, uptake and subsequent movement of CNS-acting chemicals to active sites within the patient, could instead be employed in CNS-weaponization for targeted delivery. Developments in aerosol dispersal technologies and nanotechnology could also facilitate more effective particulate agent dissemination on a widescale for both benign or malign application. Such advances could now be coupled with the contemporary commercial development of 'wide area' dual-use aerosolisation and agent delivery systems promoted for agricultural and other civilian purposes. A more direct concern has been the development of an extensive range of delivery systems primarily promoted for the 'wide area' law enforcement dispersal of RCAs. These systems include those designed for indoor or perimeter use; foggers, sprayers and water cannons; multiple barrel and automatic projectile launchers; large projectiles; and drones, unmanned vehicles and unmanned weapon systems. In its 2023 report the SAB warned that these capabilities "increasingly resemble military equipment" and that "these systems could be repurposed and filled with other chemicals including CWAs, CNS-acting chemicals and bioregulators.".

Undeniably, these multiple strands of research and the associated technological advances will yield major health and wider societal benefits. However, because of their intrinsic dual or rather multi-faceted applicability, they could instead be exploited for the weaponization of CNS-acting chemical agents, ICAs, or for other malign manipulation of core human physiological systems.

12.3 HAC Stage Two: Potentially Applicable International Law, Arms Control, Disarmament and Other Instruments and Attendant Control Regimes

12.3.1 The Biological and Toxin Weapons Convention (BTWC)

As discussed in Chapter 6, the BTWC under its Article 1 General Purpose Criterion, and the Extended Understandings agreed at successive Review Conferences, makes clear that the Convention is comprehensive in scope, covering both naturally occurring and synthetic toxins and bioregulators, and that this coverage is essentially open-ended extending to substances yet to be discovered or synthesised. Under the Convention, development, production, acquisition, stockpiling, and use of toxins and bioregulators other than for prophylactic, protective, or other peaceful purposes are prohibited. Consequently, the development and use "for hostile purposes or in armed conflict" of toxin and bioregulator weapons designed to attack the human CNS or other core physiological systems of the body would clearly be prohibited. This prohibition would appear to cover all forms of such weapons whether they were labelled as lethal, 'less lethal' or 'non-lethal'.

However, there are a number of long-standing ambiguities and limitations in the Convention that its States Parties have failed to address. Furthermore, it is not simply a question of the treaty's negotiators at the time not defining certain terms or rules in the text, but of States subsequently failing to agree (or avoid attempting to clarify) that there are issues that can (or do) create ambiguity and hence, would necessitate development of common guidance to address. These include failing to: define "toxins" and clarify the range of substances covered by this term; failures to define or provide guidance on the scope and nature of "hostile purposes", "armed conflict" and "prophylactic, protective, or other peaceful purposes". Consequently, there is uncertainty whether any toxin and bioregulator weapons can be developed and employed for use in law enforcement and if so, what types, in what circumstances and under what constraints. There is further uncertainty on how dual-use research can be regulated in accordance with the Convention to ensure it does not directly facilitate or is not employed in the development of toxins and bioregulators for prohibited purposes. These continuing ambiguities have had and are now having real world consequences, as illustrated throughout the book. For example, did Soviet scientists and officials believe, as reported by whistleblowers, that research and development of bioregulator weapons was not prohibited under the BTWC? Does the current Russian administration believe that "genetic weapons" are a legitimate line of research permitted by the Convention? What did Indian defence scientists mean when they described their work on plant toxins as being exploited for "the development of novel multisystem targeted agents for self-defence" and do they consider such activities as being consistent with the BTWC and CWC?

In addition, there are important long-standing and systemic weaknesses that severely limit the value of the BTWC as a tool to prevent and address the development of toxin and bioregulator weapons attacking the CNS or other core physiological systems. These include inadequate monitoring and review of advances in relevant science and technology; limited and patchily implemented Confidence Building Measures; and the current lack of effective transparency, verification, and compliance mechanisms. Even the limited recent proposals agreed by BTWC States seeking to address such weakness, notably through establishing a body to review S&T developments, appear to have stalled.

The structural and systemic weaknesses are further exacerbated by the absence of a sufficiently resourced international organisation empowered to coordinate such activities and to facilitate effective implementation by States Parties. Finally, since the commencement of the Russia-Ukraine war in February 2021, the discourse and interactions between many BTWC States Parties have become increasingly polarised and antagonistic with consequent inability of States Parties to collectively agree, or move forward on, much needed measures to strengthen the Convention and its implementation. The potentially far-reaching damaging consequences of the current dysfunctional environment was highlighted by the inability of BTWC States Parties to satisfactorily collectively respond to the unevidenced allegations by Russia in 2022 against the US and Ukraine, which many States Parties viewed as part of a deliberate disinformation campaign.

12.3.2 The Chemical Weapons Convention (CWC)

Chapter 7 examines the CWC, its attendant control regime and its implementing body, the Organisation for the Prohibition of Chemical Weapons (OPCW), which has been the primary framework and forum where most inter-governmental discourse regarding measures to prevent weaponization of CNS-acting chemicals and other ICAs, has taken place. Article 1 of the CWC prohibits development, production, stockpiling, transfer or use of "chemical weapons" and requires destruction of all existing chemical weapon stockpiles and production facilities.

As "chemical weapons" are not defined in terms of lethality but rather toxicity, their prohibition covers all forms of such weapons – including CNS-acting chemical agent weapons and other incapacitating chemical agent weapons – whether they are labelled as lethal, 'less lethal' or 'non-lethal'. The CWC through its definitions of "toxic chemicals" and "chemical weapons" incorporating a General Purpose Criterion, ensures a comprehensive and essentially open coverage of substances within its scope, incorporating all existing and yet to be discovered or synthesised "toxic chemicals" including CNS-acting chemicals – whether they be pharmaceutical chemicals, toxins and bioregulators.

Although the Convention is essentially open-ended in terms of the chemicals its covers, it specifically lists one CNS-acting chemical (3-Quinuclidinyl benzilate, BZ) that had previously been weaponised by at least one State (the US) and two of its immediate precursors (3-Quinuclidinol and Benzilic Acid), in its Schedules of chemicals requiring specific regulation and that are subject to routine verification measures. Although the Schedules can in theory be amended so that further toxic chemicals, including CNS-acting chemical agents or other ICAs, of particular concern can be added and thereby be subject to greater regulation and oversight, such amendments are rare and in practice require political consensus.

While the CWC prohibits the development and use of all chemical weapons against animals and humans, it permits the use of certain toxic chemicals for a range of "purposes not prohibited" notably including law enforcement, but only "as long as the types and quantities are consistent with such purposes." While there has been long-standing agreement that RCAs could be used for law enforcement, notably in domestic riot control, there has been a lack of clarity as to whether and if so which CNS-acting chemical agents or other ICAs could be employed for law enforcement, in what circumstances and under what constraints. This has been a highly disputed area amongst CWC States Parties, exacerbated by failures to define or provide sufficient guidance relating to key terms in the Convention, notably "law enforcement" and "domestic riot control". Although the CSP-26 in 2021 adopted a Decision clarifying that use of aerosolised CNS-acting chemicals was inconsistent with law enforcement as a 'purpose not prohibited' under the Convention, the Decision was limited, uncertain in scope, and its legitimacy was challenged by certain States.

The application of the Convention in other relevant areas is also contested including the appropriate regulation of dual-use research for 'protective purposes', while increasingly important but 'difficult' issues, notably regulation of wide area RCA means of delivery which could alternatively be employed for CNS-acting weapon delivery systems, have not been formally raised by States Parties in any OPCW policy-making organ.

The current situation, has in part arisen through a failure of States Parties to collectively address the long-standing ambiguities contained in the CWC text at the time of its negotiation. In addition, there has been a lack of common willingness of the States Parties to apply the existing CWC provisions to new/evolving situations and contexts in light of the treaty's objectives. These failures have been played out through, and further exacerbated by, the limitations in the structure, working practices and culture of the OPCW, notably the long-standing limited ability of the organisation to address certain important or pressing issues, even if delays in action could seriously weaken the effectiveness of the Convention. Factors contributing to producing this situation have included the organisation's culture of decision-making by consensus and the consequent avoidance of difficult or controversial issues, the wide disparity in resources, and scientific and technical expertise available to State Party delegations and national authorities, the limitations on the autonomy of the OPCW's Technical Secretariat including its inability to formally receive and act on open-source information, and the limited transparency and accountability of the OPCW to civil society. In recent years, however, the OPCW has increasingly found innovative ways to work around organisational obstacles and has been attempting to address some of its systemic failings. Notable advances have included its work to facilitate and verify Syrian destruction of declared CW facilities and stockpiles, establishment of a mechanism to identify chemical weapons perpetrators, and adoption of the "understanding" prohibiting aerosolised use of CNS-acting chemicals in law enforcement.[4]

One longstanding systemic concern has been the failure of CWC States Parties to employ the existing formal multilateral consultation, investigation, and fact-finding mechanisms under the CWC to raise and address activities of other CWC States that may breach the Convention. The US, for example, has publicly alleged, and published varying amounts of supporting evidence, that China, Iran and Russia have been undertaking research, development and/or use of weapons employing CNS-acting chemical agents which it calls pharmaceutical based agents (PBAs). Despite such allegations, neither the US nor any other CWC States Party has attempted to resolve these issues through the CWC multilateral processes. Continued failure to utilise the formal collective measures risks allowing States engaged in CNS-acting chemical agent weapons development to more easily ignore or dismiss well-founded concerns as being groundless politically motivated attempts to undermine or tarnish the international reputation of the accused States. Failure to use the formal CWC mechanisms has the additional broad corrosive effect of undermining the system of resolving compliance issues and also weakens confidence in the control regime as a whole.

Russia has also made allegations that the US, the UK and 'NATO States' have developed and transferred CNS-acting chemical agent weapons (BZ) to Ukraine. Despite the circulation of these allegations and attendant information to all CWC States Parties, Russia has not sought to initiate formal CWC consultation, clarification and fact-finding mechanisms to resolve such matters. Considering the nature of the Russian allegations and the lack of credible publicly available information that Russia has provided to justify their claims, the Russian activities appear to be part of a disinformation campaign against the US, 'NATO States' and Ukraine, rather than an attempt at raising legitimate concerns over CNS-acting chemical agent weapons. They are potentially highly damaging for the stability of the OPCW.

From this analysis, it is clear that the CWC suffers from significant ambiguities affecting the regulation of CNS-acting chemicals and ICAs that have not been sufficiently addressed by the States Parties. Furthermore, the organisation has systemic limitations which have been exacerbated by increased polarisation and antagonism between certain CWC States Parties as a result of the Russia-Ukraine War. However, it is also evident that despite its limitations, the CWC does contain the existing normative framework and a range of existing mechanisms that could be strengthened and effectively applied to regulate CNS-acting chemicals and ICAs and prevent their weaponization, if States Parties were willing.

12.3.3 International Human Rights Law

It is clear from the review, conducted in Chapter 8, that international human rights law (IHRL) is applicable to the employment of CNS-acting chemical agent weapons, as it regulates the use of force by law enforcement officials and other agents of the State. IHRL is particularly relevant as it covers the full 'use of force' spectrum from law enforcement activities through to armed conflict, including 'grey areas' such as counter-terrorist, counter-insurgency and military operations outside armed conflict. It also covers and constrains the forced application of CNS-acting chemical agents in other contexts such as to detainees or patients confined in mental institutions and in other medical contexts. While several human rights norms may be applicable to the regulation of CNS-acting chemical agent weapons and to preventing other malign employment of these substances, the rights to life; to freedom from torture and other cruel, inhuman or degrading treatment or punishment; to engage in 'peaceful protest'; to liberty and security; and to health; together with attendant obligations on the restraint of force, are the most relevant.

An important positive aspect of human rights law is the existence of a number of international and regional mechanisms to monitor adherence to relevant treaties. However, whilst there is monitoring and reporting, the response by the international community to these bodies' findings is often weak and ineffective. In contrast, certain treaties provide for the possibility of individual petition which can result in legally binding judgements for those States party to the relevant instrument (as was seen in the European Court of Human Rights rulings on the use of a CNS-acting chemical agent weapon by Russia to end the Moscow theatre siege in 2002). It must be recognised that these monitoring and enforcement mechanisms have limited preventative value (though they may have some deterrent effect), as they would only be initiated after a potential misuse of such weapons has occurred. Furthermore, there are currently no internationally accepted procedures, under IHRL, for evaluating new weapons, or for monitoring their subsequent use at a national level.

A further recent application of IHRL, of growing importance, has been the development and implementation of human rights-based national, regional and international trade control measures which could be applied to constrain, and in some cases prevent, the transfer of CNS-acting chemical agents to end-users who would employ them for torture and other ill-treatment, 'lethal injection' executions or for the development of CNS-acting chemical agent weapons.

12.3.4 International Humanitarian Law

Chapter 9 analysed the constraints upon the development, acquisition and use in armed conflict of CNS-acting chemical agent weapons, that are imposed by international humanitarian law (IHL) instruments (principally the Four Geneva Conventions and Additional Protocols) and customary IHL, and the consequent implications for State actions in this area. Indeed, the use of CNS-acting chemical agent weapons in armed conflict would potentially breach several relevant IHL prohibitions (such as those forbidding superfluous injury or unnecessary suffering, indiscriminate weapons, and deliberate attacks on civilians). IHL also provides protection to prisoners of war or others considered *hors de combat*, for example, prohibiting the use of toxic chemicals (including CNS-acting chemical agents) or the malign application of neuroscience and neuropharmacology against them in forced interrogations, torture, other ill-treatment or human experimentation. Further potential constraint on CNS-acting chemical agent weapons development and use would derive from States' obligations under Article 36 of Additional Protocol 1 to conduct reviews of any new weapon developed or acquired, to determine its compatibility with the principles of IHL.

However, the potential utility of IHL for preventing and addressing the misuse of these agents is curtailed due to limitations in investigation and enforcement procedures, and extremely low levels of State implementation of Article 36 reviews. Furthermore, IHL is only applicable to situations of armed conflict. Whilst much relevant IHL would extend to non-international armed conflicts, there may be disagreements as to whether a particular situation is indeed a non-international armed conflict, with the relevant State instead claiming to be involved in law enforcement activities against criminals or terrorist organisations.

12.3.5 International Drug Controls

Chapter 10 examined the UN Drug Control instruments, specifically the Single Convention on Narcotic Drugs (SCND) and the UN Convention on Psychotropic Substances (CPS), and over-arching UN regulatory regime and oversight bodies, including the International Narcotics Control Board (INCB), established to combat the illicit trade and abuse of such drugs whilst safeguarding the legitimate employment of these substances.

The narcotic drugs covered under the scope of the SCND, elaborated in four regularly reviewed Schedules, include a number that have previously been explored or utilised by States as CNS-acting weapons including fentanyl and its derivatives, while morphine for example has previously been employed in drug-assisted interrogations. Similarly, the psychotropic drugs covered by the CPS and listed in its Schedules include a number previously explored by States as potential CNS-acting weapons including LSD, mescalin and midazolam; explored or previously utilised in drug-assisted interrogations such as amobarbital, secobarbital, and temazepam; or used in 'lethal injection' judicial executions, for example midazolam.

Both Conventions oblige States Parties to restrict the manufacture, export, import, distribution, stockpiling, trade in, possession, and use of the Scheduled narcotic and psychotropic drugs to "medical and scientific" purposes. There are however certain exemptions, notably with the SCND allowing States to hold undefined "special stocks" of narcotic chemicals for "special government purposes and to meet exceptional circumstances".

The Conventions incorporate limited reporting obligations upon States to provide annual figures; in the SCND on Scheduled drug requirements and drug production, in the CPS on quantities of Scheduled drugs manufactured, imported and exported, and details of certain stocks. While this information is published in INCB Annual Reports, States are not required to detail the intended use of such chemicals. Although the ICNB is empowered to investigate potential circumstances of concern, such activities are extremely circumscribed, and in the first instance, confidential.

Both Conventions were developed as crime control instruments and not as disarmament and arms control mechanisms; to date, their applicability and utility for constraining State development and use of CNS-acting chemical weapons or employment of Scheduled chemicals in interrogation or judicial executions does not appear to have been tested by States Parties.

12.3.6 Further Considerations Regarding International Instruments and Control Regimes

The preceding analysis of international law and instruments that could constrain, prevent or address weaponisation of CNS-acting chemicals has not been exhaustive. Instead, we sought to examine the most relevant and promising regulatory mechanisms and measures currently available. Consequently, there are other laws, regulations and mechanisms that may also be relevant, and that a fully comprehensive HAC analysis would need to examine. Some of these have been explored by the current authors in other publications and include: international criminal law; international environmental law; international trade controls and sanctions, and regulations governing transportation of goods; and UN investigatory measures and sanctions.[5]

It must also be recognised that our HAC Stage Two analysis concentrated upon the core prohibitions, constraints and requirements established in the relevant international law and instruments examined, and did not examine in detail how such high-level obligations are implemented by individual States through national legislation and associated measures. Although critical to how these norms are translated into action in the messy real world, the level of specificity and complexity this would require – with attendant contradictions, inconsistencies and contestations – far exceeds the scope of this book.

Finally, while the HAC Stage Two analysis found albeit varying levels of activity within some of the relevant individual international legal frameworks, instruments and attendant regimes, there appears to be very little formal communication, consultation, coordination and cooperation *between* the relevant institutional control regime bodies when it came to preventing or addressing weaponization of CNSACs and other ICAs. The negative consequences of such siloisation go far beyond the issue of CNSAC/ICA weapons, detrimentally affecting the ability of the international community to prevent and address the misuse of chemical and life sciences and associated technologies more broadly. The causes of such siloisation are complex and specific to the relevant control regimes and reflect, *inter alia,* the nature of the founding international instruments, composition and interaction of State Parties, institutional mandates and working practices, differing technical expertise, resources and prioritisations. There have been some limited examples of fruitful communication and cooperation, for example between the OPCW SAB and CWC States Parties on the one hand, and the BTWC ISU and States Parties on the other, with regard to promotion of codes of conduct, education and monitoring of convergent science and technology. And it is hoped that these can be built upon and may indicate working practices and adjacent issue areas where further engagement is possible, as will be explored in HAC Stage Three.

12.3.7 The Potential Role of Civil Society

Chapter 11 examined the advances made towards fostering a 'culture of responsibility' amongst the chemical, life science and medical communities and how this can help prevent involvement of scientists and physicians in the weaponization or other malign application of CNS-acting chemical agents and ICAs, and the broader misuse of neuropharmacology and neuroscience.

The chapter explored the growing recognition from the early 2000s onwards, notably following the landmark *Fink* and *Lemon Relman Reports*, of the 'dual-use' dilemma, and subsequent introduction of oversight structures and processes notably by US and European governments, scientific bodies, academic institutions, funders and publishers to review 'dual-use' life science research of potential concern, which appear to have been patchily implemented with limited effectiveness. As illustrative case examples, the contemporary varied degree of 'dual-use' regulation of the large-scale State Human Brain Projects in China, Europe and the US were compared, and concerns, particularly regarding Chinese military neuroscience research, highlighted.

The drafting and promulgation by national and international chemical and life science associations and organisations of non-binding aspirational codes of ethics and codes of conduct to promote ethical practice and prevent engagement in chemical and biological weapons development and production was explored. For the chemical science community, this culminated in the drafting, through the auspices of the International Union of Pure and Applied Chemistry (IUPAC) and the OPCW, of *the Hague Guidelines,* which have subsequently been endorsed and promulgated by the IUPAC, International Council of Chemical Associations and the International Chemical Trade Association. A similar process undertaken by the life science community has resulted in the *Tianjin Biosecurity Guidelines for Codes of Conduct for Scientists,* authored by experts from Tianjin University, Johns Hopkins Center for Health Security, and the Inter-Academy Partnership (IAP), and now championed by China and Pakistan, the IAP and the International Union of Biochemistry and Molecular Biology.

Work on such aspirational codes has been underpinned and complemented by chemical and life science education initiatives raising the communities' awareness of the 'dual-use' research dilemma. For chemists this has been largely led by IUPAC and the OPCW through joint projects with educationalists and working scientists resulting in both in-person and e-learning modules. Since the early 2010s, the various components of the OPCW, including National Authorities, have taken a more active role, which has been facilitated since 2014 by the OPCW's Advisory Board on Education and Outreach. Without a similar strong organisation facilitating BTWC implementation measures, life science education initiatives have been more fractured and limited, largely being undertaken by University-based life scientists and academics.

Proponents of aspirational codes and associated education assert their potential utility in helping to sensitise chemical and life scientists to the dangers of 'dual use' research, and to reinforce the importance of, and promulgate, ethical 'red lines' where the legal prohibitions or normative taboos are already clearly defined and widely accepted. However, they have certain limitations. Firstly, the codes are voluntary and non-binding with no clearly identified penalties elaborated for those individuals who breach the prohibitions, or mechanisms established to monitor and enforce such prohibitions. Secondly, no existing codes and little of the parallel education, specifically addresses contentious issues, or attempts to prevent chemical and life science engagement in State weaponization of CNS-acting chemical agents or ICAs, or malign application of neuroscience. The chapter consequently emphasised the responsibility of chemical and life scientists to refrain from participation in these activities and instead 'blow the whistle' upon them, recognising the challenges and dangers of such ethical behaviour and examining current national legislative and peer support for such whistleblowers.

The chapter examined how the international medical community, primarily through the World Medical Association (WMA), has developed a comprehensive framework of ethical standards that guide the conduct of health professionals in relation to their patients and the wider society, in times of peace and in times of armed conflict. Many of these standards directly or indirectly constrain medical involvement in the development, testing or use of weapons employing toxic chemicals. Of particular importance are the WMA Declarations and Regulations that oppose biological and chemical weapons development programmes; requirements that health professionals prevent human rights violations and breaches of international humanitarian law; and guidance that strictly regulates medical and scientific experimentation upon human subjects.

However, despite the action taken by certain national medical associations, notably the British Medical Association and the Health Professions

Council of South Africa, to develop and enforce relevant ethical standards, there are, to date, no international medical guidelines specifically prohibiting medical involvement in development, testing and use of CNS-acting chemical agent/ICA weapons. One important recent development has been the publication by the World Health Organisation of the *Global guidance framework for the responsible use of the life sciences: mitigating bio-risks and governing dual-use research* and the initiation of a programme to implement educational activities to support the framework.

12.4 HAC Stage Three: Analysis and Recommendations

The HAC analytical framework seeks to actively explore the utility of employing a multiplicity of mechanisms and to facilitate the active engagement of a variety of relevant actors in the regulation of the weapons or weapons-related technology under review. Consequently, it is an axiomatic principle of HAC that a number of regulatory processes elaborated as part of existing regimes (as well as those independent of such regimes) can be pursued in parallel. However, it is clearly apparent that relevant actors – be they international organizations, States or civil society bodies – have limited human, financial and temporal resources to devote to these issues, and competing demands to address. When analysing the potential responsiveness of particular regimes, factors such as the existing political environment and previous negotiating history, relevant regulatory mechanisms and their application, and the informal and formal regime forum schedules and agenda space, will have to be considered by the relevant actors.

From the review of the regimes conducted under HAC Stage Two, it is clear that the CWC (and OPCW) is the most appropriate and probably the most receptive forum, at least in the short term, for the discussion of concerns relating to CNSAC and other ICA weapons. However, the results of such discussions are by no means certain, and parallel processes should also be established to explore alternative mechanisms with the BTWC, UN Drugs Conventions, IHL and IHRL potentially yielding positive results in the next five- to ten-year period. In light of the current siloisation of relevant control regimes, and cognisant of the difficulties in breaching these institutional and 'cultural' walls, we have proposed certain measures to encourage consultation and cooperation between relevant control regime bodies, where this seemed feasible.

Since the international instruments and attendant regimes previously examined are State-centric mechanisms, HAC Stage Three is drafted primarily in terms of policy proposals for States that are party to the relevant mechanism. However, recognising the critical roles the chemical, life science and medical communities can play in promoting, facilitating and implementing action to prevent and address the misuse of relevant chemical and life sciences and technologies, this section concludes with recommendations for complementary civil society activity in these areas.

12.4.1 Chemical Weapons Convention

12.4.1.1 Clarify Application of CSP-26 Understanding and Develop a Further Understanding on ICAs

Following the adoption in December 2021 of the "Understanding" by the 26th Conference of States Parties (CSP-26), the aerosolised use of CNS-acting chemical agents would not be consistent with law enforcement purposes as set out in the CWC. It should be emphasised that this "Understanding", formally adopted by a CSP, is binding on all States Parties, and should be implemented by all CWC States Parties. To facilitate effective and uniform implementation, technical guidance in certain areas should be developed as soon as possible, notably including a definition of "law enforcement purposes", a definition of "CNS-acting chemicals" and a clarification of the range of chemicals that would be covered by the "Understanding". Following the technical intervention by the Scientific Advisory Board (SAB) in its report to the 2023 5th CWC Review Conference that the "Understanding" as drafted did not cover 'gaseous' CNS-acting chemical agents employed for law enforcement purposes, the CWC States Parties should at a minimum clarify through a suitable mechanism – such as an amendment or revision to the Understanding – that the use of such 'gaseous' agents for law enforcement purposes would also be prohibited. Suitable draft guidance clarifying these issues should be developed by the Technical Secretariat, in consultation with the SAB, for consideration by the CWC States Parties at a forthcoming CSP.

Additional clarification of the application of the Convention with regard to the use of other incapacitating chemical agents (ICAs) for law enforcement purposes is also required. This could take the form of a second complementary "Understanding" adopted by a forthcoming CSP that would clarify that any existing or future weapons that use the broader category of incapacitating chemical agents *i.e.*, toxic chemicals (including pharmaceutical chemicals, toxins and bioregulators) that act on core human physiological processes, including and beyond the CNS, would not be consistent with law enforcement purposes as set out in the CWC. This "Understanding" should further clarify that not only aerosolised and gaseous ICAs (including CNS-acting chemical agents) but *all* weaponised use of ICAs (including CNS-acting chemical agents) for law enforcement purposes, no matter how they could be delivered, are prohibited under the CWC. The development of this second "Understanding" should be informed, at an early stage, by an independent review, undertaken by the SAB, of current developments in relevant chemical, life and associated sciences and technologies and the risks of their misuse for ICA weaponization or malign application.

12.4.1.2 Establish SAB Temporary Working Group on CNS Acting Chemicals and ICAs

In its S&T Report to the 5th CWC Review Conference, the SAB recommended that the Director General consider establishing a SAB Temporary Working Group to "to study current developments concerning CNS-acting chemicals" and that it "include the participation and expertise of other relevant international institutions (*e.g.*, the United Nations Commission on Narcotic Drugs)". Although this was supported by certain States, notably the US, this TWG has yet to be established. We believe this should be a priority for the organisation and such a body established at the earliest opportunity, when human and financial resources allow – and even if OPCW funding is constrained, the Organisation might still be able to undertake this process through voluntary financial contributions from certain concerned States Parties. The TWG would necessarily need to take a broad view of the advances in neuroscience and associated technologies such as AI and ML and Synthetic Biology. However, based on the evidence we have reviewed here, we suggest that the TWG should pay particular attention to the implications of: the ongoing elucidation of neuronal circuits and receptors of the lower brain that underly survival behaviours; the growing understanding of the impact of psychedelics on the functions of the higher CNS; and the impact of the immune system malfunctions on the operations of the higher CNS.

Depending on when it is formed and its mandate, the SAB TWG should firstly provide technical assistance to the Technical Secretariat in its preparation of guidance to States to facilitate implementation of the CSP-26 "Understanding" as discussed above. We also recommend that the SAB TWG while concentrating upon CNS-acting chemical agents, should have a more extensive mandate also encompassing the broader category of ICAs. This would enable the SAB TWG to undertake a wide-ranging survey and analysis of developments in chemical, life and associated sciences relevant to the weaponization or other malign use of ICAs (including CNS-acting agents) that undermine the object and purpose of the Convention.

In addition to providing scientific advice to the Technical Secretariat in the development of its guidance to States Parties on implementing the CSP-26 Understanding, the SAB TWG could be tasked with making recommendations:

- As to which if any CNS-acting chemical agents or other ICAs that could be readily weaponized, pose a significant risk to the CWC, and therefore merit inclusion into one of the existing CWC Schedules of chemicals. At present only one pharmaceutical chemical (BZ) and two of its precursors are listed in Schedule 2. This is despite the fact that certain States previously weaponized or attempted to weaponize a range of pharmaceutical chemicals, toxins and bioregulators attacking the CNS or other core physiological systems.
- On how to adapt the routine verification measures to cover appropriate sites of CNS-acting chemical agent/ICA production. One area requiring

particular attention is how to appropriately regulate bioregulator production. One potential route advanced by Tucker would be to modify the provisions in the CWC Verification Annex that cover other chemical production facilities (OCPFs) which produce by synthesis more than 200 metric tons per year of organic chemicals not listed by name in the CWC. Tucker proposed developing a subcategory for those facilities that manufacture peptides with a production threshold of between 50 and 100 kilograms per year, thereby capturing OCPFs that could support State-level bioregulator weapons production.[6]
- On adaptations and improvements to the Technical Secretariat's ability to conduct challenge inspections and inspections of alleged use where CNS-acting chemical agents/ICAs may be present, including its and the OPCW designated laboratories' combined capabilities to detect and identify CNS-acting chemical agents/ICAs, and provision of appropriate personal protective equipment.
- To inform the development of guidance material on the range of ICAs and associated delivery mechanisms that could be encompassed under a second "Understanding" prohibiting use of ICAs for law enforcement purposes.

12.4.1.3 Strengthen OPCW Science and Technology Monitoring and Review Capabilities

Advances in science and technology can significantly alter the implementation environment of the CWC, and they may even affect the scope of its prohibitions. Consequently, it is crucial that the OPCW conduct in-depth, systematic, and holistic analysis of relevant science and technology advances, including those related to toxins and bioregulators, as part of the activities of the five-yearly Review Conferences. To facilitate such activity, it is essential that independent technical advice is provided to the States Parties concerning developments in relevant science and technology and their potential impacts upon the operation of the Convention. While such technical expertise is provided to CWC States Parties through the SAB and the OPCW Technical Secretariat with the long-standing cooperation of the International Union of Pure and Applied Chemistry (IUPAC), measures for ensuring greater involvement of independent chemical and life scientists, including neuroscientists and neuropharmacologists, as well experts in related and converging disciplines notably AI and nanotechnology, should be developed.

CWC States Parties should further enhance the Technical Secretariat's in-house capability to monitor advances in science and technology of potential concern to the organisation; suitable mechanisms should also be established to allow the Technical Secretariat to bring relevant concerns *proactively* to the attention of the Executive Council and the States Parties, in a timely manner, and not be restricted to only responding to their requests.

12.4.1.4 Addressing Allegations of Malign Use

If a CWC State Party has information indicating that another State has been developing, acquiring, stockpiling or using CNS-acting chemical agent/ICA weapons they should bring this information to the attention of the CWC States Parties through the appropriate institutional mechanisms. Following the CSP-26 "Understanding", this would include any use of aerosolised CNS-acting chemical agents for law enforcement. We would argue that States should also bring information concerning malign use of CNSAC and ICAs for torture, other ill-treatment and other violations of IHRL and breaches of IHL, to the attention of CWC States Parties, as such use would not be consistent with "law enforcement" or any other "purposes not prohibited" as covered by the CWC, and consequently would breach the Convention.

In the first instance, States can seek to resolve such concerns through bilateral consultation. If these consultations are successful and the relevant States Parties agree, in order to dispel misperceptions, the results of such consultations should be made known to all States Parties and, if appropriate, to the wider public. If bilateral consultations with the relevant States Parties do not prove fruitful, concerned States Parties should initiate, under Article IX of the CWC, the formal multilateral consultation, investigation and fact-finding mechanisms which include on-site challenge inspections of any facility or location in the territory or in any other place under the jurisdiction or control of the State Party of potential concern. States should avoid making allegations without being prepared to fully substantiate them and allow them to be investigated through the relevant institutional mechanisms. Where appropriate such information should be placed in the public domain.

12.4.2 Biological and Toxin Weapons Convention

12.4.2.1 Resolving Ambiguities in the BTWC and Its Application

BTWC States Parties should develop guidance, through for example Extended Understandings of Meetings of States Parties (MSPs) or Review Conferences, resolving existing ambiguities in the Convention, including by defining "toxins" and determining the range of substances covered by this term; defining the scope of "hostile purposes" and making clear it encompasses the use of biological agents, toxins and bioregulators for breaches of IHL and IHRL including torture and other ill-treatment; and determining whether any toxin and bioregulator weapon can be legitimately developed and employed for use in law enforcement and if so which toxins and bioregulators can be so employed, in what circumstances and under what constraints; and establish how 'dual use' research, including when employed for bio-defence, is constrained under the Convention to ensure it does not facilitate or is not employed for malign purposes.

12.4.2.2 Addressing Allegations of Malign Use

BTWC States Parties with concerns that other States malignly employed biological agents, toxins and bioregulators (including CNS-acting chemical agents/ICAs), should employ the problem-solving provisions under Article V. If initial bilateral consultations prove ineffective, States should initiate the Article V provisions for multilateral engagement, that is, a formal BTWC consultative meeting, as employed previously by Cuba (1997) and Russia (2024). Where States have evidence of such BTWC non-compliance that has not been resolved they should employ the Article VI provision to bring a complaint to the United Nations Security Council which may request the UN Secretary General to investigate the issue.

12.4.2.3 Strengthening Science and Technology Monitoring and Review

Despite the failure of BTWC States Parties to move forward with the establishment of an independent expert body to review and assess science and technology of relevance to implementation of the BTWC, concerned States Parties should continue advocacy for these much-needed measures at forthcoming Meetings of Experts (MXs) and MSPs. Consequently, States should support the S&T review mechanism proposed in the May 2025 Non-paper by the Chairperson, Vice-Chairs and Friends of the Chair of the Working Group on the Strengthening of the Convention.[7] States should also consider advocating for additional S&T oversight measures proposed in this non-paper, including the establishment of "an annual foresight report...to do horizon-scanning across AI, bioengineering, *neurotech*, quantum and new emerging tech" (emphasis added).[8]

In the meantime, BTWC States Parties should utilise existing, albeit limited, opportunities to collectively review and discuss science and technology at MXs, MSPs and Review Conferences. This should include examination of the rapid developments in neuroscience and neuropharmacology, particularly those related to toxin and bioregulator research, to determine potential threats of weaponization and other malign application, and bring forward proposals to mitigate such threats.

The States Parties and the relevant organisations of the BTWC and the CWC should collectively discuss how to improve coordination and cooperation – including science and technology monitoring and analysis – to better address threats to both Conventions from the convergence of the chemical and life sciences and related technologies, including potential weaponization of CNS-acting chemical agents/ICAs of biological origin and their synthetic analogues.

12.4.2.4 Promoting Responsible Use of Science and Technology

The BTWC States Parties should formally endorse the *Tianjin Biosecurity Guidelines* as soon as possible at a forthcoming MSP and call on all relevant

State and civil society life science bodies to similarly endorse and promulgate its measures. Such activities need to be reinforced by State support of appropriate awareness-raising and education programmes and materials designed to inform life science students from secondary level onwards of their obligations to utilise their knowledge and expertise in conformity with the BTWC. Such education and outreach activities conducted by BTWC States Parties and the life science community should be coordinated where possible with those undertaken by the chemical science community and CWC States Parties, and this could potentially be facilitated through the OPCW's Advisory Board on Education and Outreach (ABEO).

12.4.3 International Humanitarian Law

All States should explore the existing restrictions and prohibitions arising from international humanitarian law (IHL) instruments and customary IHL on the development, acquisition and use of CNS-acting chemical agent/ICA weapons in armed conflict, and constrain the activities of their military accordingly. At a minimum States should not develop, acquire or employ CNS-acting chemical agent/ICA weapons that disperse agent over wide-areas (no matter how such dispersal was achieved), or that cause severe "abnormal physiological" or "abnormal psychological" states as these would automatically breach the prohibitions on indiscriminate weapons or superfluous injury or unnecessary suffering. Even use of CNS-acting chemical agent/ICA weapons targeting individuals such as dart guns should be severely constrained and for example never be employed on civilians or any persons considered *hors de combat.* States should ensure that these substances are never developed or employed by military forces for torture and other ill-treatment under any circumstances, including such practices as forced interrogations, or the abusive 'chemical restraint' of detainees, patients or others held in secure medical facilities under their authority.

All States should fulfil their obligations under Article 36 of Additional Protocol I to conduct reviews of any new weapon – including CNS-acting chemical agent/ICA weapons – developed or acquired, to determine their compatibility with the principles and rules of IHL. To date the effectiveness of Article 36 reviews to regulate new weapons is uncertain and contested. This is compounded by limited transparency in this area, with the number of States carrying out Article 36 reviews, the nature of the review process, and the results of such reviews largely unknown.

Although not explicitly obliged by Article 36, we believe if a State carrying out an Article 36 review subsequently determines that a new weapon – for example one utilising CNS-acting chemical agents/ICAs – would be in breach of the Additional Protocol or "any other rule of international law applicable", that State should then inform all other High Contracting Parties (HCPs) of their findings. Although the determination by one State that the employment of a particular weapon is prohibited would not be binding upon any other State, it can be argued that the exchange of such information between

HCPs could alert the international community to, and thereby help to 'close down', possible avenues of research that might – unchecked – lead to the development, proliferation and use of weapons that would be contrary to international law.

Consequently, we recommend that HCPs explore measures to standardise the Article 36 review process, facilitate the establishment of national review bodies and promote greater transparency including an obligation to inform other HCPs of weapons they have prohibited and why. One potential forum for these discussions and development of greater cooperative measures could be the International Conference of the Red Cross and Red Crescent, the supreme deliberative body of the movement, where the High Contracting Parties to the Geneva Conventions come together with the components of the movement to discuss key matters of humanitarian concern and to make joint commitments. The next (35th) International Conference will be held in 2028.

12.4.4 International Human Rights Law

All States should explore the existing restrictions arising from international and regional human rights instruments and customary international human rights law (IHRL) on the development, acquirement, and employment of CNS-acting chemical agent/ICA weapons and constrain the activities of their military and law enforcement bodies accordingly. At a minimum, this would severely restrict the use of CNS-acting chemical agent/ICA weapons, limiting them to use only in extreme law enforcement situations to preserve life and only when all other means would prove ineffective. States should ensure that these substances are never developed, acquired or employed for torture and other ill-treatment, including such practices as forced interrogations, or for the abusive 'chemical restraint' of detainees, patients or others held in secure medical facilities. As developing best practice, States should halt all use of such substances for 'lethal injection' executions.

States and civil society organisations should bring cases of reported CNS-acting chemical agent/ICA weapons use or other malign application of these substances by law enforcement officials or other State agents to the attention of the appropriate human rights mechanisms, including: UN Special Procedures (notably the UN Special Rapporteurs on Torture; or Extra-Judicial Executions; or Counter-Terrorism and Human Rights; Freedom of Association and Assembly; Right to Health); the UN Human Rights Council; and relevant international and regional treaty bodies (*e.g.* UN Human Rights Committee and the UN Committee against Torture).

If appropriate, cases could be brought before the relevant regional judicial mechanisms capable of delivering binding legal judgements regarding violations of regional treaties (*e.g.*, European Court of Human Rights, Inter-American Court, African Commission on Human and Peoples Rights). Since a number of such regional judicial mechanisms are potentially open to individual petition, victims and their families can also directly seek redress in cases of the CNS-acting chemical agent/ICA weapons employment or the misuse of such agents for torture and other ill-treatment; and civil society organisations can attempt to employ such mechanisms to develop human rights case law on these issues.

The UN Human Rights Council should request a suitable UN Special Procedure or group of UN Special Procedures (*e.g.*, UN Special Rapporteurs on Torture; Freedom of Association and Assembly; and/or the Right to Health) to conduct a study on the use and misuse of toxic chemicals for law enforcement purposes. The body conducting this study should be informed by and engage with international organisations with technical competency in associated relevant areas notably including the OPCW, the BTWC ISU and the WHO. This study would specifically examine law enforcement weaponization or other applications of CNS-acting chemical agents and ICAs (as well as other chemicals, notably riot control agents and malodorants). The study would examine use in prisons and other places of detention as well as in non-custodial contexts notably in crowd control. It would explore the implications of advances in chemical, life and associated sciences and technologies, and bring forward recommendations to the UN Human Rights Council on whether and if so, how such substances can be legitimately employed in law enforcement, under what circumstances and with what constraints.

Whilst certain States, regional and plurilateral organisations have developed instruments to regulate the transfer of certain CNS-acting agents to prevent their weaponization (notably the Australia Group and Wassenaar Arrangement) or use in the death penalty (notably the European Union (EU) and Council of Europe (CoE)); most States are not covered by such controls. Individual States, regional and plurilateral organisations should establish or adapt existing non-proliferation and human rights-based national, regional and plurilateral transfer controls that restrict the import and export of relevant CNS-acting agents and other ICAs to prevent their weaponization and use in torture, other ill-treatment or the death penalty. States and civil society organisations currently advocating the development of an international legally binding Torture-Free Trade Treaty to prohibit transfer of all inherently abusive torture equipment, and control the transfer of law enforcement equipment that could be misused for torture, should consider incorporating CNS-acting chemical agents, other ICAs and related means of delivery within the scope of any future instrument proposed. In all such non-proliferation and human rights-based transfer controls, provision should be made to ensure that the trade in CNS-acting chemicals/ICAs for medical and other legitimate purposes is not detrimentally affected.

12.4.5 UN Drug Control Conventions

Both the Single Convention on Narcotic Drugs (SCND) and the UN Convention on Psychotropic Substances (CPS) restrict the legitimate use of a range of CNS-acting chemical agents to "medical and scientific purposes". Although they have not done so to date, States Parties to these Conventions could seek to formally establish the implications of these restrictions upon the development, stockpiling, transfer and use of CNS-acting chemical agent weapons intended for law enforcement or military purposes. If appropriate, and required, State Parties could bring forward clarificatory amendments (or agree on common understandings) through the appropriate Convention mechanisms to explicitly prohibit or constrain such activities.

Given the SCND and CPS focus upon preventing the criminal misuse of narcotic and psychotropic drugs by non-State actors, there may be greater potential capability in the instrument mechanisms and more immediate willingness amongst States Parties to explore the two Conventions' applicability to preventing and addressing CNS-acting chemical agent weapons development, stockpiling, transfer and use when undertaken by non-State actors including criminal organisations, terrorists, paramilitaries, private military and security companies, and armed opposition forces. If this proves to be the case, then advances should be made in this area as an important first step.

The reporting (and potentially the consultation/investigatory) mechanisms of both Conventions could be considered by States Parties as potential routes for obtaining information relevant to the regulation of CNS-acting chemical agents to prevent or address their weaponization by other States. Where a State Party has concern about development, manufacture, stockpiling, transfer or use of a narcotic drug or psychotropic substance as a CNS-acting weapon, or for use in torture, other ill-treatment or the death penalty, they could consider bringing their concerns before the relevant Drug Convention bodies, for example, the International Narcotics Control Board (INCB), or the Commission on Narcotic Drugs (UN CND) of the Economic and Social Council of the UN.

Although the relevance and potential utility of the SCNP and CPS for constraining CNS-acting chemical agent weaponization was previously raised by the ICRC and the current authors, until recently these mechanisms had not been publicly addressed by States in the context of arms control and disarmament. However, in recent years, certain CWC States Parties, notably Russia, as well as the OPCW's SAB have recognised the relevance of the UN Drugs Conventions and associated bodies to this discourse. Consequently, engagement between the OPCW and the UN Drug Convention bodies should be an element of the OPCW's measures to facilitate and verify implementation of the CSP-26 Understanding on aerosolised CNS-acting chemical agents, and of more wide-ranging OPCW measures to prevent and address CNS-acting chemical agent and other ICA weaponization. Such engagement may lead to greater formal consultation, cooperation and collaboration at a technical level between the OPCW, the INCB and UN CND.

12.4.6 Chemical, Life Science and Medical Communities

12.4.6.1 Monitor and Respond to Science and Technology Developments of Concern

Building on previous work by bodies such as the British Medical Association, the UK Royal Society, Bradford University, the Harvard Sussex Programme, and the Sunshine Project, independent scientific and medical bodies should engage in technology monitoring to identify technologies and activities of potential concern. Particular attention should be given to research and development of CNS-acting chemical agents, ICAs and associated means of delivery conducted under the auspices of research establishments controlled, directly or indirectly, by defence, security or law enforcement organisations, or that receive funding from such organisations. Independent chemical, biomedical and life scientists should also undertake technology horizon-scanning to identify and predict likely research trajectories in neuropharmacology, neuroscience and relevant chemical, life and associated scientific disciplines and related technologies, highlighting potential applications for weapons development and other malign use. Where activities of concern are uncovered, the relevant CWC and BTWC National Authorities should be informed and requests for clarification sought. Where appropriate, the relevant international organisations should also be notified *i.e.*, the OPCW Technical Secretariat and the BTWC ISU.

Independent scientific and medical bodies and broader associations and organisations – notably IUPAC, the Inter-Academy Partnership, and the World Health Organisation (WHO) – should constructively engage with the relevant OPCW policy-making organs and subsidiary bodies, notably the Scientific Advisory Board, the BTWC ISU, and individual BTWC and CWC States Parties, to highlight existing limitations in the CWC and BTWC and attendant control regimes with regard to CNS-acting chemicals/ICAs and associated means of delivery, and to develop and promote possible science-informed policy responses. Where appropriate, broader scientific and civil society coordination bodies, notably the CWC Coalition and the Joint NGO BTWC coordination group should be involved in these processes.

Calls have recently been made for the potential malign applications of neurotechnologies, being developed now for benign purposes, to be seriously addressed.[9] Clearly, an important actor in this regard would be the World Health Organisation. Indeed, the WHO's own horizon scan of 2021[10] concluded that within 10 years and beyond there would be the possible malign misuse of neurobiology, with particular concerns raised about "neuroscientific research assessing or modifying human thought, emotions and actions and means to affect the nervous system and alter cognitive states, behaviour and functions for performance enhancement and degradation". In view of the evidence becoming available about these growing capabilities it is surely incumbent on the WHO to now make a new systematic assessment of these growing dangers and propose comprehensive and holisitic measures to prevent such malign future applications.

12.4.6.2 *Development, Promotion and Implementation of Scientific Ethical Standards*

Chemical and life science associations should continue to foster a 'culture of responsibility' in science through their development, strengthening and promulgation of national codes of conduct and ethics. Given the currently under-recognised dangers of the malign application of modern neuroscience and neuropharmacology, specific attention should be given to engaging relevant neuroscience associations and neuro-ethicists. At the international level, chemical and life science associations should continue endorsement and promulgation respectively of the *Hague Ethical Guidelines* and the *Tianjin Biosecurity Guidelines*.

Given the previous focus upon preventing dual-use research access to malign non-State actors; greater attention should now be given to norm-building initiatives, including incorporation into existing codes, of measures explicitly constraining or halting involvement of chemical and life scientists in State CNS-acting chemical/ICA weapons development programmes, or other malign State utilisation of these agents and technologies.

The chemical and life community should promote and protect the rights of whistleblowers and support colleagues who 'blow the whistle' on CNSAC/ICA weapons programmes, whether conducted by States or non-State actors. In addition, the chemical and life science community should encourage the OPCW to establish an appropriate mechanism to receive and appropriately respond to information provided by whistleblowers. Such a function or body could be established either within the OPCW itself or be an independent international body within the UN framework that has authority to talk to the OPCW. This body could be authorised to act as an "international ombudsman" and a recipient where individuals could file complaints and report issues. This body would then evaluate the information and if found credible launch a formal procedure within the OPCW framework to clarify what is going on. Such a body could also be responsible for protecting the rights and safety of whistle-blowers who had submitted information to it.

12.4.6.3 Education and Awareness Raising

Education about the dual-use research dilemma and ethical responsibilities of scientists should form a mandatory element for all those trained in the chemical and life sciences from secondary level onwards. This will require sustained and systematic engagement with educationalists, professional associations, States and the OPCW and ISU BTWC. Chemical and life science associations – notably IUPAC and the IAP – should work with educationalists, and the OPCW/BTWC ISU to develop education and awareness-raising materials and programmes targeted at scientific communities, State officials and the general public. Of particular importance in this regard is the potential impact of the work of the *WHO Technical Advisory Group on the responsible use of the life sciences and dual-use research* and its sub-groups to assist implementation of the WHO Global Guidance Framework. Given the widespread historical misuse of medical research in malign efforts to attack or control the mind, it is particularly important that this issue is a particular focus of the WHO's efforts in its continuing educational work on the responsible use of the life sciences and the problem of dual use. Coordination of such initiatives will be vital, firstly to avoid any 'mixed messaging' in this area where the BTWC and CWC overlap, and secondly to ensure the most effective use of limited financial, temporal, and personnel resources. The OPCW's Advisory Board on Education and Outreach may prove a useful forum for such coordination.

12.4.6.4 Guidance and Regulatory Measures to Prevent Malign Application of Dual-use Research

For nearly 20 years, since the US Academy of Sciences *Fink and Lemon-Relman reports,* governments and scientific associations have sought to develop, promulgate and apply appropriate guidance and regulatory measures for those conducting dual-use research in the life and associated sciences. To date, however these largely concentrated on preventing malign application of genetic engineering of pathogens, despite the *Lemon-Relman report's* warning of risks from malign application of a far broader 'threat spectrum'. This has left scientists working in other relevant, fast evolving, fields of research—such as neuroscience, neuropharmacology, as well as the broader study of toxins, bioregulators and associated physiological systems—in difficulties judging just how much dual-use concern there might be about their work and what they should do to help minimise the risks of its malign application. National and international scientific associations should establish mechanisms to explore these issues, incorporating involvement and active engagement between, security experts and researchers from the relevant scientific disciplines, to inform both communities and collectively develop appropriate guidance for all those engaged in undertaking, funding, publishing, and regulating dual-use research in these areas. One important focus for such ethical and regulatory development should be the largescale Human Brain Projects now being undertaken by a number of States.

12.4.6.5 Development and Application of Medical Ethics

National medical, and other health professional associations, should follow the example of the British Medical Association (BMA) and the Health Professions Council of South Africa (HPCSA), and clarify that it is unethical for their members to be involved in the development or use of CNS-acting chemical agent/ICA weapons or other malign applications of neuroscience and neuropharmacology. The relevant national association's ethics boards should investigate any well-founded reports that members were engaged in such practices, and where appropriate sanction those found guilty of such action, including through disbarment.

Efforts should be made to develop binding international prohibitions on medical involvement in CNS-acting chemical agent/ICA weaponization, by the World Medical Association (WMA) and associations representing other relevant health professionals such as anaesthetists and medical toxicologists. Given their long-standing active opposition to medical involvement in torture and other ill-treatment, national medical associations and ultimately the WMA and other health professional associations, should clarify application of these existing obligations to prevent medical involvement in the application of CNS-acting chemical agents or ICAs in abusive "chemical restraint", enforced interrogations and other malign application of such chemicals upon detainees for torture and other ill-treatment, and also explicitly prohibit all medical involvement in 'lethal injection' executions employing such substances.

References

1. J. Ngai, BRAIN 2.0: Transforming neuroscience, *Cell*, 2022, **185**, 4–7.
2. (a) N. S. Hsu, *et al.*, The promise of the BRAIN initiative NIH strategies for understanding neural circuit function, *Curr. Opin. Neurobiol.*, 2020, **66**, 162–166; (b) J. Ngal, BRAIN@ 10: A decade of innovation, *Neuron*, 2024, **112**, 3003–3006.
3. In this book we only touch on one particular aspect of the very recent revolution that is taking place in our understanding of the immune system and its relationship to the brain, behaviour and the microbes of our gut. See M. Lyman, *The Immune Mind: The new science of health*, Penguin Books, London, 2024, for a concise overview for the non-specialist.
4. Certain adaptations necessitated collective decisions of the States Parties that clarified the understanding of the scope of the treaty in relevant areas, and these decisions were taken by a vote at the relevant CSP or Review Conference. This has the unfortunate side effect that, although these decisions were binding for all States Parties, certain States Parties will in practice question the validity of said decisions and may ignore/not implement them. This creates a different type of uncertainty about the way the treaty is being applied.
5. (a) M. Crowley, *Chemical Control: Regulation of Incapacitating Chemical Agent Weapons, Riot Control Agents and their Means of Delivery*, Palgrave, Basingstoke, 2016, pp. 142–155; 190–198, 199–223; (b) *Preventing Chemical Weapons; Arms Control and Disarmament as the Sciences Converge*, ed. M. Crowley, M. Dando and L. Shang, Royal Society of Chemistry, London, 2018, pp. 101–145; 146–192; 418–473.
6. J. Tucker, The body's own bioweapons, *Bull. At. Sci.*, 2008, **64**(1), 16–22.
7. United Nations, Possible elements of specific and effective measures, including possible legally-binding measures to strengthen and institutionalize the Biological Weapons Convention in all its aspects, Non-paper by the Chairperson, Vice-Chairs and Friends of the Chair of the Working Group on the Strengthening of the Convention, Geneva, 6 May 2025.
8. United Nations, Possible elements of specific and effective measures, including possible legally-binding measures to strengthen and

institutionalize the Biological Weapons Convention in all its aspects, Non-paper by the Chairperson, Vice-Chairs and Friends of the Chair of the Working Group on the Strengthening of the Convention, Geneva, 6 May 2025, point 7.

9. L. J. Meier, *Mind Control: Past and Future*, Carr Center Discussion Paper, 2025-1, Harvard Kennedy Center, Harvard University, 2025.
10. World Health Organisation, *Emerging technologies and dual-use concerns: a horizon scan for global public health*, WHO, Geneva, 2021.

Subject Index

2,2-diphenyl-2-hydroxyacetic acid, 263
3-quinuclidinol, 278
3-quinuclidinyl benzilate, 39
1925 Geneva Protocol, 13

α-bungarotoxin, 174

acetylcholine, 107, 136
active placebo, 34
aerosols, 183
alpha 2 adrenoceptor agonists, 10
AlphaFold, 166
AlphaFold3, 167
Alzheimer's disease, 100
amphetamines, 385
anaesthesia, 48, 65, 108
anaesthetic agents, 40
animal model, 407
anticholinergics, 39
anti-NMDA receptor encephalitis, 140–144
antipsychotics, 40
anti-torture trade instruments, 324
    Council of Europe Committee of Ministers Recommendation CM/REC, 325–326
    EU Anti Torture Regulation, 324–325
    International Measures to Promote Torture-free Trade, 326–327
anxiolytics, 40
Applied Biology Committee (ABC), 431
armed conflict, 233
Article 36, 360, 362, 363, 364, 365, 367–368
artificial intelligence (AI), 158, 166–167
    short primer on, 163
autoimmune attack, 140
autoimmune diseases, 146
autoimmune encephalitis (AE), 140
    anti-NMDA receptor encephalitis, 140–144
awareness raising, 242–244

basolateral amygdala, 112
BBB. *See* blood–brain barrier (BBB)
benzilic acid, 278
benzodiazepines, 10, 41
bioautomation, 92
biocatalysts, 172
biodegradable nano-systems, 180
biological agents, 121, 226
Biological and Toxin Weapons Convention (BTWC), 13–14, 91, 121, 225, 468–469
    addressing allegations of malign use, 482
    CNS-acting chemicals and scope of coverage, 227–232
    general obligations under, 226–227
    measures to facilitate implementation, verification and compliance, 234

confidence building measures, 235–236
education and awareness raising, 242–244
implementation support unit, 240
non-compliance measures, 240–242
review of science and technology, 236–240
promoting responsible use of science and technology, 482–483
resolving ambiguities in BTWC and application, 481
scope of activities regulated, 232–234
strengthening science and technology monitoring and review, 482
Biological Weapons Prevention Project (BWPP), 238
biology-oriented synthesis (BIOS), 168
biomaterials, 91
bioregulators, 11, 159, 244, 403, 438
biotechnology, 168
biotoxins, 21, 173
blood–brain barrier (BBB), 12, 94, 180, 183
brain, 87
bomb, 8
structures, 89
BRAIN project, 409
BTWC. *See* Biological and Toxin Weapons Convention (BTWC)

Caenorhabditis elegans, 150
calmatives, 43
cannabinoids, 9
carbon nanoparticles, 179
catecholamines, 90
cell-free protein synthesis (CFPS), 173
Central Intelligence Agency (CIA), 60
central nervous system (CNS), 4, 89, 91, 96, 121, 158, 230, 270, 406
immune system and
disruption of, 128
basic immunology, 130–133
COVID and disruption of nervous system, 128–130
knowledge of autoimmune disease mechanisms, 135–139
some history of weapons research during Cold War, 133–134
immunopsychiatry, 139
autoimmune aspects of neurodegenerative diseases, 144–147
autoimmune encephalitis, 140–144
psychedelic disruption, 122–127
wider context of scientific advances in neuroscience, 147–151
Chemical and Biological Defense Command (CBDCOM), 42
Chemical Defence Experimental Establishment (CDEE), 30
chemical incapacitation agents, 272
chemical language models, 164
chemical, life science and medical communities, 486
development and application of medical ethics, 489–490
development, promotion and implementation of scientific ethical standards, 487–488
education and awareness raising, 488–489
guidance and regulatory measures to prevent malign application of dual-use research, 489

chemical, life science and medical communities (*continued*)
monitor and respond to science and technology developments, 486–487
chemical neurotransmitters, 90
chemical warfare agents (CWAs), 269
chemical weapons, 277
Chemical Weapons Convention (CWC), 13–17, 91, 121, 253, 469–472
addressing allegations of malign use, 481
allegations by US and other NATO countries, 270
alleged Chinese research of CNS-acting chemical agent weapons, 270–271
alleged Iranian research and development of CNS-acting weapons, 271–272
use of CNS-acting weapons and alleged subsequent research and development, 272–274
clarifying application of CSP-26 understanding and developing further understanding on ICAs, 478
definitional lacunae and their potential consequences, 256–257
definition of chemical weapons, 255–256
establishing SAB Temporary Working Group on CNS acting chemicals and ICAs, 479–480
general obligations under, 254
measures to facilitate implementation, verification and compliance, 262–264
regulating CNS-acting chemical research, development and use, 257
Russian allegations, 274–277
scope of coverage, 254–255
strengthening OPCW science and technology monitoring and review capabilities, 480
chemogenetics, 98, 99
chemokines, 136
China Brain Project (CBP), 406
Chinese Academy of Sciences (CAS), 406
Chinese military institute, 270
cholecystokinin (CCK), 43, 111, 112
chronic inflammation, 146
CNS. *See* central nervous system (CNS)
CNS-acting chemical agents (CNSAC), 8, 15, 29, 158
development of CNS-acting agent weapons employing limited area delivery mechanisms, 174
Chinese development and use of CNS-acting weapons for military and law enforcement, 174–175
Iranian military university development of CNS-agent grenades, 177–178
Israeli development and use of trans-dermal CNS-acting weapon, 175–176
US development of fentanyl felt pad projectiles, 176–177
developments in CNS-agent uptake and targeting brain and CNS, 178

aerosolised delivery of therapeutic drugs and vaccines to individuals, 181–183
commercial dual use wide area delivery aerosol devices employed for delivery of, 186–188
concerns about malign employment of aerosol drug delivery, 183–186
nanotechnology and drug delivery and uptake, 178–180
transporting CNS-acting chemicals to brain, 180–181
wide area aerosolised CNS-agent weapons in enclosed spaces, 188–189
wide area RCA dissemination devices, 190–199
discovery and synthesis, 159
biotechnology techniques employed for development and synthesis of novel agents, 168–173
chemical synthesis and production of biological agents, 173–174
from discovery to rationale engineering, 165
identifying retro-synthesis pathways, 167–168
screening for novel chemical agents, 159–165
systems biology, AI and rational drug design, 166–167
CNS-acting chemicals, 391
cocaine, 37
cognitive functions, 407
Cold War, 174
Committee on the Safety of Human Experimentation (COHSE), 430
Conference of States Parties (CSP), 15, 260, 264
Confidence Building Measures (CBMs), 235–236
Convention on Certain Conventional Weapons (CCCW), 367
convergence, 3
corticotropin-releasing factor (CRF), 43
counter-terrorism, 40
COVID-19 pandemic, 226, 245, 405
Cumulative Link Models (CLMs), 164
CWC. *See* Chemical Weapons Convention (CWC)
Cyclone Riot Control Drone System, 194
cytokines, 136
Czech Republic, 48–49

deep learning (DL), 163, 164
de facto toxins, 13
default mode network (DMN), 464
Defence Research & Development Establishment (DRDE), 49
Defence Research Policy Committee (DRPC), 31
Defence Science and Technology Laboratory (DSTL), 273
degradation market, 66
delta-9 tetrahydrocannabinol (THC), 9
depatterning process, 63
Development, Concepts and Doctrine Centre (DCDC), 366
diffusion model, 167
dimethyl sulfoxide (DMSO), 176
Directorate of Forensic Sciences (DFS), 66
dirty tricks program, 176
disorientation substances, 36
distinction, 354
dopamine, 90, 107, 137

dorsal raphe (DR), 139
Drone Tear Smoke Launcher System, 194
Drosophila melanogaster, 104
drug delivery, 178–180
drug development, 230
drug discovery, 166
dual-use dilemma, 402
dual-use neuroscience, 406
 China Brain Project and potential dual-use concern, 406–407
 European Brain Project, 410–412
 US BRAIN project and potential dual-use concern, 407–410
dual-use problem, 4

Edgewood Chemical Biological Center (ECBC), 39, 176, 195
education, 242–244
electroencephalogram (EEG), 106
endocrine cybernetics, 100
endorphins, 46
enkephalins, 46
epitope mapping, 92
Epstein–Barr virus, 136
Escherichia coli, 167
ESMFold, 166, 167
etorphine, 383
EU Anti Torture Regulation, 324–325
European Brain Project, 410–412
European Union (EU), 96
Executive Council (EC), 264

facial recognition technology, 187
fear circuits, 109–113
feed-forward regulation, 105
fentanyl, 176, 184, 383
fiber photometry, 99
Fourth Industrial Revolution, 164

gain-of-function experiments, 112
galanin, 107
gamma-aminobutyric acid (GADA), 10, 107
gene editing techniques, 168
General Purpose Criterion (GPC), 13–14, 255, 271
genetic engineering, 138, 169
Genome Project-write (GP-write), 173
G protein coupled receptors, 100

HAC. *See* holistic arms control (HAC)
herpes simplex virus encephalitis (HSE), 141
high-throughput screening (HTS), 161
holistic arms control (HAC), 17–19, 459
 advances in neuroscience, 461
  disrupting circuits of higher central nervous system, 463–465
  other relevant science and technology, 465–467
  understanding basic circuits of brain, 461–463
 Biological and Toxin Weapons Convention, 468–469
  addressing allegations of malign use, 482
  promoting responsible use of science and technology, 482–483
  resolving ambiguities in BTWC and application, 481
  strengthening science and technology monitoring and review, 482
 Chemical Weapons Convention, 469–472
  addressing allegations of malign use, 481
  clarifying application of CSP-26 understanding and developing further understanding on ICAs, 478

establishing SAB Temporary Working Group on CNS acting chemicals and ICAs, 479–480
strengthening OPCW science and technology monitoring and review capabilities, 480
further considerations regarding international instruments and control regimes, 474–475
international drug controls, 473–474
international humanitarian law, 472–473
international human rights law, 472
potential role of civil society, 475–477
weapons development and use, 460–461
Human Brain Project (HBP), 409
Human Genome Project, 98
humanitarian law, 292
human leukocyte antigen (HLA), 138
human rights law, 327
hypocretin, 108, 114, 463
hypocretinergic neurons, 108
hypothalamic peptides, 94
hypothalamus, 107

ICAs. *See* incapacitating chemical agents (ICAs)
IHL. *See* international humanitarian law (IHL)
IHRL. *See* international human rights law (IHRL)
immune dysfunction, 142
immune systems, 138, 151
Implementation Support Unit (ISU), 230, 240
incapacitating chemical agents (ICAs), 8, 38, 41, 158, 227, 291
incapacitation agents, 272
Indian Border Security Force (BSF), 194
indiscriminate attacks, 353
indiscriminate weapons, 353
informatics, 172
information exchange, 365–366
information technology, 166, 401
innate behaviours, 103–105
International Biological Security Education Network (IBSEN), 426
International Committee of the Red Cross (ICRC), 44, 351, 401
international humanitarian law (IHL), 21, 291, 350, 483–484
implementation of, 368
nature of scope of, 350–352
over-arching IHL obligations constraining development and use of weapons, 352
discrimination and distinction, 352–354
identification and protection of persons considered *Hors de Combat,* 356–358
Martens clause, 359
preventing malign application of medicine, 358
respect of, 358–359
superfluous injury or unnecessary suffering, 354–356
reviewing legality of 'new' weapons, 360
applicability of human rights considerations, 363
Article 36 implementation, 367–368
health considerations, 363–364
mechanism of review, 365–366
restrictions on the use of new weapons, 364–365
scope of new weapons, 360–362

international human rights law (IHRL), 21, 291, 352, 484–485
considerations regarding means of delivery and dispersal of CNS-acting chemical agents, 315–316
health considerations and the right to health, 312
application of CNS-acting chemicals for law enforcement chemical restraint, 314
obligations regarding design, review, training and monitoring use of 'less lethal' weapons, 317–320
prohibition against torture and other ill-treatment, 304
non-consensual application of CNS-acting chemical agents to detainees in prison/law enforcement context, 305–311
torture in non-custodial contexts, 311–312
protection of the right to life, 294
application to CNS acting chemical agent weapons, 295–302
restrictions on the use of force, 292–294
rights to freedom of association and assembly, 302–303
rights to liberty and security, 303–304
trade agreements incorporating IHRL and IHL criteria, 320
anti-torture trade instruments, 324–327
conventional arms and dual use goods, 320–324
International Narcotics Control Board (INCB), 381, 382
International Nuclear Security Education Network (INSEN), 426
interrogation
contemporary malign application of CNS-acting chemicals for, 64
India, 65–66
Iran, 65
Intersessional Process, 239, 243
investigation of alleged use (IAU), 267
in vitro high-throughput screening assays, 162
in vivo animal testing, 159
in vivo optogenetics, 108
in vivo screening techniques, 161
Institute for Non-Lethal Defense Technologies (INLDT), 40
Israeli security forces (ISF), 196

Joint Non-Lethal Weapons Directorate (JNLWD), 195

ketamine, 48, 142
Knobel's account, 36
knock in gene, 103
knock out gene, 103

lateral amygdala (LA), 112
Lemon Relman Committee, 403
lipids, 4
locus coeruleus (LC) system, 42, 460
loss-of-function experiments, 112
lysergic acid diethylamide (LSD), 9, 31
lysergic acid ethylamide (LAE), 31

machine learning (ML), 92, 163
macrophages, 136
malign application of neuroscience
application of CNS-acting chemicals and neuroscience for torture and ill-treatment, 51

of CNS-acting chemicals for interrogation, 60–64
contemporary malign application of CNS-acting chemicals for interrogation, 64–66
development of psychopharmacology for benign purposes, 51–53
of psychopharmacology, 53–60
Chinese pharmaceutic CNS-acting chemical agent weapons development and use, 47
illustrative examples of past and contemporary dual use neuroscience and related research of potential concern, 48
Czech Republic, 48–49
India, 49–50
Iran, 50–51
South Africa pharmaceutical CNS-acting chemical agent weapons development, 35–38
Soviet Union/Russia CNS-acting agent weapon development and use, 44
bioregulator and toxin CNS-acting weapons and other ICA weapons, 46–47
pharmaceutical CNS-acting agent research and weaponization, 44–45
UK pharmaceutical CNS-acting chemical agent weapons research, 30
physical incapacitating agents, 33–35
psychological incapacitating agents, 31–33
US CNS-acting chemical agent weapons development, 38
research and development of pharmaceutical CNS-acting chemical agent weapons, 38–41
research of toxin and bioregulator incapacitating chemical agents, 41–43
US rejection of CNS-acting weapons development, 43–44
mammalian neuropeptides, 102
Marshall Islands, 42
Martens clause, 359
massive convoluted cerebral hemispheres, 88
mass spectrometry, 102
medetomidine, 48
Meeting of States Parties (MSP), 239
mental illnesses, 20
mesolimbic dopamine system, 94
messenger RNA, 109
methaqualone, 35, 37
microbial agents, 226
microglia, 136
microRNAs (miRNAs), 109
midazolam, 48
mind-and mood-altering effects, 65
MKULTRA sub-projects, 61–62
Modular Advanced Armed Robotic System (MAARS), 193–194
monoamines, 100
Montreal experiments, 62–64
morphine, 10, 383
multiple sclerosis (MS), 136–137, 465
myasthenia gravis (MG), 135–136, 465
history of investigations of, 135
myelin basic protein (MBP), 46

nalorphine, 34
nanocarrier-based delivery systems, 179

nanomaterials, 180
nanoparticles, 178, 181
nanotechnology, 158, 178–180
narcoanalysis, 65
narcolepsy, 114, 137–139, 465
nasal delivery, 182
nasal route, 182
National Academy of Sciences (NAS), 185
National Institute of Justice (NIJ), 39
National Intelligence Service (NIS), 38
nervous systems, 104
neurodegenerative and psychiatric diseases (NPDs), 145
neurodegenerative diseases, 115
 autoimmune aspects of, 144–147
neuroleptic anaesthetic agents, 10
neuromodulatory function, 90
neuronal receptors, 90
neurons, 115, 122
neuropeptides
 in 2000, 93–95
 importance of, 91–93
 neuronal circuit research in 2023, 95–100
 research in 2025, 100–103
 simple circuits and higher levels of CNS functions and behaviours, 113–115
 understanding simpler circuits, 103
  fear circuits, 109–113
  innate behaviours, 103–105
  sleep/wake cycle, 105–109
neuropsychopharmacology, 144
neuroscience, 3, 461
 advancing and converging chemical and life sciences, 3–5
 current limited regulation of CNS-acting agents under CBW arms control and disarmament treaties, 12
  1925 Geneva Protocol, 13
  Biological and Toxin Weapons Convention, 13–14
  Chemical Weapons Convention, 14–17
  holistic arms control, 17–19
 disrupting circuits of higher central nervous system, 463–465
 other relevant science and technology, 465–467
 overview of advances in, 5
  implications for monitoring dual use research and preventing weaponization of CNS-acting chemical agents, 11–12
  potential dual-use concern, 7–11
 understanding basic circuits of brain, 461–463
 wider context of scientific advances in, 147–151
neurotechnology, 409
neurotransmitters, 91, 115
N-methyl-4-piperidyl-isopropyl-phenyl glycollate (MPIPG), 32
N-methyl-D-aspartate receptor (NMDAr), 141
 behavioural change in, 142
non-compliance measures, 240–242
non-human primates (NHPs), 406
non-lethal (NL), 40
 agents, 266
 application, 39
 weapons, 230, 463
nonlethal weapons, 121
non-rapid eye movement (NREM), 106, 137
noradrenaline, 90, 107, 137
North Industries Corporation (NORINCO), 47

opiate receptors (ORs), 45
opioid analgesics, 40
opioid peptide systems, 94
opioids, 10, 48
optogenetics, 98, 99, 138, 150
orexin, 107, 108, 408
Organisation for the Prohibition of Chemical Weapons (OPCW), 254, 480
  'culture' and working practices, 262

panic attacks (PAs), 111
panic disorder (PD), 111
Panksepp's basic circuits, 114
Parkinson's disease, 100, 145–147
patchwork drug-control legal framework, 381
pathological process, 136
Pelak-e 100, 65
pelargonic acid vanillylamide (PAVA), 192
peptides, 95, 173
peptide synthesis, 91, 92
pharmaceutical-based agent (PBA) program, 270, 273
pharmaceutical industry, 31, 230
pharmaceuticals, 159
pharmacological cocktails, 48
pharmacological non-lethal weapons, 49
phencyclidine (PCP), 142
physical incapacitating agents, 33–35
political, security, intelligence and military (PSIM), 411
poly-(lactic-*co*-glycolic acid), 179
post traumatic stress disorder (PTSD), 113
preventative arms control, 17
primary self-antigens (pSAgs), 146
Prisoner of War (POW) camps, 351
proteins, 173
psilocybin, 9
psychedelic disruption, 122–127
psychedelic drugs, 9, 122
psychological incapacitating agents, 31–33
psychological techniques, 357
psychoneuroimmunology, 139
pyrexal, 33, 34

quinuclidin-3-ol, 263

rapid eye movement (REM), 106, 137
rational drug design, 166–167
receptors, 122
reconnaissance, surveillance and target acquisition (RSTA), 193
recreational drugs, 381
recurrent loops, 100
remifentanil, 383
Research Development and Engineering Center (RDEC), 42
retro-synthesis techniques, 167
reverse genetics, 174
riot control agents (RCAs), 159
  drones, unmanned ground vehicles and unmanned weapons systems, 193–195
  foggers, sprayers and water cannon, 191–192
  indoor and perimeter RCA dispersion system, 190–191
  large RCA projectiles, 195–196
  multiple barrel and automatic launchers, 192–193
RoseTTAFold, 166
Russian allegations, 274
  alleged "western" transfer of CNS-acting agent to Ukrainian forces, 276–277
  use of CNS acting agent rather than novichok in assassination attempt in UK, 274–275

Salmonella abortus equi, 33
schizophrenia, 63
schizophrenic symptomatology, 63
Science and Technology (S&T), 236–240

Scientific Advisory Board (SAB), 264
scientific and medical communities, 401
  education and outreach for chemical and life science communities, 420
    education and outreach to chemists, 420–424
    education and outreach to life science communities, 424–426
  medical ethics and ethical practice, 429
    developing and applying medical ethical standards to regulation of the life sciences, 437–439
    in development and use of CNS-acting chemical agent weapons, 435–437
    prohibiting medical involvement in chemical and biological weapons development, 432–435
    regulating scientific research on human subjects, 429–431
    upholding IHRL and IHL, 431–432
  oaths, codes and pledges for chemical and life science communities, 412
    constraints on effectiveness of codes, 419–420
    ethical codes for chemists, 416–419
    ethical codes for the life sciences, 413–415
  regulating 'dual-use' research, 402
    dual-use neuroscience, 406–412
  whistle-blowing, 426–429
second hypothesis, 106
Secret Intelligence Service (SIS), 31
sedative-hypnotic agents, 40
serotonin, 90, 137, 464
Simplified Molecular Input Line Entry Systems (SMILES), 164
single-cell transcriptomics, 108
single convention on narcotic drugs (SCND), 382
    civil society and state discourse concerning potential relevancy of, 388–392
    UN Convention on Psychotropic Substances, 386–388
skeletal muscle relaxants, 40
Skunk Riot Control Copter, 194–195
sleep disorders, 108
sleep/wake cycle, 105–109
smokeless method, 407
sodium pentothal, 65
South African Defence Force (SADF), 35
spatial RNASeq, 149
special stocks, 384, 385
Staphylococcal enterotoxin B (SEB), 42, 228
structure–activity relationships, 31
sublaterodorsal tegmental nucleus (SLD), 138, 139
sufentanil, 383
superfluous injury or unnecessary suffering (SIRUS), 354–356
suprachiasmatic nucleus (SCN), 107
swarming capability, 187
synaptic connectome, 150
synthetic biology, 92, 168, 171, 172
systems biology, 166–167

temporary working groups (TWGs), 92, 239, 265
therapeutic drugs, 182
titanium oxide nanoparticles, 179
Torture-Free Trade Treaty (TFTT), 328
toxic chemicals, 277

toxins, 13, 159, 227, 244
transformative projects, 97
transparency, 365–366
treaty ratification, 350
Truth and Reconciliation Commission (TRC), 35, 434
truth drugs, 60, 65
truth serums, 60
TWGs. *See* temporary working groups (TWGs)
type 1 narcolepsy, 138, 463

UK Chemical Weapons Development and Human Experimentation, 430–431
UK's Defence Science and Technology Laboratory (UK DSTL), 45
UN Convention on Psychotropic Substances (CPS), 386–388
UN drug control conventions, 381–393, 485–486
UN Economic and Social Council (ECOSOC), 382
United Nations Security Council (UNSC), 240
unmanned ground vehicle (UGV), 187, 193
UN Secretary General's Mechanism (USGM), 240–241
US Brain Research Through Advancing Innovative Neurotechnologies (BRAIN), 97, 407–409

ventromedial medulla (VMM), 138–139
vertebrate novelty, 101
vesicular monoamine transporter-2 (VMAT2), 146
virtual HTS (vHTS), 162
voluntary muscles, 138

whistle-blowing, 426–429
wiring diagram, 98

Yersinia pseudotuberculosis, 46

zebrafish, 406